D0605988

THE HANDBOOK OF NANOTECHNOLOGY

NANOMETER STRUCTURES

Theory, Modeling, and Simulation

THE HANDBOOK OF NANOTECHNOLOGY

NANOMETER STRUCTURES

Theory, Modeling, and Simulation

AKHLESH LAKHTAKIA, EDITOR

A Publication of ASME
New York, New York USA

SPIE PRESS
A Publication of SPIE—The International Society for Optical Engineering
Bellingham, Washington USA

Library of Congress Cataloging-in-Publication Data

Nanometer structures: theory, modeling, and simulation / editor: Akhlesh Lakhtakia.
p. cm. – (Handbook of nanotechnology)
Includes bibliographical references and index.
ISBN 0-8194-5186-X
1. Nanotechnology–Handbooks, manuals, etc. I. Lakhtakia, A. (Akhlesh), 1957- II. Series.

T174.7.N353 2004
620'.5–dc22 2004041716

Published by

SPIE—The International Society for Optical Engineering
P.O. Box 10
Bellingham, Washington 98227-0010 USA
Phone: +1 360 676 3290
Fax: +1 360 647 1445
Email: spie@spie.org
Web: www.spie.org

ASME
Three Park Ave.
New York, New York 10016-5990 USA
Phone: +1 800 843 2763
Fax: +1 212 591 7292
Email: infocentral@asme.org
Web: www.asme.org

Copublished in the United Kingdom by Professional Engineering Publishing Ltd.
Northgate Avenue, Bury St Edmunds, Suffolk, IP32 6BW, www.pepublishing.com
UK ISBN 1-86058-458-6

Printed in the United States of America.

About the cover: The images shown are part of a simulation studying the formation of complex junction structures in metals undergoing work-hardening induced by tensile strain. The work was produced by Farid Abraham of IBM Almaden Research in collaboration with Lawrence Livermore National Laboratory (LLNL) personnel Mark Duchaineau and Tomas Diaz De La Rubia. The images are screenshots from a movie depicting a billion-atom dislocation simulation in copper. Further information on this simulation can be found at www.llnl.gov/largevis/atoms/ductile-failure/ and in Reference 56 in Chapter 7 of this book. Special thanks are due to the University of California, LLNL, and the U.S. Department of Energy, under whose auspices the work was performed.

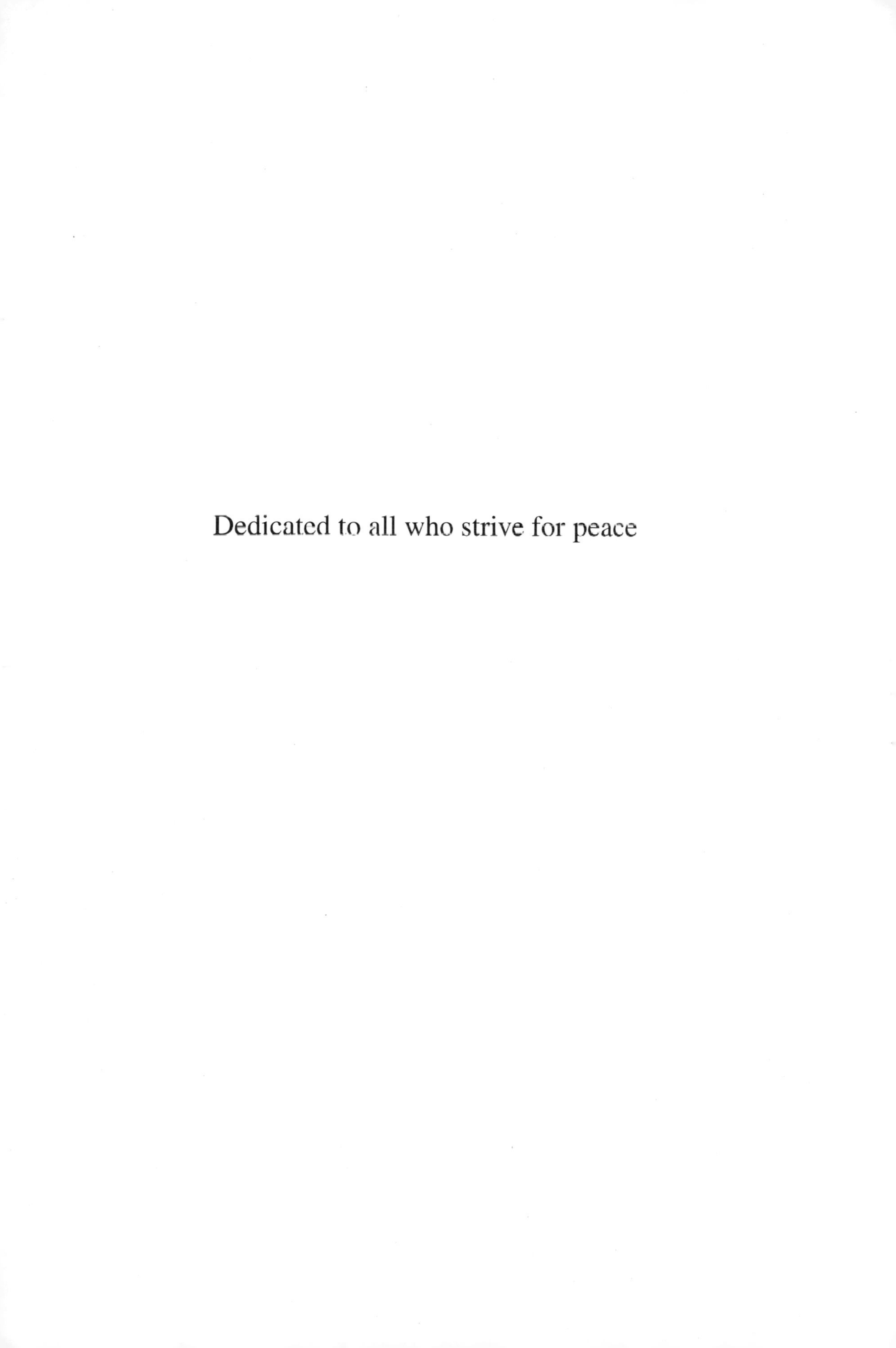

Dedicated to all who strive for peace

Order of Chapters

Foreword / ix
Brian J. Thompson

Preface / xi

List of Contributors / xiii

1. Editorial / 1
Akhlesh Lakhtakia

2. Sculptured Thin Films / 5
Akhlesh Lakhtakia and Russell Messier

3. Photonic Band Gap Structures / 45
Joseph W. Haus

4. Quantum Dots: Phenomenology, Photonic and Electronic Properties, Modeling and Technology / 107
Fredrik Boxberg and Jukka Tulkki

5. Nanoelectromagnetics of Low-Dimensional Structures / 145
Sergey A. Maksimenko and Gregory Ya. Slepyan

6. Atomistic Simulation Methods / 207
Pierre A. Deymier, Vivek Kapila and Krishna Muralidharan

7. Nanomechanics / 255
Vijay B. Shenoy

8. Nanoscale Fluid Mechanics / 319
Petros Koumoutsakos, Urs Zimmerli, Thomas Werder and Jens H. Walther

9. Introduction to Quantum Information Theory / 395
Mary Beth Ruskai

Index / 465

Foreword

It is both a rare privilege and a distinct challenge to prepare a short foreword to this volume of the *Handbook of Nanotechnology*. So, why me and why did I agree? The answer to that is certainly not the usual answer. Traditionally, someone pre-eminent in the field of nanometer structures would be asked to provide a short overview of this subfield, its importance, and its trajectory. Obviously, I am not an expert in this particular branch of science and technology; the fact is that I am intellectually challenged by the material in its totality even though I feel comfortable and at home with a significant fraction of that totality as stand-alone components.

The answer to "why me?" is perhaps because I have always championed the integrated approach to science and engineering, specifically optical science and engineering. This approach involves the integration of theory, modeling, setting up and evaluating specific examples, testing those examples, and applying the results to specific experimental and engineering studies. The resultant knowledge is then used to devise new technology, implement that technology, and apply it to problem solving and to the development of new components and systems. The final step is to design and create new instruments and products to serve the local world in which we live.

Having now taken the time to accept the challenge of working through this volume, I can certainly report that it was well worth the effort. Those readers who follow my example will find that it will provide a significant stimulation to those already working in the field and encourage others to make an intellectual investment in moving nanotechnology forward.

This handbook is not presenting a fully developed theoretical model, but is presenting significant theory based on sound physical laws augmented by other approaches to provide a framework to test ideas and make progress. We have all learned over the years that there are a number of valuable ways to approach the mathematical description of physical observations: modeling, simulation, algorithms, interactive processes, transformations to other spaces and coordinates, curve fitting, and statistical methods, to name a few. The reader will find many of these techniques used in the text.

There is no doubt that nanotechnology will play a very important role in the coming years in a variety of areas that are listed in Professor Lakhtakia's preface and in the table of contents. These areas will certainly be interdisciplinary between science and engineering, but also interdisciplinary in the traditional sense between optical science, optical engineering mechanics, electronics, material science, etc.

It is not without significance that this volume is published as a joint venture between SPIE—The International Society for Optical Engineering and ASME, The American Society of Mechanical Engineering.

My expectation (and hence my prediction) is that this volume may well become a milestone volume for some time to come with perhaps new editions in the future as the field progresses. I hope the editor will ask someone more qualified than I am to prepare the foreword to future editions!

Brian J. Thompson
University of Rochester
May 2004

Preface

The *Handbook of Nanotechnology* series is intended to provide a reference to researchers in nanotechnology, offering readers a combination of tutorial material and review of the state of the art. This volume focuses on modeling and simulation at the nanoscale. Being sponsored by both SPIE—The International Society for Optical Engineering and the American Society of Mechanical Engineering, its coverage is confined to optical and mechanical topics.

The eight substantive chapters of this volume—entitled *Nanometer Structures: Theory, Modeling, and Simulation*—cover nanostructured thin films, photonic bandgap structures, quantum dots, carbon nanotubes, atomistic techniques, nanomechanics, nanofluidics, and quantum information processing. Modeling and simulation research on these topics has acquired a sufficient degree of maturity as to merit inclusion. While the intent is to serve as a reference source for expert researchers, there is sufficient content for novice researchers as well. The level of presentation in each chapter assumes a fundamental background at the level of an engineering or science graduate.

I am appreciative of both SPIE and ASME for undertaking this project at a pivotal point in the evolution of nanotechnology, just when actual devices and applications seem poised to spring forth. My employer, Pennsylvania State University, kindly provided me a sabbatical leave-of-absence during the Spring 2003 semester, when the major part of my editorial duties were performed.

All contributing authors cooperated graciously during the various phases of the production of this volume and its contents, and they deserve the applause of all colleagues for putting their normal research and teaching activities aside while writing their chapters for the common good. Tim Lamkins of SPIE Press coordinated the production of this volume promptly and efficiently. I consider myself specially privileged to have worked with all of these fine people.

Akhlesh Lakhtakia
University Park, Pennsylvania
May 2004

List of Contributors

Fredrik Boxberg
Helsinki University of Technology, Finland

Pierre A. Deymier
University of Arizona, USA

Joseph W. Haus
The University of Dayton, USA

Vivek Kapila
University of Arizona, USA

Petros Koumoutsakos
Institute of Computational Science
Swiss Federal Institute of Technology,
Switzerland

Akhlesh Lakhtakia
Pennsylvania State University, USA

Sergey A. Maksimenko
Belarus State University, Belarus

Russell Messier
Pennsylvania State University, USA

Krishna Muralidharan
University of Arizona, USA

Mary Beth Ruskai
Tufts University, USA

Vijay B. Shenoy
Indian Institute of Science, India

Gregory Ya. Slepyan
Belarus State University, Belarus

Jukka Tulkki
Helsinki University of Technology, Finland

Jens Walther
Institute of Computational Science
Swiss Federal Institute of Technology,
Switzerland

Thomas Werder
Institute of Computational Science
Swiss Federal Institute of Technology,
Switzerland

Urs Zimmerli
Institute of Computational Science
Swiss Federal Institute of Technology,
Switzerland

Chapter 1

Editorial

Akhlesh Lakhtakia

1.1 Introduction

Can any community of researchers remain unaware of the idea of nanotechnology today? Consider that the U.S. National Science Foundation launched the National Nanotechnology Initiative in 2002, accompanied by a website[1] with a special section for kids and a projected annual funding that exceeds $600M. Consider also that copies of Michael Crichton's 2002 book *Prey: A Novel*, in which he introduces the notion of predatory nanobots, have been lapped up members of both sexes at $27 per volume. Not surprisingly, pundits have pronounced on the future of nanotechnology in numerous publications.[2–6] Real as well as virtual journals on nanotechnology have sprouted, and not a week passes by when either a new conference on nanotechnology is not announced or a new book on nanotechnology is not published. Nanotechnology is shaping up as a megaideology—for the solution of *any* problem afflicting humanity—in the minds of many researchers as well as those who control research funds; and it could very well become a gigaideology when fully coupled in the United States with the theme of homeland security.

Skepticism about nanotechnology as a panacea has also been offered, on economic,[7] environmental,[8] as well as ethical[9] grounds. Indeed, beginning in the Iron Age and perhaps even earlier, our history provides numerous instances of false promises and unexpectedly deleterious outcomes of technological bonanzas. Yet, there is no doubt that we are materially better off than our great-grandparents were, leave aside our immediate evolutionary precursor species—and mostly because of technological progress. Therefore, even though nanotechnology may be a double-edged sword, we may be able to wield it in such a way as to cause the least harm all around.

Nanotechnology spans a vast mindscape in the world of academic, industrial, and governmental research; and I must stress that it is still in an embryonic stage despite a history that, some researchers say, spans two decades. The decision by both SPIE and ASME to launch the *Handbook of Nanotechnology* series therefore came at a very appropriate time. It will provide guidance on the state of the art to burgeoning ranks of nanotechnology researchers, and thus shape the contours of both experimental and theoretical research.

A huge fraction of nanotechnology research output is focused on synthesis and characterization of materials. Considerable attention is paid to potential and primitive devices as well, chiefly for biomedical applications and nanoelectromechanical systems. Reported research on modeling and simulation in nanotechnology, the scope of this volume, is scantier—as becomes evident on perusing the tables of contents of relevant journals and conference proceedings.

In part, the preponderance of experimental research over theoretical research in nanotechnology is due to the natural excitement about potentially revolutionary devices. In part also, the relative paucity of attention bestowed on modeling and simulation in nanotechnology derives from the Janusian characteristic of the nanoscale. Both macroscopic and molecular aspects apply at the nanoscale, sometimes simultaneously, sometimes not; and it becomes difficult to either handle together or decide between macroscopic and molecular approaches. This attribute of theoretical nanotechnology is clearly evident in the following eight chapters.

1.2 Coverage

Solid slabs and crystals have long been the workhorse materials of optics. Their nanotechnological counterparts today are thin solid films with engineered nanostructure and photonic crystals. In Chapter 2, A. Lakhtakia and R. Messier summarize developments regarding sculptured thin films (STFs). These films with unidirectionally varying properties can be designed and realized in a controllable manner using physical vapor deposition. The ability to virtually instantaneously change the growth direction of their columnar morphology through simple variations in the direction of the incident vapor flux leads to a wide spectrum of columnar forms. These forms can be 2D and 3D. Nominal nanoscopic-to-continuum models provide a way to extract structure-property relationships.

J. W. Haus describes, in Chapter 3, the optical properties of two- and three-dimensionally periodically nonhomogeneous materials called photonic band gap (PBG) structures. Analogous to crystals in some ways, a PBG structure enables the transmission of light through it in certain frequency bands, but not in others. Analytical, semianalytical, and numerical methods are presented along with programs for the reader to explore the band structure.

The last decade has witnessed an explosion in research on quantum dots. Progress in semiconductor technology, chiefly on epitaxial growth and lithography, has made it possible to fabricate structures wherein electrons are confined in dots that are 1 to 2 nm in diameter. In Chapter 4, F. Boxberg and J. Tulkki discuss the physical principles as well as experiments along with the first expected commercial applications of quantum dots.

In Chapter 5, S. A. Maksimenko and G. Ya. Slepyan formulate the nanoelectromagnetics of low-dimensional structures exemplified by carbon nanotubes and quantum dots. A wide range of theoretical results on the electromagnetic properties of carbon nanotubes as quasi-1D structures is presented in the first part of this chapter, spanning linear electrodynamics, nonlinear optical effects, and foundations of their quantum electrodynamics. In the second part of this chapter, a quantum dot is modeled as a spatially localized, two-level quantum oscillator illuminated by either classical or quantum light.

The availability of powerful supercomputers during the last decade has led to a proliferation of numerical studies on atomistic methods, such as molecular dynamics and Monte Carlo methods, which are grounded in classical statistical mechanics. Given a model for interaction between the discrete interacting units—howsoever small—of a material system, an energy formulation can be undertaken, and the microscopic states of that system can be sampled either deterministically or stochastically. P. A. Deymier, V. Kapila, and K. Muralidharan describe both classes of methods in Chapter 6.

In addition to electromagnetic modeling, mechanical modeling of devices is necessary for both fabrication and operation. In Chapter 7, therefore, V. B. Shenoy undertakes a discussion of mechanics at the nanoscale. The multiscale methods described in this chapter are meant to model the nanoscale mechanical behavior of materials as well as the mechanical behavior of nanostructures. Traditional continuum approaches having severe limitations at the nanoscale, atomistic methods must be resorted to. But atomistic methods are computationally intensive, which has engendered the emergence of hybrid methods.

The great potential of nanotechnology for biomedical applications has led to massive interest in nanofluidics. In Chapter 8, P. Koumoutsakos, U. Zimmerli, T. Werder, and J. H. Walther present a detailed account of nanoscale fluid mechanics. While discussing computational issues, the authors emphasize the choices of molecular interaction potentials and simulation boundary conditions, which critically control the physics of fluids. A careful review of experimental research is also provided.

The unremitting increase of device density in semiconductor chips brings quantum effects into the picture. Control of these quantum effects could be exploited to build quantum computers that would be more efficient than classical computers for some tasks. Whereas quantum computing devices are best described as barely embryonic, the mathematics of quantum information processing is progressing by leaps and bounds. A comprehensive account of quantum information processing is provided in Chapter 9 by M. B. Ruskai.

1.3 Concluding remark

The eight substantive chapters of *Nanometer Structures: Theory, Modeling, and Simulation* address those topics in nanotechnology that have acquired a reasonable degree of theoretical maturity in my opinion. No doubt, so rapid is the pace of progress in nanotechnology that later editions of this volume, not to mention volumes produced in the future by others, will offer coverage of topics neglected here.

In the meanwhile, I tender my apologies to any reader who feels that his or her area of theoretical research, modeling, and simulation suffered from editorial myopia.

References

1. http://www.nano.gov/
2. M. P. Frank and T. F. Knight, Jr., "Ultimate theoretical models of nanocomputers," *Nanotechnology* **9**, 162–176 (1998).
3. C. Hu, "Silicon nanoelectronics for the 21st century," *Nanotechnology* **10**, 113–116 (1999).
4. R. Tsu, "Challenges in nanoelectronics," *Nanotechnology* **12**, 625–628 (2001).
5. M. L. Cohen, "Nanotubes, nanoscience, and nanotechnology," *Mater. Sci. Eng. C* **15**, 1–11 (2001).
6. A. M. Stoneham, "The challenges of nanostructures for theory," *Mater. Sci. Eng. C* **23**, 235–241 (2003).
7. J. J. Gilman, "Nanotechnology," *Mater. Res. Innovat.* **5**, 12–14 (2001).
8. G. Brumfiel, "A little knowledge . . .," *Nature* **424**, 246–248 (2003).
9. A. Mnyusiwalla, A. S. Daar, and P. A. Singer, " 'Mind the gap': science and ethics in nanotechnology," *Nanotechnology* **14**, R9–R13 (2003).

Chapter 2

Sculptured Thin Films

Akhlesh Lakhtakia and Russell Messier

2.1. Introduction 6
2.2. Genesis 7
  2.2.1. Columnar thin films 7
  2.2.2. Primitive STFs with nematic morphology 9
  2.2.3. Chiral sculptured thin films 9
  2.2.4. Sculptured thin films 10
2.3. Electromagnetic fundamentals 11
  2.3.1. Linear constitutive relations 11
  2.3.2. From the nanostructure to the continuum 13
  2.3.3. Electromagnetic wave propagation 16
  2.3.4. Reflection and transmission 17
2.4. Dielectric STFs 21
  2.4.1. Relative permittivity dyadics 22
  2.4.2. Local homogenization 23
  2.4.3. Wave propagation 24
2.5. Applications 26
  2.5.1. Optical filters 26
  2.5.2. Optical fluid sensors 29
  2.5.3. Chiral PBG materials 29
  2.5.4. Displays 30
  2.5.5. Optical interconnects 30
  2.5.6. Optical pulse shapers 30
  2.5.7. Biochips 30
  2.5.8. Other applications 31
2.6. Directions for future research 32
References 33
List of symbols 42

2.1 Introduction

Sculptured thin films (STFs) are nanostructured materials with unidirectionally varying properties that can be designed and realized in a controllable manner using century-old techniques of physical vapor deposition (PVD).[1–4] The ability to virtually instantaneously change the growth direction of their columnar morphology through simple variations in the direction of the incident vapor flux leads to a wide spectrum of columnar forms. These forms can be (i) 2D, ranging from the simple slanted columns and chevrons to the more complex C- and S-shaped morphologies; and (ii) 3D, including simple helixes and superhelixes. A few examples of STFs are presented in Figs. 2.1 and 2.2.

For most optical applications envisioned, the column diameter and the column separation normal to the thickness direction of any STF should be constant. The column diameter can range from about 10 to 300 nm, while the density may lie between its theoretical maximum value to less than 20% thereof. The crystallinity must be at a scale smaller than the column diameter. The chemical composition is essentially unlimited, ranging from insulators to semiconductors to metals. Despite the fact that precursors of STFs have been made for over a century,[5–12] systematic

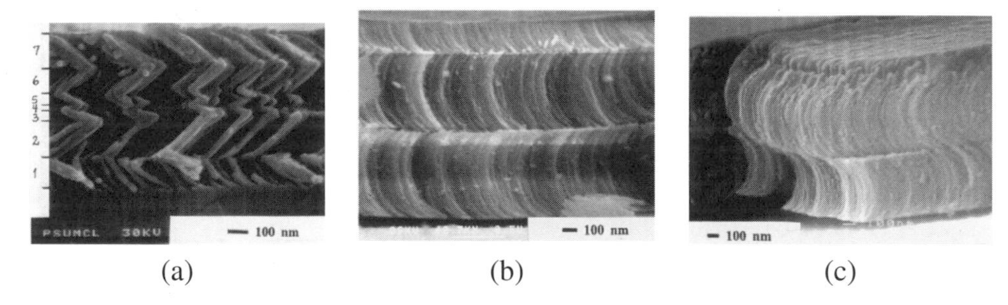

Figure 2.1 Scanning electron micrographs of sculptured thin films made of magnesium fluoride (MgF_2) with 2D morphologies: (a) 7-section zigzag, (b) C shaped, and (c) S shaped.

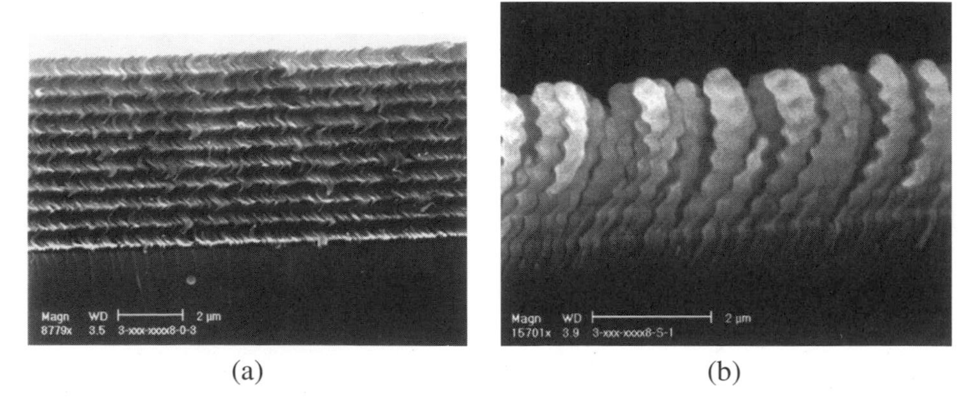

Figure 2.2 Scanning electron micrographs of sculptured thin films with 3D morphologies: (a) helical, made of silicon oxide (SiO), and (b) superhelical, made of MgF_2.

exploration of the science and technology of STFs began only during the mid-1990s.[3,4,13,14]

At visible and infrared wavelengths, a single-section STF is a unidirectionally nonhomogeneous continuum with direction-dependent properties. Several sections can be grown consecutively into a multisection STF, which can be conceived of as an optical circuit that can be integrated with electronic circuitry on a chip. Being porous, a STF can act as a sensor of fluids and also can be impregnated with liquid crystals for switching applications. Application as low-permittivity barrier layers in electronic chips has also been suggested. The first optical applications of STFs saw the light of the day in 1999.

This chapter is organized as follows: Sec. 2.2 traces the genesis of STFs from the columnar thin films first grown in the 1880s to the emergence of the STF concept during the 1990s. Section 2.3 describes STFs as unidirectionally nonhomogeneous, bianisotropic continuums at optical wavelengths; provides a nominal model to connect the nanostructure to the macroscopic electromagnetic response properties; and presents a matrix method to handle boundary value problems. Dielectric STFs are described in Sec. 2.4, followed by a survey of optical as well as other applications of STFs in Sec. 2.5. Directions for future research are suggested in Sec. 2.6.

A note on notation: Vectors are in boldface; dyadics (Ref. 15, Ch. 1) are in normal face and double underlined; column vectors and matrixes are in boldface and enclosed within square brackets. A dyadic can be interpreted as a 3×3 matrix throughout this chapter. The position vector is denoted by $\mathbf{r} = x\mathbf{u}_x + y\mathbf{u}_y + z\mathbf{u}_z$; the z axis is parallel to the thickness direction of all films; and an $\exp(-i\omega t)$ time dependence is implicit for all electromagnetic fields.

2.2 Genesis

2.2.1 Columnar thin films

Chronologically as well as morphologically, it is sensible to begin with the so-called columnar thin films (CTFs). Vapor from a source boat is directed towards a substrate in PVD, as shown in Fig. 2.3. Both sputtering and evaporation PVD techniques[16] deposit films at sufficiently low vapor pressures, so that the adatoms move toward the growing film surface with ballistic trajectories for which an average direction of arrival can be defined. At a low substrate temperature ($\lesssim$0.3 of the melting point of the depositing material), the arriving adatoms move very little on condensation. Instead, clustering at the 1- to 3-nm level occurs. The clusters evolve into clusters of clusters, which in turn evolve into expanding cones that compete with their neighbors for growth.[17,18] The surviving columns grow in the direction of the vapor flux, albeit somewhat closer to the substrate normal, as shown in Fig. 2.4.

The growth of nonnormal CTFs by the evaporation PVD technique at oblique angles is usually credited to Kundt[5] in 1885. It was the anisotropy of the optical properties of the films that focused interest on the columnar morphology.

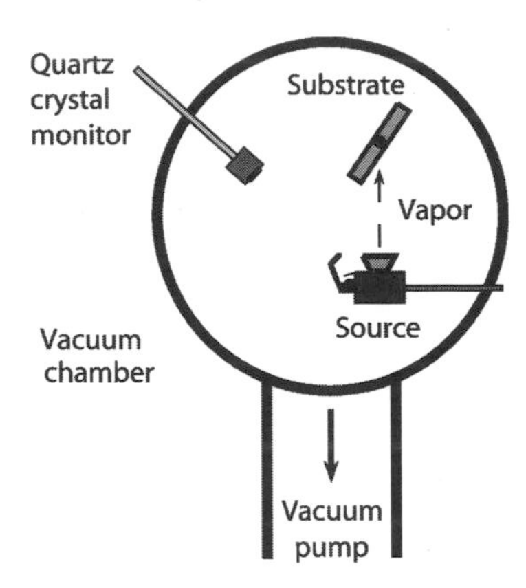

Figure 2.3 Schematic of the basic system for physical vapor deposition of columnar thin films on planar substrates. Although an electron-beam evaporation point source is shown, distributed directional sources—such as those used in sputter deposition—can be used to similar effect.

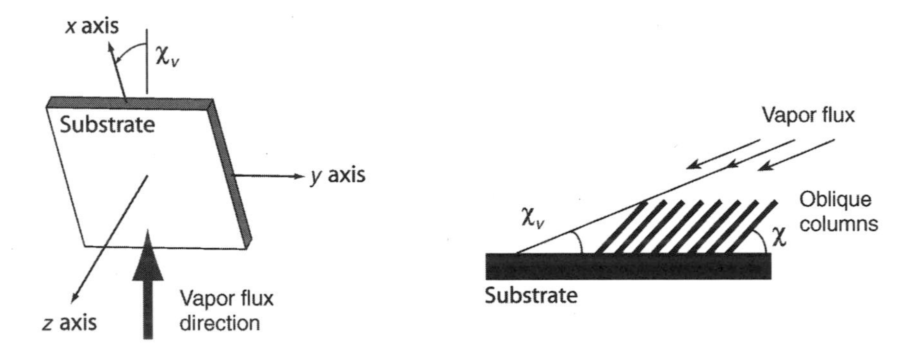

Figure 2.4 Coordinate system, the vapor incidence angle χ_v, and the column inclination angle χ.

The addition of ion bombardment during growth—either in sputtering or ion-assisted evaporation techniques—can eliminate columns, thereby yielding dense, smooth and stable thin films that meet the stringent requirements for laser-based applications of optical coatings.[19]

Significantly, an intermediate state occurs between columnar expansion and the elimination of the columns. In that state, competition between neighboring columns is frustrated[20] and stable columns grow. This CTF morphology is achieved either through intermediate levels of ion bombardment or simply by depositing the films at oblique angles.[21] The columns thus grow at a controllable angle $\chi \geq 25$ deg to the substrate, while the average direction of the incident vapor flux is delineated by the angle $\chi_v \leq \chi$ in Fig. 2.4.

In an extensive review of both experimental and ballistic aggregation modeling studies of obliquely deposited CTFs, van Kranenburg and Lodder[22] concluded that elongated clusters and columns generally pointing in the direction of the incoming vapor flux are a direct consequence of the adatomic self-shadowing process;

furthermore, when viewed from directly overhead, the length of the long axis relative to the width of the cluster increases markedly for $\chi_v < 30$ deg. The columns become separated and begin to grow as noncompeting cylinders—with elliptical cross sections due to anisotropy in self-shadowing[21,22]—as χ_v is decreased further. The columns become more separated in the vapor incidence direction due to the increased shadowing effect in the longitudinal direction (parallel to the vapor incidence plane), while shadowing in the transverse direction is unaffected by changes in χ_v. This leads to a higher material density in the transverse direction.

As the columnar cross-sectional dimensions are less than or equal to 150 nm for a large variety of CTFs, these films can be considered effectively as homogeneous orthorhombic continuums in the visible and infrared regimes, depending on the constitutive parameters of the deposited material.[23] Generally thought of as dielectric materials, their optical birefringence has long been appreciated and exploited.[24,25]

2.2.2 Primitive STFs with nematic morphology

A seminal event occurred in 1966 that eventually led to the emergence of the STF concept in 1994.[1] While a CTF was growing, Nieuwenhuizen and Haanstra deliberately altered χ_v to prove that columnar morphology "cannot be a result of the method of preparation itself."[11] The resulting change in χ was accomplished while the film thickness grew by just ~3 nm, the transition thus being practically abrupt in comparison to optical wavelengths. Some two decades later, Motohiro and Taga demonstrated that χ can be abruptly altered many times during growth,[12] which is the basis for realizing STFs with bent nematic morphologies.

Thus, primitive STFs with zigzag and chevronic morphologies came into existence. The similarity of CTFs to crystals had long been noticed in the optical literature,[24] so that the primitive STFs with nematic morphology can be considered as stacked crystalline plates. This has been astutely exploited for designing, fabricating, and testing various optical devices.[25,26] Furthermore, serial as well as simultaneous bideposition of CTFs and chevronic STFs are now routine in the manufacture of wave plates for the automobile industry.[27,28]

2.2.3 Chiral sculptured thin films

Another seminal event toward the emergence of the STF concept had already occurred in 1959. Although it had evidently been ignored then, all credit for periodic STFs with chiral (i.e., handed) morphology should be accorded to Young and Kowal.[8] Without actually seeing the anisotropic morphology of CTFs via scanning electron microscopy or otherwise, but surmising it from the well-known effects of anisotropy on optical response characteristics, these two pioneers consciously rotated the substrate about the z axis constantly during growth to create thin films with morphology predicted to display transmission optical activity. Most likely, they were the first researchers to deliberately engineer thin-film morphology for producing a nontrivial STF—one with a fully 3D morphology.

Remarkably, Young and Kowal inferred that "the [optical] activity of a helically deposited film could be due to the co-operative action of a helically symmetrical arrangement of crystallites, crystal growth or voids." Furthermore, they conjectured that the columnar direction could change virtually instantaneously and continuously with changes in the position and the orientation of the substrate. Happily, the Young–Kowal technique of rotating the substrate, the helicoidal morphology realized thereby, and the transmission optical activity of chiral STFs, were rediscovered in the last decade.[29–31]

2.2.4 Sculptured thin films

Recognition came during the 1990s that a very wide variety of columnar morphologies is possible, and that preparation-property-application connections can be truly engineered by coupling theoretical and experimental results.[1,2]

STFs are modifications of CTFs in which the column direction can be changed almost abruptly and often, even continuously, during growth. When CTFs are obliquely deposited, a wide variety of STF morphologies tailored at the nanoscale are realizable by simple variations of two fundamental axes of rotation, either separately or concurrently.[19,29–40] These fundamental axes lead to two canonical classes of STFs that have been termed

1. sculptured nematic thin films (SNTFs)[33] and
2. thin-film helicoidal bianisotropic mediums (TFHBMs).[1,29]

More complex shapes and even multisections, in which either the material or the shape or both are changed from section to section along the z axis, have been executed.[41,42]

SNTF morphologies include such simple 2D shapes as slanted columns, chevrons, and zigzags as well as the more complex C and S shapes; see Fig. 2.1. The substrate must be rotated about the y axis, which lies in the substrate plane and is perpendicular to the vapor incidence direction, while χ_v is varied either episodically or continuously.[33] One concern with this approach is related to the fact that the density of a CTF is highly dependent[4,33,35] on χ_v and, therefore, density variations are expected as a SNTF grows. The compensation of these variations is an area of future research.

TFHBMs are fabricated by tilting the substrate at some oblique angle to the incident vapor flux (i.e., $\chi_v \leq 90$ deg), followed by substrate rotation about the z axis. Helicoidal morphologies result for constant rotational velocity about the z axis.[8,29] By varying the rotational velocity in some prescribed manner throughout a rotational cycle, a slanted helicoidal structure occurs with the slant angle controllable over all χ above its minimum value for static glancing angle deposition. Furthermore, it is possible to engineer a wide range of superhelixes with controlled handedness.[4,19] The mass density as a function of film thickness is expected to remain constant since χ_v is fixed for TFHBMs, so long as the columns attain a steady-state diameter in the early nucleation and growth stages.

2.3 Electromagnetic fundamentals

2.3.1 Linear constitutive relations

The macroscopic conception of STFs at optical wavelengths is as unidirectionally nonhomogenous continuums, with the constitutive relations

$$\left.\begin{aligned} \mathbf{D}(\mathbf{r},\omega) &= \epsilon_0\big[\underline{\underline{\epsilon}}_r(z,\omega)\cdot\mathbf{E}(\mathbf{r},\omega)+\underline{\underline{\alpha}}_r(z,\omega)\cdot\mathbf{H}(\mathbf{r},\omega)\big] \\ \mathbf{B}(\mathbf{r},\omega) &= \mu_0\big[\underline{\underline{\beta}}_r(z,\omega)\cdot\mathbf{E}(\mathbf{r},\omega)+\underline{\underline{\mu}}_r(z,\omega)\cdot\mathbf{H}(\mathbf{r},\omega)\big] \end{aligned}\right\}, \tag{2.1}$$

indicating that the z axis of the coordinate system is aligned parallel to the direction of nonhomogeneity. These relations model the STF as a bianisotropic continuum,[43] with $\epsilon_0 = 8.854\times10^{-12}\ \mathrm{F\,m^{-1}}$ and $\mu_0 = 4\pi\times10^{-7}\ \mathrm{H\,m^{-1}}$ as the constitutive parameters of free space (i.e., vacuum). Whereas the relative permittivity dyadic $\underline{\underline{\epsilon}}_r(z,\omega)$ and the relative permeability dyadic $\underline{\underline{\mu}}_r(z,\omega)$ represent the electric and magnetic properties, respectively, the dyadics $\underline{\underline{\alpha}}_r(z,\omega)$ and $\underline{\underline{\beta}}_r(z,\omega)$ delineate the magnetoelectric properties.[44] These four constitutive dyadics have to be modeled with guidance from the STF morphology.

All of the columns in a single-section STF are nominally parallel to each other, and can be assumed to be rectifiable curves. A tangential unit vector can be prescribed at any point on a curves,[45] as shown in Fig. 2.5. Differential geometry can then be used to prescribe an osculating plane for the curve, leading to the identification of a normal unit vector. A third unit vector, called the binormal unit vector, is perpendicular to the first two unit vectors. These vectors may be written as $\underline{\underline{S}}(z)\cdot\mathbf{u}_\tau$, $\underline{\underline{S}}(z)\cdot\mathbf{u}_n$, and $\underline{\underline{S}}(z)\cdot\mathbf{u}_b$, for any particular column in the chosen STF. The rotation dyadic $\underline{\underline{S}}(z)$ incorporates the locus of points on the axis of the column; while the unit vectors $\mathbf{u}_\tau$, $\mathbf{u}_n$, and $\mathbf{u}_b$ should be chosen with the columnar cross section in mind. The rotation dyadic is some composition of the following three

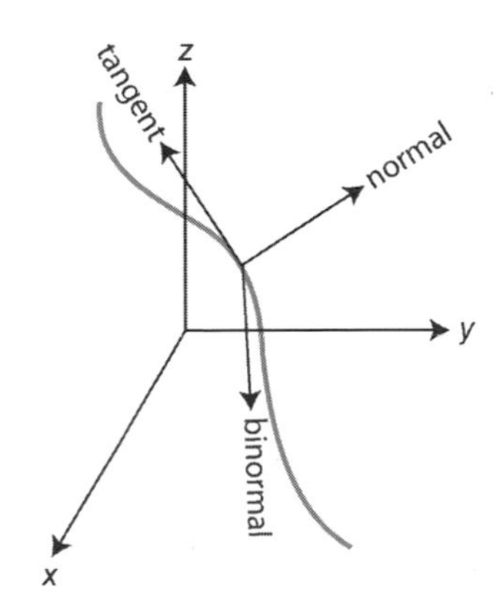

Figure 2.5 Tangential, normal, and binormal unit vectors at a point on a curve.

elementary rotation dyadics:

$$\underline{\underline{S}}_x(z) = \mathbf{u}_x\mathbf{u}_x + (\mathbf{u}_y\mathbf{u}_y + \mathbf{u}_z\mathbf{u}_z)\cos\xi(z) + (\mathbf{u}_z\mathbf{u}_y - \mathbf{u}_y\mathbf{u}_z)\sin\xi(z), \quad (2.2)$$

$$\underline{\underline{S}}_y(z) = \mathbf{u}_y\mathbf{u}_y + (\mathbf{u}_x\mathbf{u}_x + \mathbf{u}_z\mathbf{u}_z)\cos\tau(z) + (\mathbf{u}_z\mathbf{u}_x - \mathbf{u}_x\mathbf{u}_z)\sin\tau(z), \quad (2.3)$$

$$\underline{\underline{S}}_z(z) = \mathbf{u}_z\mathbf{u}_z + (\mathbf{u}_x\mathbf{u}_x + \mathbf{u}_y\mathbf{u}_y)\cos\zeta(z) + (\mathbf{u}_y\mathbf{u}_x - \mathbf{u}_x\mathbf{u}_y)\sin\zeta(z). \quad (2.4)$$

The angles $\xi(z)$, $\tau(z)$, and $\zeta(z)$ can be prescribed piecewise. The choice

$$\mathbf{u}_\tau = \mathbf{u}_x\cos\chi + \mathbf{u}_z\sin\chi, \quad (2.5)$$

$$\mathbf{u}_n = -\mathbf{u}_x\sin\chi + \mathbf{u}_z\cos\chi, \quad (2.6)$$

$$\mathbf{u}_b = -\mathbf{u}_y, \quad (2.7)$$

recalls the column inclination angle χ of CTFs, and is most appropriate for STFs.

Accordingly, the linear constitutive relations of a single-section STF are set up as[3,46]

$$\mathbf{D}(\mathbf{r},\omega) = \epsilon_0\,\underline{\underline{S}}(z)\cdot\left[\underline{\underline{\epsilon}}_{\mathrm{ref}}(\omega)\cdot\underline{\underline{S}}^T(z)\cdot\mathbf{E}(\mathbf{r},\omega) + \underline{\underline{\alpha}}_{\mathrm{ref}}(\omega)\cdot\underline{\underline{S}}^T(z)\cdot\mathbf{H}(\mathbf{r},\omega)\right], \quad (2.8)$$

$$\mathbf{B}(\mathbf{r},\omega) = \mu_0\,\underline{\underline{S}}(z)\cdot\left[\underline{\underline{\beta}}_{\mathrm{ref}}(\omega)\cdot\underline{\underline{S}}^T(z)\cdot\mathbf{E}(\mathbf{r},\omega) + \underline{\underline{\mu}}_{\mathrm{ref}}(\omega)\cdot\underline{\underline{S}}^T(z)\cdot\mathbf{H}(\mathbf{r},\omega)\right]. \quad (2.9)$$

The dyadics $\underline{\underline{\epsilon}}_{\mathrm{ref}}(\omega) = \underline{\underline{S}}^T(z)\cdot\underline{\underline{\epsilon}}_r(z,\omega)\cdot\underline{\underline{S}}(z)$, etc., are called the *reference* constitutive dyadics, because $\underline{\underline{S}}(z_0) = \underline{\underline{I}}$ in some reference plane $z = z_0$. Here and hereafter, $\underline{\underline{I}} = \mathbf{u}_x\mathbf{u}_x + \mathbf{u}_y\mathbf{u}_y + \mathbf{u}_z\mathbf{u}_z$ is the identity dyadic.

The foregoing equations reflect the fact that the morphology of a single-section STF in any plane $z = z_1$ can be made to nominally coincide with the morphology in another plane $z = z_2$ with the help of a suitable rotation. In conformity with the requirement that $\mathbf{u}_y\cdot\underline{\underline{S}}(z) \equiv \mathbf{u}_y\ \forall z$, the choice $\underline{\underline{S}}(z) = \underline{\underline{S}}_y(z)$ is appropriate for STFs with nematic morphology. For TFHBMs, the correct choice is $\underline{\underline{S}}(z) = \underline{\underline{S}}_z(z)$. Although a helicoidal STF need not be *periodically* nonhomogeneous along the z axis, it is easy to fabricate such films with periods chosen anywhere between 50 and 2000 nm. Chiral STFs are generally analyzed as periodic dielectric TFHBMs with $\zeta(z) = \pi z/\Omega$ in Eq. (2.4), where 2Ω is the structural period.[13,47] More complicated specifications of $\underline{\underline{S}}(z)$ are possible—to wit, slanted chiral STFs.[48,49]

The choice

$$\underline{\underline{\sigma}}_{\mathrm{ref}}(\omega) = \sigma_a(\omega)\mathbf{u}_n\mathbf{u}_n + \sigma_b(\omega)\mathbf{u}_\tau\mathbf{u}_\tau + \sigma_c(\omega)\mathbf{u}_b\mathbf{u}_b \quad (2.10)$$

is in accord with the local orthorhombicity of STFs. The density anisotropy occurring during deposition is thus taken into account. For magneto-optics, gyrotropic terms such as $i\sigma_g(\omega)\mathbf{u}_\tau\times\underline{\underline{I}}$ can be added to the right side of Eq. (2.10).[50]

A multisection STF is a cascade of single-section STFs fabricated in an integrated manner.[2] Substrate rotational dynamics may be chosen differently for each section, and the rotation dyadic $\underline{\underline{S}}(z)$ then must be specified sectionwise. The deposited material(s) and/or the vapor incidence angle may also be changed from section to section, so that the constitutive dyadics $\underline{\underline{\epsilon}}_{ref}(\omega)$, $\underline{\underline{\mu}}_{ref}(\omega)$, $\underline{\underline{\alpha}}_{ref}(\omega)$, and $\underline{\underline{\beta}}_{ref}(\omega)$ are different for each section. Furthermore, the constitutive dyadics will be affected by the substrate rotational dynamics in each section. Since renucleation clusters are 3 to 5 nm in diameter, the transition between two consecutive sections is virtually abrupt and, therefore, optically insignificant.[41,42]

2.3.2 From the nanostructure to the continuum

Implicit in the constitutive relations of Eq. (2.1) is the assumption of a STF as a continuous medium. The relationship of the nanostructure to the macroscopic constitutive dyadics must be modeled adequately for intelligent design and fabrication of STF devices.

As any STF can be viewed as a composite material with two different constituent materials, the constitutive dyadics $\underline{\underline{\epsilon}}_r(z,\omega)$, etc., must emerge from both composition and morphology. The mathematical process describing this transition from the microscopic to the continuum length scales is called *homogenization*. It is very commonly implemented in various forms for random distributions of electrically small inclusions in an otherwise homogeneous host material (Ref. 23 and Ref. 51, Ch. 4); and homogenization research continues to flourish.[52,53]

But, as the inclusions are randomly distributed, the *effective* constitutive dyadics computed with any particular homogenization formalism are independent of position. In contrast, a STF is effectively a nonhomogeneous continuum, because the orientation of inclusions of the deposited material must depend on z. This is a serious difficulty, when devising structure-property relationships.

If the aim is just to construct a control model to span the nanostructure-continuum divide for manufacturing STFs with desirable optical response characteristics, the homogenization procedure can be *localized*.[54] In a nominal model being presently developed,[54–56] the deposited material as well as the voids are to be thought of as parallel ellipsoidal inclusions in any thin slice of the STF parallel to substrate plane. Each slice is homogenized in the local homogenization procedure. But any two consecutive slices in a single-section STF are identical, except for a small rotation captured by $\underline{\underline{S}}(z)$. This dyadic is presumably known, either from examination of scanning electron micrographs or because it was programmed into the fabrication process. Therefore, in this nominal model, the aim of the local homogenization procedure for a STF changes from estimating $\underline{\underline{\epsilon}}_r(z,\omega)$, etc., in Eq. (2.1) to estimating $\underline{\underline{\epsilon}}_{ref}(\omega)$, etc., in Eqs. (2.8) and (2.9).

Suppose that the chosen single-section STF is made of a bianisotropic material whose bulk constitutive relations are specified as

$$\left.\begin{aligned} \mathbf{D}(\mathbf{r},\omega) &= \epsilon_0\big[\underline{\underline{\epsilon}}_s(\omega)\cdot\mathbf{E}(\mathbf{r},\omega)+\underline{\underline{\alpha}}_s(\omega)\cdot\mathbf{H}(\mathbf{r},\omega)\big] \\ \mathbf{B}(\mathbf{r},\omega) &= \mu_0\big[\underline{\underline{\beta}}_s(\omega)\cdot\mathbf{E}(\mathbf{r},\omega)+\underline{\underline{\mu}}_s(\omega)\cdot\mathbf{H}(\mathbf{r},\omega)\big] \end{aligned}\right\}. \tag{2.11}$$

The voids in the STF are taken to be occupied by a material with the following bulk constitutive relations:

$$\left.\begin{aligned}\mathbf{D}(\mathbf{r},\omega) &= \epsilon_0\big[\underline{\underline{\epsilon}}_v(\omega)\cdot\mathbf{E}(\mathbf{r},\omega) + \underline{\underline{\alpha}}_v(\omega)\cdot\mathbf{H}(\mathbf{r},\omega)\big]\\ \mathbf{B}(\mathbf{r},\omega) &= \mu_0\big[\underline{\underline{\beta}}_v(\omega)\cdot\mathbf{E}(\mathbf{r},\omega) + \underline{\underline{\mu}}_v(\omega)\cdot\mathbf{H}(\mathbf{r},\omega)\big]\end{aligned}\right\}. \tag{2.12}$$

The voids may not necessarily be vacuous; in fact, scanning electron microscopy shows that voids should be considered as low-density regions. The nominal porosity of the STF is denoted by f_v, $(0 \le f_v \le 1)$, which is actually the void volume fraction.

Each column in the chosen STF is represented as a string of ellipsoids in the nominal model, as shown in Fig. 2.6. In the thin slice containing the reference plane $z = z_0$—defined by the condition $\underline{\underline{S}}(z_0) = \underline{\underline{I}}$—the surface of a particular ellipsoid is delineated by the position vectors

$$\mathbf{r}(\vartheta,\varphi) = \delta_s \underline{\underline{U}}_s \cdot (\sin\vartheta\cos\varphi\,\mathbf{u}_n + \cos\vartheta\,\mathbf{u}_\tau + \sin\vartheta\sin\varphi\,\mathbf{u}_b),$$
$$\vartheta \in [0,\pi], \quad \varphi \in [0,2\pi], \tag{2.13}$$

with respect to the ellipsoidal centroid. In this equation, δ_s is a linear measure of size and the shape dyadic

$$\underline{\underline{U}}_s = \mathbf{u}_n\mathbf{u}_n + \gamma_\tau^{(s)}\mathbf{u}_\tau\mathbf{u}_\tau + \gamma_b^{(s)}\mathbf{u}_b\mathbf{u}_b. \tag{2.14}$$

Setting the shape factors $\gamma_\tau^{(s)} \gg 1$ and $\gamma_b^{(s)} \gtrsim 1$ will make each ellipsoid resemble a needle with a slight bulge in its middle part. The voids in the reference thin slice can also be represented by similarly aligned ellipsoids whose shape dyadic is

$$\underline{\underline{U}}_v = \mathbf{u}_n\mathbf{u}_n + \gamma_\tau^{(v)}\mathbf{u}_\tau\mathbf{u}_\tau + \gamma_b^{(v)}\mathbf{u}_b\mathbf{u}_b. \tag{2.15}$$

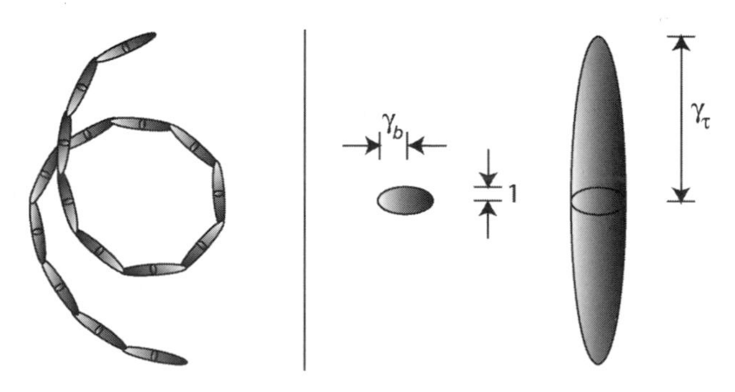

Figure 2.6 A column modeled as a string of electrically small ellipsoids, and the shape factors γ_τ and γ_b of an ellipsoid.

The use of 6×6 matrixes provides notational simplicity for treating electromagnetic fields in bianisotropic materials. Let us therefore define the 6×6 constitutive matrixes

$$[\mathbf{C}]_{\mathrm{ref},s,v} = \left[\begin{array}{c|c} \epsilon_0[\epsilon]_{\mathrm{ref},s,v} & \epsilon_0[\alpha]_{\mathrm{ref},s,v} \\ \hline \mu_0[\beta]_{\mathrm{ref},s,v} & \mu_0[\mu]_{\mathrm{ref},s,v} \end{array}\right], \tag{2.16}$$

where $[\epsilon]_{\mathrm{ref}}$ is the 3×3 matrix equivalent to $\underline{\underline{\epsilon}}_{\mathrm{ref}}$, etc. The ω dependences of various quantities are not explicitly mentioned in this and the following equations for compactness. Many homogenization formalisms can be chosen to determine $[\mathbf{C}]_{\mathrm{ref}}$ from $[\mathbf{C}]_s$ and $[\mathbf{C}]_v$, but the Bruggeman formalism[52,55] appears particularly attractive because of its simplicity as well as its widespread use in optics.[23]

For this purpose, the 6×6 polarizability density matrixes

$$[\mathbf{A}]_{s,v} = ([\mathbf{C}]_{s,v} - [\mathbf{C}]_{\mathrm{ref}}) \cdot \left\{[\mathbf{I}] + i\omega[\mathbf{D}]_{s,v} \cdot ([\mathbf{C}]_{s,v} - [\mathbf{C}]_{\mathrm{ref}})\right\}^{-1} \tag{2.17}$$

are set up, where $[\mathbf{I}]$ is the 6×6 identity matrix. The 6×6 depolarization matrixes $[\mathbf{D}]_{s,v}$ must be computed via 2D integration as follows:

$$[\mathbf{D}]_{s,v} = \frac{1}{4\pi i\omega\epsilon_0\mu_0} \int_{\varphi=0}^{2\pi} \int_{\vartheta=0}^{\pi} \frac{\sin\vartheta}{\Lambda_{s,v}} \times \left[\begin{array}{c|c} \mu_0[\mathbf{w}]_{s,v}[\mu]_{\mathrm{ref}}[\mathbf{w}]_{s,v} & -\epsilon_0[\mathbf{w}]_{s,v}[\alpha]_{\mathrm{ref}}[\mathbf{w}]_{s,v} \\ \hline -\mu_0[\mathbf{w}]_{s,v}[\beta]_{\mathrm{ref}}[\mathbf{w}]_{s,v} & \epsilon_0[\mathbf{w}]_{s,v}[\epsilon]_{\mathrm{ref}}[\mathbf{w}]_{s,v} \end{array}\right] d\vartheta\, d\varphi. \tag{2.18}$$

In the foregoing equation, the scalars

$$\Lambda_{s,v} = \left(\mathbf{v}_{s,v} \cdot \underline{\underline{\epsilon}}_{\mathrm{ref}} \cdot \mathbf{v}_{s,v}\right)\left(\mathbf{v}_{s,v} \cdot \underline{\underline{\mu}}_{\mathrm{ref}} \cdot \mathbf{v}_{s,v}\right) - \left(\mathbf{v}_{s,v} \cdot \underline{\underline{\alpha}}_{\mathrm{ref}} \cdot \mathbf{v}_{s,v}\right)\left(\mathbf{v}_{s,v} \cdot \underline{\underline{\beta}}_{\mathrm{ref}} \cdot \mathbf{v}_{s,v}\right), \tag{2.19}$$

the 3×3 matrixes $[\mathbf{w}]_{s,v}$ are equivalent to the dyads

$$\underline{\underline{w}}_{s,v} = \mathbf{v}_{s,v}\mathbf{v}_{s,v}, \tag{2.20}$$

and

$$\mathbf{v}_{s,v} = \underline{\underline{U}}_{s,v}^{-1} \cdot (\sin\vartheta\cos\varphi\,\mathbf{u}_n + \cos\vartheta\,\mathbf{u}_\tau + \sin\vartheta\sin\varphi\,\mathbf{u}_b). \tag{2.21}$$

The Bruggeman formalism requires the solution of the matrix equation

$$f_v[\mathbf{A}]_v + (1 - f_v)[\mathbf{A}]_s = [\mathbf{0}], \tag{2.22}$$

with $[\mathbf{0}]$ as the 6×6 null matrix. This equation has to be numerically solved for $[\mathbf{C}]_{\text{ref}}$, and a Jacobi iteration technique is recommended for that purpose.[52,56]

The solution of Eq. (2.22) represents the homogenization of an ensemble of objects of microscopic linear dimensions into a continuum. The quantities entering $\underline{\underline{S}}(z)$ are to be fixed prior to fabrication, as also are $[\mathbf{C}]_s$ and $[\mathbf{C}]_v$. To calibrate the nominal model presented, the shape dyadics $\underline{\underline{U}}_s$ and $\underline{\underline{U}}_v$ can be chosen by comparison of the predicted $[\mathbf{C}]_{\text{ref}}$ against measured data.[57]

2.3.3 Electromagnetic wave propagation

Electromagnetic wave propagation in a STF is best handled using 4×4 matrixes and column vectors of size 4. At any given frequency, with the transverse wave number κ and the angle ψ fixed by excitation conditions, the following spatial Fourier representation of the electric and the magnetic field phasors is useful:

$$\left.\begin{aligned} \mathbf{E}(\mathbf{r}, \omega) &= \mathbf{e}(z, \kappa, \psi, \omega) \exp[i\kappa(x \cos\psi + y \sin\psi)] \\ \mathbf{H}(\mathbf{r}, \omega) &= \mathbf{h}(z, \kappa, \psi, \omega) \exp[i\kappa(x \cos\psi + y \sin\psi)] \end{aligned}\right\}. \tag{2.23}$$

Substitution of the foregoing representation into the source-free Maxwell curl postulates $\nabla \times \mathbf{E}(\mathbf{r}, \omega) = i\omega\mathbf{B}(\mathbf{r}, \omega)$ and $\nabla \times \mathbf{H}(\mathbf{r}, \omega) = -i\omega\mathbf{D}(\mathbf{r}, \omega)$, followed by the use of the constitutive relations, leads to four ordinary differential equations and two algebraic equations. The phasor components $e_z(z, \kappa, \psi, \omega)$ and $h_z(z, \kappa, \psi, \omega)$ are then eliminated to obtain the 4×4 matrix ordinary differential equation (MODE)[46]

$$\frac{d}{dz}[\mathbf{f}(z, \kappa, \psi, \omega)] = i[\mathbf{P}(z, \kappa, \psi, \omega)][\mathbf{f}(z, \kappa, \psi, \omega)]. \tag{2.24}$$

In this equation,

$$[\mathbf{f}(z, \kappa, \psi, \omega)] = \begin{bmatrix} e_x(z, \kappa, \psi, \omega) \\ e_y(z, \kappa, \psi, \omega) \\ h_x(z, \kappa, \psi, \omega) \\ h_y(z, \kappa, \psi, \omega) \end{bmatrix} \tag{2.25}$$

is a column vector, and $[\mathbf{P}(z, \kappa, \psi, \omega)]$ is a 4×4 matrix function of z that can be easily obtained using symbolic manipulation programs. The 4×4 system can reduce to two autonomous 2×2 systems in special cases, e.g., for propagation in the morphologically significant planes of single-section SNTFs.[58]

Analytic solution of Eq. (2.24) can be obtained, provided $[\mathbf{P}(z,\kappa,\psi,\omega)]$ is not a function of z {i.e., $[\mathbf{P}(z,\kappa,\psi,\omega)] = [\mathbf{P}_{\mathrm{con}}(\kappa,\psi,\omega)]$}—which happens for CTFs. Exact analytic solution of Eq. (2.24) has been obtained also for axial propagation (i.e., $\kappa = 0$) in periodic TFHBMs and chiral STFs.[47,59,60] A solution in terms of a convergent matrix polynomial series is available for nonaxial propagation (i.e., $\kappa \neq 0$) in periodic TFHBMs.[61–64]

More generally, only a numerical solution of Eq. (2.24) can be obtained. If the matrix $[\mathbf{P}(z,\kappa,\psi,\omega)]$ is a periodic function of z, a perturbative approach[65] can be used to obtain simple results for weakly anisotropic STFs;[61,66] coupled-wave methods can come in handy, if otherwise.[48,67–69]

But if $[\mathbf{P}(z,\kappa,\psi,\omega)]$ is not periodic, the constitutive dyadics can be assumed as piecewise constant—i.e., constant over slices of thickness Δz—and the approximate transfer equation[46]

$$[\mathbf{f}(z+\Delta z,\kappa,\psi,\omega)] \simeq \exp\left\{ i \left[\mathbf{P}\left(z+\frac{\Delta z}{2},\kappa,\psi,\omega \right) \right] \Delta z \right\} [\mathbf{f}(z,\kappa,\psi,\omega)] \tag{2.26}$$

can be suitably manipulated with appropriately small values of Δz. This numerical technique has been applied to chiral STFs.[48,62]

Regardless of the method used to solve Eq. (2.24), it can be used to formulate a matrizant. Defined via the transfer equation

$$[\mathbf{f}(z,\kappa,\psi,\omega)] = [\mathbf{M}(z,\kappa,\psi,\omega)]\,[\mathbf{f}(0,\kappa,\psi,\omega)], \tag{2.27}$$

the matrizant $[\mathbf{M}]$ is the solution of the differential equation

$$\frac{d}{dz}[\mathbf{M}(z,\kappa,\psi,\omega)] = i[\mathbf{P}(z,\kappa,\psi,\omega)]\,[\mathbf{M}(z,\kappa,\psi,\omega)]. \tag{2.28}$$

Only one boundary value of the matrizant is needed to determine it uniquely, and that boundary value is supplied by Eq. (2.27) as

$$[\mathbf{M}(0,\kappa,\psi,\omega)] = [\mathbf{I}], \tag{2.29}$$

where $[\mathbf{I}]$ is the 4×4 identity matrix.

Finally, quasi-static solutions of Eq. (2.24) can be obtained in the same ways, after taking the limit $\omega \to 0$ *ab initio*.[70] These are useful if applications of STFs in the microwave and lower-frequency regimes are desired—for example, as interlayer dielectrics in integrated electronic circuits[71,72] and for humidity sensors that rely on capacitance change induced by altered humidity.[73]

2.3.4 Reflection and transmission

The quintessential problem for optics is that of the reflection and transmission of a plane wave by a STF of thickness L. Suppose that the half-spaces $z \leq 0$ and $z \geq L$

are vacuous. An arbitrarily polarized plane wave is obliquely incident on the STF from the half-space $z \leq 0$. As a result, there is a reflected plane wave in the same half-space, as well as a transmitted plane wave in the half-space $z \geq L$.

The propagation vector of the obliquely incident plane wave makes an angle $\theta \in [0, \pi/2)$ with respect to the $+z$ axis, and is inclined to the x axis in the xy plane by an angle $\psi \in [0, 2\pi]$, as shown in Fig. 2.7. Accordingly, the transverse wave number

$$\kappa = k_0 \sin\theta, \tag{2.30}$$

where $k_0 = \omega\sqrt{\epsilon_0\mu_0}$ is the free-space wave number. Evanescent plane waves can be taken into account as well by making the angle θ complex-valued.[74]

The incident plane wave is conventionally represented in terms of linear polarization components in the optics literature. An equivalent description in terms of circular polarization components is more appropriate for chiral STFs. Thus, the incident plane wave is delineated by the phasors

$$\left.\begin{aligned} \mathbf{e}_{\mathrm{inc}}(z) &= \begin{cases} (a_s\mathbf{s} + a_p\mathbf{p}_+)e^{ik_0 z\cos\theta} \\ \left(a_L\dfrac{i\mathbf{s} - \mathbf{p}_+}{\sqrt{2}} - a_R\dfrac{i\mathbf{s} + \mathbf{p}_+}{\sqrt{2}}\right)e^{ik_0 z\cos\theta} \end{cases} \\ \mathbf{h}_{\mathrm{inc}}(z) &= \begin{cases} \eta_0^{-1}(a_s\mathbf{p}_+ - a_p\mathbf{s})e^{ik_0 z\cos\theta} \\ -i\eta_0^{-1}\left(a_L\dfrac{i\mathbf{s} - \mathbf{p}_+}{\sqrt{2}} + a_R\dfrac{i\mathbf{s} + \mathbf{p}_+}{\sqrt{2}}\right)e^{ik_0 z\cos\theta} \end{cases} \end{aligned}\right\}, \quad z \leq 0, \tag{2.31}$$

where $\eta_0 = \sqrt{\mu_0/\epsilon_0}$ is the intrinsic impedance of free space; a_s and a_p are the amplitudes of the perpendicular- and parallel-polarized components, respectively; a_L and a_R are the amplitudes of the left and right circularly polarized (LCP and

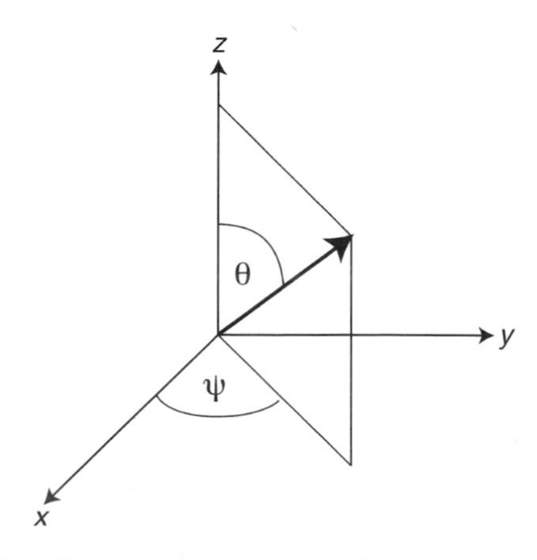

Figure 2.7 Propagation direction of incident plane wave.

RCP) components; and the plane-wave polarization vectors

$$\mathbf{s} = -\mathbf{u}_x \sin\psi + \mathbf{u}_y \cos\psi, \tag{2.32}$$

$$\mathbf{p}_\pm = \mp(\mathbf{u}_x \cos\psi + \mathbf{u}_y \sin\psi)\cos\theta + \mathbf{u}_z \sin\theta \tag{2.33}$$

are of unit magnitude. For notational simplicity, the dependences on κ, ψ, and ω are explicitly mentioned from this point onward only if necessary.

The electromagnetic field phasors associated with the reflected and transmitted plane waves, respectively, are expressed by

$$\left.\begin{aligned} \mathbf{e}_{\mathrm{ref}}(z) &= \begin{cases} (r_s\mathbf{s} + r_p\mathbf{p}_-)e^{-ik_0 z\cos\theta} \\ \left(-r_L\dfrac{i\mathbf{s} - \mathbf{p}_-}{\sqrt{2}} + r_R\dfrac{i\mathbf{s} + \mathbf{p}_-}{\sqrt{2}}\right)e^{-ik_0 z\cos\theta} \end{cases} \\ \mathbf{h}_{\mathrm{ref}}(z) &= \begin{cases} \eta_0^{-1}(r_s\mathbf{p}_- - r_p\mathbf{s})e^{-ik_0 z\cos\theta} \\ i\eta_0^{-1}\left(r_L\dfrac{i\mathbf{s} - \mathbf{p}_-}{\sqrt{2}} + r_R\dfrac{i\mathbf{s} + \mathbf{p}_-}{\sqrt{2}}\right)e^{-ik_0 z\cos\theta} \end{cases} \end{aligned}\right\}, \quad z \le 0, \tag{2.34}$$

and

$$\left.\begin{aligned} \mathbf{e}_{\mathrm{tr}}(z) &= \begin{cases} (t_s\mathbf{s} + t_p\mathbf{p}_+)e^{ik_0(z-L)\cos\theta} \\ \left(t_L\dfrac{i\mathbf{s} - \mathbf{p}_+}{\sqrt{2}} - t_R\dfrac{i\mathbf{s} + \mathbf{p}_+}{\sqrt{2}}\right)e^{ik_0(z-L)\cos\theta} \end{cases} \\ \mathbf{h}_{\mathrm{tr}}(z) &= \begin{cases} \eta_0^{-1}(t_s\mathbf{p}_+ - t_p\mathbf{s})e^{ik_0(z-L)\cos\theta} \\ -i\eta_0^{-1}\left(t_L\dfrac{i\mathbf{s} - \mathbf{p}_+}{\sqrt{2}} + t_R\dfrac{i\mathbf{s} + \mathbf{p}_+}{\sqrt{2}}\right)e^{ik_0(z-L)\cos\theta} \end{cases} \end{aligned}\right\}, \quad z \ge L. \tag{2.35}$$

The amplitudes $r_{s,p}$ and $t_{s,p}$ indicate the strengths of the perpendicular- and parallel-polarized components of the reflected and transmitted plane waves, both of which are elliptically polarized in general. Equivalently, the amplitudes $r_{L,R}$ and $t_{L,R}$ indicate the strengths of the LCP and RCP components.

The transfer matrix of a STF of thickness L is $[\mathbf{M}(L,\kappa,\psi,\omega)]$, because the relationship

$$[\mathbf{f}(L,\kappa,\psi,\omega)] = [\mathbf{M}(L,\kappa,\psi,\omega)]\,[\mathbf{f}(0,\kappa,\psi,\omega)] \tag{2.36}$$

between the two boundary values of $[\mathbf{f}(z,\kappa,\psi,\omega)]$ follows from Eq. (2.27). As the tangential components of $\mathbf{E}(\mathbf{r},\omega)$ and $\mathbf{H}(\mathbf{r},\omega)$ must be continuous across the planes $z = 0$ and $z = L$, the boundary values $[\mathbf{f}(0,\kappa,\psi,\omega)]$ and $[\mathbf{f}(L,\kappa,\psi,\omega)]$ can be fixed by virtue of Eqs. (2.31) to (2.35). Hence,

$$[\mathbf{f}(0,\kappa,\psi,\omega)] = [\mathbf{K}(\theta,\psi)] \begin{bmatrix} a_s \\ a_p \\ r_s \\ r_p \end{bmatrix} = \frac{[\mathbf{K}(\theta,\psi)]}{\sqrt{2}} \begin{bmatrix} i(a_L - a_R) \\ -(a_L + a_R) \\ -i(r_L - r_R) \\ r_L + r_R \end{bmatrix}, \tag{2.37}$$

$$[\mathbf{f}(L,\kappa,\psi,\omega)] = [\mathbf{K}(\theta,\psi)] \begin{bmatrix} t_s \\ t_p \\ 0 \\ 0 \end{bmatrix} = \frac{[\mathbf{K}(\theta,\psi)]}{\sqrt{2}} \begin{bmatrix} i(t_L - t_R) \\ -(t_L + t_R) \\ 0 \\ 0 \end{bmatrix}, \tag{2.38}$$

where the 4×4 matrix

$$[\mathbf{K}(\theta,\psi)] = \begin{bmatrix} -\sin\psi & -\cos\psi\cos\theta & -\sin\psi & \cos\psi\cos\theta \\ \cos\psi & -\sin\psi\cos\theta & \cos\psi & \sin\psi\cos\theta \\ -\eta_0^{-1}\cos\psi\cos\theta & \eta_0^{-1}\sin\psi & \eta_0^{-1}\cos\psi\cos\theta & \eta_0^{-1}\sin\psi \\ -\eta_0^{-1}\sin\psi\cos\theta & -\eta_0^{-1}\cos\psi & \eta_0^{-1}\sin\psi\cos\theta & -\eta_0^{-1}\cos\psi \end{bmatrix}. \tag{2.39}$$

The plane-wave reflection/transmission problem then amounts to four simultaneous, linear algebraic equation stated in matrix form as

$$\begin{bmatrix} t_s \\ t_p \\ 0 \\ 0 \end{bmatrix} = [\mathbf{K}(\theta,\psi)]^{-1}\,[\mathbf{M}(L,\kappa,\psi,\omega)]\,[\mathbf{K}(\theta,\psi)] \begin{bmatrix} a_s \\ a_p \\ r_s \\ r_p \end{bmatrix}, \tag{2.40}$$

equivalently,

$$\begin{bmatrix} i(t_L - t_R) \\ -(t_L + t_R) \\ 0 \\ 0 \end{bmatrix} = [\mathbf{K}(\theta,\psi)]^{-1}\,[\mathbf{M}(L,\kappa,\psi,\omega)]\,[\mathbf{K}(\theta,\psi)] \begin{bmatrix} i(a_L - a_R) \\ -(a_L + a_R) \\ -i(r_L - r_R) \\ r_L + r_R \end{bmatrix}. \tag{2.41}$$

These sets of equations can be solved by standard matrix manipulations to compute the reflection and transmission amplitudes when the incidence amplitudes are known.

It is usually convenient to define reflection and transmission coefficients. These appear as the elements of the 2×2 matrixes in the following relations:

$$\begin{bmatrix} r_s \\ r_p \end{bmatrix} = \begin{bmatrix} r_{ss} & r_{sp} \\ r_{ps} & r_{pp} \end{bmatrix} \begin{bmatrix} a_s \\ a_p \end{bmatrix}, \quad \begin{bmatrix} r_L \\ r_R \end{bmatrix} = \begin{bmatrix} r_{LL} & r_{LR} \\ r_{RL} & r_{RR} \end{bmatrix} \begin{bmatrix} a_L \\ a_R \end{bmatrix}, \tag{2.42}$$

$$\begin{bmatrix} t_s \\ t_p \end{bmatrix} = \begin{bmatrix} t_{ss} & t_{sp} \\ t_{ps} & t_{pp} \end{bmatrix} \begin{bmatrix} a_s \\ a_p \end{bmatrix}, \quad \begin{bmatrix} t_L \\ t_R \end{bmatrix} = \begin{bmatrix} t_{LL} & t_{LR} \\ t_{RL} & t_{RR} \end{bmatrix} \begin{bmatrix} a_L \\ a_R \end{bmatrix}. \tag{2.43}$$

Copolarized coefficients have both subscripts identical, but cross-polarized coefficients do not. The relationships between the linear and circular coefficients are as follows:

$$\left.\begin{aligned} r_{ss} &= -\frac{(r_{LL}+r_{RR})-(r_{LR}+r_{RL})}{2} \\ r_{sp} &= i\frac{(r_{LL}-r_{RR})+(r_{LR}-r_{RL})}{2} \\ r_{ps} &= -i\frac{(r_{LL}-r_{RR})-(r_{LR}-r_{RL})}{2} \\ r_{pp} &= -\frac{(r_{LL}+r_{RR})+(r_{LR}+r_{RL})}{2} \end{aligned}\right\}, \tag{2.44}$$

$$\left.\begin{aligned} t_{ss} &= \frac{(t_{LL}+t_{RR})-(t_{LR}+t_{RL})}{2} \\ t_{sp} &= -i\frac{(t_{LL}-t_{RR})+(t_{LR}-t_{RL})}{2} \\ t_{ps} &= i\frac{(t_{LL}-t_{RR})-(t_{LR}-t_{RL})}{2} \\ t_{pp} &= \frac{(t_{LL}+t_{RR})+(t_{LR}+t_{RL})}{2} \end{aligned}\right\}. \tag{2.45}$$

The square of the magnitude of a reflection/transmission coefficient is the corresponding reflectance/transmittance; thus, $R_{LR} = |r_{LR}|^2$ is the reflectance corresponding to the reflection coefficient r_{LR}, and so on. The principle of conservation of energy mandates the constraints

$$\left.\begin{aligned} R_{ss}+R_{ps}+T_{ss}+T_{ps} \le 1 \\ R_{pp}+R_{sp}+T_{pp}+T_{sp} \le 1 \\ R_{LL}+R_{RL}+T_{LL}+T_{RL} \le 1 \\ R_{RR}+R_{LR}+T_{RR}+T_{LR} \le 1 \end{aligned}\right\}, \tag{2.46}$$

with the inequalities turning to equalities only in the absence of dissipation.

2.4 Dielectric STFs

Despite the generality of Sec. 2.3, at this time it appears sufficient to model STFs as dielectric materials. The constitutive relations of a dielectric STF are as follows:

$$\begin{aligned} \mathbf{D}(\mathbf{r},\omega) &= \epsilon_0 \underline{\underline{\epsilon}}_r(z,\omega)\cdot\mathbf{E}(\mathbf{r},\omega) \\ &= \epsilon_0 \underline{\underline{S}}(z)\cdot\underline{\underline{\epsilon}}_{\mathrm{ref}}(\omega)\cdot\underline{\underline{S}}^T(z)\cdot\mathbf{E}(\mathbf{r},\omega), \end{aligned} \tag{2.47}$$

$$\mathbf{B}(\mathbf{r},\omega) = \mu_0 \mathbf{H}(\mathbf{r},\omega). \tag{2.48}$$

Description of dielectric STFs is greatly facilitated by the definition of two auxiliary rotation dyadics

$$\hat{\underline{\underline{S}}}_y(\chi) = \mathbf{u}_y\mathbf{u}_y + (\mathbf{u}_x\mathbf{u}_x + \mathbf{u}_z\mathbf{u}_z)\cos\chi + (\mathbf{u}_z\mathbf{u}_x - \mathbf{u}_x\mathbf{u}_z)\sin\chi, \tag{2.49}$$

$$\hat{\underline{\underline{S}}}_z(h,\sigma) = \mathbf{u}_z\mathbf{u}_z + (\mathbf{u}_x\mathbf{u}_x + \mathbf{u}_y\mathbf{u}_y)\cos(h\sigma) + (\mathbf{u}_y\mathbf{u}_x - \mathbf{u}_x\mathbf{u}_y)\sin(h\sigma), \tag{2.50}$$

and an auxiliary relative permittivity dyadic

$$\underline{\underline{\epsilon}}^{o}_{\text{ref}}(\omega) = \underline{\underline{\epsilon}}_{\text{ref}}(\omega)\Big|_{\chi=0} = \epsilon_a(\omega)\mathbf{u}_z\mathbf{u}_z + \epsilon_b(\omega)\mathbf{u}_x\mathbf{u}_x + \epsilon_c(\omega)\mathbf{u}_y\mathbf{u}_y. \tag{2.51}$$

In these equations, h is the structural handedness parameter, which can take one of only two values: either $+1$ for right-handedness or -1 for left-handedness. Locally uniaxial STFs are accommodated by the relations $\epsilon_c(\omega) = \epsilon_a(\omega) \neq \epsilon_b(\omega)$, but all three scalars are different for local biaxiality.

2.4.1 Relative permittivity dyadics

The simplest STFs are, of course, CTFs whose relative permittivity dyadics do not depend on z, i.e.,

$$\underline{\underline{\epsilon}}_r^{\text{CTF}}(z,\omega) = \underline{\underline{\epsilon}}_{\text{ref}}(\omega) = \hat{\underline{\underline{S}}}_y(\chi)\cdot\underline{\underline{\epsilon}}^{o}_{\text{ref}}(\omega)\cdot\hat{\underline{\underline{S}}}_y^{T}(\chi). \tag{2.52}$$

The relative permittivity dyadic of a SNTF is given by

$$\underline{\underline{\epsilon}}_r^{\text{SNTF}}(z,\omega) = \underline{\underline{S}}_y(z)\cdot\underline{\underline{\epsilon}}^{o}_{\text{ref}}(\omega)\cdot\underline{\underline{S}}_y^{T}(z), \tag{2.53}$$

where $\underline{\underline{S}}_y(z)$ is defined in Eq. (2.3). The angular function

$$\tau(z) = \frac{\pi z}{\Omega} \tag{2.54}$$

for a C-shaped SNTF, and

$$\tau(z) = \begin{cases} \dfrac{\pi z}{\Omega}, & 2m < \dfrac{z}{\Omega} < 2m+1, \quad m = 0, 1, 2, \ldots \\ -\dfrac{\pi z}{\Omega}, & 2m-1 < \dfrac{z}{\Omega} < 2m, \quad m = 1, 2, 3, \ldots, \end{cases} \tag{2.55}$$

for an S-shaped SNTF, where Ω is the thickness of a C section, as shown in Fig. 2.8. Parenthetically, if it is convenient to have the morphology in the yz plane, then $\underline{\underline{S}}_y(z)$ should be replaced by $\underline{\underline{S}}_x(z)$ of Eq. (2.2).

The relative permittivity dyadic of a dielectric TFHBM is decomposed into simple factors as

$$\underline{\underline{\epsilon}}_r^{\text{TFHBM}}(z,\omega) = \underline{\underline{S}}_z(z)\cdot\underline{\underline{\hat{S}}}_y(\chi)\cdot\underline{\underline{\epsilon}}_{\text{ref}}^{o}(\omega)\cdot\underline{\underline{\hat{S}}}_y^T(\chi)\cdot\underline{\underline{S}}_z^T(z), \tag{2.56}$$

where $\underline{\underline{S}}_z(z)$ is specified by Eq. (2.4). Chiral STFs are periodically nonhomogeneous, and their relative permittivity dyadics are better represented in the form

$$\underline{\underline{\epsilon}}_r^{\text{chiral STF}}(z,\omega) = \underline{\underline{\hat{S}}}_z\left(h,\frac{z}{\Omega}\right)\cdot\underline{\underline{\hat{S}}}_y(\chi)\cdot\underline{\underline{\epsilon}}_{\text{ref}}^{o}(\omega)\cdot\underline{\underline{\hat{S}}}_y^T(\chi)\cdot\underline{\underline{\hat{S}}}_z^T\left(h,\frac{z}{\Omega}\right), \tag{2.57}$$

where 2Ω is the structural period. The parameter h appears in Eq. (2.57) to indicate one of the two types of structural handedness illustrated in Fig. 2.9.

2.4.2 Local homogenization

The nominal model of Sec. 2.3.2 simplifies greatly for dielectric STFs. In effect, only the upper left quadrants of the constitutive matrixes $[\mathbf{C}]_{\text{ref},s,v}$, the polarizability density matrixes $[\mathbf{A}]_{s,v}$, and the depolarization matrixes $[\mathbf{D}]_{s,v}$ of Eqs. (2.16) to (2.18) must be handled. Further simplification of the Bruggeman formalism comes from assuming that the deposited material as well as the material in the voids are isotropic dielectric, albeit with ellipsoidal topology.

Therefore, let $\epsilon_{s,v}$ be the relative permittivity scalars of the two constituent materials, while the shape factors for the two types of ellipsoidal inclusions are

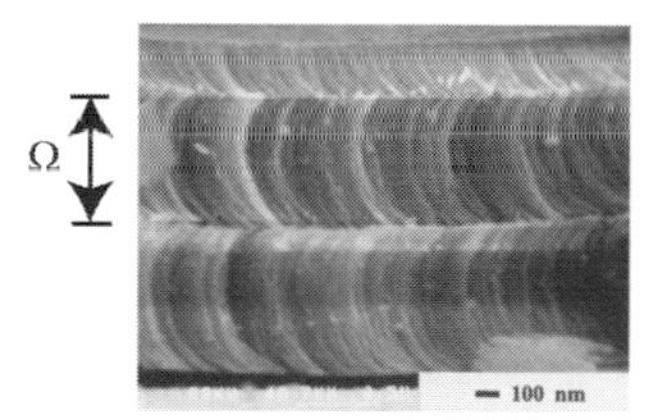

Figure 2.8 Thickness of a C section in a C-shaped SNTF.

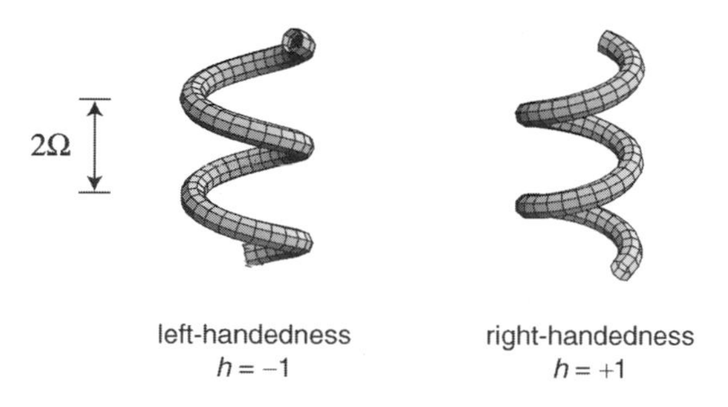

Figure 2.9 Structural handedness and period of chiral STFs.

$\gamma_{\tau,b}^{(s,v)}$. The equation to be solved is the dyadic counterpart of Eq. (2.22):

$$f_v \underline{\underline{A}}_v + (1 - f_v) \underline{\underline{A}}_s = \underline{\underline{0}}. \tag{2.58}$$

The polarizability density dyadics

$$\underline{\underline{A}}_{s,v} = \epsilon_0 \big(\epsilon_{s,v} \underline{\underline{I}} - \underline{\underline{\epsilon}}_{\text{ref}}\big) \cdot \big[\underline{\underline{I}} + i\omega\epsilon_0 \underline{\underline{D}}_{s,v} \cdot \big(\epsilon_{s,v} \underline{\underline{I}} - \underline{\underline{\epsilon}}_{\text{ref}}\big)\big]^{-1} \tag{2.59}$$

require the computation of the depolarization dyadics

$$\underline{\underline{D}}_{s,v} = \frac{2}{i\pi\omega\epsilon_0} \int_{\varphi=0}^{\pi/2} \int_{\vartheta=0}^{\pi/2} \sin\vartheta \times \frac{(\sin\vartheta\cos\varphi)^2 \mathbf{u}_n\mathbf{u}_n + [\cos\vartheta/\gamma_\tau^{(s,v)}]^2 \mathbf{u}_\tau\mathbf{u}_\tau + [\sin\vartheta\sin\varphi/\gamma_b^{(s,v)}]^2 \mathbf{u}_b\mathbf{u}_b}{(\sin\vartheta\cos\varphi)^2 \epsilon_a + [\cos\vartheta/\gamma_\tau^{(s,v)}]^2 \epsilon_b + [\sin\vartheta\sin\varphi/\gamma_b^{(s,v)}]^2 \epsilon_c} \, d\vartheta \, d\varphi \tag{2.60}$$

by an appropriate numerical integration scheme.

The devised model has been used extensively[55,56] to study the plane wave responses of dispersive chiral STFs on axial excitation, studying in particular the spectrums of various measures of transmission optical activity. The dependencies of these quantities on the column inclination angle, periodicity, porosity, and two ellipsoidal shape factors were deduced. After calibration against experimentally obtained reflectance/transmittance data,[57] the nominal model may turn out a powerful design tool and process-control paradigm. It has already been applied to assess the piezoelectric tunability of lasers and filters made of chiral STFs.[75,76]

2.4.3 Wave propagation

The matrix $[\mathbf{P}(z, \kappa, \psi, \omega)]$ of Eq. (2.24) determines the transfer of electromagnetic field phasors across a STF. This matrix is independent of z for a CTF; i.e.,

$$[\mathbf{P}(\kappa, \psi, \omega)] = \omega \begin{bmatrix} 0 & 0 & 0 & \mu_0 \\ 0 & 0 & -\mu_0 & 0 \\ 0 & -\epsilon_0\epsilon_c(\omega) & 0 & 0 \\ \epsilon_0\epsilon_d(\omega) & 0 & 0 & 0 \end{bmatrix} + \kappa \frac{\epsilon_d(\omega)[\epsilon_a(\omega) - \epsilon_b(\omega)]}{\epsilon_a(\omega)\epsilon_b(\omega)} \sin\chi\cos\chi \begin{bmatrix} \cos\psi & 0 & 0 & 0 \\ \sin\psi & 0 & 0 & 0 \\ 0 & 0 & 0 & 0 \\ 0 & 0 & -\sin\psi & \cos\psi \end{bmatrix}$$

$$
+\frac{\kappa^2}{\omega\epsilon_0}\frac{\epsilon_d(\omega)}{\epsilon_a(\omega)\epsilon_b(\omega)}\begin{bmatrix} 0 & 0 & \cos\psi\sin\psi & -\cos^2\psi \\ 0 & 0 & \sin^2\psi & -\cos\psi\sin\psi \\ 0 & 0 & 0 & 0 \\ 0 & 0 & 0 & 0 \end{bmatrix}
$$

$$
+\frac{\kappa^2}{\omega\mu_0}\begin{bmatrix} 0 & 0 & 0 & 0 \\ 0 & 0 & 0 & 0 \\ -\cos\psi\sin\psi & \cos^2\psi & 0 & 0 \\ -\sin^2\psi & \cos\psi\sin\psi & 0 & 0 \end{bmatrix}, \tag{2.61}
$$

where

$$
\epsilon_d(\omega) = \frac{\epsilon_a(\omega)\epsilon_b(\omega)}{\epsilon_a(\omega)\cos^2\chi + \epsilon_b(\omega)\sin^2\chi} \tag{2.62}
$$

is a *composite* relative permittivity scalar. The corresponding matrix for SNTFs, given by

$$
[\mathbf{P}(z,\kappa,\psi,\omega)]
= \omega\begin{bmatrix} 0 & 0 & 0 & \mu_0 \\ 0 & 0 & -\mu_0 & 0 \\ 0 & -\epsilon_0\,\epsilon_c(\omega) & 0 & 0 \\ \epsilon_0\,\varsigma_d(\omega,z) & 0 & 0 & 0 \end{bmatrix}
$$

$$
+\kappa\frac{\varsigma_d(\omega,z)[\epsilon_a(\omega)-\epsilon_b(\omega)]}{\epsilon_a(\omega)\epsilon_b(\omega)}\frac{\sin 2\tau(z)}{2}\begin{bmatrix} \cos\psi & 0 & 0 & 0 \\ \sin\psi & 0 & 0 & 0 \\ 0 & 0 & 0 & 0 \\ 0 & 0 & -\sin\psi & \cos\psi \end{bmatrix}
$$

$$
+\frac{\kappa^2}{\omega\epsilon_0}\frac{\varsigma_d(\omega,z)}{\epsilon_a(\omega)\epsilon_b(\omega)}\begin{bmatrix} 0 & 0 & \cos\psi\sin\psi & -\cos^2\psi \\ 0 & 0 & \sin^2\psi & -\cos\psi\sin\psi \\ 0 & 0 & 0 & 0 \\ 0 & 0 & 0 & 0 \end{bmatrix}
$$

$$
+\frac{\kappa^2}{\omega\mu_0}\begin{bmatrix} 0 & 0 & 0 & 0 \\ 0 & 0 & 0 & 0 \\ -\cos\psi\sin\psi & \cos^2\psi & 0 & 0 \\ -\sin^2\psi & \cos\psi\sin\psi & 0 & 0 \end{bmatrix}, \tag{2.63}
$$

is not spatially constant but depends on z instead. The auxiliary function

$$
\varsigma_d(\omega,z) = \frac{\epsilon_a(\omega)\epsilon_b(\omega)}{\epsilon_a(\omega)\cos^2\tau(z) + \epsilon_b(\omega)\sin^2\tau(z)} \tag{2.64}
$$

in Eq. (2.63) is analogous to $\epsilon_d(\omega)$. The matrixes of Eqs. (2.61) and (2.63) simplify either for propagation in morphologically significant planes (i.e., $\psi = 0$) or along the thickness direction (i.e., $\kappa = 0$), and the 4×4 MODE (2.24) then simplifies into two autonomous 2×2 MODEs.

Finally, the matrix $[\mathbf{P}(z, \kappa, \psi, \omega)]$ for a chiral STF turns out to be as follows:

$$
\begin{aligned}
&[\mathbf{P}(z,\kappa,\psi,\omega)] \\
&= \omega \begin{bmatrix} 0 & 0 & 0 & \mu_0 \\ 0 & 0 & -\mu_0 & 0 \\ h\epsilon_0[\epsilon_c(\omega)-\epsilon_d(\omega)]\cos(\frac{\pi z}{\Omega})\sin(\frac{\pi z}{\Omega}) & -\epsilon_0[\epsilon_c(\omega)\cos^2(\frac{\pi z}{\Omega})+\epsilon_d(\omega)\sin^2(\frac{\pi z}{\Omega})] & 0 & 0 \\ \epsilon_0[\epsilon_c(\omega)\sin^2(\frac{\pi z}{\Omega})+\epsilon_d(\omega)\cos^2(\frac{\pi z}{\Omega})] & -h\epsilon_0[\epsilon_c(\omega)-\epsilon_d(\omega)]\cos(\frac{\pi z}{\Omega})\sin(\frac{\pi z}{\Omega}) & 0 & 0 \end{bmatrix} \\
&+ \kappa \frac{\epsilon_d(\omega)[\epsilon_a(\omega)-\epsilon_b(\omega)]}{\epsilon_a(\omega)\epsilon_b(\omega)} \frac{\sin 2\chi}{2} \begin{bmatrix} \cos(\frac{\pi z}{\Omega})\cos\psi & h\sin(\frac{\pi z}{\Omega})\cos\psi & 0 & 0 \\ \cos(\frac{\pi z}{\Omega})\sin\psi & h\sin(\frac{\pi z}{\Omega})\sin\psi & 0 & 0 \\ 0 & 0 & h\sin(\frac{\pi z}{\Omega})\sin\psi & -h\sin(\frac{\pi z}{\Omega})\cos\psi \\ 0 & 0 & -\cos(\frac{\pi z}{\Omega})\sin\psi & \cos(\frac{\pi z}{\Omega})\cos\psi \end{bmatrix} \\
&+ \frac{\kappa^2}{\omega\epsilon_0} \frac{\epsilon_d(\omega)}{\epsilon_a(\omega)\epsilon_b(\omega)} \begin{bmatrix} 0 & 0 & \cos\psi\sin\psi & -\cos^2\psi \\ 0 & 0 & \sin^2\psi & -\cos\psi\sin\psi \\ 0 & 0 & 0 & 0 \\ 0 & 0 & 0 & 0 \end{bmatrix} \\
&+ \frac{\kappa^2}{\omega\mu_0} \begin{bmatrix} 0 & 0 & 0 & 0 \\ 0 & 0 & 0 & 0 \\ -\cos\psi\sin\psi & \cos^2\psi & 0 & 0 \\ -\sin^2\psi & \cos\psi\sin\psi & 0 & 0 \end{bmatrix}.
\end{aligned}
\tag{2.65}
$$

2.5 Applications

Although optical, electronic, acoustic, thermal, chemical, and biological applications of STFs were forecast early on,[2] the potential of these nanostructured materials has been most successfully actualized in linear optics thus far. Several types of optical filters, sensors, photonic band gap (PBG) materials, and electrically addressable displays are in various stages of development but are definitely past their embryonic stages.

2.5.1 Optical filters

Chiral STFs display the circular Bragg phenomenon in accordance with their periodic nonhomogeneity along the z axis.[62] Briefly, a structurally right/left-handed chiral STF only a few periods thick almost completely reflects normally incident RCP/LCP plane waves with wavelength lying in the so-called Bragg regime; while the reflection of normally incident LCP/RCP plane waves in the same regime is very little. Figure 2.10 presents the measured and the predicted transmittances of a structurally left-handed chiral STF made of titanium oxide, showing the almost complete blockage of an incident LCP plane wave and the high transmission of an incident RCP plane wave at free-space wavelengths in the neighborhood of 620 nm.

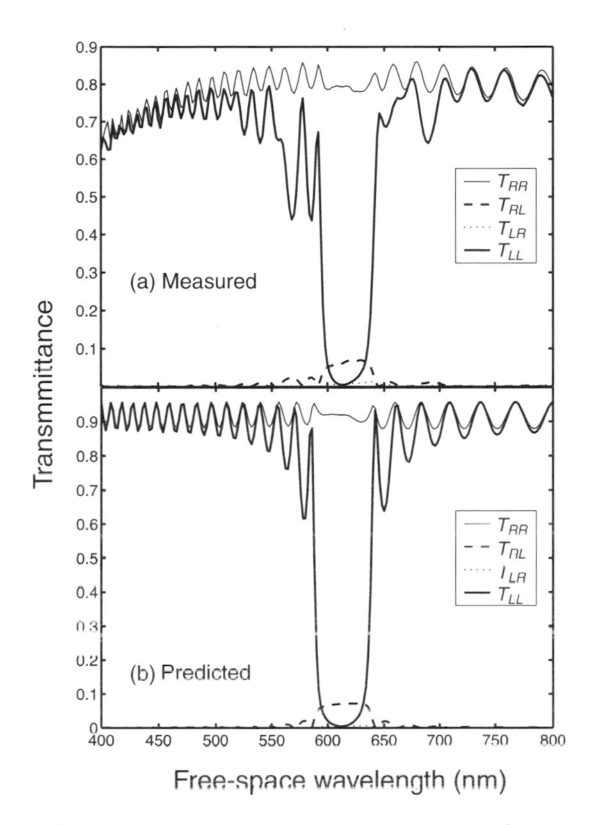

Figure 2.10 Measured and predicted transmittance spectrums of a structurally left-handed chiral STF for normal incidence ($\kappa = 0$). The transmittance T_{LR} is the intensity of the LCP component of the transmitted plane wave relative to the intensity of the RCP component of the incident plane wave, etc. Dispersion (i.e., frequency-dependence of constitutive parameters) was not taken into account when predicting the transmittances from the solution of Eq. (2.24). (Adapted from Wu et al.[78])

The bandwidth of the Bragg regime and the peak reflectance therein first increase with the thickness of the chiral STF, and then saturate. Once this saturation has occurred, further thickening of the film has negligible effects on the reflection spectrum. The Bragg regime is also marked by high levels of optical activity,[13,30,38] which, however, does not scale with the film thickness and is also highly dependent on the orientation of the incident electric field phasor.[77]

More than one Bragg regime is possible when a plane wave is obliquely incident (i.e., $\kappa \neq 0$),[62] but it is the normal-incidence case that appears to be of the greatest value in the context of planar technology. The major successes reported are as follows:

- *Circular polarization filters.* The circular Bragg phenomenon can be employed to realize circular polarization filters. A normally incident, circularly polarized plane wave of one handedness can be reflected almost completely, while that of the other handedness is substantially transmitted, if absorption is small enough and the film is sufficiently thick, in the Bragg regime. This has been experimentally demonstrated.[78] As of now, the Bragg regime can be positioned at virtually any free-space wavelength between 450 and 1700 nm.

Calculations show that polarization–insensitivity, for application in laser mirrors, can be realized with a cascade of two otherwise identical chiral STFs but of opposite structural handedness.[79,80] Furthermore, stepwise chirping can widen the bandwidth,[81] and tightly interlaced chiral STFs may be attractive for bandwidth engineering.[82] Finally, dispersive characteristics can allow more than one Bragg regime,[83] as exemplified by the calculated reflectance spectrums shown in Fig. 2.11.

A handedness inverter for light of only one of the two circular polarization states was designed[84] and then fabricated as well as tested.[85] As the first reported two-section STF device, it comprises a chiral STF and a CTF functioning as a half waveplate. Basically, it almost completely reflects, say, LCP light, while it substantially transmits incident RCP light after transforming it into LCP light, in the Bragg regime.

- *Spectral hole filters.* A two-section STF was proposed as a spectral hole filter.[86] Both sections are chiral STFs of the same structural handedness and identical thickness L. Their structural periods $2\Omega_1$ and $2\Omega_2$ are chosen such that $2L(\Omega_2^{-1} - \Omega_1^{-1}) = 1$. A narrow transmission band then appears for circular polarized plane waves of the same handedness as the two chiral STF sections.

  A more robust three-section STF was also proposed as a spectral reflection hole filter. Its first and third sections are identical chiral STFs, whereas

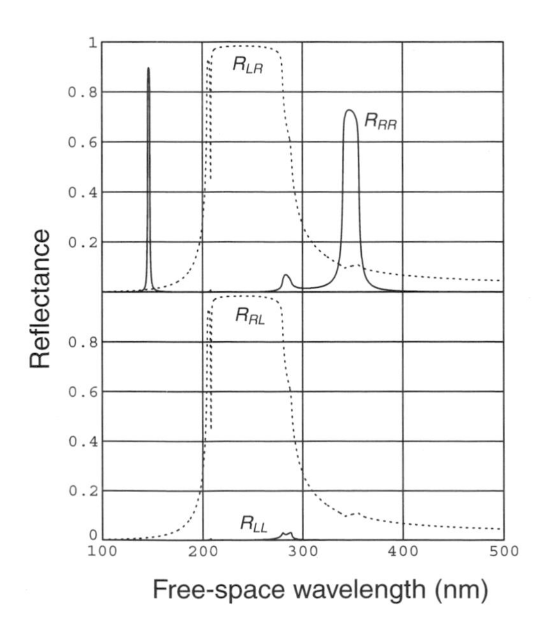

Figure 2.11 Calculated reflectance spectrums of a structurally right-handed chiral STF half-space for normal incidence ($\kappa = 0$). Dispersion is responsible for the circular-polarization-sensitive Bragg regimes centered at 147 and 349 nm wavelengths. (Adapted from Wang et al.[83])

the thin middle section is a homogeneous layer which acts like a phase defect.[87,88] This design was implemented to obtain a 11-nm-wide spectral hole centered at a free-space wavelength of 580 nm in the reflectance spectrum.[89]

An even better design became available shortly thereafter, wherein the middle layer was eliminated, but the lower chiral STF was twisted by 90 deg with respect to the upper chiral STF about the z axis. The twist performed satisfactorily as the required phase defect.[41] With much thicker chiral STFs on either side of the phase defect, calculations show that ultranarrow spectral holes ($\lesssim 0.1$ nm bandwidth) can be obtained in the transmittance spectrum;[49,90] but their performance could be impaired by attenuation within the thick sections.

Most recently, slanted chiral STFs have been introduced[48] to couple the circular Bragg phenomenon to the Rayleigh-Wood anomalies exhibited by surface-relief gratings.[91] This coupling occurs when the helicoidal axis is inclined with respect to the z axis, and suggests the use of these new types of STFs as narrowband circular polarization beamsplitters.

SNTFs can also be pressed into service as optical filters—for linearly polarized plane waves. Rugate filters have been realized as piecewise uniform SNTFs to function as narrow-band reflectors.[14] Šolc filters of the *fan* and the *folded* types are also possible with the same technology.[25,92] The major issue for further research and development is the control of mass density and, hence, $\underline{\underline{\epsilon}}_{\,ref}(\omega)$ with χ_v when fabricating continuously nonhomogeneous SNTFs.

The future of multisection STF devices in optics appears bright because of the recent feat of Suzuki and Taga[42] in being able to deposit a cascade of six different sections of combined thickness $\sim 2\ \mu$m.

2.5.2 Optical fluid sensors

The porosity of STFs makes them attractive for fluid-concentration-sensing applications,[93,94] because their optical response properties must change in accordance with the number density of infiltrant molecules. In particular, theoretical research has shown that the Bragg regime of a chiral STF must shift accordingly, thereby providing a measure of the fluid concentration.[93] Qualitative support for this finding is provided by experiments on wet and dry chiral STFs.[95]

Furthermore, STF spectral hole filters can function as highly sensitive fluid concentration sensors. Proof-of-concept experiments with both circularly polarized and unpolarized incident light have confirmed the redshift of spectral holes on exposure to moisture.[96]

2.5.3 Chiral PBG materials

Chiral STFs have been grown on regular lattices by lithographically patterning the substrates.[39,97] Whereas slow substrate rotation results in the growth of arrays of nano- and micro-helixes spaced as close as 20 nm from their nearest neighbors, faster rotation yields arrays of increasingly denser pillars.[19] Such STFs are essentially PBG materials in the visible and the infrared regimes,[98,99] and the possibility of fabricating them on cheap polymeric substrates is very attractive.[100]

2.5.4 Displays

Liquid crystals (LCs) can be electrically addressed and are therefore widely used now for displays.[101,102] Although STFs are not electronically addressable, the alignment of nematic LCs forced into the void regions of chiral STFs has been shown to respond to applied voltages.[103] Thus, STF-LC composite materials may have a future as robust displays.

Another interesting possibility, in the same vein, is to grow carbon (and other) nanotubes by chemical reactions involving fluid catalysts and precursors[104] inside highly porous STFs. The growing nanotubes would have to conform to the structure imposed by the STF skeleton, and the nanotube-STF composite material thus formed could be useful for field emission devices.

2.5.5 Optical interconnects

STF technology is compatible with the planar technology of electronic chips. Chiral STFs have the potential to simultaneously guide waves with different phase velocities in different directions[105,106] and could therefore function as optical interconnects, leading to efficient use of the available *real estate* in electronic chips. Furthermore, the helicoidal structure of chiral STFs would resist vertical cleavage and fracture. Simultaneous microrefrigeration enabled by the porous STFs would be a bonus.

2.5.6 Optical pulse shapers

The current explosive growth of digital optics communication has provided impetus for time-domain research on novel materials. As chiral STFs are very attractive for optical applications, the circular Bragg phenomenon is being studied in the time domain. A pulse-bleeding phenomenon has been identified as the underlying mechanism, which can drastically affect the shapes, amplitudes, and spectral components of femtosecond pulses.[107] However, narrow-band rectangular pulses can pass through without significant loss of information.[108] The application of STFs to shape optical pulses appears to be waiting in the wings.

2.5.7 Biochips

Endowed with porosity of nanoengineered texture, STFs can function as microreactors for luminescence-producing reactions involving biochemicals. Bioluminescent emission is bound to be affected by the reactor characteristics. If the reactor is a chiral STF, its helicoidal periodicity can be exploited. The structural handedness as well as the periodicity of chiral STFs have been theoretically shown to critically control the emission spectrum and intensity, while the polarization state of the emitted light is strongly correlated with the structural handedness of the embedded source filaments.[109] Optimization with respect to χ_v appears possible.[110,111]

2.5.8 Other applications

From their inception,[2] STFs were expected to have a wide range of applications, implementable after their properties came to be better understood. Their optical applications came to be investigated first. However, their high porosity—in combination with optical anisotropy and possible 2D electron confinement in the nanostructure—makes STFs potential candidates also as

1. electroluminescent devices;
2. high-speed, high-efficiency electrochromic films;
3. optically transparent conducting films sculptured from pure metals; and
4. multistate electronic switches based on filamentary conduction.

That same porosity can be harnessed in microreactors and thermal barriers, as it is accompanied by high surface area.[112–114] For the same reason, STFs may be useful as nanosieves and microsieves for the entrapment of viruses or for growing biological tissues on surfaces of diverse provenances. The potential of STFs as biosubstrates is bolstered by many reports on altered adsorption of proteins and cells on nanopatterned surfaces.[115,116]

These applications of STFs are still in their incipient stages, but some advances have been made on the following two fronts:

- *Interlayer dielectrics.* With the microelectronics industry moving relentlessly toward decreasing feature sizes and increasingly stringent tolerance levels, an urgent need exists for the use of low-permittivity materials as interlayer dielectrics. Silicon dioxide, the current material of choice, has too high a quasi-static permittivity. The porosity of STFs and nanoporous silica makes them attractive low-permittivity materials for microelectronic and electronic packaging applications.[72] However, chiral STFs are likely to have significant thermal, mechanical, as well as electrical advantages over nanoporous silica—because of (1) porosity with controllable texture and (2) helicoidal morphology. Also, STFs can be impregnated with various kinds of polymers.
- *Ultrasonic applications.* The sciences of electromagnetics and elastodynamics have an underlying mathematical unity. For that reason, many optical applications described thus far possess ultrasonic analogs. Indeed, ultrasonic wave propagation in chiral STFs is now theoretically well established,[117–119] as also is the potential for its applications.[120,121] Actual implementation would, however, require[122] the fabrication of chiral STFs with periods $\sim 20\ \mu$m, of which development is still awaited.

2.6 Directions for future research

Several of the emerging applications mentioned in Sec. 2.5 are barely past conceptualization. Considerable research on them is warranted, before they become commercially viable. Just a few of the optical applications have crossed the threshold of academic research and now require several issues to be addressed.

A key issue is that of environmental stability of STFs. The chemical stability of STFs has not yet been examined in any detail, although the susceptibility of porous thin films to moisture is known.[95] An indentation experiment on a chiral STF[123] as well as the successful deposition of six-section STFs[42] strongly indicate that mechanical stability must be investigated in depth. However, only a preliminary model for the mechanical loading of STFs exists at this time.[124,125] Due to the porosity, internal stresses, and morphological stability of STFs in the absence of external loads have to be examined carefully as well.[126,127]

Another key issue is that of efficiency. The vapor incidence angle χ_v, the bulk constitutive properties of the deposited material (responsible, e.g., for ϵ_a, ϵ_b, and ϵ_c) and the substrate rotation parameters appearing in $\underline{\underline{S}}(z)$ must be optimized to achieve desired performance characteristics. As examples, the photocatalytic efficiency of chiral STFs of tantalum oxide is known to be optimal when[114] $\chi_v = 20$ deg, efficient bioluminiscent emission has been shown[110] to require $\chi_v \lesssim 15$ deg, while χ_v could be manipulated to maximize the bandwidth of a Bragg regime.[111] A study on second-harmonic generation in uniaxial chiral STFs has underscored the criticality of χ (and therefore of χ_v) for efficiency.[128] Cross-polarized remittances are drastically reduced and the diversity in the copolarized remittances is enhanced by the incorporation of index-matched layers at the entry and the exit pupils of circular polarization filters.[78] Further improvements may require the simultaneous deposition of different types of materials to reduce absorption and dispersion in desired wavelength regimes in optical filters based on the STF concept.

Nonlinear optics with STFs is practically uncharted territory, despite two reported forays into second-harmonic generation.[128,129] Due to the numerous classes of nonlinearity,[130] the delineation of nanocrystallinity in STFs will be of primary importance. Likewise, understanding of nanodomains in magnetic STFs, as well as of magnetoelectric effects in bianisotropic STFs, are topics of future research.

Although the demonstrated successes of the STF concept and technology are few as yet, the electromagnetic and elastodynamic frameworks for STFs are reasonably mature. But for STF research and use to be truly widespread, economical production must be enabled. Any satisfactory production technique must be rapid and deliver high yields, so that large-scale fabrication must become possible. The latter appears feasible with the adaptation of ion-thruster technology.[131] Furthermore, the films will have to be laterally uniform with growth evolution, and χ may have to be lower than 20 deg. If PVD (or any variant) is to be industrially successful, then new architectures for the evaporant flux source—whether discrete or continuous, single, or multiple—must be developed to deposit STFs on large substrates. Reliability of deposition uniformity would be facilitated by computer-controlled source architectures. In turn, they will require the development of *in situ* monitoring of the deposition process and appropriate control models. These and related avenues for manufacturing research must be opened up. Some progress has been recently made.[132]

References

1. A. Lakhtakia and R. Messier, "The key to a thin film HBM: the Motohiro–Taga interface," *Proceedings of Chiral '94: 3rd International Workshop on Chiral, Bi-Isotropic and Bi-Anisotropic Media*, F. Mariotte and J.-P. Parneix, Eds., pp. 125–130, Périgueux, France (1994).
2. A. Lakhtakia, R. Messier, M. J. Brett, and K. Robbie, "Sculptured thin films (STFs) for optical, chemical and biological applications," *Innovat. Mater. Res.* **1**, 165–176 (1996).
3. A. Lakhtakia and R. Messier, "Sculptured thin films—I. Concepts," *Mater. Res. Innovat.* **1**, 145–148 (1997).
4. R. Messier and A. Lakhtakia, "Sculptured thin films—II. Experiments and applications," *Mater. Res. Innovat.* **2**, 217–222 (1999).
5. A. Kundt, "Ueber Doppelbrechung des Lichtes in Metallschichten, welche durch Zerstäuben einer Kathode hergestellt sind," *Ann. Phys. Chem. Lpz.* **27**, 59–71 (1886).
6. H. König and G. Helwig, "Über die Struktur schräg aufgedampfter Schichten und ihr Einfluß auf die Entwicklung submikroskopischer Oberflächenrauhigkeiten," *Optik* **6**, 111–124 (1950).
7. L. Holland, "The effect of vapor incidence on the structure of evaporated aluminum films," *J. Opt. Soc. Am.* **43**, 376–380 (1953).
8. N. O. Young and J. Kowal, "Optically active fluorite films," *Nature* **183**, 104–105 (1959).
9. D. O. Smith, M. S. Cohen, and G. P. Weiss, "Oblique-incidence anisotropy in evaporated permalloy films," *J. Appl. Phys.* **31**, 1755–1762 (1960).
10. W. Metzdorf and H. E. Wiehl, "Negative oblique-incidence anisotropy in magnetostriction-free permalloy films," *Phys. Status Solidi* **17**, 285–294 (1966).
11. J. M. Nieuwenhuizen and H. B. Haanstra, "Microfractography of thin films," *Philips Tech. Rev.* **27**, 87–91 (1966).
12. T. Motohiro and Y. Taga, "Thin film retardation plate by oblique deposition," *Appl. Opt.* **28**, 2466–2482 (1989).
13. V. C. Venugopal and A. Lakhtakia, "Sculptured thin films: conception, optical properties and applications," *Electromagnetic Fields in Unconventional Materials and Structures*, O. N. Singh and A. Lakhtakia, Eds., 151–216, Wiley, New York (2000).

14. I. J. Hodgkinson and Q. H. Wu, "Inorganic chiral optical materials," *Adv. Mater.* **13**, 889–897 (2001).
15. H. C. Chen, *Theory of Electromagnetic Waves*, TechBooks, Fairfax, VA (1992).
16. D. M. Mattox, "Physical vapor deposition (PVD) processes," *Vac. Technol. Coat.* **3**(7), 60–62 (2002).
17. R. Messier, A. P. Giri, and R. A. Roy, "Revised structure zone model for thin film physical structure," *J. Vac. Sci. Technol. A* **2**, 500–503 (1984).
18. R. A. Roy and R. Messier, "Evolutionary growth development in SiC sputtered films," *MRS Symp. Proc.* **38**, 363–370 (1985).
19. R. Messier, V. C. Venugopal, and P. D. Sunal, "Origin and evolution of sculptured thin films," *J. Vac. Sci. Technol. A* **18**, 1538–1545 (2000).
20. J. M. García-Ruiz, A. Lakhtakia, and R. Messier, "Does competition between growth elements eventually eliminate self-affinity?" *Speculat. Sci. Technol.* **15**, 60–71 (1992).
21. R. Messier, P. Sunal, and V. C. Venugopal, "Evolution of sculptured thin films," *Engineered Nanostructural Films and Materials*, A. Lakhtakia and R. F. Messier, Eds., *Proc. SPIE* **3790**, 133–141 (1999).
22. H. van Kranenberg and C. Lodder, "Tailoring growth and local composition by oblique-incidence deposition: a review and new experimental data," *Mater. Sci. Eng. R* **11**, 295–354 (1994).
23. A. Lakhtakia, Ed., *Selected Papers on Linear Optical Composite Materials*, SPIE Press, Bellingham, WA (1996).
24. O. S. Heavens, *Optical Properties of Thin Solid Films*, Butterworths, London (1955).
25. I. J. Hodgkinson and Q.-h. Wu, *Birefringent Thin Films and Polarizing Elements*, World Scientific, Singapore (1997).
26. H. A. Macleod, *Thin-Film Optical Filters*, 3rd ed., Institute of Physics, Bristol (2001).
27. M. Suzuki, S. Tokito, and Y. Taga, "Review of thin film technology in automobile industry," *Mater. Sci. Eng. B* **51**, 66–71 (1998).
28. M. Suzuki, T. Ito, and Y. Taga, "Recent progress of obliquely deposited thin films for industrial applications," *Engineered Nanostructural Films and Materials*, A. Lakhtakia and R. F. Messier, Eds., *Proc. SPIE* **3790**, 94–105 (1999).
29. K. Robbie, M. J. Brett, and A. Lakhtakia, "First thin film realization of a helicoidal bianisotropic medium," *J. Vac. Sci. Technol. A* **13**, 2991–2993 (1995).
30. K. Robbie, M. J. Brett, and A. Lakhtakia, "Chiral sculptured thin films," *Nature* **384**, 616 (1996).
31. P. I. Rovira, R. A. Yarussi, R. W. Collins, R. Messier, V. C. Venugopal, A. Lakhtakia, K. Robbie, and M. J. Brett, "Transmission ellipsometry of a thin-film helicoidal bianisotropic medium," *Appl. Phys. Lett.* **71**, 1180–1182 (1997).

32. K. Robbie and M. J. Brett, "Sculptured thin films and glancing angle deposition: Growth mechanics and applications," *J. Vac. Sci. Technol. A* **15**, 1460–1465 (1997).
33. R. Messier, T. Gehrke, C. Frankel, V. C. Venugopal, W. Otaño, and A. Lakhtakia, "Engineered sculptured nematic thin films," *J. Vac. Sci. Technol. A* **15**, 2148–2152 (1997).
34. J. C. Sit, D. Vick, K. Robbie, and M. J. Brett, "Thin film microstructure control using glancing angle deposition by sputtering," *J. Mater. Res.* **14**, 1197–1199 (1999).
35. K. Robbie, J. C. Sit, and M. J. Brett, "Advanced techniques for glancing angle deposition," *J. Vac. Sci. Technol. B* **16**, 1115–1122 (1998).
36. O. R. Monteiro, A. Vizir, and I. G. Brown, "Multilayer thin-films with chevron-like microstructure," *J. Phys. D: Appl. Phys.* **31**, 3188–3196 (1998).
37. F. Liu, M. T. Umlor, L. Shen, W. Eads, J. A. Barnard, and G. J. Mankey, "The growth of nanoscale structured iron films by glancing angle deposition," *J. Appl. Phys.* **85**, 5486–5488 (1999).
38. I. Hodgkinson, Q. H. Wu, B. Knight, A. Lakhtakia, and K. Robbie, "Vacuum deposition of chiral sculptured thin films with high optical activity," *Appl. Opt.* **39**, 642–649 (2000).
39. M. Malac and R. F. Egerton, "Observations of the microscopic growth mechanism of pillars and helices formed by glancing-angle thin-film deposition," *J. Vac. Sci. Technol. A* **19**, 158–166 (2001).
40. Y.-P. Zhao, D.-X. Ye, G.-C. Wang, and T.-M. Lu, "Novel nano-column and nano-flower arrays by glancing angle deposition," *Nano Lett.* **2**, 351–354 (2002).
41. I. J. Hodgkinson, Q. H. Wu, K. E. Thorn, A. Lakhtakia, and M. W. McCall, "Spacerless circular-polarization spectral-hole filters using chiral sculptured thin films: theory and experiment," *Opt. Commun.* **184**, 57–66 (2000).
42. M. Suzuki and Y. Taga, "Integrated sculptured thin films," *Jpn. J. Appl. Phys. Part 2* **40**, L358–L359 (2001).
43. W. S. Weiglhofer, "Constitutive characterization of simple and complex mediums," in *Introduction to Complex Mediums for Optics and Electromagnetics*, W. S. Weiglhofer and A. Lakhtakia, Eds., 27–61, SPIE Press, Bellingham, WA (2003).
44. T. H. O'Dell, *The Electrodynamics of Magneto-Electric Media*, North-Holland, Amsterdam (1970).
45. R. Aris, *Vectors, Tensors, and the Basic Equations of Fluid Mechanics*, Prentice-Hall, Englewood Cliffs, NJ (1962).
46. A. Lakhtakia, "Director-based theory for the optics of sculptured thin films," *Optik* **107**, 57–61 (1997).
47. A. Lakhtakia and W. S. Weiglhofer, "Axial propagation in general helicoidal bianisotropic media," *Microwave Opt. Technol. Lett.* **6**, 804–806 (1993).

48. F. Wang, A. Lakhtakia, and R. Messier, "Coupling of Rayleigh-Wood anomalies and the circular Bragg phenomenon in slanted chiral sculptured thin films," *Eur. Phys. J. Appl. Phys.* **20**, 91–104 (2002); Corrections: **24**, 91 (2003).
49. F. Wang and A. Lakhtakia, "Specular and nonspecular, thickness-dependent spectral holes in a slanted chiral sculptured thin film with a central twist defect," *Opt. Commun.* **215**, 79–92 (2003).
50. M. D. Pickett and A. Lakhtakia, "On gyrotropic chiral sculptured thin films for magneto-optics," *Optik* **113**, 367–371 (2002).
51. P. S. Neelakanta, *Handbook of Composite Materials*, CRC Press, Boca Raton, FL (1995).
52. B. Michel, "Recent developments in the homogenization of linear bianisotropic composite materials," *Electromagnetic Fields in Unconventional Materials and Structures*, O. N. Singh and A. Lakhtakia, Eds., 39–82, Wiley, New York (2000).
53. T. G. Mackay, "Homogenization of linear and nonlinear composite materials," in *Introduction to Complex Mediums for Optics and Electromagnetics*, W. S. Weiglhofer and A. Lakhtakia, Eds., 317–345, SPIE Press, Bellingham, WA (2003).
54. A. Lakhtakia, P. D. Sunal, V. C. Venugopal, and E. Ertekin, "Homogenization and optical response properties of sculptured thin films," in *Engineered Nanostructural Films and Materials*, A. Lakhtakia and R. F. Messier, Eds., *Proc. SPIE* **3790**, 77–83 (1999).
55. J. A. Sherwin and A. Lakhtakia, "Nominal model for structure-property relations of chiral dielectric sculptured thin films," *Math. Comput. Model.* **34**, 1499–1514 (2001); corrections: **35**, 1355–1363 (2002).
56. J. A. Sherwin and A. Lakhtakia, "Nominal model for the optical response of a chiral sculptured thin film infiltrated with an isotropic chiral fluid," *Opt. Commun.* **214**, 231–245 (2002).
57. J. A. Sherwin, A. Lakhtakia, and I. J. Hodgkinson, "On calibration of a nominal structure-property relationship model for chiral sculptured thin films by axial transmittance measurements," *Opt. Commun.* **209**, 369–375 (2002).
58. A Lakhtakia, "Linear optical responses of sculptured thin films (SNTFs)," *Optik* **106**, 45–52 (1997).
59. A. Lakhtakia and W. S. Weiglhofer, "On light propagation in helicoidal bianisotropic mediums," *Proc. R. Soc. Lond. A* **448**, 419–437 (1995); correction: **454**, 3275 (1998).
60. A. Lakhtakia, "Anomalous axial propagation in helicoidal bianisotropic media," *Opt. Commun.* **157**, 193–201 (1998).
61. A. Lakhtakia and W. S. Weiglhofer, "Further results on light propagation in helicoidal bianisotropic mediums: oblique propagation," *Proc. R. Soc. Lond. A* **453**, 93–105 (1997); correction: **454**, 3275 (1998).
62. V. C. Venugopal and A. Lakhtakia, "Electromagnetic plane-wave response characteristics of non-axially excited slabs of dielectric thin-film helicoidal bianisotropic mediums," *Proc. R. Soc. Lond. A* **456**, 125–161 (2000).

63. M. Schubert and C. M. Herzinger, "Ellipsometry on anisotropic materials: Bragg conditions and phonons in dielectric helical thin films," *Phys. Status Solidi (a)* **188**, 1563–1575 (2001).
64. J. A. Polo, Jr. and A. Lakhtakia, "Numerical implementation of exact analytical solution for oblique propagation in a cholesteric liquid crystal," *Microwave Opt. Technol. Lett.* **35**, 397–400 (2002); [Equation (16) of this paper should read as follows: $[\underline{\underline{\tilde{M}}}'(\zeta)] = [\underline{\underline{\tilde{M}}}'(\xi)][\underline{\underline{\tilde{M}}}'(1)]^{\ell}$.]
65. V. A. Yakubovich and V. M. Starzhinskii, *Linear Differential Equations with Periodic Coefficients*, Wiley, New York (1975).
66. W. S. Weiglhofer and A. Lakhtakia, "Oblique propagation in a cholesteric liquid crystal: 4×4 matrix perturbational solution," *Optik* **102**, 111–114 (1996).
67. K. Rokushima and J. Yamakita, "Analysis of diffraction in periodic liquid crystals: the optics of the chiral smectic C phase," *J. Opt. Soc. Am. A* **4**, 27–33 (1987).
68. M. W. McCall and A. Lakhtakia, "Development and assessment of coupled wave theory of axial propagation in thin-film helicoidal bianisotropic media. Part 1: reflectances and transmittances," *J. Mod. Opt.* **47**, 973–991 (2000).
69. M. W. McCall and A. Lakhtakia, "Development and assessment of coupled wave theory of axial propagation in thin-film helicoidal bi-anisotropic media. Part 2: dichroisms, ellipticity transformation and optical rotation," *J. Mod. Opt.* **48**, 143–158 (2001).
70. A. Lakhtakia, "On the quasistatic approximation for helicoidal bianisotropic mediums," *Electromagnetics* **19**, 513–525 (1999).
71. A. Lakhtakia, "Capacitance of a slab of a dielectric thin-film helicoidal bianisotropic medium," *Microwave Opt. Technol. Lett.* **21**, 286–288 (1999).
72. V. C. Venugopal, A. Lakhtakia, R. Messier, and J.-P. Kucera, "Low-permittivity materials using sculptured thin film technology," *J. Vac. Sci. Technol. B* **18**, 32–36 (2000).
73. A. T. Wu and M. J. Brett, "Sensing humidity using nanostructured SiO posts: mechanism and optimization," *Sens. Mater.* **13**, 399–431 (2001).
74. A. Boström, G. Kristensson, and S. Ström, "Transformation properties of plane, spherical and cylindrical scalar and vector wave functions," *Field Representations and Introduction to Scattering*, V. V. Varadan, A. Lakhtakia, and V. K. Varadan, Eds., 165–210, North-Holland, Amsterdam (1991).
75. F. Wang, A. Lakhtakia, and R. Messier, "On piezoelectric control of the optical response of sculptured thin films," *J. Mod. Opt.* **50**, 239–249 (2003).
76. F. Wang, A. Lakhtakia, and R. Messier, "Towards piezoelectrically tunable chiral sculptured thin film lasers," *Sens. Actuat. A: Phys.* **102**, 31–35 (2002).
77. V. C. Venugopal and A. Lakhtakia, "On optical rotation and ellipticity transformation by axially excited slabs of dielectric thin-film helicoidal bianisotropic mediums (TFHBMs)," *Int. J. Appl. Electromag. Mech.* **9**, 201–210 (1998).

78. Q. Wu, I. J. Hodgkinson, and A. Lakhtakia, "Circular polarization filters made of chiral sculptured thin films: experimental and simulation results," *Opt. Eng.* **39**, 1863–1868 (2000).
79. A. Lakhtakia and V. C. Venugopal, "Dielectric thin-film helicoidal bianisotropic medium bilayers as tunable polarization-independent laser mirrors and notch filters," *Microwave Opt. Technol. Lett.* **17**, 135–140 (1998).
80. A. Lakhtakia and I. J. Hodgkinson, "Spectral response of dielectric thin-film helicoidal bianisotropic medium bilayer," *Opt. Commun.* **167**, 191–202 (1999).
81. A. Lakhtakia, "Stepwise chirping of chiral sculptured thin films for Bragg bandwidth enhancement," *Microwave Opt. Technol. Lett.* **28**, 323–326 (2001).
82. A. Lakhtakia, "Axial excitation of tightly interlaced chiral sculptured thin films: 'averaged' circular Bragg phenomenon," *Optik* **112**, 119–124 (2001).
83. J. Wang, A. Lakhtakia, and J. B. Geddes III, "Multiple Bragg regimes exhibited by a chiral sculptured thin film half-space on axial excitation," *Optik* **113**, 213–221 (2002).
84. A. Lakhtakia, "Dielectric sculptured thin films for polarization-discriminatory handedness-inversion of circularly polarized light," *Opt. Eng.* **38**, 1596–1602 (1999).
85. I. J. Hodgkinson, A. Lakhtakia, and Q. H. Wu, "Experimental realization of sculptured-thin-film polarization-discriminatory light-handedness inverters," *Opt. Eng.* **39**, 2831–2834 (2000).
86. M. W. McCall and A. Lakhtakia, "Polarization-dependent narrowband spectral filtering by chiral sculptured thin films," *J. Mod. Opt.* **47**, 743–755 (2000).
87. A. Lakhtakia and M. McCall, "Sculptured thin films as ultranarrow-bandpass circular-polarization filters," *Opt. Commun.* **168**, 457–465 (1999).
88. A. Lakhtakia, V. C. Venugopal, and M. W. McCall, "Spectral holes in Bragg reflection from chiral sculptured thin films: circular polarization filters," *Opt. Commun.* **177**, 57–68 (2000).
89. I. J. Hodgkinson, Q. H. Wu, A. Lakhtakia, and M. W. McCall, "Spectral-hole filter fabricated using sculptured thin-film technology," *Opt. Commun.* **177**, 79–84 (2000).
90. V. I. Kopp and A. Z. Genack, "Twist defect in chiral photonic structures," *Phys. Rev. Lett.* **89**, 033901 (2002).
91. D. Maystre, Ed., *Selected Papers on Diffraction Gratings*, SPIE Press, Bellingham, WA (1993).
92. A. Lakhtakia, "Dielectric sculptured thin films as Šolc filters," *Opt. Eng.* **37**, 1870–1875 (1998).
93. A. Lakhtakia, "On determining gas concentrations using thin-film helicoidal bianisotropic medium bilayers," *Sens. Actuat. B: Chem.* **52**, 243–250 (1998).

94. E. Ertekin and A. Lakhtakia, "Sculptured thin film Šolc filters for optical sensing of gas concentration," *Eur. Phys. J. Appl. Phys.* **5**, 45–50 (1999).
95. I. J. Hodgkinson, Q. H. Wu, and K. M. McGrath, "Moisture adsorption effects in biaxial and chiral optical thin film coatings," *Engineered Nanostructural Films and Materials*, A. Lakhtakia and R. F. Messier, Eds., *Proc. SPIE* **3790**, 184–194 (1999).
96. A. Lakhtakia, M. W. McCall, J. A. Sherwin, Q. H. Wu, and I. J. Hodgkinson, "Sculptured-thin-film spectral holes for optical sensing of fluids," *Opt. Commun.* **194**, 33–46 (2001).
97. M. Malac and R. F. Egerton, "Thin-film regular-array structures with 10–100 nm repeat distance," *Nanotechnology* **12**, 11–13 (2001).
98. O. Toader and S. John, "Proposed square spiral microfabrication architecture for large 3D photonic band gap crystals," *Science* **292**, 1133–1135 (2001).
99. S. R. Kennedy, M. J. Brett, O. Toader, and S. John, "Fabrication of tetragonal square spiral photonic crystals," *Nano Lett.* **2**, 59–62 (2002).
100. B. Dick, J. C. Sit, M. J. Brett, I. M. N. Votte, and C. W. M. Bastiaansen, "Embossed polymeric relief structures as a template for the growth of periodic inorganic microstructures," *Nano Lett.* **1**, 71–73 (2001).
101. S. D. Jacobs, Ed., *Selected Papers on Liquid Crystals for Optics*, SPIE Press, Bellingham, WA (1992).
102. H. Kawamoto, "The history of liquid crystal displays," *Proc. IEEE* **90**, 460–500 (2002).
103. J. C. Sit, D. J. Broer, and M. J. Brett, "Liquid crystal alignment and switching in porous chiral thin films," *Adv. Mater.* **12**, 371–373 (2000).
104. R. Vajtai, B. Q. Wei, Z. J. Zhang, Y. Jung, G. Ramanath, and P. M. Ajayan, "Building carbon nanotubes and their smart architectures," *Smart Mater. Struct.* **11**, 691–698 (2002).
105. A. Lakhtakia, "Towards sculptured thin films (STFs) as optical interconnects," *Optik* **110**, 289–293 (1999).
106. E. Ertekin and A. Lakhtakia, "Optical interconnects realizable with thin-film helicoidal bianisotropic mediums," *Proc. R. Soc. Lond. A* **457**, 817–836 (2001).
107. J. B. Geddes III and A. Lakhtakia, "Reflection and transmission of optical narrow-extent pulses by axially excited chiral sculptured thin films," *Eur. Phys. J. Appl. Phys.* **13**, 3–14 (2001); corrections: **16**, 247 (2001).
108. J. B. Geddes III and A. Lakhtakia, "Pulse-coded information transmission across an axially excited chiral sculptured thin film in the Bragg regime," *Microwave Opt. Technol. Lett.* **28**, 59–62 (2001).
109. A. Lakhtakia, "On bioluminescent emission from chiral sculptured thin films," *Opt. Commun.* **188**, 313–320 (2001).
110. A. Lakhtakia, "Local inclination angle: a key structural factor in emission from chiral sculptured thin films," *Opt. Commun.* **202**, 103–112 (2002); correction: **203**, 447 (2002).

111. A. Lakhtakia, "Pseudo-isotropic and maximum-bandwidth points for axially excited chiral sculptured thin films," *Microwave Opt. Technol. Lett.* **34**, 367–371 (2002).
112. K. D. Harris, M. J. Brett, T. J. Smy, and C. Backhouse, "Microchannel surface area enhancement using porous thin films," *J. Electrochem. Soc.* **147**, 2002–2006 (2000).
113. K. D. Harris, D. Vick, E. J. Gonzalez, T. Smy, K. Robbie, and M. J. Brett, "Porous thin films for thermal barrier coatings," *Surf. Coat. Technol.* **138**, 185–191 (2001).
114. M. Suzuki, T. Ito, and Y. Taga, "Photocatalysis of sculptured thin films of TiO_2," *Appl. Phys. Lett.* **78**, 3968–3970 (2001).
115. M. Riedel, B. Müller, and E. Wintermantel, "Protein adsorption and monocyte activation on germanium nanopyramids," *Biomaterials* **22**, 2307–2316 (2001).
116. C. D. W. Wilkinson, M. Riehle, M. Wood, J. Gallagher, and A. S. G. Curtis, "The use of materials patterned on a nano- and micro-metric scale in cellular engineering," *Mater. Sci. Eng. C* **19**, 263–269 (2002).
117. A. Lakhtakia, "Wave propagation in a piezoelectric, continuously twisted, structurally chiral medium along the axis of spirality," *Appl. Acoust.* **44**, 25–37 (1995); corrections: **44**, 385 (1995).
118. A. Lakhtakia, "Exact analytic solution for oblique propagation in a piezoelectric, continuously twisted, structurally chiral medium," *Appl. Acoust.* **49**, 225–236 (1996).
119. C. Oldano and S. Ponti, "Acoustic wave propagation in structurally helical media," *Phys. Rev. E* **63**, 011703 (2000).
120. A. Lakhtakia and M. W. Meredith, "Shear axial modes in a PCTSCM. Part IV: bandstop and notch filters," *Sens. Actuat. A: Phys.* **73**, 193–200 (1999).
121. A. Lakhtakia, "Shear axial modes in a PCTSCM. Part VI: simpler transmission spectral holes," *Sens. Actuat. A: Phys.* **87**, 78–80 (2000).
122. A. Lakhtakia, K. Robbie, and M. J. Brett, "Spectral Green's function for wave excitation and propagation in a piezoelectric, continuously twisted, structurally chiral medium," *J. Acoust. Soc. Am.* **101**, 2052–2059 (1997).
123. M. W. Seto, K. Robbie, D. Vick, M. J. Brett, and L. Kuhn, "Mechanical response of thin films with helical microstructures," *J. Vac. Sci. Technol. B* **17**, 2172–2177 (1999).
124. A. Lakhtakia, "Axial loading of a chiral sculptured thin film," *Model. Simul. Mater. Sci. Eng.* **8**, 677–686 (2000).
125. A. Lakhtakia, "Microscopic model for elastostatic and elastodynamic excitation of chiral sculptured thin films," *J. Compos. Mater.* **36**, 1277–1298 (2002).
126. R. Knepper and R. Messier, "Morphology and mechanical properties of oblique angle columnar thin films," *Complex Mediums*, A. Lakhtakia, W. S. Weiglhofer, and R. F. Messier, Eds., *Proc. SPIE* **4097**, 291–298 (2000).

127. M. Suzuki, T. Ito, and Y. Taga, "Morphological stability of TiO_2 thin films with isolated columns," *Jpn. J. Appl. Phys. Part 2* **40**, L398–L400 (2001).
128. V. C. Venugopal and A. Lakhtakia, "Second harmonic emission from an axially excited slab of a dielectric thin-film helicoidal bianisotropic medium," *Proc. R. Soc. Lond. A* **454**, 1535–1571 (1998); corrections: **455**, 4383 (1999).
129. A. Lakhtakia, "On second harmonic generation in sculptured nematic thin films (SNTFs)," *Optik* **105**, 115–120 (1997).
130. R. W. Boyd, *Nonlinear Optics*, Academic Press, San Diego, CA (1992).
131. S. G. Bilén, M. T. Domonkos, and A. D. Gallimore, "Simulating ionospheric plasma with a hollow cathode in a large vacuum chamber," *J. Spacecraft Rockets* **38**, 617–621 (2001).
132. M. W. Horn, M. D. Pickett, R. Messier, and A. Lakhtakia, "Blending of nanoscale and microscale in uniform large-area sculptured thin-film architectures," *Nanotechnology* **15**, 243–250 (2004).

List of symbols

$\underline{\underline{0}}$	null dyadic
$[\mathbf{0}]$	null matrix
$a_{L,R}$	circular amplitudes of incident plane wave
$a_{s,p}$	linear amplitudes of incident plane wave
$\underline{\underline{A}}_{s,v}$	polarizability density dyadics
$[\mathbf{A}]_{s,v}$	6×6 polarizability density matrixes
$\mathbf{B}$	primitive magnetic field phasor
$[\mathbf{C}]_{\mathrm{ref},s,v}$	6×6 constitutive matrixes
$\mathbf{D}$	induction electric field phasor
$\underline{\underline{D}}_{s,v}$	depolarization dyadics
$[\mathbf{D}]_{s,v}$	6×6 depolarization matrixes
$e_{x,y,z}$	Cartesian components of $\mathbf{e}$
$\mathbf{e}$, $\mathbf{E}$	primitive electric field phasor
f_v	void volume fraction, porosity
$[\mathbf{f}]$	column vector of size 4
h	structural handedness parameter
$h_{x,y,z}$	Cartesian components of $\mathbf{h}$
$\mathbf{h}$, $\mathbf{H}$	induction magnetic field phasor
i	$= \sqrt{-1}$
$\underline{\underline{I}}$	identity dyadic
$[\mathbf{I}]$	identity matrix
k_0	free-space wave number
L	film thickness
$[\mathbf{M}]$	4×4 matrizant
$\mathbf{p}_{\pm}$	plane-wave polarization vectors
$[\mathbf{P}]$	4×4 matrix function
$\mathbf{r}$	position vector
$r_{L,R}$	circular amplitudes of reflected plane wave
$r_{LL,LR,RL,RR}$	circular reflection coefficients
$r_{s,p}$	linear amplitudes of reflected plane wave
$r_{ss,sp,ps,pp}$	linear reflection coefficients
$R_{LL,LR,RL,RR}$	circular reflectances
$R_{ss,sp,ps,pp}$	linear reflectances
$\mathbf{s}$	plane-wave polarization vector
$\underline{\underline{S}}$	rotation dyadic
$\underline{\underline{S}}_{x,y,z}$	elementary rotation dyadics
$\underline{\underline{\hat{S}}}_{y,z}$	rotation dyadics
t	time
$t_{L,R}$	circular amplitudes of transmitted plane wave
$t_{LL,LR,RL,RR}$	circular transmission coefficients
$t_{s,p}$	linear amplitudes of transmitted plane wave

$t_{ss,sp,ps,pp}$	linear transmission coefficients
$T_{LL,LR,RL,RR}$	circular transmittances
$T_{ss,sp,ps,pp}$	linear transmittances
$\mathbf{u}_{x,y,z}$	Cartesian unit vectors
$\mathbf{u}_{\tau,n,b}$	tangential, normal and binormal unit vectors
$\underline{\underline{U}}_{s,v}$	ellipsoidal shape dyadics
x, y, z	Cartesian coordinates
$\underline{\underline{\alpha}}_{\rm ref}$	reference relative magnetoelectricity dyadic
$\underline{\underline{\alpha}}_{r}$	relative magnetoelectricity dyadic
$\underline{\underline{\alpha}}_{s}$	relative magnetoelectricity dyadic of deposited material
$\underline{\underline{\alpha}}_{v}$	relative magnetoelectricity dyadic in the void region
$[\alpha]_{{\rm ref},s,v}$	3×3 matrix equivalents of $\underline{\underline{\alpha}}_{{\rm ref},s,v}$
$\underline{\underline{\beta}}_{\rm ref}$	reference relative magnetoelectricity dyadic
$\underline{\underline{\beta}}_{r}$	relative magnetoelectricity dyadic
$\underline{\underline{\beta}}_{s}$	relative magnetoelectricity dyadic of deposited material
$\underline{\underline{\beta}}_{v}$	relative magnetoelectricity dyadic in the void region
$[\beta]_{{\rm ref},s,v}$	3×3 matrix equivalents of $\underline{\underline{\beta}}_{{\rm ref},s,v}$
$\gamma_{\tau,b}^{(s,v)}$	ellipsoidal shape factors
$\delta_{s,v}$	ellipsoidal size measures
ϵ_0	permittivity of free space
$\epsilon_{a,b,c}$	relative permittivity scalars
ϵ_d	composite relative permittivity scalar
$\underline{\underline{\epsilon}}_{r}$	relative permittivity dyadic
$\underline{\underline{\epsilon}}_{\rm ref}$	reference relative permittivity dyadic
$\underline{\underline{\epsilon}}_{\rm ref}^{o}$	auxiliary relative permittivity dyadic
$\underline{\underline{\epsilon}}_{s}$	relative permittivity dyadic of deposited material
$\underline{\underline{\epsilon}}_{v}$	relative permittivity dyadic in the void region
$[\epsilon]_{{\rm ref},s,v}$	3×3 matrix equivalents of $\underline{\underline{\epsilon}}_{{\rm ref},s,v}$
ζ	angular function
η_0	intrinsic impedance of free space
ϑ	angle
θ	angle of incidence with respect to z axis
κ	transverse wave number
μ_0	permeability of free space
$\underline{\underline{\mu}}_{r}$	relative permeability dyadic
$\underline{\underline{\mu}}_{\rm ref}$	reference relative permeability dyadic
$\underline{\underline{\mu}}_{s}$	relative permittivity dyadic of deposited material
$\underline{\underline{\mu}}_{v}$	relative permittivity dyadic in the void region
$[\mu]_{{\rm ref},s,v}$	3×3 matrix equivalents of $\underline{\underline{\mu}}_{{\rm ref},s,v}$

$\xi(z)$	angular function
φ	angle
σ	dummy variable
ς_d	composite relative permittivity function
τ	angular function
χ	column inclination angle
χ_v	vapor incidence angle
ψ	angle of incidence in xy plane
ω	angular frequency
Ω	structural period of C-shaped SNTF and structural half-period of chiral STF

Akhlesh Lakhtakia is a distinguished professor of engineering science and mechanics at the Pennsylvania State University. He has published more than 460 journal articles; has contributed chapters to eight research books; has edited, coedited, authored, or coauthored nine books and five conference proceedings; has reviewed for 78 journals; and was the editor-in-chief of the international journal *Speculations in Science and Technology* from 1993 to 1995. He headed the IEEE EMC Technical Committee on Nonsinusoidal Fields from 1992 to 1994, and served as the 1995 Scottish Amicable Visiting Lecturer at the University of Glasgow. He is a Fellow of the Optical Society of America, SPIE—the International Society for Optical Engineering, and the Institute of Physics (UK). He was awarded the PSES Outstanding Research Award in 1996. Since 1999, he has organized five SPIE conferences on nanostructured materials and complex mediums. His current research interests lie in the electromagnetics of complex mediums, sculptured thin films, and chiral nanotubes.

Russell Messier is a professor of engineering science and mechanics at the Pennsylvania State University. He received his BSEE degree from Northeastern University and his PhD degree from the Pennsylvania State University. He has worked in the area of vapor deposition for over 30 years. His doctoral research was on one of the first commercial radio-frequency sputtering systems. He has published over 200 papers, holds seven U.S. patents, edited three books, is the founding editor of the international journal *Diamond and Related Materials*, and is a fellow of the American Vacuum Society. The Institute of Scientific Information, Philadelphia, has identified him as a Highly Cited Researcher. His interest in thin-film morphology has extended over much of his research career. His current interests are in the application of thin-film morphology fundamentals to sculptured thin films, hard coatings, sensor coatings, and biomaterials.

Chapter 3

Photonic Band Gap Structures

Joseph W. Haus

3.1. Introduction 46
3.2. One-dimensional structures 47
  3.2.1. Finite periodic structures: arbitrary angles of incidence 47
  3.2.2. Brief summary of infinite periodic structures 51
  3.2.3. Finite periodic structures: perpendicular incidence 55
  3.2.4. Slowly varying envelope techniques 61
  3.2.5. Nonlinear optics in 1D PBGs 62
3.3. Higher dimensions 63
  3.3.1. Vector wave equations 64
  3.3.2. Two dimensions 65
  3.3.3. Dielectric fluctuations 68
  3.3.4. Band structure 69
  3.3.5. Band eigenfunction symmetry and uncoupled modes 71
  3.3.6. Three dimensions 73
3.4. Summary 88
3.5. Appendix A 89
3.6. Appendix B 95
References 98
List of symbols 105

3.1 Introduction

Nanotechnology is a scientific frontier with enormous possibilities. Reducing the size of objects to nanometer scale to physically manipulate the electronic or structural properties offers a fabrication challenge with a large payoff. Nanophotonics is a subfield of nanotechnology and a part of nanophotonics includes photonic band gap structures, which manipulate the properties of light to enable new applications by periodically modulating the relative permittivity. In photonic band gap (PBG) structures, the electromagnetic properties of materials, such as the electromagnetic density of states, phase, group velocities, signal velocities, field confinement, and field polarization are precisely controlled. The size scale of interest in PBG structures is typically of the order of a wavelength, which is not quite as demanding as required to observe quantum confinement effects in electronic materials. Nevertheless, photonic devices designed with nanophotonic technology enable new technology for devices and applications in sensing, characterization, and fabrication.

Even though PBG photonic devices are complex and the fabrication is often expensive, rapid progress on PBG structures has been possible because of the development of powerful numerical computation tools that provide a detailed analysis of the electromagnetic properties of the system prior to fabrication. To design photonic devices we use a variety of computational techniques that help in evaluating performance.

Several books have already been written about the optical properties of PBGs. A classic book on 1D periodic structures was written by Brillouin.[1] Yariv and Yeh's book is an excellent resource on many aspects of periodic optical media.[2] Recent books devoted to the subject include the books by Joannopoulos et al.,[3] the very thorough book by Sakoda,[4] and a recent book on nonlinear optics of PBGs by Slusher and Eggleton[5] that features results of several researchers who have contributed to the subject. In addition, many good articles on PBG structures can be found in special issues[6–8] or in summer school proceedings.[9,10]

Numerical approaches are available to completely describe the properties of electromagnetic wave propagation in PBG structures. Three methods are of general use; they are the plane wave, the transfer matrix, and the finite-difference time-domain (FDTD) methods. The results of the plane wave method with the latter two are to some degree complementary, as is demonstrated and discussed later in this chapter.

Analytical methods are also available and have been especially useful for 1D systems. For instance, the development of coupled-mode equations for propagation by using multiple scales or slowly varying amplitude methods has given researchers powerful tools for studying nonlinear effects and designing new electro-optic (EO) devices, such as tunable optical sources from the ultraviolet to the terahertz regime, EO modulators, and a new generation of sensitive bio/chem sensors.

In this chapter, several basic algorithms are introduced and results exemplifying each numerical method are presented. MATLAB programs for the 1D transfer matrix method and the 2D plane wave method are provided in the appendixes. They are the simplest to understand and implement, and do not demand a large computational effort. The reader can use these programs to explore the band structure or transmission and reflection characteristics.

Periodic structures form a special subset in the subject of inhomogeneous media. Rayleigh published early studies of optical properties of inhomogeneous media devoted to the long-wavelength regime where the effective relative permittivity was calculated using various analytical approximations. For periodic structures, accurate calculations of the relative permittivity tensor are possible by the plane wave method by using the asymptotic form of the dispersion relation for long wavelengths.[11] These properties also have potentially new applications.

In the following sections, the conceptual foundations of PBG structures are elucidated. Section 3.2 provides a brief description of 1D systems. Basic optical properties are introduced, such as the electromagnetic density of modes, group velocity dispersion, band-edge field enhancement and nonlinear optical response characteristics.

Section 3.3 is devoted to higher-dimensional PBG systems. The common numerical techniques used to explore the optical properties of these systems are presented and results illustrating the techniques are provided. The final section summarizes the theoretical points of the previous sections and provides some future research directions. Two appendixes are provided with MATLAB programs that the reader can apply to illustrate some of the concepts and explore different parameter regimes.

3.2 One-dimensional structures

A multilayered dielectric stack is the simplest material that has some properties that are identified with PBG structures. The general properties of periodic structures can be found in the books by Brillouin[1] and Yariv and Yeh.[2] Multilayer PBGs exhibit interesting phenomena, such as high reflectivity over a frequency range, a so-called forbidden band or *stop band*. The theory has a strong correspondence with the quantum theory of periodic lattice.

3.2.1 Finite periodic structures: arbitrary angles of incidence

An optical filter consists of a large number of thin layers of differing optical properties. In any one layer, relative permittivity $\epsilon(t)$ has dispersive characteristics, but is independent of position. The temporal Fourier transform of the permittivity is frequency dependent, i.e., $\tilde{\epsilon}(\omega)$. The relative permeability $\mu(t)$ may also be dispersive. Although magnetic systems are rare in practice, they have become the center of attention in recent years due to the special properties of microscopically inhomogeneous systems that can have both negative real parts of $\tilde{\epsilon}(\omega)$ and $\tilde{\mu}(\omega)$. Such systems are called "left-handed" materials and one special interfacial property is the bending of the phase front of light at an angle on the opposite side to the normal refraction angle. For zero electric current and vanishing charge density, Maxwell's equations are stated as follows:

$$\frac{\partial \mathbf{D}}{\partial t} = \nabla \times \mathbf{H}, \qquad \frac{\partial \mathbf{B}}{\partial t} = -\nabla \times \mathbf{E}, \qquad \nabla \cdot \mathbf{B} = \nabla \cdot \mathbf{D} = 0. \tag{3.1}$$

The constitutive relations for a homogeneous, isotropic, dielectric-magnetic medium are given as

$$\mathbf{B}(t) = \mu_0 \int_{-\infty}^{\infty} \mu(t-t')\mathbf{H}(t')\,dt', \qquad \mathbf{D}(t) = \epsilon_0 \int_{-\infty}^{\infty} \epsilon(t-t')\mathbf{E}(t')\,dt', \tag{3.2}$$

which must be consistent with the Kramers-Kronig relations.[12,13] For nonmagnetic materials [i.e., $\tilde{\mu}(\omega) = 1$], the foregoing equations can be reformulated to give the vector wave equation

$$\frac{\partial^2 \mathbf{D}}{\partial t^2} = c^2 \nabla^2 \mathbf{E}, \tag{3.3}$$

where the speed of light in vacuum is $c = (\epsilon_0 \mu_0)^{-1/2}$. In the remainder of this chapter, $\tilde{\epsilon}(\omega)$ and $\tilde{\mu}(\omega)$ are written simply as ϵ and μ, with their functional dependences on the angular frequency ω being implicit.

For an electromagnetic field associated with a light ray traveling in the xz plane and making an angle α (also called the angle of incidence) with the z axis, $\mathbf{E}$ is a sinusoidal plane wave,

$$\mathbf{E} = \mathrm{Re}\big[\tilde{\mathbf{E}} e^{ik(ct - z\cos\alpha - x\sin\alpha)}\big], \tag{3.4}$$

where k is the wave number and $(\sin\alpha, 0, \cos\alpha)$ is the unit vector in the direction of propagation. In the rest of this section, the permittivities are frequency-domain functions. The tilde is dropped from the notation.

Two linear polarizations are distinguished in the analysis for boundary value problems. The p-polarized wave has its electric field vector in the plane of incidence, i.e., the plane defined by the incident, reflected and transmitted wave vectors, and the s-polarization has its electric field vector confined perpendicular to the plane of incidence. The two polarizations are indistinguishable at normal incidence ($\alpha = 0$). Thus, for

- *s-polarization*: $\mathbf{E}$ is in the y direction, so $\tilde{\mathbf{E}} = (0, A, 0)$ and, correspondingly, $\tilde{\mathbf{B}} = (A/c)(-\cos\alpha, 0, \sin\alpha)$ $\{\mathbf{B} = \mathrm{Re}[\tilde{\mathbf{B}} e^{ik(ct - z\cos\alpha - x\sin\alpha)}]\}$.
- *p-polarization*: $\mathbf{E}$ lies in the x-z plane so $\tilde{\mathbf{E}} = A(\cos\alpha, 0, -\sin\alpha)$ and $\tilde{\mathbf{B}} = (A/c)(0, 1, 0)$.

In either case, A is the complex amplitude of the electric field and contains information about the phase as well as the magnitude.

At a single interface, say $z = 0$ for the present, between two media with relative permittivities ϵ_a and ϵ_b and relative permeabilities μ_a and μ_b lying in $z < 0$ and $z > 0$ respectively, an incoming wave

$$\tilde{\mathbf{E}}_I e^{ik_a(ct - z\cos\alpha - x\sin\alpha)} \tag{3.5}$$

in $z < 0$ undergoes reflection and refraction so there are both a reflected plane wave

$$\tilde{\mathbf{E}}_R e^{ik_a(ct - z\cos\gamma - x\sin\gamma)}, \quad z < 0, \tag{3.6}$$

and a transmitted one

$$\tilde{\mathbf{E}}_T e^{ik_b(ct - z\cos\beta - x\sin\beta)}, \quad z > 0, \tag{3.7}$$

where $\gamma = \pi - \alpha$, $k_a = \sqrt{\mu_a \epsilon_a}\omega/c$, and $k_b = \sqrt{\mu_b \epsilon_b}\omega/c$.

At the interface, the tangential components of **E** and of **H** and the normal components of **B** and **D** are continuous. Application of the boundary conditions to the total field for $z < 0$ and the transmitted field in $z > 0$ gives the so-called Snell's law

$$\frac{\sin\beta}{\sin\alpha} = \sqrt{\frac{\epsilon_a \mu_a}{\epsilon_b \mu_b}} \tag{3.8}$$

for both polarizations, and the following polarization-specific relationships:

- *s-polarization*:

$$\frac{A_R}{A_I} = \frac{\sqrt{\epsilon_a/\mu_a}\cos\alpha - \sqrt{\epsilon_b/\mu_b}\cos\beta}{\sqrt{\epsilon_a/\mu_a}\cos\alpha + \sqrt{\epsilon_b/\mu_b}\cos\beta}, \tag{3.9}$$

$$\frac{A_T}{A_I} = \frac{2\sqrt{\epsilon_a/\mu_a}\cos\alpha}{\sqrt{\epsilon_a/\mu_a}\cos\alpha + \sqrt{\epsilon_b/\mu_b}\cos\beta}; \tag{3.10}$$

- *p-polarization*:

$$\frac{A_R}{A_I} = \frac{\sqrt{\mu_a/\epsilon_a}\cos\alpha - \sqrt{\mu_b/\epsilon_b}\cos\beta}{\sqrt{\mu_a/\epsilon_a}\cos\alpha + \sqrt{\mu_b/\epsilon_b}\cos\beta}, \tag{3.11}$$

$$\frac{A_T}{A_I} = \frac{2\sqrt{\mu_b/\epsilon_b}\cos\alpha}{\sqrt{\mu_a/\epsilon_a}\cos\alpha + \sqrt{\mu_b/\epsilon_b}\cos\beta}. \tag{3.12}$$

A PBG structure is a multilayer extension of the preceding results. The process of reflection and refraction repeatedly occurs at each internal boundary, which generates a pair of internal forward- and backward-propagating plane wave fields in each layer whose amplitudes are uniquely determined from the boundary conditions. Because $z \neq 0$ at the interfaces, some care must be exercised with phases.

The transmission and reflection relations are used to determine equations for the complex amplitudes of the waves in the positive and negative z directions in layer m (A_m and C_m, respectively). At the boundary $z = z_m$ between layers $m-1$ and m ($m = 1, \ldots, M$, with $z < z_1$ being the exterior of the filter), using s-polarization:

$$
\begin{aligned}
C_{m-1}e^{ik_m z_m \cos\alpha_{m-1}} &= \frac{\sqrt{\epsilon_{m-1}/\mu_{m-1}}\cos\alpha_{m-1} - \sqrt{\epsilon_m/\mu_m}\cos\alpha_m}{\sqrt{\epsilon_{m-1}/\mu_{m-1}}\cos\alpha_{m-1} + \sqrt{\epsilon_m/\mu_m}\cos\alpha_m} A_{m-1} \\
&\quad \times e^{-ik_m z_m \cos\alpha_{m-1}} \\
&\quad + \frac{2\sqrt{\epsilon_m/\mu_m}\cos\alpha_m}{\sqrt{\epsilon_{m-1}/\mu_{m-1}}\cos\alpha_{m-1} + \sqrt{\epsilon_m/\mu_m}\cos\alpha_m} C_m \\
&\quad \times e^{ik_m z_m \cos\alpha_m}
\end{aligned}
\tag{3.13}
$$

[see Eqs. (3.9) and (3.10)]; and

$$
\begin{aligned}
A_m e^{-ik_m z_m \cos\alpha_m} &= \frac{2\sqrt{\epsilon_{m-1}/\mu_{m-1}}\cos\alpha_{m-1}}{\sqrt{\epsilon_{m-1}/\mu_{m-1}}\cos\alpha_{m-1} + \sqrt{\epsilon_m/\mu_m}\cos\alpha_m} A_{m-1} \\
&\quad \times e^{-ik_m z_m \cos\alpha_{m-1}} \\
&\quad + \frac{\sqrt{\epsilon_m/\mu_m}\cos\alpha_m - \sqrt{\epsilon_{m-1}/\mu_{m-1}}\cos\alpha_{m-1}}{\sqrt{\epsilon_{m-1}/\mu_{m-1}}\cos\alpha_{m-1} + \sqrt{\epsilon_m/\mu_m}\cos\alpha_m} C_m \\
&\quad \times e^{ik_m z_m \cos\alpha_m}.
\end{aligned}
\tag{3.14}
$$

The extra terms on the right sides are due to the incoming wave moving down from $z > z_m$. The factors $e^{ik_m z_m \cos\alpha_{m-1}}$ etc. account for the nonzero value of z at the interface $z = z_m$.

There are now $2M$ equations for $A_1, \ldots, A_M, C_0, \ldots, C_{M-1}$, given A_0, the incoming wave, and assuming that $C_M = 0$, i.e., that no light is returned from the filter (or C_M is otherwise specified). The other polarization can be similarly handled. (It is clear that, because of the different transmission and reflection coefficients for the two polarizations when $\alpha \neq 0$, light not arriving normally complicates matters.)

The foregoing amplitude equations are solved by matrix methods, which we will refer to as the transfer matrix method. The MATLAB program in Appendix A implements the 1D transfer matrix method; it was used to illustrate several results presented in this chapter. A compact interference filter can be designed using the transfer matrix program. Most parameters are annotated in the main program. An interpolation function is used for dispersive dielectric properties, but this feature will not be demonstrated here. The program has been tested using metal layers, an extreme case, where the imaginary part of the refractive index exceeds the real part in magnitude.

Two figures are generated from the transfer matrix program in Appendix A. For simplicity, the layers are a quarter-wavelength thick at the free-space (vacuum) wavelength $\lambda_0 = 2\pi c/\omega = 1.5\ \mu$m. The refractive indices of the materials in one layer pair are $n_1 = \sqrt{\epsilon_a} = 3.5$ and $n_2 = \sqrt{\epsilon_b} = 1.5$, both materials being nonmagnetic (i.e., $\mu_a = \mu_b = 1$); the first is typical of a semiconductor material, while the second is typical of a wide electronic band gap insulator. No attempt was made to put in specific material parameters in the program, which has the capacity to apply complex index parameters and to interpolate from a table of data. The superstrate and substrate materials are assumed to have the same electromagnetic properties as vacuum. Five periods of the two layers are sufficient to create a large transmission stop band that covers wavelengths ranging from 1.1 to about 2 μm. This filter, which is less than 2 μm thick, is very compact indeed.

The larger issue in designing an interference filter is the dependence of the transmission on the angle of incidence of the radiation. Large variations may foil the rejection wavelength range for the filter. In Fig. 3.1, the p-polarization transmission function is shown for two angles of incidence $\alpha = 0$ and $\pi/4$. The edges of the stop bands shift by about 10% over this range. For larger angles, the shift becomes even larger. The s-polarization displayed in Fig. 3.2 has a similar angular dependence with a 10% shift of the band edge near 1.1 μm. Both have a larger shift of the gap to a wavelength near 500 nm. Evaluation of the angular dependence highlights a problem for the operation of the interference filters.

3.2.2 Brief summary of infinite periodic structures

Consider the simplest form of an infinite periodic dielectric structure whose index is defined by a periodic step function, where the steps have index values of n_1 and n_2 for widths a and b, respectively, with $d = a + b$ being the period of the lattice, and $m = 0, 1, 2, \ldots$, being the translation factor (Fig. 3.3). The permittivity and the permeability are related to the indices.

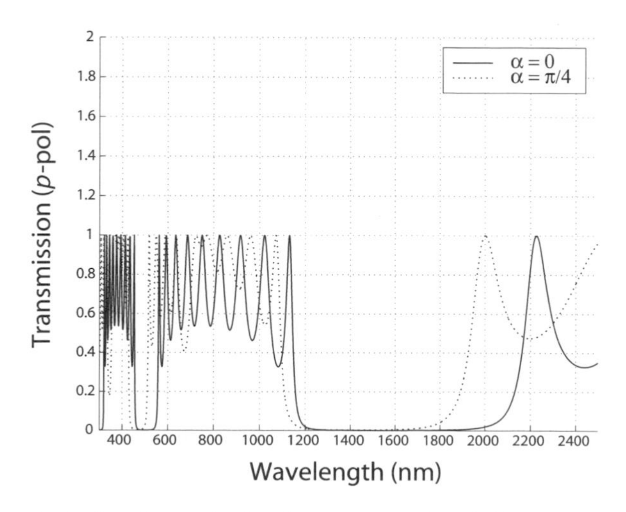

Figure 3.1 A p-polarization transmission for $\alpha = 0$ and $\pi/4$. Parameters are given in the text. The free-space (vacuum) wavelength $\lambda_0 = 2\pi c/\omega$ is the independent variable.

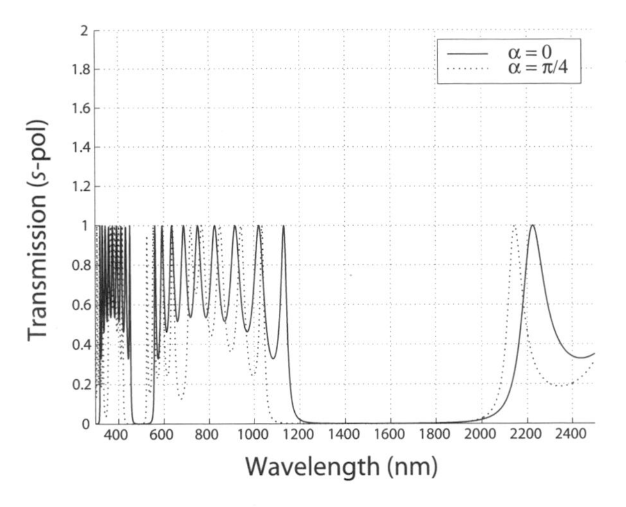

Figure 3.2 Same as Fig. 3.1, but for s-polarization.

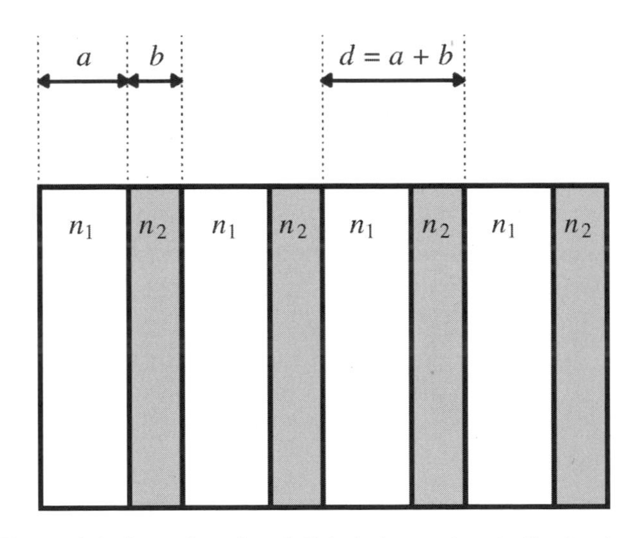

Figure 3.3 A section of an infinitely layered periodic structure.

The electromagnetic plane waves are assumed to propagate perpendicular to the layers. The electric field, which is perpendicular to the propagation direction, satisfies the scalar wave equation for a plane wave with frequency ω propagating in the z direction; thus,

$$\frac{d}{dz}\frac{1}{\mu(z)}\frac{dA}{dz} + \left(\frac{\omega}{c}\right)^2 \epsilon(z)A = 0. \tag{3.15}$$

This second-order ordinary differential equation has two independent solutions, which we denote as $A_1(z)$ and $A_2(z)$. The Wronskian, defined as the determinant of the solutions and their first derivative,

$$W(z) = \begin{vmatrix} A_1(z) & A_2(z) \\ A_1'(z) & A_2'(z) \end{vmatrix}, \tag{3.16}$$

is nonzero when the solutions are independent. Moreover, if n_1 and n_2 are kept constant within each unit cell, the Wronskian is also a constant—i.e., the value of the Wronskian is equal in the layers with identical physical properties. The solutions translated by a lattice constant can be written as a linear combination of the original solutions

$$\begin{bmatrix} A_1(z+d) \\ A_2(z+d) \end{bmatrix} = \mathbf{M} \begin{bmatrix} A_1(z) \\ A_2(z) \end{bmatrix}, \tag{3.17}$$

where $\mathbf{M}$ is the 2×2 transfer matrix. The Wronskian of the translated solutions is identical, i.e., $W(z+d) = W(z)$. From the constancy of the Wronskian, we conclude that the determinant of $\mathbf{M}$ must be equal to unity, i.e.,

$$M_{11}M_{22} - M_{12}M_{21} = 1. \tag{3.18}$$

Note that $\mathbf{M}$ has eigenvalues λ of the form

$$\lambda_{\pm} = e^{\pm ikd}, \tag{3.19}$$

where k is the wave number. Thus, we have

$$A_\ell(z+d) = e^{ikd} A_\ell(z), \quad \ell = 1, 2, \tag{3.20}$$

which is a manifestation of the Floquet–Bloch theorem. The amplitudes of the functions $A_\ell(z)$ called Floquet–Bloch functions are strictly periodic in z. The general solution in each region can be written as

$$\begin{aligned} A_1 &= C_1 e^{ik_1 z} + D_1 e^{-ik_1 z}, \\ A_2 &= C_2 e^{ik_2(z-a)} + D_2 e^{-ik_2(z-a)}. \end{aligned} \tag{3.21}$$

The solution A_1 is valid in the region $z \in (0, a)$ and all other regions displaced from this by md, where $(m = 1, 2, \ldots)$, and the wave number is related to the frequency by

$$k_1 = n_1 \frac{\omega}{c} = \sqrt{\epsilon_1 \mu_1} \frac{\omega}{c}. \tag{3.22}$$

The solution A_2 is valid in the region $z \in (a, d)$ and all its translations by md, and the wave number for this region is similarly defined as

$$k_2 = n_2 \frac{\omega}{c} = \sqrt{\epsilon_2 \mu_2} \frac{\omega}{c}. \tag{3.23}$$

The wave number k in Eq. (3.20) is normally restricted to the first Brillouin zone, since values outside this zone are redundant. To solve the eigenvalue problem and determine the connection (i.e., dispersion relation) between the wave number k and the frequency ω, the boundary conditions must be applied.

The two matrix coefficients M_{11} and M_{22} are

$$M_{11} = M_{22}^{*} = \frac{(1+Z_{12})^2}{4Z_{12}} e^{i(k_1 a + k_2 b)} - \frac{(Z_{12}-1)^2}{4Z_{12}} e^{i(k_2 b - k_1 a)}, \quad (3.24)$$

where $Z_{12} = n_1/n_2$ is the impedance ratio and the asterisk denotes complex conjugation. The dispersion equation is

$$\cos(kd) = \frac{(1+Z_{12})^2}{4Z_{12}} \cos(k_1 a + k_2 b) - \frac{(Z_{12}-1)^2}{4Z_{12}} \cos(k_2 b - k_1 a). \quad (3.25)$$

By plotting this transcendental equation, we deduce that the band structure of the 1D lattice always possesses a stop band, no matter how close Z_{12} is to unity. Figure 3.4 displays a strong distortion of the dispersion equation from a straight line. This is an example of the strong dispersion introduced by a PBG.

3.2.2.1 Electromagnetic mode density

The electromagnetic mode density is the number of electromagnetic modes per unit frequency range. The density of modes (DOM) is given by[14]

$$\rho(\omega) = \int_0^{\pi/d} \delta[\omega - \omega(k)]\, dk, \quad (3.26)$$

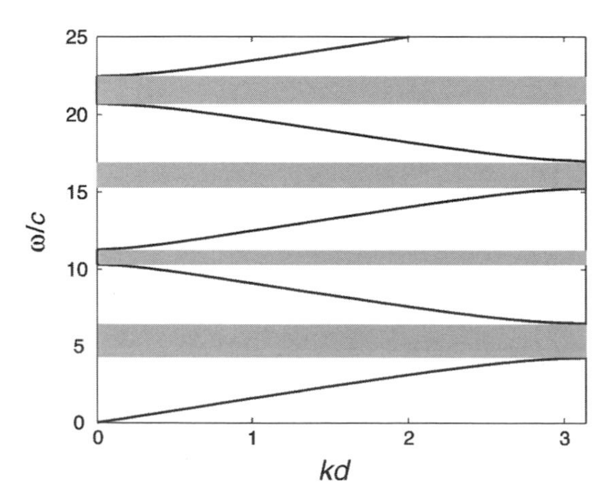

Figure 3.4 Dispersion curve for an infinite 1D lattice. The parameters are $n_1 = 2$, $n_2 = 1$, $a = 1/6$, and $b = 1/4$. The edge of the first Brillouin zone is $kd = \pi$ in the plotted units. The band gaps occur where k is complex-valued, signaling thereby that the wave is no longer propagating.

where the Dirac delta function $\delta(\cdot)$ is used as a sifting function. The DOM is given by

$$\rho = \frac{1}{v_g} = \frac{1}{|d\omega/dk|}, \tag{3.27}$$

where v_g is the group velocity of the wave.

The DOM of a periodic structure is normalized with respect to the DOM of a homogeneous medium of the same length and same average refractive index. The wave number in a homogeneous medium of refractive index n_{homo} is $k = n_{\text{homo}}\omega/c$. Thus the DOM ρ_{homo} in such a medium is

$$\rho_{\text{homo}} = \frac{n_{\text{homo}}}{c}. \tag{3.28}$$

The normalized DOM for a periodic structure is then given by

$$\rho_{\text{homo}}^{\text{norm}} = \frac{\rho}{\rho_{\text{homo}}} = \frac{n_{\text{homo}}}{|d\omega/dk|c}. \tag{3.29}$$

The DOM in one dimension diverges at the band edge, as the group velocity vanishes for an infinite lattice.

3.2.3 Finite periodic structures: perpendicular incidence

In analyzing a finite periodic structure, the effect of the finite boundaries on the internal field structure is considered. We begin with the simplest case of a unit cell composed of two layers with refractive indices n_1 and n_2 and widths a and b as before, with $a + b = d$.

After normalizing the input field at $z = 0$ to unity, and defining r_1 and t_1 as the complex reflectance and transmittance for the unit cell, the solutions to the field in column vector form at the boundaries are

$$\mathbf{A}(0) = \begin{pmatrix} 1 \\ r_1 \end{pmatrix}, \quad \mathbf{A}(d) = \begin{pmatrix} t_1 \\ 0 \end{pmatrix}. \tag{3.30}$$

In this case, the 2×2 transfer matrix[15] is defined as

$$\mathbf{A}(0) = \mathbf{M}\mathbf{A}(d). \tag{3.31}$$

Using the boundary conditions at $z = a$, and the fact that for real-valued refractive index profiles, the behavior of the field must be invariant under time reversal, yields

$$\mathbf{M} = \begin{pmatrix} 1/t_1 & r_1^*/t_1^* \\ r_1/t_1 & 1/t_1^* \end{pmatrix}. \tag{3.32}$$

The eigenvalue equation of **M**, with eigenvalues λ, is therefore

$$\lambda^2 - 2\lambda \mathrm{Re}(1/t_1) + 1 = 0, \tag{3.33}$$

where Re(·) denotes the real part. Rewriting λ in a convenient form as $\lambda = e^{i\theta_0}$, we obtain the useful relation

$$\mathrm{Re}(1/t_1) = \cos\theta_0. \tag{3.34}$$

The Cayley–Hamilton theorem states that every matrix obeys its own eigenvalue equation; hence,

$$\mathbf{M}^2 - 2\mathbf{M}\cos\theta_0 + \mathbf{I} = 0, \tag{3.35}$$

where **I** is the 2×2 identity matrix. It follows by induction that the transfer matrix for an N-layered structure is

$$\mathbf{M}^N = \mathbf{M}\frac{\sin N\theta_0}{\sin\theta_0} - \mathbf{I}\frac{\sin(N-1)\theta_0}{\sin\theta_0}. \tag{3.36}$$

The general form of the transfer matrix **M** shown in Eq. (3.32) for a unit cell can be applied to N unit cells, as well. Thus,

$$\mathbf{M}^N = \begin{pmatrix} 1/t_N & r_N^*/t_N^* \\ r_N/t_N & 1/t_N^* \end{pmatrix}, \tag{3.37}$$

where r_N and t_N are the complex reflection and transmission amplitudes for an N-period PBG. The transmission for an N-period structure is therefore

$$\frac{1}{T_N} = \frac{1}{|t_N|^2} = 1 + \frac{\sin^2 N\theta_0}{\sin^2\theta_0}\left(\frac{1}{T_1} - 1\right), \tag{3.38}$$

where $T_1 = |t_1|^2$ is the transmittance for a single unit cell. To explicitly calculate T_1, the boundary conditions are applied at each interface. Following the same procedure as for the infinitely layered structure, we can calculate the transfer matrix for the field as it goes from the first layer to the second layer explicitly as

$$\mathbf{M}_{12} = \begin{bmatrix} (1+Z_{12})e^{-ik_1 a} & (1-Z_{12})e^{-ik_1 a} \\ (1-Z_{12})e^{ik_1 a} & (1+Z_{12})e^{ik_1 a} \end{bmatrix}. \tag{3.39}$$

The transfer matrix for the field propagation through the layer pair making up the unit cell is $\mathbf{M} = \mathbf{M}_{ab}\mathbf{M}_{ba}$. Defining T_{12} and R_{12} as $T_{12} = (4n_1n_2)/(n_1+n_2)^2$ and $R_{12} = 1 - T_{12}$, the transmission for a single layer denoted as b is

$$t_b = \frac{T_{12}e^{i(p+q)}}{1 - R_{12}e^{-2iq}}, \tag{3.40}$$

where $p = n_1 a\omega/c$ and $q = n_2 b\omega/c$. The field within an N-layered structure can be calculated by applying the transfer matrix to the input field at each layer boundary. The phase θ_0 in Eq. (3.38) plays a very important role in the behavior of the transmission curve of a finite period structure. In the passbands, θ_0 is real valued and T_N varies periodically with θ_0. When θ_0 is complex-valued, this behavior changes to a hyperbolic exponential form, giving rise to band gaps. The quantity θ_0 is real-valued when $|\mathrm{Re}(1/t_1)| \leqslant 1|$ and complex-valued otherwise, as seen from Eq. (3.34). For real values of θ_0, Eq. (3.38) shows that T_N is periodic in θ_0 with a period π/N resulting in N oscillations in each passband interval of θ_0-length π. Reference 15 is a detailed exposition of the properties of T_N. For large values of N, the transfer matrix method requires several repeated matrix multiplications.

3.2.3.1 Density of modes

A general expression for the DOM for a finite periodic structure was derived using cavity quantum electrodynamics (QED) in Refs. 15 and 16. For an N-period structure, the DOM is defined as

$$\rho_N = \frac{dk_N}{d\omega}, \tag{3.41}$$

and the group velocity as

$$v_N = \frac{1}{\rho_N} = \frac{d\omega}{dk_N}. \tag{3.42}$$

Given that the complex-valued transmission amplitude t_N is available from the transfer matrix method, the DOM can be calculated as in Ref. 15, i.e.,

$$\rho_N = \frac{1}{d}\frac{\mathrm{Im}(t_N)'\mathrm{Re}(t_N) - \mathrm{Re}(t_N)'\mathrm{Im}(t_N)}{\mathrm{Re}(t_N)^2 + \mathrm{Im}(t_N)^2}, \tag{3.43}$$

where the prime indicates differentiation with respect to ω, while $\mathrm{Im}(\cdot)$ denotes the imaginary part. As demonstrated in Fig. 3.5, the DOM mimics the behavior of the transmission curve. The maxima and minima of the DOM and transmission seem to line up, but there is a slight offset between the extreme values of the two curves. This offset becomes rapidly negligible with increasing number of periods N. The DOM has its largest values at the band edge resonance, which means that the group velocity is the smallest at the band edge.

3.2.3.2 Effective refractive index

The spatial variation of the refractive index within the periodic structure has a square wave profile, as inferred from Fig. 3.3. The refractive index $n_{\mathrm{eff}}(\omega)$ for the structure as a whole, which we shall call the effective refractive index, can be calculated by a simple yet elegant method.[16] The effective refractive index $n_{\mathrm{eff}}(\omega)$ is

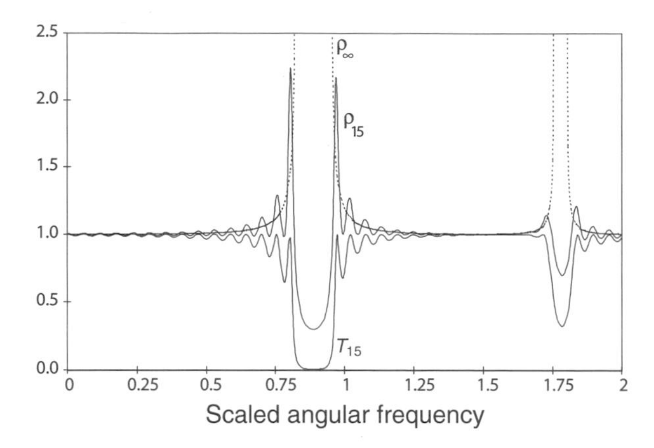

Figure 3.5 Transmission T_{15} and density of states ρ_{15} versus a scaled angular frequency for a layered structure comprising 15 two-layer periods with the same thickness for each dielectric layer. For comparison, ρ_∞ (the density of states for an infinite lattice) is also shown. Refractive indices are $n_1 = 1$ and $n_2 = 1.25$.

a very useful tool in understanding the phase-matching concerns that are essential for good conversion efficiency in parametric frequency conversion processes.

The transmission amplitude at the output of a periodic structure of length L can be written as

$$\hat{t}(\omega) = A(L)e^{ik_{\text{eff}}(\omega)L}, \tag{3.44}$$

where A is the position-dependent amplitude, and k_{eff} is the effective wave number of the whole structure defined as

$$k_{\text{eff}}(\omega) = n_{\text{eff}}(\omega)\frac{\omega}{c}. \tag{3.45}$$

These two equations yield the expression

$$n_{\text{eff}}(\omega) = \frac{c}{\omega}\frac{1}{L}\tan^{-1}\left[\frac{\text{Im}(\hat{t})}{\text{Re}(\hat{t})}\right]. \tag{3.46}$$

The actual value of the transmission amplitude $\hat{t}$ for a given refractive-index profile can be easily determined using the transfer matrix method.

The behavior of the effective refractive index as a function of frequency for a dielectric structure with 10 equal two-layer periods ($n_1 = 1$, $n_2 = 1.5$) is seen in Fig. 3.6. The transmission through this structure is also plotted on the same graph for comparison. Within the first band gap, the effective refractive index falls sharply as a function of frequency, while n_{eff} varies very slowly with ω outside that regime.

The behavior of n_{eff} is analogous to that of a Lorentzian atom under the influence of a sinusoidally varying electric field. The normal dispersion region for the Lorentzian atom, associated with an increase in the real part of the refractive index with angular frequency, corresponds to the region *outside* the band gaps in the periodic structure. The anomalous dispersion regions then correspond to the band gaps in the transmission curve in Fig. 3.6. In the anomalous dispersion region, when the imaginary part of n_{eff} is large, there is a large dissipation of energy into the medium, i.e. there is resonant absorption in these regions. However, in the case of the periodic structures, the corresponding band gap regions represent reflection of the electromagnetic wave from the structure, either partially or completely.

3.2.3.3 Field profiles

The transmitted (forward) and reflected (backward) field amplitudes within the periodic structures are computed by repeated applications of the transfer matrix method and by retaining the amplitudes generated in each layer. The fields are functions of both position z and angular frequency ω.

The field amplitude is largest at the first transmission resonance below the band gap. The first transmission resonances on either side of the band gap are called the lower and the upper band edge transmission resonance, respectively. The field amplitude at the resonance frequency has one maximum as would be found for the lowest transmission in a Fabry–Pérot étalon. The difference is that the Fabry–Pérot resonance is a half wavelength, while for the Bragg grating, the field amplitude varies slowly over the scale of a wavelength. The maximum field amplitude is larger than the input field value. For transmission resonances farther from the band gap edge, the field profiles exhibit an increasing number of maxima, again similar to higher-order modes in the Fabry–Pérot étalon. The field profile shows two maxima at the second transmission resonance, three maxima at the third transmission resonance, etc. However, note that the field values at the band edge transmission resonance are the largest. This follows directly from the fact that the DOM has its largest values at the band edge transmission resonances, as seen in Fig. 3.5. This selective enhancement of the fundamental field at the band edge is a significant fact

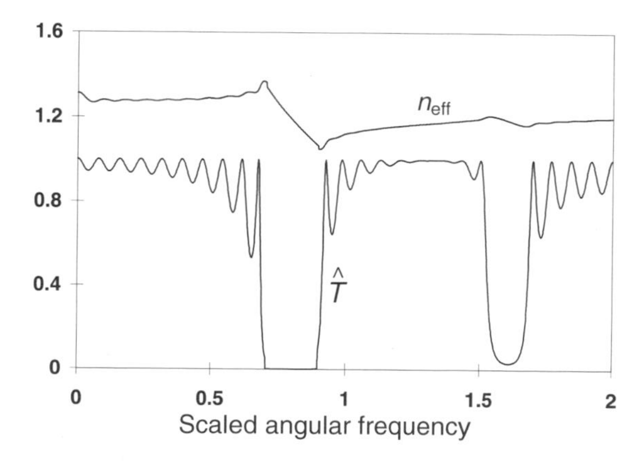

Figure 3.6 Effective refractive index n_{eff} and transmission $\hat{T} = |\hat{t}|^2$ versus a scaled angular frequency for a periodic dielectric structure of 10 two-layer periods, with refractive indices $n_1 = 1$ and $n_2 = 1.5$ and the same thickness for each layer.

that can be exploited in designing efficient nonlinear optical devices. The behavior of the fields on the short-wavelength side of the band gap is analogous to that on the long-wavelength side. The presence of the transmission maxima at the band edge is also critical to the selective enhancement of the fundamental field. The maxima are washed out when the grating is apodized or chirped or when absorption is present.

3.2.3.4 Absorption

Absorption in a material medium is quantified by the absorption coefficient, which is defined in terms of the imaginary part of the refractive index. There is a distinction depending on which material (high or low index) has absorption. To illustrate the effects of absorption, we consider the case where the high index is complex. Figure 3.7(a) represents the transmission spectrum about the center of the first band gap for two cases: lossless dielectrics and dielectrics with equal absorption in the two media. The transmission is symmetrically lowered around the band gap region.

In Fig. 3.7(b), the transmission spectra indicate that the absorption is not uniformly distributed. The dashed curve represents the spectrum when the complex refractive index is concentrated in the high-index medium, while the solid curve represents the case where the lower index medium is absorptive. The transmission through such a structure can then be calculated following the usual transfer matrix recipe. The introduction of absorption causes a fall in the transmitted intensity, but more importantly, the asymmetry in the absorption spectrum reveals where the electric field is concentrated. It is mainly in the high-index medium on the low-frequency side of the band gap and vice versa on the high-frequency side. The size of the band gap, however, does not change. In addition to a general drop in the transmission, a complex index smoothes out the oscillations on each side of the

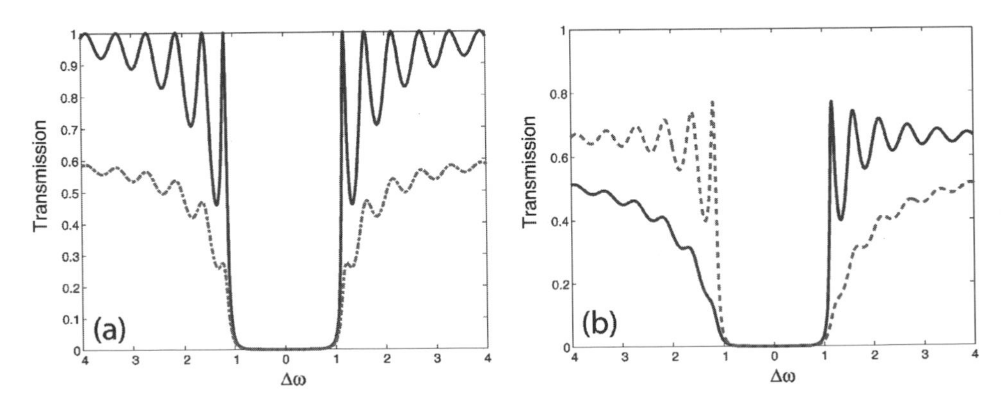

Figure 3.7 Transmission versus $\Delta\omega$ (detuning from the angular frequency at the center of the band gap) for lossy and nonlossy dielectrics. (a) The solid line is the lossless dielectric case and the dash-dotted line represents lossy dielectric with equal absorption in each dielectric. (b) One dielectric is lossy. The lossy medium is the higher permittivity medium for the dashed line and it is the lower permittivity medium for the solid line.

band gap. The oscillations become progressively smaller with increasing values of the absorption. The fundamental field intensity at the band edge also falls as a result of the absorption and the smoothing out of the transmission resonances. Absorption limits the ability of the structure to enhance the fundamental field at the band edge, by smoothing out the transmission resonances.

The transfer matrix method is also used for metal/dielectric layers that have demonstrated surprizing transmission and reflection propereties when placed together.[17,18]

3.2.4 Slowly varying envelope techniques

Scalora and Crenshaw[19] developed a generalization of the beam propagation algorithm that has found wide application in nonlinear optics. This method has many advantages over the other beam propagation methods because it handles forward and backward propagating waves and is simple to implement. It can be used for nonlinear media and is useful to describe pulse propagation. This method is a powerful numerical procedure for studies of 1D and 2D photonic band structures.

The method, called the slowly varying envelope approximation in time (SVEAT), is discussed here for 1D lattices. The wave is incident perpendicular to the interface and satisfies the scalar wave equation:

$$\frac{\partial^2 E}{\partial z^2} - \left(\frac{1}{c}\right)^2 \frac{\partial^2 E}{\partial t^2} = \frac{4\pi}{c^2}\frac{\partial^2 P}{\partial t^2}, \tag{3.47}$$

where P is the polarization of the medium. This can be as simple as an expression proportional to the electric field or as complicated as a contribution that is a nonlinear function of the electric field.

The equation is approximated by a slowly varying envelope expansion, but only the rapid time variable is approximated. Let the field be represented by

$$E(z,t) = 2\mathrm{Re}(\hat{E}e^{-i\omega t}), \tag{3.48}$$

where $\hat{E}$ is an envelope function. The polarization is also decomposed into rapid and slow varying contributions. After expressing the polarization envelope through a linear susceptibility χ as

$$\hat{P} = \chi\hat{E}, \tag{3.49}$$

the wave equation is approximated as

$$\frac{\partial\hat{E}}{\partial\tau} - i\mathcal{D}\hat{E} = i\frac{1+4\pi\chi}{4\pi}\hat{E}. \tag{3.50}$$

Therein, the spatial variable has been scaled to the wavelength $\xi = z/\lambda_0$, time has been scaled by the oscillation frequency to $\tau = \nu t$, and the operator

$$\mathcal{D} \equiv \frac{1}{4\pi} \frac{\partial^2}{\partial \xi^2} \tag{3.51}$$

has been used.

The solution of Eq. (3.50) is formally written as

$$\hat{E}(\xi, \tau) = \mathcal{T} e^{i \int_0^\tau [\mathcal{D} + (1/4\pi) + \chi] d\tau} \hat{E}(\xi, 0), \tag{3.52}$$

where $\mathcal{T}$ is the time-ordering operator. The differential operator $\mathcal{D}$ can be diagonalized by Fourier transformation, but the susceptibility χ may be a complicated function of the field and the spatial and temporal variables. Therefore, this equation is solved by a spectral method called the split-step propagation algorithm or beam propagation method. A second-order version of this algorithm is as follows:

$$\hat{E}(\xi, \tau) = e^{i\mathcal{D}\tau/2} e^{i \int_0^\tau [(1/4\pi) + \chi] d\tau} e^{i\mathcal{D}\tau/2} \hat{E}(\xi, 0). \tag{3.53}$$

It is solved by applying the fast Fourier transform technique to diagonalize the operator $\mathcal{D}$ and then solving for the susceptibility in the original space.

This method has been applied to a wide variety of problems, including pulse reshaping and dispersion in transmission through[20] a PBG and emission rates of dipoles embedded[21] in a PBG. Recent work on pulse propagation in nonlinear metallodielectrics, i.e., stacks containing alternate layers of metals and insulating dielectrics, has elucidated the optical limiting properties of the complex, nonlinear systems.[22]

The method has been generalized to cover forward-backward coupled-mode equations to describe nonlinear media with space-time effects. Such methods continue to find uses as we explore new parameter regimes and have led to potential novel applications of PBGs.

3.2.5 Nonlinear optics in 1D PBGs

One-dimensional PBGs constitute a large portion of the research effort for obvious reasons. They are often simple to fabricate with good control over layer thickness and surface smoothness, they are cost-effective, and the experiments are simpler to design. What is perhaps surprising is the control that can be exercised over the optical properties of 1D systems to enhance a system's nonlinear response.

Among the many interesting nonlinear phenomena that are predicted (and often observed) in 1D PBGs are gap solitons,[23] optical limiting and switching,[22,24–27] optical parametric generation,[28–30] optical diodes,[31] photonic band edge lasers,[32] Raman gap solitons,[33] and superfluorescence.[34] To illustrate the usefulness of nonlinear effects in 1D systems, this section concludes with a highlight of two nonlinear effects in PBGs: gap solitons and second-harmonic generation.

Gap solitons were first reported by Chen and Mills[23] for materials with a third-order nonlinear susceptibility, i.e., so-called Kerr nonlinear media, and quickly thereafter, a large number of papers appeared elucidating the gap soliton's properties. Other examples of solitons near the gap also followed. This chapter limits the citations to a few papers[35,36] and refers the reader to de Sterke and Sipe's review article.[37] Several experiments were designed to explore the nonlinear response in different regimes. Experimental tests of general nonlinear optical effects have been attempted in several systems, including semiconductor wave guides,[38] colloidal solutions,[39] and fiber Bragg gratings.[40,41]

Bragg solitons, which have the laser frequency tuned near but outside the band gap, were studied by Eggleton et al.[40,41] Gap solitons have the incident laser spectrum contained mostly in the gap region; they require that the nonlinearity be large enough to shift the band gap away from the laser spectrum, thus creating transparency. Taverner et al.[42] used narrow-band gap fiber Bragg gratings driven by a narrow-band source to achieve formation of the gap soliton.

Second-harmonic generation was among the first observed nonlinear optical effects reported after the invention of the laser. It is also a success story, since there have long been laser systems using the second harmonic to transform light from one wavelength to another. Parametric processes are related second-order phenom ena that are also now applied to a wide range of systems to generate near- and far-infrared and even terahertz radiation.

Large enhancement of second-harmonic generation in PBG systems was discussed in the context of band edge electric field enhancement and slow group velocity phenomena by the groups of Scalora and Haus.[28,29] Calculations showing enhanced second harmonic generation in waveguides were reported by Pezzetta et al.[30] In all cases, a careful theoretical analysis of finite periodic systems shows that several order of magnitude enhancement could be expected when the PBG was designed to include these effects and phase matching. The nonlinear conversion efficiency is predicted to increase as L^6, where L is the length of the PBG. By comparison, in a perfectly phase-matched sample, the conversion efficiency is proportional to L^2. With the additional enhancement from a PBG, compact samples measuring only a few wavelengths in thickness can be used and modest incident pump powers can be applied.

Several experimental tests of enhanced second-harmonic generation were performed on multilayer stacks. Balakin et al.[43] used alternated layers of ZnS and SrF. Dumeige and coworkers[44,45] reported second-harmonic experiments in molecular-beam-epitaxy grown AlGaAs and AlAs multilayers. More than an order of magnitude enhancement was observed in each case. In each case, the experiments were guided by theoretical calculations.

3.3 Higher dimensions

The qualitative character of PBGs is changed in higher dimensions. The off-axis diffraction of the waves leading to coupling between plane wave modes in different directions leads to a number of new phenomena that can be exploited. First of all, the bands are nonmonotonic, leading to changes in the group velocity direction as well as its magnitude; and the Bloch waves of the structure have a particular transverse symmetry that leads to uncoupled modes in the structure. An excellent pedagogical and research book covering advanced topics in photonic crystals has been recently published by Sakoda.[4] A review of plane wave calculations in 3D systems was published by Haus.[46]

Plane wave methods are developed for infinite lattices. Plane wave techniques have been extended to explore the properties of lattices containing defects, but they rely on an expansion of periodic functions. They reduce the problem to finding the eigenvalues and eigenvectors of a generalized eigenproblem. The number of plane waves is N. The matrixes are not sparse and many diagonalization schemes demand a large amount of the computational time and memory allocation. Moreover, convergence is difficult to achieve, due to high-spatial-frequency terms in the relative permittivity; the convergence goes as $N^{-1/D}$, where D is the lattice dimension. Hence, for higher dimensions, the matrixes must be made as large as possible to be certain that the band structure features are correctly reproduced.

3.3.1 Vector wave equations

With the convenient assumption that the permittivity is not a function of the angular frequency, Maxwell's equations for $\mathbf{E}$ and $\mathbf{H}$ lead to the following vector wave equations:

$$\nabla \times \nabla \times \mathbf{E}(\mathbf{x}, t) + \frac{1}{c^2}\frac{\partial^2}{\partial t^2}\epsilon(\mathbf{x})\mathbf{E}(\mathbf{x}, t) = \mathbf{0}, \tag{3.54}$$

$$\nabla \times \eta(\mathbf{x})\nabla \times \mathbf{H}(\mathbf{x}, t) + \frac{1}{c^2}\frac{\partial^2}{\partial t^2}\mathbf{H}(\mathbf{x}, t) = \mathbf{0}. \tag{3.55}$$

Here and hereafter, $\eta(\mathbf{x}) \equiv 1/\epsilon(\mathbf{x})$ and the relative permittivity

$$\epsilon(\mathbf{x}) = \epsilon_b \Theta_b(\mathbf{x}) + \epsilon_a \Theta_a(\mathbf{x}) \tag{3.56}$$

is linear, locally isotropic, positive-definite, and periodic with lattice vectors $\mathbf{R}$; whereas ϵ_b is the relative permittivity of the background material, and ϵ_a is that of the inclusion material. The Heaviside functions $\Theta_\ell(\mathbf{x})$ are unity in the region occupied by the material ℓ and vanish otherwise. For nonoverlapping spheres of radius r, the spatial Fourier transform of $\epsilon(\mathbf{x})$ is

$$\epsilon(\mathbf{G}) = \frac{1}{V_{\text{cell}}}\int_{\text{WS cell}} d^3x e^{-i\mathbf{G}\cdot\mathbf{x}}\epsilon(\mathbf{x}) = \epsilon_b \delta_{\mathbf{G},\mathbf{0}} + 3f(\epsilon_a - \epsilon_b)\frac{j_1(Gr)}{Gr}. \tag{3.57}$$

The function $j_1(Gr)$ is the spherical Bessel function of order 1, while $\delta_{\mathbf{G},\mathbf{G}'}$ is the 3D Kronecker delta. The eigenfunctions of Eqs. (3.54) and (3.55) are Bloch functions of the form

$$\mathbf{E}_{n\mathbf{k}}(\mathbf{x},t) = \exp[i(\mathbf{k}\cdot\mathbf{x} - \omega_{n\mathbf{k}}t)]\sum_{\mathbf{G}}\mathbf{E}_{n\mathbf{k}}(\mathbf{G})\exp[i(\mathbf{G}\cdot\mathbf{x})], \tag{3.58}$$

$$\mathbf{H}_{n\mathbf{k}}(\mathbf{x},t) = \exp[i(\mathbf{k}\cdot\mathbf{x} - \omega_{n\mathbf{k}}t)]\sum_{\mathbf{G}}\mathbf{H}_{n\mathbf{k}}(\mathbf{G})\exp[i(\mathbf{G}\cdot\mathbf{x})], \tag{3.59}$$

where $\mathbf{G}$ is a reciprocal lattice vector, $\mathbf{k}$ is the reduced wave vector in the first Brillouin zone (BZ), and n is the band index including the polarization.

The Fourier coefficients $\mathbf{E}_{\mathbf{G}} \equiv \mathbf{E}_{n\mathbf{k}}(\mathbf{G})$ and $\mathbf{H}_{\mathbf{G}} \equiv \mathbf{H}_{n\mathbf{k}}(\mathbf{G})$ satisfy, respectively, the infinite-dimensional matrix equations

$$(\mathbf{k}+\mathbf{G}) \times [(\mathbf{k}+\mathbf{G}) \times \mathbf{E}_{\mathbf{G}}] + \frac{\omega^2}{c^2}\sum_{\mathbf{G}'}\epsilon_{\mathbf{G}\mathbf{G}'}\mathbf{E}_{\mathbf{G}'} = \mathbf{0}, \tag{3.60}$$

$$(\mathbf{k}+\mathbf{G}) \times \left[\sum_{\mathbf{G}'}\eta_{\mathbf{G}\mathbf{G}'}(\mathbf{k}+\mathbf{G}') \times \mathbf{H}_{\mathbf{G}'}\right] + \frac{\omega^2}{c^2}\mathbf{H}_{\mathbf{G}} = \mathbf{0}, \tag{3.61}$$

with $\epsilon_{\mathbf{G}\mathbf{G}'} \equiv \epsilon(\mathbf{G}-\mathbf{G}')$ and $\eta_{\mathbf{G}\mathbf{G}'} \equiv \eta(\mathbf{G}-\mathbf{G}')$. Their respective solutions constitute the E and the H methods, respectively. The choice of other fields to express the wave equation is redundant to these two, at least for nonmagnetic materials. For instance, Zhang and Satpathy[47] used the displacement field, which satisfies the wave equation

$$\nabla \times \nabla \times \eta(\mathbf{x})\mathbf{D}(\mathbf{x},t) + \frac{1}{c^2}\frac{\partial^2}{\partial t^2}\mathbf{D}(\mathbf{x},t) = \mathbf{0}, \tag{3.62}$$

but their method is identical to the H-method.[48]

The solution procedure required the truncation of the infinite set of reciprocal lattice vectors to just N lattice vectors, which produces matrixes of size $3N \times 3N$ from Eqs. (3.60) and (3.61). Using $\nabla \cdot \nabla \times \mathbf{E} = 0$ and $\nabla \cdot \mathbf{H} = 0$, only $2N \times 2N$ matrixes turn out to be necessary.

Although the two methods yield the same spectrum when an infinite number of plane waves are included, their truncated forms yield, in general, very different spectra even when N equals a few thousands.

In the following subsections, results for 2D and 3D PBGs are highlighted. All lengths are in units of $a/2\pi$, and the magnitudes of the wave vectors $\mathbf{k}$ and $\mathbf{G}$ are in units of $2\pi/a$, where a is the side of the real space conventional cubic unit cell for the relevant Bravais lattice.

3.3.2 Two dimensions

The scalar wave equation has limited application in photonic band structures. Two-dimensional periodic structures, such as rods in a lattice arrangement (Fig. 3.8),

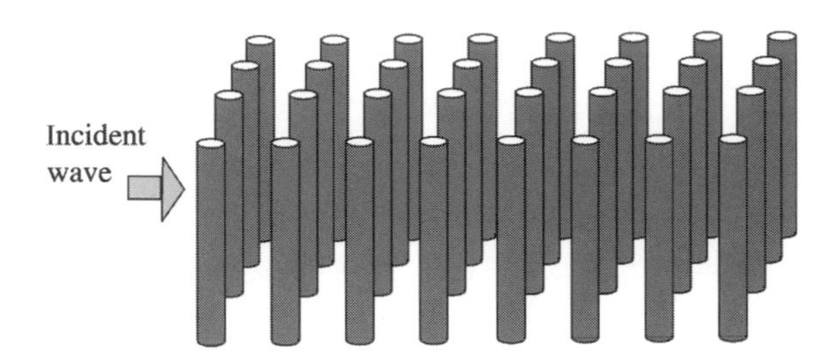

Figure 3.8 A 2D lattice of rods. The incident light propagates in the plane to which the rods are perpendicular.

with the field polarized parallel to the rod axis satisfy scalar wave equations.[49–54] In this section, it is relevant to consider for a moment the analysis of the simpler scalar problem and contrast the results with those of the vector equations in Sec. 3.3.6.

The derivation of the appropriate scalar wave equation depends upon the chosen field polarization. When the electric field is oriented along the symmetry axis, the equation is as follows:

$$\nabla^2 E - \frac{1}{c^2}\epsilon(\mathbf{x})\frac{\partial^2 E}{\partial t^2} = 0. \tag{3.63}$$

This leads to a generalized eigenvalue problem because the frequency eigenvalues, obtained by Fourier transformation in the time coordinate, are multiplied by the periodic relative permittivity. For this reason, it is not precisely equivalent to the Schrödinger equation, since the eigenvalue multiplies the "potential." In 1D periodic materials with the field propagating perpendicular to the surfaces, this is equivalent to the Kronig-Penney model.[55,56]

When the magnetic field is oriented along the symmetry axis, the scalar wave equation

$$\nabla \cdot \eta(\mathbf{x})\nabla H - \frac{1}{c^2}\frac{\partial^2 H}{\partial t^2} = 0 \tag{3.64}$$

emerges. This equation yields an ordinary eigenvalue problem for the frequency that is equivalent to the Schrödinger equation with a periodically varied mass.

As discussed in Sec. 3.3.1, several methods can be applied to solving the two foregoing equations. The simplest, the most widely used, and the most tractable method is derived from the Bloch wave analysis of periodic structures. The eigenvalues and eigenvectors can be found by introducing the Bloch functions

$$\phi_{n\mathbf{k}}(\mathbf{x}, t) = e^{i(\mathbf{k}\cdot\mathbf{x}-\omega_{n\mathbf{k}}t)}\sum_{\mathbf{G}}\phi_{\mathbf{k}}(\mathbf{G})e^{i\mathbf{G}\cdot\mathbf{x}}, \tag{3.65}$$

where $\mathbf{G}$ is the reciprocal lattice vector for the chosen lattice and the wave vector $\mathbf{k}$ lies within the first BZ. The index n labels the band for a particular wave vector $\mathbf{k}$.

A triangular lattice and its reciprocal lattice are depicted in Fig. 3.9. A lattice is decomposed into unit cells that repeat and tile the space. Each lattice point can be reached from the origin by a linear combination of two primitive basis vectors, which we denote by $(\mathbf{b}_1, \mathbf{b}_2)$. The corresponding reciprocal lattice vectors are defined by the relation

$$\mathbf{G}_\ell \cdot \mathbf{b}_m = 2\pi \delta_{i\ell,m}, \tag{3.66}$$

where $\delta_{\ell,m}$ is the Kronecker delta function. For example, the triangular lattice in Fig. 3.9 has $\mathbf{b}_1 = a\hat{\mathbf{e}}_x$ and $\mathbf{b}_2 = (a/2)\hat{\mathbf{e}}_x + (\sqrt{3}a/2)\hat{\mathbf{e}}_y$ as its basis vectors. The corresponding reciprocal lattice vectors are $\mathbf{G}_1 = (2\pi/a)[\hat{\mathbf{e}}_x - (1/\sqrt{3})\hat{e}_y]$ and $\mathbf{G}_2 = (4\pi/\sqrt{3}a)\hat{\mathbf{e}}_y$, as also depicted in the same figure. The reciprocal lattice is constructed by summing combinations of the two basis reciprocal lattice vectors.

Also shown in Fig. 3.9 is the construction for the first BZ: perpendicular bisectors of the reciprocal lattice points are drawn and where they intersect, the edge of the first BZ is formed. The first BZ is a hexagon; the point at the center of the zone is called the Γ point; two other important points are shown, the X point at $(2\pi/\sqrt{3}a)\hat{\mathbf{e}}_y$ and the J point at $(2\pi/\sqrt{3}a)[(1/\sqrt{3})\hat{\mathbf{e}}_x + \hat{\mathbf{e}}_y]$. These three symmetry points in the lattice appear in the following sections in which the band structure is presented.

The equations are transformed into matrix equations when this expansion is inserted into the wave equations, along with the spatial Fourier transform of either the relative permittivity

$$\epsilon(\mathbf{x}) = \sum_{\mathbf{G}} \epsilon(\mathbf{G}) e^{i\mathbf{G}\cdot\mathbf{x}}, \tag{3.67}$$

or of its inverse

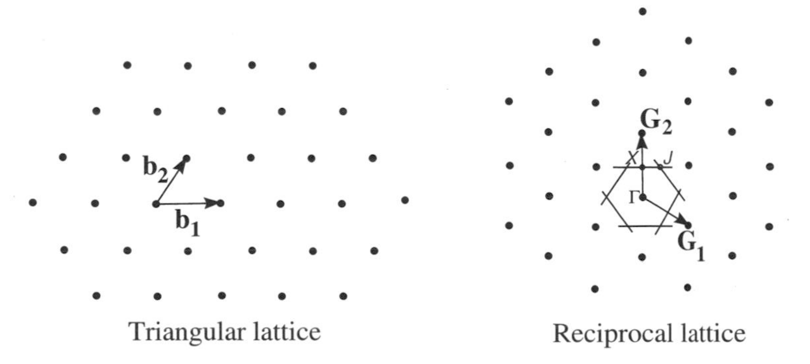

Figure 3.9 A triangular lattice constructed from basis vectors $\mathbf{b}_1$ and $\mathbf{b}_2$. The reciprocal lattice is also shown with the corresponding reciprocal lattice basis vectors $\mathbf{G}_1$ and $\mathbf{G}_2$ and the boundary of the first Brillouin zone (BZ). The Γ point is at the center of the zone, while the X and J points lie on the boundary.

$$\eta(\mathbf{x}) = \sum_{\mathbf{G}} \eta(\mathbf{G}) e^{i\mathbf{G}\cdot\mathbf{x}}. \tag{3.68}$$

The Fourier transform of Eq. (3.63) is

$$-|\mathbf{k}+\mathbf{G}|^2 E_{\mathbf{G}} + \left(\frac{\omega}{\mathbf{c}}\right)^2 \sum_{\mathbf{G}'} \epsilon(\mathbf{G}-\mathbf{G}') E_{\mathbf{G}'} = 0, \tag{3.69}$$

and similarly, the Fourier transform of Eq. (3.64) is

$$-(\mathbf{k}+\mathbf{G}) \cdot \sum_{\mathbf{G}'} (\mathbf{G}-\mathbf{G}') \eta(\mathbf{G}-\mathbf{G}') H_{\mathbf{G}'} + \left(\frac{\omega}{c}\right)^2 H_{\mathbf{G}} = 0. \tag{3.70}$$

Determination of the eigenvalues again proceeds by truncating the series of amplitudes and diagonalizing the matrixes. Truncation has its pitfalls and extra care must be exercised to obtain accurate and reliable results.[48] The relative permittivity has large disconinuities, which means that there are important Fourier components at large reciprocal lattice vectors. The convergence of the Fourier amplitudes proceeds slowly and the number of terms required for a specified accuracy increases as the power of the lattice dimensionality. The MATLAB program in Appendix B can be used to study the eigenvalue spectra of Eqs. (3.69) and (3.70) for circular cylinders.

3.3.3 Dielectric fluctuations

The dispersion curves are a result of the spatial variation of the relative permittivity. The homogeneous medium has a linear relation between the wave number and the frequency; but in a inhomogeneous medium, the waves are scattered and interfere with one another and quantitative results require numerical computations. The resulting dispersion curve is complicated. Nevertheless, we can crudely learn about the size of the perturbation by considering the fluctuations of the relative permittivity.[48,46]

The relative fluctuation of $\epsilon(\mathbf{x})$ from its spatial average provides a measure of the spectrum's deviation from the homogeneous medium. The relative permittivity is expressed as

$$\epsilon(\mathbf{x}) = \bar{\epsilon}[1 + \epsilon_r(\mathbf{x})], \tag{3.71}$$

where $\bar{\epsilon}$ is the spatial average of $\epsilon(\mathbf{x})$ over the unit cell. As $\bar{\epsilon}$ is an overall scaling factor, $\epsilon_r(\mathbf{x})$ is the relative ripple of the relative permittivity.

The perturbation parameter is related to the variance of the ripple

$$\langle \epsilon_r^2 \rangle \equiv \langle \epsilon_{\text{fluc}}^2 \rangle / \bar{\epsilon}^2,$$

the average is taken with respect to the unit cell volume. For more than one space dimension, we find significant deviations from the linear dispersion relation when $\langle\epsilon_r^2\rangle \sim 1$ or larger; and nonperturbative effects, such as band gaps for both linear polarizations, begin to appear.

For any two-component medium [i.e., where $\epsilon(\mathbf{x})$ can assume only two values: ϵ_a and ϵ_b], with the a-type medium occupying a volume fraction f of space, the ripple variance

$$\langle\epsilon_r^2\rangle = \frac{\langle\epsilon^2\rangle}{\bar{\epsilon}^2} - 1 = \frac{f\epsilon_a^2 + (1-f)\epsilon_b^2}{[f\epsilon_a + (1-f)\epsilon_b]^2} - 1. \tag{3.72}$$

For fixed f, the ripple saturates in value as either of the two relative permittivities goes to infinity. Given ϵ_a and ϵ_b, the value of f that maximizes $\langle\epsilon_r^2\rangle$ is

$$f_{\max} = \frac{\epsilon_b}{\epsilon_a + \epsilon_b}. \tag{3.73}$$

The corresponding maximum variance of the ripple is

$$\langle\epsilon_r^2\rangle = \frac{(\epsilon_a - \epsilon_b)^2}{4\epsilon_a\epsilon_b}. \tag{3.74}$$

The computed relative bandwidth of the gap $\delta\omega/\omega$ as a function of ϵ_a and/or f peaks roughly where $\langle\epsilon_r^2\rangle$ does. For scalar waves in three dimensions, the quantitative agreement is excellent,[57,58] while for vector waves the competition from the effects of dielectric connectivity and complexity of the polarization eigenstate shifts the transition region.

The ratio of the relative permittivities must be large, about 6, for large enough perturbations to open a gap. Also, the low-permittivity material has a high volume fraction (e.g., $\epsilon_b = 6$ and $f_{\max} \approx 0.86$) when the ripple variance is unity. These numbers provide a guide to search for the band gaps. This simple analysis is in accord with the observations that a band gap for both polarizations only exists when the contrast between the two relative permittivities is large and the low-permittivity material occupies most of the unit cell; also, increasing the ratio of relative permittivities eventually leads to a saturation of the band structure features, such as, the bandwidth of the gap.

3.3.4 Band structure

The band structure is computed by a straightforward matrix manipulation procedure. The Fourier transform of the relative permittivity $\epsilon(\mathbf{x})$, for nonoverlapping rods of radius r and relative permittivity ϵ_a in a background relative permittivity ϵ_b, is

$$\epsilon(\mathbf{G}) = \epsilon_b\delta_{\mathbf{G},\mathbf{0}} + 2f(\epsilon_a - \epsilon_b)\frac{J_1(Gr)}{Gr}, \tag{3.75}$$

where $J_1(Gr)$ is the cylindrical Bessel function of order 1. When $\mathbf{G} = \mathbf{0}$, the average relative permittivity is $\epsilon(\mathbf{0}) = f\epsilon_a + (1 - f)\epsilon_b$.

Truncation of the Fourier series results in a rounding-off of the relative permittivity at the bimaterial interfaces. This occurs because a sharp interface represents a structure with very high Fourier components and the convergence there is not uniform. There is also the Gibbs phenomenon,[59] which persists even in the limit $N \to \infty$. Convergence can be a difficult problem indeed for the plane wave expansion. This problem is more critical in three dimensions. A representation of the truncated relative permittivity is shown in Fig. 3.10. With $N = 271$ plane waves, there are ripples in evidence; the convergence for this case is very good at least for the lower frequency bands.

The band structure is computed by choosing values of the wave vector $\mathbf{k}$ and solving Eqs. (3.69) and (3.70) as an eigenvalue problem. There are N eigenvalues for N plane waves. The lowest 10 to 20 eigenvalues are usually sufficient for most applications. The MATLAB program in Appendix B only plots the lowest six eigenvalues. The matrix subroutines are chosen for the generalized eigenvalue problem. The H field computation is similar, but the program in Appendix B applies the inverse permittivity matrix $\eta(\mathbf{G} - \mathbf{G}')$. This is a choice made for convenience only, but the user should be mindful of the slow convergence of the truncated expansion before making any conclusions.

The band structure when the H field is parallel to the rods is shown in Fig. 3.11. There is a large gap that opens up between the first and the second bands. The volume fraction of the low-permittivity material is $f = 0.906$, which means that the rods are close packed; and the ratio of relative permittivities equals 5. The

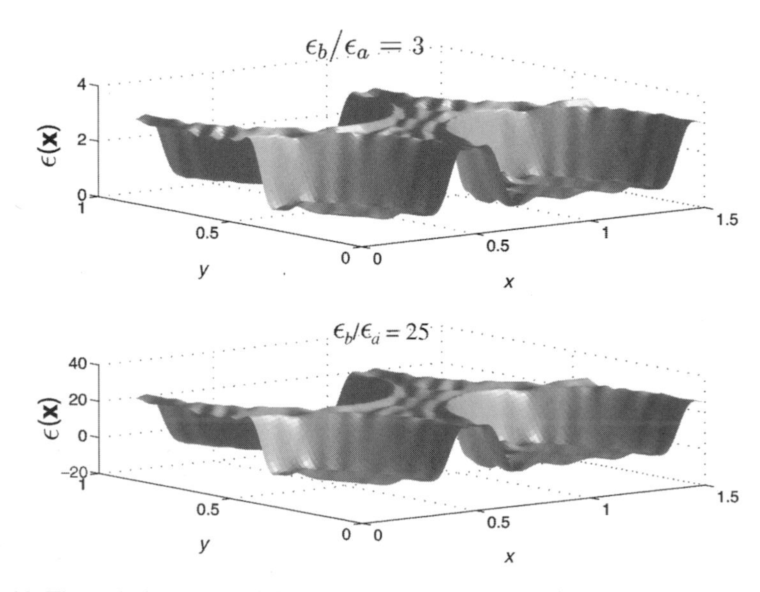

Figure 3.10 The relative permittivity $\epsilon(\mathbf{x})$ of a 2D triangular lattice, reconstructed from its spatial Fourier transform $\epsilon(\mathbf{G})$ with $N = 271$. Top: $\epsilon_b/\epsilon_a = 3$. Bottom: $\epsilon_b/\epsilon_a = 25$. The convergence in both cases is about the same.

horizontal axis represents a path in the BZ, which can be understood by reference to Fig. 3.9. The third vertical axis in Fig. 3.11 represents the J point on the zone boundary, from which point the path leads directly to the center of the zone until the Γ point is reached. Next, the path turns directly to the X point, and from there back to the J point.

The results are presented in Fig. 3.12 for the case when the electric field is parallel to the rods. The same itinerary is chosen, as in Fig. 3.11. For several values of ϵ_a, the bands transform from straight lines without dispersion in the homogeneous limit (i.e., $\epsilon_a = \epsilon_b = 1$) to a strong fluctuation limit in which large band gaps are in evidence. A gap opens up between the second and the third bands, when ϵ_b/ϵ_a is around 10. A complete gap for both polarizations is only found for very large ratios of the permittivities. This is one distinction between the 1D and 2D band structures: In two (and higher) dimensions, band gaps appear when f and ϵ_b/ϵ_a are sufficiently large.

3.3.5 Band eigenfunction symmetry and uncoupled modes

Another new property of wave propagation is found in higher dimensions. To this point, we have been concerned only with the eigenvalues of the bands, but the eigenfunctions have symmetry properties derived from the underlying lattice. Uncoupled modes in PBGs were first identified by Robertson et al.[60,61] in terahertz propagation experiments where the transmission vanishes, but the density of states does not vanish. This occurs because the uncoupled mode is antisymmetric under

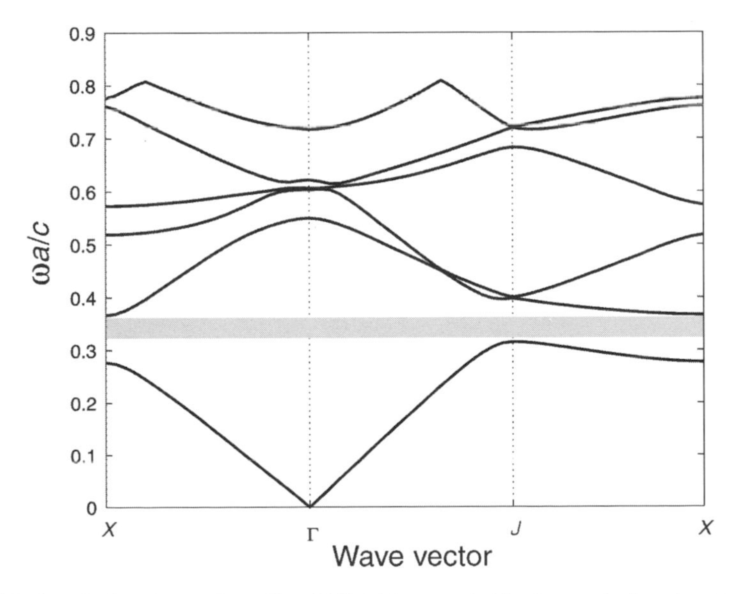

Figure 3.11 Band structure when the H field is parallel to the rods (made of air), whose volume fraction $f = 0.906$. The relative permittivity of the background material is $\epsilon_b = 5$, while $\epsilon_a = 1$.

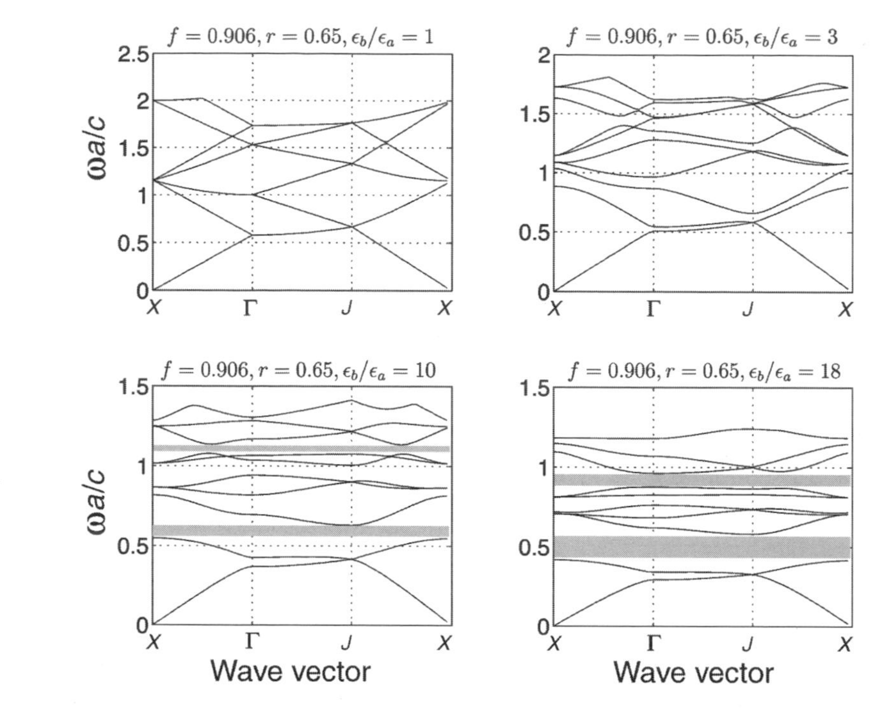

Figure 3.12 Band structure when the electric field is parallel to the rods (made of air). The permittivity ratio $\epsilon_b/\epsilon_a = 1$, 3, 10 and 18. A gap forms between the second and the third bands in the lower left frame; this is close to the value for band gap formation based on the dielectric fluctuation argument.

a mirror plane reflection, whereas the external incident wave is symmetric under this same reflection. Hence, the coupling of the two waves vanishes.

The calculation of photonic band structure is insufficient to provide a qualitative explanation of the transmission spectra. Boundary conditions and finite-size effects are important factors, and gaps in the transmission spectra are found where eigenmodes are present in the spectrum. However, the modes are uncoupled from the incident wave due to a symmetry mismatch between the incoming wave and the eigenmode of the photonic crystal. In other words, the eigenmodes of the photonic crystal are either symmetric or antisymmetric with respect to operations of the group that leave the crystal unchanged, and the incoming plane wave also has a definite symmetry with respect to the same operations; hence, when the plane wave and the eigenmode have opposite symmetries, they do not couple. Group theoretic methods can be applied to tag each band by its eigenfunction's symmetry, requiring the numerical computation of eigenfunctions only for a few cases.

For 2D lattices, Maxwell's equations can be reduced to a scalar form. Group-theoretical analysis[4,62–64] of such structures was confirmed by experiments on triangular and square lattice structures.[65,66] Uncoupled modes add a new aspect to finding novel applications for photonic crystals; they do not exist in 1D photonic crystals. Two-dimensional photonic crystals are also fabricated with micron or submicron lattice constants,[67–70] which makes them candidates for further study to develop a number of applications using these new concepts.

For 2D structures, both theoretical analyses and experimental results are available from the microwave to the visible regimes. In the microwave regime, the lattice constant is machined on the order of millimeters.[51,60,61,71] Two-dimensional hollow-rod structures in glass were fabricated with a lattice constants of a micron or less.[67,69] There are potential applications of this technology to optoelectronic devices. Etching in semiconductor materials is another avenue to producing good PBG structures.

The band structure of a 2D triangular lattice composed of air holes was examined both theoretically and experimentally in the terahertz regime by Wada et al.[66] They demonstrated a transmission minimum for one polarization in a region where no band gap is expected. A larger gap was found for the uncoupled mode and the characteristics were in good agreement with calculations.

Microchannel plate samples have been developed for several new applications by filling the air holes with a specifically doped fluid material. Two example applications are optical limiting and lasing.

Laser action was reported in samples where a dye-filled solution was placed in the air holes. Lasing was found to be correlated with a flat-band feature of the photonic band spectrum.[72] Below the lasing threshold, broadband emission is expected. However, near the threshold two peaks were observed, which became spectrally narrow as the pump fluence increased. The interesting feature is the lasing peak found off the gain maximum and thus, lasing is not normally favored at this wavelength. However, the band structure has a flat band at this wavelength, which means that the group velocity is small and the effective interaction length is longer in the medium.[72] Lasing due to this feature has been identified in many different samples with fluid-modified relative permittivity.

3.3.6 Three dimensions

The 3D scalar wave equation, analogous to Eq. (3.63), has been studied in detail by Datta et al.[73] All cubic structures have been investigated with spheres around each lattice point: simple cubic (SC), body-centered cubic (BCC), and face-centered cubic (FCC), as well as the diamond structure consisting of two spheres per unit cell. When low-permittivity spheres are embedded in a high-permittivity background medium, we denote this case as the *air sphere* material. For the scalar wave equation, band gaps open up for all cases between the first and the second bands, and again between the fourth and the fifth bands. The cases of dielectric spheres and air spheres are both interesting and display common features; the gaps can be as large as about 30%, and the size of a gap saturates as the ratio of relative permittivities becomes large. This is consistent with the saturation of the dielectric fluctuations discussed in Sec. 3.3.3.

The volume fraction at which the level gap is widest is closely given by the maximum of relative permittivity fluctuations; see Eq. (3.73). The gaps begin to appear roughly when the magnitude of the dielectric fluctuations is near unity. The plane wave method has the advantage of allowing a simple check on the convergence; namely, the actual relative permittivity can be compared with the truncated one.[48]

A large body of literature already exists on vector-wave band structure calculations in three dimensions.[47,57,58,74–80] Many experiments have been designed to prove the basic principles, to learn about the structures, and to investigate other phenomena such as defect modes.[82,83] Chemists have synthesized opal structures and inverse opal structures, where the core sphere is coated by another material and then the core is removed by etching.[84] The submicron spheres have strong dispersion and band gaps in the visible or the near-infrared regimes. Semiconductor fabrication methods have been developed to build structures layer-by-layer.[85,86] Full 3D band gaps have been observed in the infrared regime from 3D PBG structures.

3.3.6.1 Three-dimensional band calculation results

The earliest treatment of 3D periodic lattices was for the FCC structure. The first experiments were performed with machined samples in the millimeter lattice constant regime. The choice of an FCC lattice was based on the idea that Brillouin zones without protruding edges were more likely to form full band gaps. The BZ for the FCC structure is quite round in shape, as depicted in the inset in Fig. 3.13.

In treating the FCC lattice, there are two situations to examine: dielectric spheres embedded in a host with the relative permittivity of air (dielectric spheres) and spherical voids in a dielectric background (air spheres). Experimentalists have examined the case of air spheres. The frequency regime was investigated in a range where the second and third bands lie and no complete gap was found; however, a pseudo-gap was identified. The pseudo-gap is evident at the W point in Fig. 3.13 as a point of a degeneracy between the second and third bands. The density of states is reduced in this region, making an appearance similar to a full band gap. The air spheres were overlapping on the lattice, so the volume fraction of air is very high ($\approx 72\%$). A full band gap was identified by Sozuer et al.[48] for the air-sphere FCC lattice between the eighth and ninth bands. This gap lies above the frequency regime region reported by the experiments and is it missing in the earliest papers due to poor convergence of the plane wave expansion. Figure 3.13 is a plot of the band structure for $\epsilon_b = 16$, $\epsilon_a = 1$, and $f = 0.74$.

Opal structures synthesized by using nanometer-sized spheres of silica or some other substance will self-assemble into FCC crystals. The structures can be infiltrated in various ways to completely cover the surface of the spheres. On sintering the spheres before infiltration, the spheres bind together enough so that the silica can be chemically extracted. This leaves the inverse opal structure. An inverse opal structure of graphite was reported by Zakhidov et al.[84] in 1998, and several research groups have since reported improvements in the synthesis of opaline structures.[87]

The study of PBG materials with the periodicity of the SC lattice was undertaken because of its geometric simplicity, which could possibly translate into

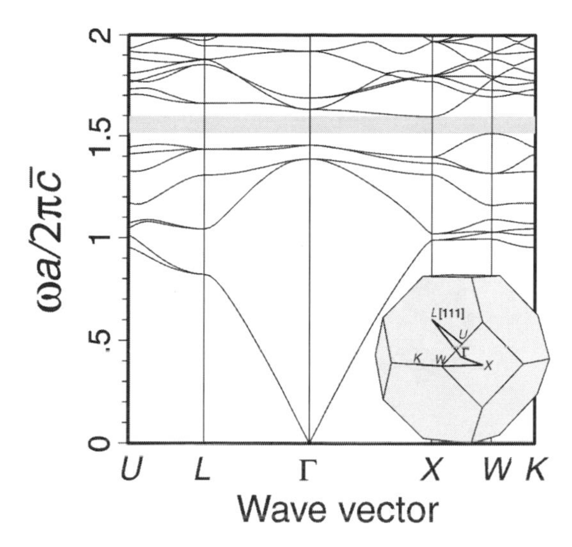

Figure 3.13 Band structure for for close-packed air spheres on a FCC lattice; $\bar{c} = c/\sqrt{\bar{\epsilon}}$; $\epsilon_b = 13$, $\epsilon_a = 1$, $f = 0.74$, and $N = 749$. Inset shows the path in the BZ. (Reprinted with permission from Ref. 48, © 1992 The American Physical Society.)

easier fabrication. The SC lattice also provides a framework in which structures with different topologies can be investigated, since the computational results obtained for the FCC structures indicate a strong relationship between connectivity and photonic band gaps. The band structures of a variety of geometries—involving nonoverlapping spheres, overlapping spheres, and the topologically equivalent structures with rods of circular and square cross section along the three Cartesian axes—were reported by Sözüer and Haus.[76]

The simplest "square-rod" structure is when the faces of the rods are oriented parallel to those of the unit cell; see Fig. 3.14. When the volume fraction of the dielectric material is 50%, the topologies of both types of media become identical. Photonic band gaps exist for these structures. The band structure corresponding to the square square-rod SC lattice is shown in Fig. 3.15. The inset shows the BZ for the SC lattice, which is the squarest of the Bravais lattice types.

3.3.6.2 Band symmetry

The simple cubic lattice of spheres has been chosen to study the details of the symmetries of the band eigenfunctions.[88] The band structure was calculated when the radii of the spheres were smaller than close-packed. Two cases are of importance: (1) dielectric spheres embedded in a host dielectric medium and (2) air holes (i.e., spherical voids) cut out of a dielectric medium. In both cases, the ratio of high to low relative permittivities must be high. The band structures can be calculated for the SC lattice using the plane wave method.

Symmetries have been assigned to the first seven bands in Fig. 3.16 based on the group theory symmetry. The angular frequency is scaled by the lattice constant a and a numerical factor including the speed of light, i.e., $a/2\pi c$. The waves prop-

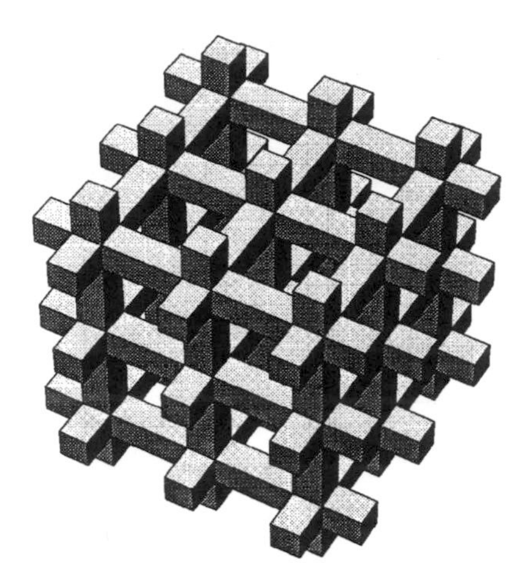

Figure 3.14 The square-rod structure. (Reprinted with permission from Ref. 76, © 1993 OSA.)

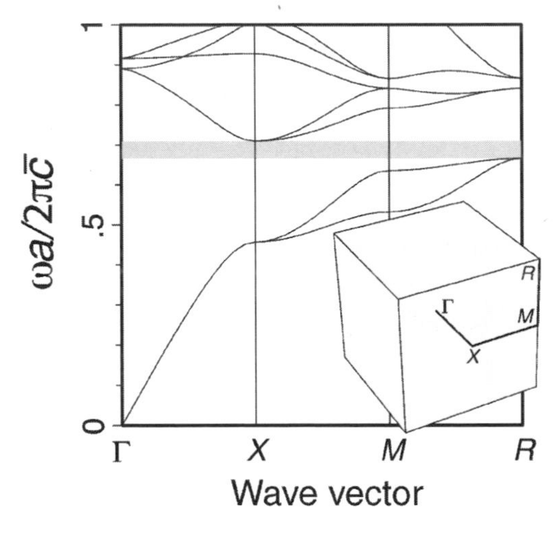

Figure 3.15 Band structure for square rods on a SC lattice; $\bar{c} \equiv c/\sqrt{\bar{\epsilon}}$; $\epsilon_b = 13$, $\epsilon_a = 1$, and $f = 0.82$. Inset shows the path in the BZ. (Reprinted with permission from Ref. 76, © 1993 OSA.)

agating in the Γ–M direction have an environment with reduced symmetry and the lowest bands for the two polarizations are nondegenerate. The wave vector is along the $(1, 1, 0)$ axis. This is a twofold symmetry direction, C_{2v}. The irreducible representations are A_1, A_2, B_1, and B_2. At the M point, the irreducible representation of the D_{4h} symmetry is A_{1g}, A_{1u}, B_{1g}, B_{1u}, A_{2g}, A_{2u}, B_{2g}, B_{2u}, E_g, and E_u. The corresponding symmetry of the H field vector has also been elucidated.[89]

The Γ–M symmetry contains invariance under two mirror reflection operations. One is the vertical plane defined by the Γ–Z and Γ–M lines; the other is the horizontal plane defined by the Γ–X and Γ–M lines. The eigenfunctions are either symmetric or antisymmetric with respect to these operations. We define the symmetry with respect to the E field vector, a complex vector field amplitude.

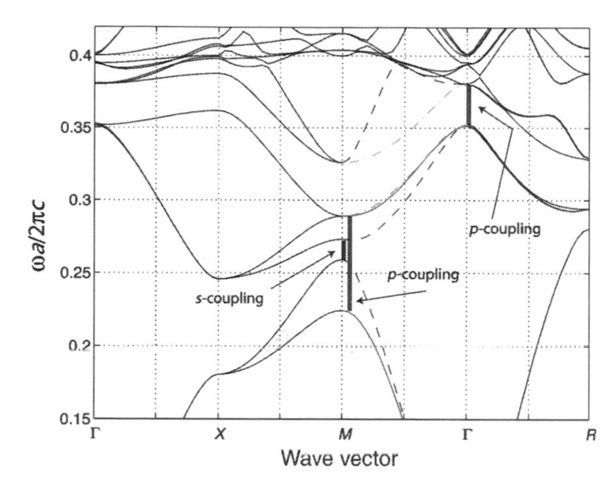

Figure 3.16 Band structure of the infinite simple cubic lattice of air holes. The E method calculations use $N = 729$ plane waves for each polarization and the cubic symmetry was deformed by 1% for direct comparison with the transmittance. The scaled radius of the air holes is 0.495 and the relative permittivity ratio is 13. The dashed lines for modes 2 and 3 in the Γ–M direction couple only to the s-polarization; the solid lines for bands 1 and 4 couple only to the p-polarization. A p-polarization gap exists between bands 4 and 5. The vertical bars show the positions of the gaps.

Note that B_1 is symmetric with respect to the horizontal plane and antisymmetric with respect to the vertical plane. It can be coupled to an incident s-polarized wave, which is polarized parallel to the horizontal plane; this mode is colored blue and is a dashed line. By contrast, B_2 (denoted by the solid line) is symmetric with respect to the vertical plane and antisymmetric with respect to the horizontal plane; it can couple with a p-polarized wave, whereas A_1 (dashed dotted line) is symmetric with respect to both planes, and A_2 (also a dashed-dotted line) is antisymmetric in both planes. These uncoupled modes are not excited by incident plane waves.

3.3.6.3 Transfer matrix method

Transfer matrix methods are useful for systems where the sample has an infinite transverse extent, but a finite thickness. A program developed by Pendry and coworkers is available for basic transfer matrix computations. This program has been rewritten by Andrew Reynolds to incorporate a graphics user interface (GUI). Several freeware programs that are useful for PBG calculations can be downloaded from http://www.pbglink.com/software.html.

The transfer matrix is much more useful for device development than the plane wave method because, as already demonstrated for 1D layers, boundary conditions change the wave amplitudes inside the PBG; these features are present in the transmission results. Calculations are automatically done with a group of matrix diagonalization subroutines. In the transfer matrix method, the computational time is proportional to N, where N is the number of grid points for a single layer. The transfer matrix method becomes very efficient as layers are doubled, thus allowing calculations of thick crystals. The grid spacing is typically about 10 to 15 points per lattice period.

The transmission spectrum is calculated by applying the transfer matrix method. The transfer matrix method is a generalization of the algorithm used to solve propagation in 1D layered materials to complex multidimensional structures. In analogy with its 1D counterpart, the transfer matrix method incorporates all wave vectors at fixed angular frequency, even the ones that have complex-valued components leading to decay of the wave amplitude through the medium. The incoming wave is assumed to be a plane wave that, in general, may be obliquely incident.

This section gives only an outline of the approach developed by Pendry.[90] It is based on a finite difference scheme that divides space into fine cells. The transverse directions are assumed to be periodic and extend to infinity. Sakoda has used a plane wave expansion technique that gives equivalent results with the same degree of accuracy.[4]

The transfer matrix method begins with Maxwell's equations in the frequency domain. Loss in a medium is expressed through a complex-valued relative permittivity; we do not consider magnetic media here. Faraday's law and Ampere's law are extracted from Eqs. (3.1) in the form

$$-i\omega\epsilon_0\epsilon(\mathbf{x})\mathbf{E} = \nabla \times \mathbf{H}, \quad -i\omega\mathbf{B} = -\nabla \times \mathbf{E}. \tag{3.76}$$

The sample is oriented with its end faces perpendicular to the z axis. The z components of the fields can be algebraically eliminated, leaving four field components to be solved. The field values in some plane with defined z are known, and values in the new plane $z + \Delta z$ are sought. This requires further rewriting the four equations to separate the z derivatives from the rest. The difference rule is the same in all cases; thus,

$$\frac{\partial E_x(x, y, z)}{\partial z} \approx \frac{E_x(x, y, z + \Delta a) - E_x(x, y, z)}{\Delta a}, \tag{3.77}$$

etc.

For simplicity, we assume the discretization is performed on a cubic lattice of side Δa. Finally, the following scaling transformation is performed on the H field $\mathbf{H}' = (i/\omega\Delta a)\mathbf{H}$ (where the prime does not indicate differentiation). The E field equations are expressed in terms of the E and H fields in the previous plane as follows:

$$\begin{aligned}
E_x(x, y, z + \Delta a) = E_x(\mathbf{x}) &+ \frac{\Delta a^2\omega^2}{c^2} H'_y(\mathbf{x}) + \frac{1}{\epsilon(\mathbf{x})}[H'_y(x - \Delta a, y, z) - H'_y(\mathbf{x}) \\
&- H'_x(x, y - \Delta a, z) + H'_x(x - \Delta a, y, z)] \\
&- \frac{1}{\epsilon(x + \Delta a, y, z)}[H'_y(\mathbf{x}) - H'_y(x + \Delta a, y, z)
\end{aligned}$$

$$- H_x'(x+\Delta a, y-\Delta a, z) + H_x'(x+\Delta a, y, z)], \tag{3.78}$$

$$\begin{aligned}E_y(x,y,z+\Delta a) = E_y(\mathbf{x}) - \frac{\Delta a^2\omega^2}{c^2}H_x'(\mathbf{x}) + \frac{1}{\epsilon(\mathbf{x})}[H_y'(x-\Delta a, y, z)\\ - H_y'(\mathbf{x}) - H_x'(x, y-\Delta a, z) + H_x'(x-\Delta a, y, z)]\\ - \frac{1}{\epsilon(x, y+\Delta a, z)}[H_y'(x-\Delta a, y+\Delta a, z)\\ - H_y'(x, y+\Delta a, z)\\ - H_x'(\mathbf{x}) + H_x'(x+\Delta a, y+\Delta a, z)].\end{aligned} \tag{3.79}$$

Subsequently, the H fields are expressed in terms of the H fields in the previous plane and the E fields in the same plane as follows:

$$\begin{aligned}H_x'(x,y,z+\Delta a) = H_x'(\mathbf{x}) + \epsilon(x,y,z+\Delta a)E_y(x,y,z+\Delta a)\\ - \frac{c^2}{\Delta a^2\omega^2}[E_y(x-\Delta a, y, z+\Delta a)\\ - E_y(x-\Delta a, y, z+\Delta a)\\ - E_x(x-\Delta a, y+\Delta a, z+\Delta a) + E_x(x-\Delta a, y, z+\Delta a)]\\ + \frac{c^2}{\Delta a^2\omega^2}[E_y(x+\Delta a, y, z+\Delta a) - E_y(x, y, z+\Delta a)\\ - E_x(x, y+\Delta a, z+\Delta a) + E_x(x, y, z+\Delta a)],\end{aligned} \tag{3.80}$$

$$\begin{aligned}H_y'(x,y,z+\Delta a) = H_y'(\mathbf{x}) + \epsilon(x,y,z+\Delta a)E_y(x,y,z+\Delta a)\\ - \frac{c^2}{\Delta a^2\omega^2}[E_y(x+\Delta a, y-\Delta a, z+\Delta a)\\ - E_y(x, y-\Delta a, z+\Delta a)\\ - E_x(x, y, z+\Delta a) + E_x(x-\Delta a, y-\Delta a, z+\Delta a)]\\ + \frac{c^2}{\Delta a^2\omega^2}[E_y(x+\Delta a, y, z+\Delta a) - E_y(x, y, z+\Delta a)\\ - E_x(x, y+\Delta a, z+\Delta a) + E_x(x, y, z+\Delta a)].\end{aligned} \tag{3.81}$$

For a structure with $N \times N \times N$ cells, the transfer matrixes are of dimension $4N^2$, yielding a very efficient representation of the computational problem. The matrixes are also sparse, which makes the solution of even large matrix equations fast when specialized matrix methods are used.

In the following, we exemplify the use of the transfer matrix methods and demonstrate the interplay that can exist between the plane wave method and the transfer matrix method. For computating the transmission, the computer program developed by Pendry's group[91] was used.

3.3.6.4 Transmission spectra

The transfer matrix method is a powerful tool for analyzing the properties of PBG structures. Here, results are provided for a simple cubic lattice of spheres. For propagation along the Γ–M direction with a crystal that is 32 periods thick, transmission data is plotted in Fig. 3.17. There is a considerable shift in the width and the depth of the band gaps as the sample thickness is increased, but 32 layers provide a clear determination of the band gap positions. The oscillations at low frequencies are of the Fabry–Pérot type arising due to reflections from opposite planes. These oscillations are strongly affected by the sample thickness; indeed, at low frequencies where only one band gap is found, the number of oscillations is used by us to verify the number of layers. Oscillations also occur at the higher frequencies, but they are difficult to interpret because of the strong dispersion in the bands and the existence of multiply excited bands with different dispersion.

To apply Pendry's method to the Γ–M direction, the unit cell is slightly deformed. The separation of the sphere centers is $\sqrt{2}$ in the propagation direction, but is unity in the transverse directions. The unit cell's geometry is modified to make the lateral to longitudinal length ratio 10:14. This creates a deformed SC geometry, a contraction of 1% along the longitudinal direction making the lattice parameters 0.99:0.99:1.0. This distortion does not noticeably affect the band structure, however.

The band structure presented in Fig. 3.16 was calculated by the E method for the deformed SC lattice of air holes with scaled radii $r = 0.495$ (i.e., they are nearly touching). The symmetries of the lowest seven bands in the Γ–M direction were assigned. A direct gap opens between the second and the third bands, but there is no common gap over all directions. The direct gap is found with both the E and H methods. The Γ–M direction is distinguished by the nondegeneracy of the bands.

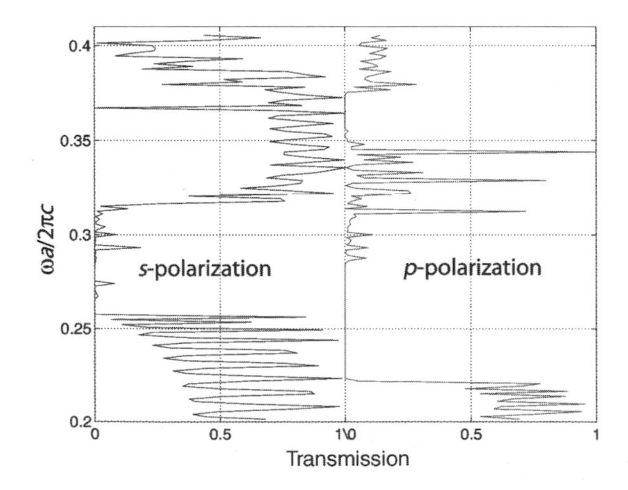

Figure 3.17 The p- and s-polarization transmission spectra along the Γ–M direction for a SC lattice of air holes in a sample that is 32 periods thick. The scaled radius of the air holes is 0.495 and the relative permittivity ratio is 13.

By examining the symmetry of each band, we determine whether an incoming wave will be coupled to it.

The validity of the analysis is checked by the structure of the transmission spectra. From the band structure calculations and group analysis, two gaps are identified for p-polarization and one for the s-polarization. The positions of the gaps are indicated in Fig. 3.16. The corresponding transmission spectra for p- and s-polarization are plotted in Fig. 3.17. On comparison, the gaps extracted from the computations are close to the band structure results from Fig. 3.16.

To further demonstrate the importance of band symmetry on transmission, we present results for dielectric spheres with scaled radii $r = 0.297$. Then the volume fraction of spheres is low enough that no direct gaps are observed in the spectrum. The band structure and transmission spectra are available in Figs. 3.18 and 3.19. There is strong dispersion in the band structure, including the appearance of distinct nonmonotonic bands. The density of states is nonzero over the entire frequency range, which makes this case a good candidate to demonstrate the correspondence between band symmetry and the transmission features.

Figure 3.18 predicts large, distinct gaps in transmission spectra, as expected from group-theory arguments. In each case, the appearance of the uncoupled A_1 or A_2 modes spans a portion of the gap region. The lowest two bands for each linear polarization are in good quantitative agreement with the transmission spectra. Although the volume fraction is small, the appearance of large gaps due entirely to predicted uncoupled modes means that the device design parameters based on these features are not stringent. Indeed, the features appear over a wide range of volume fractions.

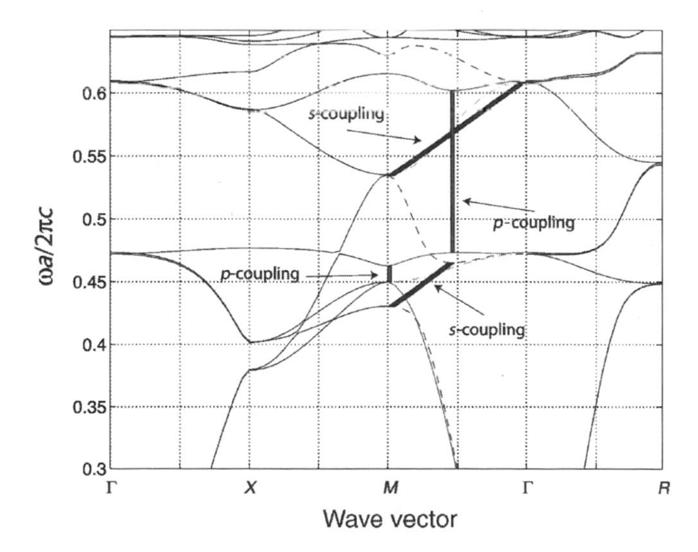

Figure 3.18 Band structure of the infinite simple cubic lattice of dielectric spheres. The E method calculations use $N = 729$ plane waves for each polarization and the cubic symmetry was deformed by 1% for direct comparison with the transmittance. The scaled radius of the air holes is 0.297 and the relative permittivity ratio is 13.

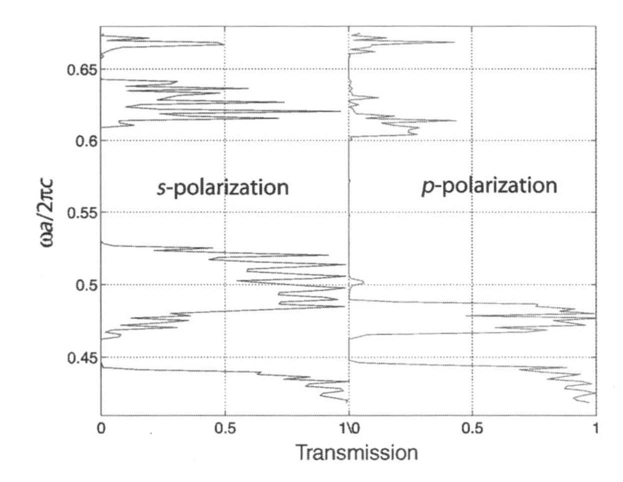

Figure 3.19 The p- and s-polarization transmission spectra along the $\Gamma{-}M$ direction for a SC lattice of dielectric spheres in a sample that is 32 periods thick. The scaled radius of the spheres is 0.297 and the relative permittivity ratio is 13.

3.3.6.5 Finite-difference time-domain method

A powerful but computationally intensive method for solving Maxwell's equations is the FDTD method. As the somewhat lengthy, but descriptive title implies, this method discretizes Maxwell's equations in the spatiotemporal domain, whether or not dispersion in the medium is accounted for. Actually, what is called the FDTD method depends on which form of Maxwell's equations is discretized.

The most popular FDTD algorithm is called the Yee method,[92] which is based on solving Maxwell's curl equations, rather than the wave equation. The differencing scheme is consistent with Faraday's and Ampere's laws. The Yee method is formulated using a technique called the leap-frog method. The **E** and **H** fields are evaluated in alternate half steps. This technique and the corresponding algorithm can be found in Taflove's book.[93]

The FDTD method is a computationally intensive method, but can be applied to complex, finite, 3D geometries and transient sources. It has been employed for the accurate determination of localized defect-mode frequencies and Q-factors. Important are the grid density—typically, 15 to 20 points per wavelength (or per lattice constant)—and appropriate absorbing boundary conditions. Lattice size is ultimately limited by memory (without swapping the grid). The memory size is reducible by taking advantage of the mode symmetry, thus reducing the required number of sites in the lattice. The algorithm is amenable to high-performance computation, as the lattice can be distributed among many processors.

In its simplest form—for systems with high symmetry—the electromagnetic equations are reducible to a scalar wave equation, which can be discretized by using explicit and implicit schemes. Even the vector wave equations Eqs. (3.54) and (3.55) can be discretized for an accurate description of one particular vector field.

For 2D systems, when the electric field is aligned parallel to the symmetry axis, Eq. (3.63) must be addressed. According to an explicit discretization using steps of Δx, Δy, and Δt, the derivatives are approximated by a central difference scheme as follows:

$$\frac{\partial^2 E(x,y,t)}{\partial t^2} \approx \frac{E(x,y,t+\Delta t)+E(x,y,t-\Delta t)-2E(x,y,t)}{(\Delta t)^2}, \tag{3.82}$$

$$\frac{\partial^2 E(x,y,t)}{\partial x^2} \approx \frac{E(x+\Delta x,y,t)+E(x-\Delta x,y,t)-2E(x,y,t)}{(\Delta x)^2}, \tag{3.83}$$

$$\frac{\partial^2 E(x,y,t)}{\partial y^2} \approx \frac{E(x,y+\Delta y,t)+E(x,y-\Delta y,t)-2E(x,y,t)}{(\Delta y)^2}. \tag{3.84}$$

Each time step solves for time $t+\Delta t$ given the information about the previous two times and nearest-neighbor spatial points. Either the field is prescribed at an initial time or an oscillating dipole source term can be incorporated to radiate into the PBG structure.

Several types of boundary conditions can be implemented. The simplest way perhaps is to keep the endpoints of the lattice far enough so that the field does not reflect back to the region of interest; periodic boundary conditions can be used with the same caveat. As these schemes waste memory space, better schemes continue to be reported.

Transparent boundary conditions are used to match the outgoing wave to a one-way propagator. Paraxial wave equations can be discretized to determine the boundary fields. Naturally, the uncertainty in the matching condition and the variations in the angle of incidence provide a small backward wave, which generates spurious interference effects.

A perfectly matched layer (PML) is a region at the end of the computational domain, introduced so that the incoming wave is impedance-matched at the boundary. The dielectric and magnetic constitutive parameters of PMLs are complex-valued, resulting in absorption of the wave without a reflection. The PML method is very effective in suppressing boundary effects over a wide range of angles of incidence.

Maxwell's equations in vector form can be discretized using a simple differencing scheme. The simplest discretization, with Δt as the time step and Δa as the step, of Maxwell's curl equations is as follows:

$$\begin{aligned}&E_x(\mathbf{x},t+\Delta t)\\&\quad = E_x(\mathbf{x},t)+\frac{\Delta t}{\Delta a}\frac{H_z(\mathbf{x},t)-H_z(x,y-\Delta a,z,t)-H_y(\mathbf{x},t)+H_y(x,y,z-\Delta a,t)}{\epsilon_0\epsilon(\mathbf{x})},\end{aligned} \tag{3.85}$$

$$\begin{aligned}&E_y(\mathbf{x},t+\Delta t)\\&\quad = E_y(\mathbf{x},t)+\frac{\Delta t}{\Delta a}\frac{H_x(\mathbf{x},t)-H_x(x,y,z-\Delta a,t)-H_z(\mathbf{x},t)+H_z(x-\Delta a,y,z,t)}{\epsilon_0\epsilon(\mathbf{x})},\end{aligned} \tag{3.86}$$

$$\begin{aligned}&E_z(\mathbf{x},t+\Delta t)\\&\quad = E_z(\mathbf{x},t)+\frac{\Delta t}{\Delta a}\frac{H_y(\mathbf{x},t)-H_y(x-\Delta a,y,z,t)-H_x(\mathbf{x},t)+H_x(x,y-\Delta a,z,t)}{\epsilon_0\epsilon(\mathbf{x})},\end{aligned} \tag{3.87}$$

$$H_x(\mathbf{x}, t+\Delta t) = H_x(\mathbf{x}, t) - \frac{\Delta t}{\Delta a} \frac{E_z(\mathbf{x}, t) - E_z(x, y-\Delta a, z, t) - E_y(\mathbf{x}, t) + E_y(x, y, z-\Delta a, t)}{\mu_0}, \quad (3.88)$$

$$H_y(\mathbf{x}, t+\Delta t) = H_y(\mathbf{x}, t) - \frac{\Delta t}{\Delta a} \frac{E_x(\mathbf{x}, t) - E_x(x, y, z-\Delta a, t) - E_z(\mathbf{x}, t) + E_z(x-\Delta a, y, z, t)}{\mu_0}, \quad (3.89)$$

$$H_z(\mathbf{x}, t+\Delta t) = H_z(\mathbf{x}, t) - \frac{\Delta t}{\Delta a} \frac{E_y(\mathbf{x}, t) - E_y(x-\Delta a, y, z, t) - E_x(\mathbf{x}, t) + E_x(x, y-\Delta a, z, t)}{\mu_0}. \quad (3.90)$$

More elaborate schemes are commonly available.[93,94]

3.3.6.6 FDTD results for defect modes

The vector electromagnetic field in a 2D photonic lattice, with its plane of periodicity designated as the xy plane, can be decoupled into two independent modes:

- Transverse electric (TE) modes: the E field is perpendicular to the plane of periodicity, and the nonzero field components are $\mathbf{E}_z$, $\mathbf{H}_x$, and $\mathbf{H}_y$); and
- Transverse magnetic (TM) modes: the H field is perpendicular to the plane of periodicity, and the nonzero field components are $\mathbf{E}_x$, $\mathbf{E}_y$, and $\mathbf{H}_z$.

A line of dipoles at the center of the defect may be introduced for the TE case. The dipoles are oriented perpendicular to the plane of periodicity. Such TE defect modes have been thoroughly investigated in 2D photonic lattices (square and triangular) by Sakoda using the scalar FDTD method.[95,98] The TE defect mode in a square lattice is localized close to the defect site, as shown in Fig. 3.20. There is very good agreement (1) between numerical results obtained from either the discretized scalar wave equation method or the full vectorial FDTD method and (2) between experimental results of McCall et al.,[51] on a 2D square photonic lattice with circular dielectric rods immersed in air for a large permittivity contrast.

The lattice defect is formed by the removal of a dielectric rod from the center of the lattice. A resonance frequency is clearly identified after 20 periods of oscillation and continues to sharpen as energy continues to build up in the defect mode. The plot of the electric field profile along the x axis in Fig. 3.20 shows that the field is concentrated close to the defect. The field is confined to a region around the defect extending out to about three lattice constants. The difference between numerical and experimental results is less than 1%. Similarly favorable comparison of theoretical results has been reported for TE modes on a triangular lattice with experimental results.[51]

The square lattice does not have a band gap for TM modes, but a triangular lattice containing air holes in a dielectric background medium does show a sizable gap. Calculations were made for the lattice geometry depicted in Fig. 3.21: ϵ_a is the relative permittivity of the rods of radius r, and ϵ_b is the relative permittivity of the background medium. A defect in the form of an air rod with a modified radius r_{defect} was introduced in the center of the lattice. To couple with different defect-mode symmetries, the orientation of the oscillating dipole can be changed. Depending on the dipole excitation and the size of the defect rod, different modes will appear.

The band structure for TM modes is shown in Fig. 3.22. The calculation was based on the plane wave method with $N = 919$ basis vectors. The relative error

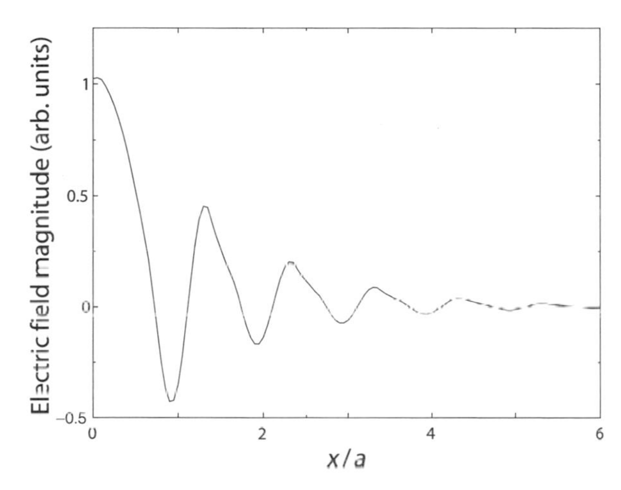

Figure 3.20 The electric field as a function of the distance from a dipole in square photonic lattice, after 100 periods of oscillations, when $\omega a/2\pi c = 0.407$. The dipole models a defect. (Reprinted with permission from Ref. 99, © 2001 The American Physical Society.)

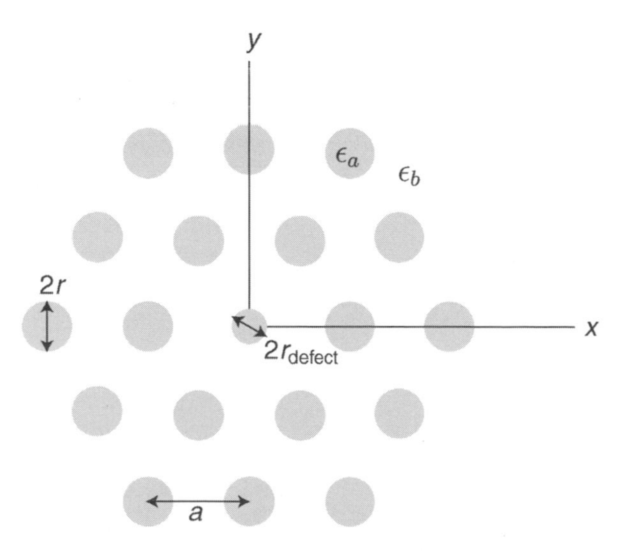

Figure 3.21 The top view of a 2D triangular array of circular rods used for examining TM defect modes; ϵ_a and ϵ_b denote the relative permittivities of the rods and the background medium, respectively; r and r_{defect} are the radii of the lattice rods and the defect rod, respectively, while a denotes the lattice constant. (Reprinted with permission from Ref. 99, © 2001 The American Physical Society.)

was determined by comparing the results for different numbers of plane waves with the asymptotic value. The error depends on the band number and increases from less than 1% for the first few bands up to 6% for the eighth band. The first gap in Fig. 3.22 exists between the first and second bands, while the second gap lies between the seventh and eight bands, for the chosen parameters.

To illustrate that the TM modes have well-defined resonance frequencies, a line of dipoles is placed at the center of the defect rod. The ratio of the defect radius to unit cell dimension is $r_{\text{defect}}/a = 0.35$, while $r/a = 0.48$. The dipoles are driven at different frequencies and the radiated energy is computed. Figure 3.23 shows the electromagnetic energy radiated as a function of the oscillation frequency. The peak in the radiated energy spectrum at $\omega a/2\pi c = 0.461$ represents the eigenfrequency of the defect mode. This peak is well established after 25 oscillation periods and it continues to grow and narrow as more time elapses. After 100 oscillation periods, the full width at half maximum width of the resonance is about 0.005, which corresponds to a Q-factor of around 100. As saturation of the peak width is not observed, the Q-factor is larger than that observed after 100 oscillation periods.

The chosen photonic crystal has C_{6v} symmetry and, therefore, six irreducible representations.[4,99] By changing the radius of the defect rod and the dipole orientation, different defect modes with different symmetries appear. The H field is concentrated in the regions with larger relative permittivity, just as for the E field. Sakoda and Shiroma[100] demonstrated that the spatial variation of the electric fields is faster for the modes in the second gap than for those in the first gap.

A typical plot of the the defect-mode resonance frequency versus defect rod radius is presented in Fig. 3.24. The horizontal lines in the figure represent the

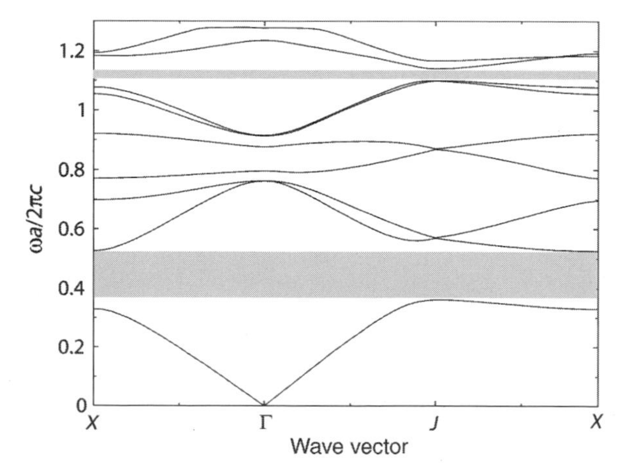

Figure 3.22 Band structure of a regular triangular lattice for TM modes, calculated by the plane wave method using 919 basis functions. The following parameters were assumed: $r/a = 0.48$, $r_{\text{defect}}/a = 0.48$, $\epsilon_a = 1$, and $\epsilon_b = 13$. A large band gap exists between $\omega a/2\pi c = 0.375$ and 0.52 in the normalized units. The Γ–X direction is along the second-nearest-neighbor lines through the crystal (i.e., the y axis in Fig. 3.21), and the Γ–X direction is along the nearest neighbor lines through the crystal (i.e., the x axis). (Reprinted with permission from Ref. 99, © 2001 The American Physical Society.)

boundaries of the band gap. As the ratio r_{defect}/a increases, the eigenfrequencies of the photonic crystal modes tend to rise monotonically and linearly due to the larger air fraction and resulting lower average index. As shown in Refs. 100 and 101, the eigenfrequency is proportional to $1/\sqrt{\epsilon_{r_{\text{defect}}}}$. With decreasing defect radius, the effective relative permittivity at the defect increases proportionally to r^2_{defect}. Hence, with constant relative permittivity and variable defect radius, the eigenfrequency versus defect radius relation becomes $\omega_{r_{\text{defect}}} \sim r_{r_{\text{defect}}}$.

The energy accumulated in the defect unit cell can be used to determine the localization properties for different defect-mode frequencies (i.e., ratios r_{defect}/a). The defect mode is localized within a few unit cells of the defect position.

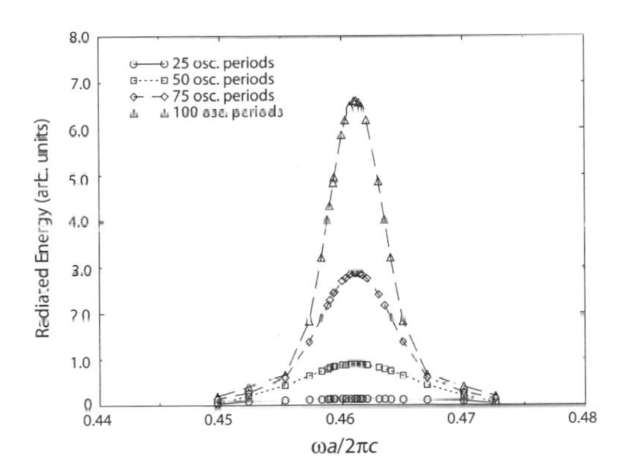

Figure 3.23 The frequency dependence of the total radiated energy for a TM mode radiated by an oscillating dipole at the center of a defect rod after 25, 50, 75 and 100 oscillation periods; $r/a = 0.48$ and $r_{\text{defect}}/a = 0.35$. (Reprinted with permission from Ref. 99, © 2001 The American Physical Society.)

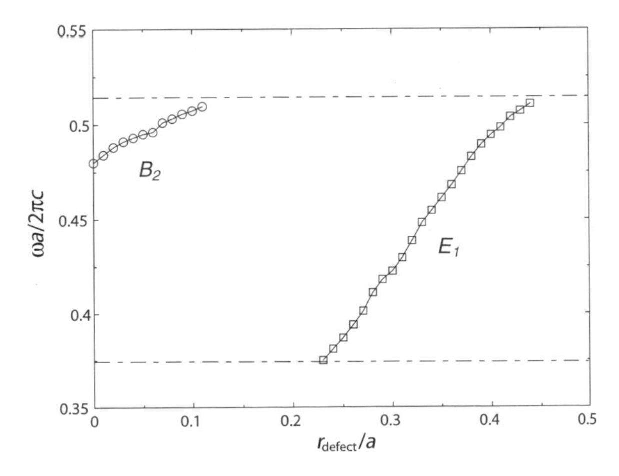

Figure 3.24 The eigenfrequencies of two localized defect modes as functions of the normalized radius of the defect rod, where $r/a = 0.48$. The two horizontal lines represent the boundaries of the photonic band gap calculated from Fig. 3.22. (Reprinted with permission from Ref. 99, © 2001 The American Physical Society.)

Further applications of the FDTD method are commonplace. Sakoda's book[4] has a detailed analysis of the symmetry properties of defect modes in various photonic crystal geometries. The real behavior of the devices is complex, but computations have identified favorable geometries and placed limits on the losses. Results for point and line defects clearly identify the resonance frequencies and the defect-mode symmetries. Lin et al.[102] reported a series of millimeter-wave experiments on samples with line or point defects. They showed that the Q-factor increases by adding rows of dielectric rods in the lateral direction around a line defect. A plateau for the spectral width of the defect mode was reached after eight layers in their study. Numerical studies with a small dielectric loss added largely explain the experiments[103] with additional losses incurred due to the finite lattice size in the third dimension.

PBG slabs with finite height have been investigated by many researchers, because such structures confine the field and control the diffraction losses. For instance, Painter et al.[101] explored the coupling of light to leaky modes, which further reduced the Q-factors of point defects in PBG cavities. Paddon and Young[104] as well as Ochiai and Sakoda[105] examined the role of mode symmetry on the coupling of light to leaky modes. Two-dimensional slabs can have reduced diffraction losses because the internal modes are forbidden by symmetry to couple with the radiation modes. This property is related to the symmetry forbidden transmission of uncoupled modes previously discussed for 2D and 3D PBGs.

3.4 Summary

Computational techniques for PBGs are available to provide detailed results. For instance, if the band gap or mode symmetry is desired, the simple plane wave method can be applied to give basic design information, i.e., the existence of band gaps for different polarizations and the eigenfunction symmetry. Using the plane wave method, researchers have shown that many different geometries—including the face-centered cubic lattice,[48] the diamond lattice,[75] the simple cubic lattice,[76] and intersecting rod geometries[83]—possess full band gaps. The eigenfunctions can be used to develop a set of coupled-mode equations for application to finite system geometries.

However, as this chapter indicated, the plane wave method is primarily restricted to information about infinite systems. More powerful methods can be recruited to study more complex media and device geometries, such as the transfer matrix method and the FDTD method. Systems with finite thickness are amenable to solution by the transfer matrix method. This has been very useful in determining the transmission spectra, and comparison with experiments have been very favorable. The FDTD method is costly in terms of computation time and memory, and problems can very easily exceed the capabilities of current computers. However, the FDTD method is an excellent simulation tool that has provided a great deal of insight into the propagation of light through PBG structures. The beam propagation method can cover forward- and backward-wave coupling, is simpler to implement, and is less demanding on computational resources than the FDTD method.

Based on the computational simulations, several applications have been identified. Filtering and waveguiding are the most obvious applications. In addition, by careful design low-threshold lasing is predicted in PBGs due to the a variety of effects such as high Q due to uncoupled modes, low group velocity due to flattening of the dispersion bands, and field confinement near a defect or by transmission resonances from the end faces.

3.5 Appendix A

Program for the plane-wave calculation of the transmission and reflection coefficients in a 1D layered structure.

Main Program:

```
clear;
% Calculate change of a intensity profile after
% transmitting a multiple thin film system
% Written by Feiran Huang on 05/31/99. Update
% JW Haus 11/16/99
% Updated and corrected by JW Haus on 8/21/2002
%---------------------------
% given parameters for film system, n,k,d of
% various wavelength given expression of input
% intensity profile in time domain and
% center wavelength
  v=[500 2100 0 1];
%film thickness information
  d1=1000/2/1.4285714; d2=1000/4/1.;
% number of periods
  m=10;
% angle of incidence
  fi0=0/180*pi;
% superstrate index
  n00=[1. 1. 1. 1. 1.];
% wavelength vector
  lambda=v(1):.5:v(2);
% wavelengths used for dielectric data
  ld0=[180 500 600 800 6100];
% real part of the complex-valued refractive index
% of film 1
  n10=[1.4285714 1.4285714 1.4285714 1.4285714
        1.4285714];
% imaginary part of the complex-valued refractive
% index of film 1
```

```
  k10=[0.0 0.0 0.0 0.0 0.0];
% real part of the complex-valued refractive index
% of film 2
  n20=[1.0 1.00 1.00 1.00 1.00];
% imaginary part of the complex-valued refractive
% index of film 2
  k20=[0.0 0.0 0.0 0.0 0.0];
% substrate refractive index (real-valued only)
  n30=[1. 1. 1. 1. 1.];
  n0=interp1(ld0,n00,lambda); n1=interp1(ld0,n10,
                               lambda);
  k1=interp1(ld0,k10,lambda); n2=interp1(ld0,n20,
                               lambda);
  k2=interp1(ld0,k20,lambda); n3=interp1(ld0,n30,
                               lambda);
% Reflection and transmission coefficients are
% calculated
  [Tp,Rp,Ts,Rs]=f_2mlyr(n0,n1,k1,n2,k2,n3,d1,d2,m,fi0,
                          lambda);
  figure(1);
  plot(lambda, Tp,'b',lambda, Ts,'r');
  figure(2);
  plot(lambda, Rp,'b',lambda, Rs,'r');
  ylabel('Transmittance');
  xlabel('Wavelength (nm)'); grid on; axis(v);
% field amplitude calculation, a single wavelength
% is used.
  lambdae=1405.26;
  n0e=interp1(ld0,n00,lambdae); n1e=interp1(ld0,n10,
                                 lambdae);
  k1e=interp1(ld0,k10,lambdae); n2e=interp1(ld0,n20,
                                 lambdae);
  k2e=interp1(ld0,k20,lambdae); n3e=interp1(ld0,n30,
                                 lambdae);
% The field amplitudes from the boundary conditions
% are calculated
  [Ees,Eep,fi1,fi2,fi3]=f_2mlyrAmp(n0e,n1e,k1e,n2e,k2e,
                          n3e,d1,d2,m,fi0,...lambdae);
% The data are plotted after constructing fields
% inside the PBG
  count=1;
  dz1=d1/10; dz2=d2/10;
  for jj=1:m  %check for the initial value 1 or 2
% layer 1
```

```
  for kk=0:9
  Eys(count)=(((Ees(1,2*jj)*exp(-i*2*pi*(n1e-i*k1e)
                *cos(fi1)/lambdae*kk*dz1))...
  +(Ees(2,2*jj)*exp(i*2*pi*(n1e-i*k1e)*cos(fi1)/
                      lambdae*kk*dz1))));
  Exp(count)=(((Eep(1,2*jj)*exp(-i*2*pi*(n1e-i*k1e)
                *cos(fi1)/lambdae*kk*dz1))+...
  (Eep(2,2*jj)*exp(i*2*pi*(n1e-i*k1e)*cos(fi1)/lambdae
                     *kk*dz1)))*cos(fi1));
  count=(count+1);
  end
% layer 2
  for kk=0:9
  Eys(count)=(((Ees(1,2*jj+1)*exp(-i*2*pi*(n2e-i*k2e)
                 *cos(fi2)/lambdae*kk*dz2))...
  +(Ees(2,2*jj+1)*exp(i*2*pi*(n2e-i*k2e)*cos(fi2)/
                         lambdae*kk*dz2))));
  Exp(count)=(((Eep(1,2*jj+1)*exp(-i*2*pi*(n2e-i*k2e)
                 *cos(fi2)/lambdae*kk*dz2))...
  +(Eep(2,2*jj+1)*exp(i*2*pi*(n2e-i*k2e)*cos(fi2)/
   lambdae*kk*dz2))).*cos(fi2));
  count=(count+1);
  end
  end
    figure(3)
    plot(abs(Exp).^2)
  figure(4)
    plot(abs(Eys).^2)
```

Function called from the main program to compute the transmission and reflection coefficients

```
function
[Tp,Rp,Ts,Rs]=f_2mlyr(n0,n1,k1,n2,k2,n3,d1,d2,m,fi0,
                     lambda)
% calculate transmittance and reflectance of a
% multi-layer system
% Written by Feiran Huang 05/31/99.
% Last Update 08/13/2002 Amplitudes of the fields
% added.
%-------------------------
%
for n=1:length(lambda)
  fi1=asin(sin(fi0)*n0(n)/(n1(n)+i*k1(n)));
  phi1(n)=fi1;
  fi2=asin(sin(fi0)*n0(n)/(n2(n)+i*k2(n)));
```

```
phi2(n)=fi2;
fi3=asin(sin(fi0)*n0(n)/n3(n));
phi3(n)=fi3;
[r01p,t01p,r01s,t01s]=f_rtamp(n0(n),0,n1(n),k1(n),
                                fi0,fi1);
[r12p,t12p,r12s,t12s]=f_rtamp(n1(n),k1(n),n2(n),
                                k2(n),fi1,fi2);
[r21p,t21p,r21s,t21s]=f_rtamp(n2(n),k2(n),n1(n),
                                k1(n),fi2,fi1);
[r23p,t23p,r23s,t23s]=f_rtamp(n2(n),k2(n),n3(n),0,
                                fi2,fi3);
delta01=0;
delta12=2*pi*(n1(n)-i*k1(n))*d1*cos(fi1)/lambda(n);
delta21=2*pi*(n2(n)-i*k2(n))*d2*cos(fi2)/lambda(n);
delta23=2*pi*(n2(n)-i*k2(n))*d2*cos(fi2)/lambda(n);
C01p=[1 r01p; r01p 1];
C12p=[exp(i*delta12) r12p*exp(i*delta12); ...
r12p*exp(-i*delta12) exp(-i*delta12)];
C21p=[exp(i*delta21) r21p*exp(i*delta21); ...
r21p*exp(-i*delta21) exp(-i*delta21)];
C23p=[exp(i*delta23) r23p*exp(i*delta23); ...
r23p*exp(-i*delta23) exp(-i*delta23)];
Ap=C01p*C12p;
tp=t01p*t12p;
for j=1:m-1
  Ap=Ap*C21p*C12p;
  tp=tp*t21p*t12p;
end
Ap=Ap*C23p;
tp=tp*t23p;
Rp(n)=abs(Ap(2,1)/Ap(1,1))^2;
Tp(n)=n3(n)*cos(fi3)/n0(n)/cos(fi0)*abs(tp/
                                      Ap(1,1))^2;
C01s=[1 r01s; r01s 1];
C12s=[exp(i*delta12) r12s*exp(i*delta12); ...
r12s*exp(-i*delta12) exp(-i*delta12)];
C21s=[exp(i*delta21) r21s*exp(i*delta21); ...
r21s*exp(-i*delta21) exp(-i*delta21)];
C23s=[exp(i*delta23) r23s*exp(i*delta23); ...
r23s*exp(-i*delta23) exp(-i*delta23)];
As=C01s*C12s;
ts=t01s*t12s;
for j=1:m-1
  As=As*C21s*C12s;
```

```
    ts=ts*t21s*t12s;
  end
  As=As*C23s;
  ts=ts*t23s;
  Rs(n)=abs(As(2,1)/As(1,1))^2;
  Ts(n)=n3(n)*cos(fi3)/n0(n)/cos(fi0)*abs(ts/
                                      As(1,1))^2;
end
```

Function called to compute the field amplitudes.

```
function
[Ees,Eep,fi1,fi2,fi3]=f_2mlyrAmp(n0,n1,k1,n2,k2,n3,d1,
d2,m,...fi0,lambda)
% calculate transmittance and reflectance of a
% multi-layer
% system. JW Haus 08/13/2002 Amplitudes of the
% fields added.
%------------------------
  fi1=asin(sin(fi0)*n0/(n1-i*k1));
  phi1=fi1;
  fi2=asin(sin(fi0)*n0/(n2-i*k2));
  phi2=fi2;
  fi3=asin(sin(fi0)*n0/n3);
  phi3=fi3;
  [r01p,t01p,r01s,t01s]=f_rtamp(n0,0,n1,k1,fi0,fi1);
  [r12p,t12p,r12s,t12s]=f_rtamp(n1,k1,n2,k2,fi1,fi2);
  [r21p,t21p,r21s,t21s]=f_rtamp(n2,k2,n1,k1,fi2,fi1);
  [r23p,t23p,r23s,t23s]=f_rtamp(n2,k2,n3,0,fi2,fi3);
  delta01=0;
  delta12=2*pi*(n1-i*k1)*d1*cos(fi1)/lambda;
  delta21=2*pi*(n2-i*k2)*d2*cos(fi2)/lambda;
  delta23=2*pi*(n2-i*k2)*d2*cos(fi2)/lambda;
  C01p=[1 r01p; r01p 1];
  C12p=[exp(i*delta12) r12p*exp(i*delta12); ...
  r12p*exp(-i*delta12) exp(-i*delta12)];
  C21p=[exp(i*delta21) r21p*exp(i*delta21); ...
  r21p*exp(-i*delta21) exp(-i*delta21)];
  C23p=[exp(i*delta23) r23p*exp(i*delta23); ...
  r23p*exp(-i*delta23) exp(-i*delta23)];
% field amplitude calculations p-polarization
  % transmitted field amplitude
  Temp=[1 0]';
  Eep(:,2*m+2)=Temp;
  % Amplitude in the last layer
  Temp=C23p*Temp/t23p;
```

```
  Eep(:,2*m+1)=Temp;
  for j=0:m-2
    Temp=C12p*Temp/t12p;
    Eep(:,2*(m-j))= Temp;
    Temp=C21p*Temp/t21p;
    Eep(:,2*(m-j)-1)= Temp;
  end
  Temp=C12p*Temp/t12p;
  Eep(:,2)=Temp;
  Temp=C01p*Temp/t01p;
  Eep(:,1)=Temp;
  % NORMALIZING AMPLITUDE
    Eep=Eep/Eep(1,1);
  C01s=[1 r01s; r01s 1];
  C12s=[exp(i*delta12) r12s*exp(i*delta12); ...
  r12s*exp(-i*delta12) exp(-i*delta12)];
  C21s=[exp(i*delta21) r21s*exp(i*delta21); ...
  r21s*exp(-i*delta21) exp(-i*delta21)];
  C23s=[exp(i*delta23) r23s*exp(i*delta23); ...
  r23s*exp(-i*delta23) exp(-i*delta23)];
% field amplitude calculations s-polarization
  % transmitted field amplitude
  Temp=[1 0]';
  Ees(:,2*m+2)=Temp;
  % Amplitude in the last layer
  Temp=C23s*Temp/t23s;
  Ees(:,2*m+1)=Temp;
  for j=0:m-2
    Temp=C12s*Temp/t12s;
    Ees(:,2*(m-j))= Temp;
    Temp=C21s*Temp/t21s;
    Ees(:,2*(m-j)-1)= Temp;
  end
  Temp=C12s*Temp/t12s;
  Ees(:,2)=Temp;
  Temp=C01s*Temp/t01s;
  Ees(:,1)=Temp;
% NORMALIZING AMPLITUDE
    Ees=Ees/Ees(1,1);
end
```

Function called by the preceding function programs to compute the Fresnel coefficients

```
function [rp,tp,rs,ts]=f_rtamp(n1,k1,n2,k2,fi1,fi2)
%calculate amplitude of transmission and reflection
```

```
%Written by Feiran Huang on 05/31/99. Last
%Update 05/31/99
%-------------------------
%given index and absorption
n1=n1-i*k1; n2=n2-i*k2;
  rp=(n1*cos(fi2)-n2*cos(fi1))/(n1*cos(fi2)+n2
                                   *cos(fi1));
  rs=(n1*cos(fi1)-n2*cos(fi2))/(n1*cos(fi1)+n2
                                   *cos(fi2));
  tp=2*n1*cos(fi1)/(n1*cos(fi2)+n2*cos(fi1));
  ts=2*n1*cos(fi1)/(n1*cos(fi1)+n2*cos(fi2));
```

3.6 Appendix B

Plane-wave method programs. The E method calculates the eigenvalues for the electric field polarized parallel to the rods. Comments are inserted that change the program to the H method.

```
% Program for calculating the E-polarized
% band structure of the triangular lattice.
% Lines are inserted to convert it to band
% structure computation for the H-polarized waves.
% A parameter determining the number of plane waves
  N1 = 9;
  pi = acos(-1.0);
% The lattice constant is scaled to unity. The radius
% of the rod.
  d=1.; a=.48;
% Volume fraction of rod in the unit cell.
  bbeta =3.14159*a*a/d/d*2.0/sqrt(3.0);
% relative permittivities.
  EPSA=1.; % ROD DIELECTRIC
  EPSB=10.;% MATRIX DIELECTRIC
% The major symmetry points in the Brillouin zone are
% defined
  gxg = 0.0; gyg = 0.0;
  gxx = 0.0; gyx = 2.0*pi/sqrt(3.0);
  gxj =2.0*pi/3.0; gyj = 2.0*pi/sqrt(3.0);
% The itinerary around the Brillouin zone is laid out.
  gxi(1) = gxx;
  gyi(1) = gyx;
  gxi(2) = gxg;
  gyi(2) = gyg;
  gxi(3) = gxj;
```

```
  gyi(3) = gyj;
  gxi(4) = gxx;
  gyi(4) = gyx;
% Construct the wave vectors used in the plane
% wave calculation
      L=0;
    gx1 = 2.0*pi;
    gy1 = -2.0*pi/sqrt(3.0);
    gx2 = 0.0;
    gy2 = 2.0*pi*2.0/sqrt(3.0);
% generate all the wave vectors within a hexagon of
% diameter 2*n1.
for ii=-N1:N1;
  for j=-N1:N1;
    if ( (-(N1+1)<(ii+j)) & ((ii+j)<(N1+1))),
      L=L+1;
      gx(L) = gx1*ii + gx2*j;
      gy(L) = gy1*ii + gy2*j;
    end;
  end;
end;
% Construct the dielectric matrix.
% The volume of the unit cell is defined
    vcel = sqrt(3.0)/2.0*d*d;
  for ii=1:L;
      EPS(ii,ii) = bbeta*EPSA + (1.0-bbeta)*EPSB;
%% H-method insert the following line to replace
% the preceding line.
% Construct the inverse-dielectric matrix
%     EPS(ii,ii)= bbeta/EPSA+(1.0-bbeta)/EPSB;
  i1= ii + 1;
  for j=i1:L;
    x1 = gx(ii) - gx(j);
    x2 = gy(ii) - gy(j);
    x = sqrt(x1*x1+x2*x2)*a/d;
    EPS(ii,j)= 2.0*pi*(EPSA-EPSB)*a*a/(vcel)
                *besselj(1,x)/x;
%% H-method insert the following line to replace the
%  preceding line.
% Construct the inverse-dielectric matrix
%   EPS(ii,j)= 2.d0*pi*(1/EPSA-1/EPSB)*a*a/(vcel)
%               *besselj(1,x)/x;
% The matrix is symmetric
      EPS(j,ii)= EPS(ii,j);
```

```
    end;
  end
% Use the itinerary constructed above to move around
% the Brillouin zone
  n=L;
% Counter for the number of points in the itinerary
  nknt = 1;
for nk = 1:3;
      nk1 = nk+1
     nmax= 20;
  for l=1:nmax;
      dkx =(l-1)*(gxi(nk1)-gxi(nk))/(nmax) + gxi(nk);
      dky =(l-1)*(gyi(nk1)-gyi(nk))/(nmax) + gyi(nk);
    for ii=1:n;
        a(ii,ii) = ((dkx+gx(ii))^2 + (dky+gy(ii))^2)/
                   4.0/pi/pi;
        b(ii,ii) = EPS(ii,ii);
        for j=ii+1:n;
            a(ii,j) = 0.0;
            a(j,ii) = 0.0;
            b(ii,j) = EPS(ii,j);
            b(j,ii) = EPS(ii,j);
        end
    end
%% H-method insert the following lines and replace the
%  preceding 10 lines
%  for ii=1:n;
%   a(ii,ii)= EPS(ii,ii)*((dkx+gx(ii))^2
             +(dky+gy(ii))^2)/4.0/pi/pi;
%    for j=ii+1:n;
%     a(ii,j)= EPS(ii,j)*((dkx+gx(ii))*(dkx+gx(j))
              +(dky+gy(ii))*...
%             (dky+gy(j)))/4.0/pi/pi;
%     a(j,ii)= a(ii,j);
%    end
%   end
% Find the eigenvalues
     v=eig(a,b);
% Find the eigenvalues for the H-method. Replace
% the preceding line.
%    v=eig(a);
% Store the lowest 10 eigenvalues for plotting
     for li=1:10;
       rr(nknt,li)=v(li);
```

```
        kxx(nknt)=nknt-1;
      end

    nknt = nknt + 1;
  end

end

% Plot lowest six eigenvalues

figure(1)
 hold on;
 plot(kxx,rr(:,1)); plot(kxx,rr(:,2));
 plot(kxx,rr(:,3)); plot(kxx,rr(:,4));
 plot(kxx,rr(:,5));plot(kxx,rr(:,6));
```

References

1. L. Brillouin, *Wave Propagation in Periodic Structures*, Wiley, New York (1946).
2. A. Yariv and P. Yeh, *Optical Waves in Layered Media*, Wiley, New York (1988).
3. J. D. Joannopoulos, R. D. Meade, and J. N. Winn, *Photonic Crystals*, Princeton Univ. Press, Princeton, NJ (1995).
4. K. Sakoda, *Optical Properties of Photonic Crystals*, Springer, Berlin (2002).
5. R. E. Slusher and B. J. Eggleton, Eds., *Nonlinear Photonic Crystals*, Springer, Berlin (2003).
6. See articles in C. M. Soukoulis, Ed., *Photonic Band Gaps and Localization*, Plenum, New York (1993).
7. C. M. Bowden, J. D. Dowling, and H. Everitt, Eds., "Development and applications of materials exhibiting photonic band gaps," *J. Opt. Soc. Am. B* **10**, 279–413 (1993).
8. G. Kurizki and J. W. Haus, Eds., "Photonic band structures," *J. Mod. Opt.* **41**, 171–404 (1994).
9. M. Bertolotti, C. M. Bowden, and C. Sibilia, Eds., *Nanoscale Linear and Nonlinear Optics*, AIP Conference Proceedings **560** (2001).
10. J. W. Haus, "Photonic band structures," in *Quantum Optics of Confined Systems*, M. Ducloy and D. Bloch, Eds., pp. 101–142, Kluwer, Dordrecht, The Netherlands (1996).
11. P. Halevi, A. A. Krokhin and J. Arriaga, "Photonic crystal optics and homogenization of 2D periodic composites," *Phys. Rev. Lett.* **82**, 719–722 (1999).
12. E. J. Post, *Formal Structure of Electromagnetics*, Dover Press, New York (1997).

13. W. S. Weiglhofer and A. Lakhtakia, Eds., *Introduction to Complex Mediums for Electromagnetics and Optics*, SPIE Press, Bellingham, WA (2003).
14. N. W. Ashcroft and N. D. Mermin, *Solid State Physics*, Harcourt, Orlando, FL (1976).
15. J. M. Benedickson, J. P. Dowling, and M. Scalora, "Analytic expressions for the electromagnetic mode density in finite, 1D, photonic band-gap structures," *Phys. Rev. B* **53**, 4107–4121 (1996).
16. M. Centini, C. Sibilia, M. Scalora, G. D'Aguanno, M. Bertolotti, M. J. Bloemer, C. M. Bowden, and I. Nefedov, "Dispersive properties of finite, 1D photonic band gap structures: applications to nonlinear quadratic interactions," *Phys. Rev. E* **60**, 4891–4898 (1999).
17. M. Scalora, M. J. Bloemer, and C. M. Bowden, "Transparent, metallo-dielectric, 1D, photonic band-gap structures," *J. Appl. Phys.* **83**, 2377–2383 (1998).
18. M. J. Bloemer, M. Scalora, and C. M. Bowden, "Transmissive properties of Ag/MgF_2 photonic band gaps," *Appl. Phys. Lett.* **72**, 1676–1678 (1998).
19. M. Scalora and M. Crenshaw, "A beam propagation method that handles reflections," *Opt. Commun.* **108**, 191–196 (1994).
20. M. Scalora, J. D. Dowling, A. S. Manka, C. M. Bowden, and J. W. Haus, "Pulse-propagation near highly reflective surfaces—applications to photonic band-gap structures and the question of superluminal tunneling times," *Phys. Rev. A* **52**, 726–734 (1995).
21. M. Scalora, M. Tocci, M. J. Bloemer, C. M. Bowden, and J. W. Haus, "Dipole emission rates in 1D photonic band-gap materials," *Appl. Phys. B* **60**, S57–S61 (1995).
22. M. C. Larciprete, C. Sibilia, S. Paoloni, M. Bertolotti, F. Sarto, and M. Scalora, "Accessing the optical limiting properties of metallo-dielectric photonic band gap structures," *J. Appl. Phys.* **93**, 5013–5017 (2003).
23. W. Chen and D. L. Mills, "Gap solitons and the nonlinear response of superlattices," *Phys. Rev. Lett.* **58**, 160–163 (1987).
24. M. Scalora, J. P. Dowling, C. M. Bowden, and M. J. Bloemer, "Optical limiting and switching of ultrashort pulses in nonlinear photonic band gap materials," *Phys. Rev. Lett.* **73**, 1368–1371 (1994).
25. J. W. Haus, B. Y. Soon, M. Scalora, M. J. Bloemer, C. M. Bowden, C. Sibilia, and A. Zheltikov, "Spatio-temporal instabilities for counter-propagating waves in periodic media," *Opt. Expr.* **10**, 114–121 (2002).
26. J. W. Haus, B. Y. Soon, M. Scalora, C. Sibilia, and I. V. Mel'nikov, "Coupled mode equations for Kerr media with periodically modulated linear and nonlinear coefficients," *J. Opt. Soc. Am. B* **19**, 2282–2291 (2002).
27. B. Y. Soon, M. Scalora, and C. Sibilia, "One-dimensional photonic crystal optical limiter," *Opt. Expr.* **11**, 2007–2018 (2003).
28. M. Scalora, M. J. Bloemer, A. S. Manka, J. P. Dowling, C. M. Bowden, R. Viswanathan, and J. W. Haus, "Pulsed second-harmonic generation in nonlinear, 1D periodic structures," *Phys. Rev. A* **56**, 3166–3174 (1997).

29. J. W. Haus, R. Viswanathan, A. Kalocsai, J. Cole, M. Scalora, and J. Theimer, "Enhanced second-harmonic generation in weakly periodic media," *Phys. Rev. A* **57**, 2120–2128 (1998).
30. D. Pezzetta, C. Sibilia, M. Bertolotti, J. W. Haus, M. Scalora, M. J. Bloemer, and C. M. Bowden, "Photonic band-gap structures in planar nonlinear waveguides: application to second harmonic generation," *J. Opt. Soc. Am. B* **18**, 1326–1333 (2001).
31. M. D. Tocci, M. J. Bloemer, M. Scalora, J. P. Dowling, and C. M. Bowden, "Thin-film nonlinear optical diode," *Appl. Phys. Lett.* **66**, 2324–2327 (1995).
32. J. P. Dowling, M. Scalora, M. J. Bloemer, A. S. Manka, J. P. Dowling, and C. M. Bowden, "The photonic band edge laser: a new approach to gain enhancement," *J. Appl. Phys.* **75**, 1896–1899 (1994).
33. H. G. Winful and V. Perlin, "Raman gap solitons," *Phys. Rev. Lett.* **84**, 3586–3589 (2000).
34. K. Sakoda and J. W. Haus, "Superradiance in photonic crystals with pencil-like excitation," *Phys. Rev. A* **68**, 053809 (2003).
35. C. M. de Sterke and J. E. Sipe, "Possibilities for the observation of gap solitons in wave-guide geometries," *J. Opt. Soc. Am. A* **6**, 1722–1725 (1989).
36. C. M. de Sterke and J. E. Sipe, "Extensions and generalizations of an envelope function approach for the electrodynamics of nonlinear periodic structures," *Phys. Rev. A* **39**, 5163–5178 (1989).
37. C. M. de Sterke and J. E. Sipe, "Gap solitons," in *Progress in Optics*, Vol. 33, E. Wolf, Ed., Elsevier, Amsterdam (1994).
38. D. F. Prelewitz and T. G. Brown, "Optical limiting and free-carrier dynamics in a periodic semiconductor waveguide," *J. Opt. Soc. Am. B* **11**, 304–312 (1994).
39. C. J. Herbert and M. S. Malcuit, "Optical bistability in nonlinear periodic structures," *Opt. Lett.* **18**, 1783–1785 (1993).
40. B. J. Eggleton, C. M. de Sterke, and R. E. Slusher, "Bragg solitons in the nonlinear Schrodinger limit: experiment and theory," *J. Opt. Soc. Am. B* **16**, 587–599 (1999).
41. B. J. Eggleton, G. Lenz, and N. M. Litchinitser, "Optical pulse compression schemes that use nonlinear Bragg gratings," *Fiber Integr. Opt.* **19**, 383–421 (2000).
42. D. Taverner, N. G. R. Broderick, D. J. Richardson, M. Ibsen, and R. I. Laming, "Nonlinear self-switching and multiple gap soliton formation in a fibre Bragg grating," *Opt. Lett.* **23**, 328–330 (1998).
43. A. V. Balakin, D. Boucher, V. A. Bushuev, N. I. Koroteev, B. I. Mantsyzov, P. Masselin, I. A. Ozheredov, and A. P. Shkurinov, "Enhancement of second-harmonic generation with femtosecond laser pulses near the photonic band edge for different polarizations of incident light," *Opt. Lett.* **24**, 793–795 (1999).
44. Y. Dumeige, P. Vidakovic, S. Sauvage, I. Sagnes, J. A. Levenson, C. Sibilia, M. Centini, G. D'Aguanno, and M. Scalora, "Enhancement of second-harmonic generation in a 1D semiconductor photonic band gap," *Appl. Phys. Lett.* **78**, 3021–3023 (2001).

45. G. D'Aguanno, M. Centini, M. Scalora, C. Sibilia, Y. Dumeige, P. Vidakovic, J. A. Levenson, J. W. Haus, M. J. Bloemer, C. M. Bowden, and M. Bertolotti, "Photonic band edge effects in finite structures and applications to $\chi^{(2)}$ interactions," *Phys. Rev. E* **64**, 16609 (2001).
46. J. W. Haus, "A brief review of theoretical results for photonic band structures," *J. Mod. Opt.* **41**, 195–207 (1994).
47. Z. Zhang and S. Satpathy, "Electromagnetic wave propagation in periodic structures: Bloch wave solution of Maxwell's equations," *Phys. Rev. Lett.* **65**, 2650–2653 (1990).
48. H. S. Sözüer, J. W. Haus, and R. Inguva, "Photonic bands: convergence problems with the plane-wave method," *Phys. Rev. B* **45**, 13962–13972 (1992).
49. M. Plihal and A. A. Maradudin, "Photonic band structure of 2D systems: the triangular lattice," *Phys. Rev. B* **44**, 8565–8571 (1991).
50. M. Plihal, A. Shambrook, A. A. Maradudin, and P. Sheng, "Two-dimensional photonic band structures," *Opt. Commun.* **80**, 199–205 (1991).
51. S. L. McCall, P. M. Platzman, R. Dalichaouch, D. Smith, and S. Schultz, "Microwave propagation in 2D dielectric lattices," *Phys. Rev. Lett.* **67**, 2017–2020 (1991).
52. R. D. Meade, K. D. Brommer, A. M. Rappe, and J. D. Joannopoulos, "Existence of a photonic band gap in two dimensions," *Appl. Phys. Lett.* **61**, 495–498 (1992).
53. A. A. Maradudin and A. R. McGurn, "Photonic band structure of a truncated, 2D, periodic dielectric medium," *J. Opt. Soc. Am. B* **10**, 307–314 (1993).
54. T. K. Gaylord, G. N. Henderson, and E. N. Glytsis, "Application of electromagnetics formalism to quantum-mechanical electron-wave propagation in semiconductors," *J. Opt. Soc. Am. B* **10**, 333–342 (1993).
55. J. Dowling and C. M. Bowden, "Atomic emission rates in inhomogeneous media with applications to photonic band structures," *Phys. Rev. A* **46**, 612–622 (1992).
56. J. Dowling and C. M. Bowden, "Beat radiation from dipoles near a photonic band edge," *J. Opt. Soc. Am. B* **10**, 353–355 (1993).
57. S. Satpathy, Z. Zhang, and M. R. Salehpour, "Theory of photon bands in 3D periodic dielectric structures," *Phys. Rev. Lett.* **64**, 1239–1242 (1990).
58. K. M. Leung and Y. F. Liu, "Photonic band structures: the plane-wave method," *Phys. Rev. B* **42**, 10188–10190 (1990).
59. T. W. Körner, *Fourier Analysis*, 62–66, Cambridge University Press, Cambridge (1988).
60. W. M. Robertson, G. Arjavalingam, R. D. Meade, K. D. Brommer, A. M. Rappe, and J. D. Joannopoulos, "Measurement of photonic band structure in a 2D periodic dielectric array," *Phys. Rev. Lett.* **68**, 2023–2026 (1992).
61. W. M. Robertson, G. Arjavalingam, R. D. Meade, K. D. Brommer, A. M. Rappe, and J. D. Joannopoulos, "Measurement of the photon dispersion relation in 2D ordered dielectric arrays," *J. Opt. Soc. Am. B* **10**, 322–330 (1993).

62. K. Sakoda, "Optical transmittance of a 2D triangular photonic lattice," *Phys. Rev. B* **51**, 4672–4675 (1995).
63. K. Sakoda, "Symmetry, degeneracy, and uncoupled modes in 2D photonic lattices," *Phys. Rev. B* **52**, 7982–7986 (1995).
64. K. Sakoda, "Transmittance and Bragg reflectivity of 2D photonic lattices," *Phys. Rev. B* **52**, 8992–9002 (1995).
65. M. Wada, K. Sakoda, and K. Inoue, "Far-infrared spectroscopy study of an uncoupled mode in a 2D photonic lattice," *Phys. Rev. B* **52**, 16297–16300 (1995).
66. M. Wada, Y. Doi, K. Inoue, and J. W. Haus, "Far-infrared transmittance and band-structure correspondence in 2D air-rod photonic crystals," *Phys. Rev. B* **55**, 10443–10450 (1997).
67. K. Inoue, M. Wada, K. Sakoda, A. Yamanaka, M. Hayashi, and J. W. Haus, "Fabrication of 2D photonic band structure with near-infrared band gap," *J. Appl. Phys.* **33**, L1463–L1466 (1994).
68. K. Inoue, M. Wada, K. Sakoda, M. Hayashi, T. Fukushima, and A. Yamanaka, "Near-infrared photonic band gap of 2D triangular air-rod lattices as revealed by transmittance measurement," *Phys. Rev. B* **53**, 1010–1013 (1996).
69. A. Rosenberg, R. J. Tonucci, H.-B. Lin, and A. J. Campillo, "Near-infrared 2D photonic band-gap materials," *Opt. Lett.* **21**, 830–833 (1996).
70. H.-B. Lin, R. J. Tonucci, and A. J. Campillo, "Observation of 2D photonic band behavior in the visible," *Appl. Phys. Lett.* **68**, 2927–2930 (1996).
71. R. D. Meade, K. D. Brommer, A. M. Rappe, and J. D. Joannopoulos, "Existence of a photonic band gap in two dimensions," *Appl. Phys. Lett.* **61**, 495–498 (1992).
72. K. Inoue, M. Sasada, J. Kawamata, K. Sakoda, and J. W. Haus, "A 2D photonic crystal laser," *Jpn. J. Appl. Phys.* **38**, L157–L159 (1999).
73. S. Datta, C. T. Chan, K. M. Ho, and C. M. Soukoulis, "Photonic band gaps in periodic dielectric structures: the scalar-wave approximation," *Phys. Rev. B* **46**, 10650–10656 (1992).
74. K. M. Leung and Y. F. Liu, "Full vector wave calculation of photonic band structures in face-centered-cubic dielectric media," *Phys. Rev. Lett.* **65**, 2646–2649 (1990).
75. K. M. Ho, C. T. Chan, and C. M. Soukoulis, "Existence of a photonic gap in periodic dielectric structures," *Phys. Rev. Lett.* **65**, 3152–3155 (1990).
76. H. S. Sözüer and J. W. Haus, "Photonic bands: the simple cubic lattice," *J. Opt. Soc. Am. B* **10**, 296–301 (1993).
77. I. H. H. Zabel and D. Stroud, "Photonic band structures of optically anisotropic periodic arrays," *Phys. Rev. B* **48**, 13962–13969 (1992).
78. C. T. Chan, K. M. Ho, and C. M. Soukoulis, "Photonic band gaps in experimentally realizable periodic dielectric structures," *Europhys. Lett.* **16**, 563–567 (1991).
79. J. W. Haus, H. S. Sözüer, and R. Inguva, "Photonic bands: ellipsoidal dielectric atoms in an FCC lattice," *J. Mod. Opt.* **39**, 1991–1998 (1991).

80. H. S. Sözüer and J. P. Dowling, "Photonic band calculations for woodpile structures," *J. Mod. Opt.* **41**, 231–240 (1994).
81. K. M. Ho, C. T. Chan, C. M. Soukoulis, R. Biswas, and M. Sigalas, "Photonic band gaps in three dimensions: new layer-by-layer periodic structures," *Solid State Commun.* **89**, 413–417 (1994).
82. E. Yablonovitch and T. J. Gmitter, "Photonic band structure: the face-centered-cubic case," *Phys. Rev. Lett.* **63**, 1950–1953 (1989).
83. E. Yablonovitch, T. J. Gmitter, and K. M. Leung, "Photonic band structure: the face-centered-cubic case employing nonspherical atoms," *Phys. Rev. Lett.* **67**, 2295–2298 (1991).
84. A. A. Zakhidov, R. H. Baughman, Z. Iqbal, C. Cui, I. Khayrullin, S. O. Dantas, J. Marti, and V. Ralchenko, "Carbon structures with 3D periodicity at optical wavelengths," *Science* **282**, 897–901 (1998).
85. S. Noda, K. Tomoda, N. Yamamoto, and A. Chutinan, "Full 3D photonic crystals at near-infrared wavelengths," *Science* **289**, 604–606 (2000).
86. S.-Y. Lin, J. G. Fleming, D. L. Hetherington, B. K. Smith, R. Biswas, K. M. Ho, M. M. Sigalas, W. Zubrzycki, S. R. Kurtz, and J. Bur, "A 3D photonic crystal operating at infrared wavelengths," *Nature* **394**, 251–253 (1998).
87. Y. A. Vlasov, M. Deutsch, and D. J. Norris, "Single-domain spectroscopy of self-assembled photonic crystals," *Appl. Phys. Lett.* **76**, 1627–1629 (2000).
88. Z. Yuan, J. W. Haus, and K. Sakoda, "Eigenmode symmetry for simple cubic lattices and the transmission spectra," *Opt. Expr.* **3**, 19–26 (1998).
89. K. Sakoda, "Group-theoretical classification of eigenmodes in 3D photonic lattices," *Phys. Rev. B* **55**, 15345–15348 (1997).
90. J. B. Pendry, "Photonic band structures," *J. Mod. Opt.* **41**, 209–229 (1994).
91. P. M. Bell, J. B. Pendry, L. M. Moreno, and A. J. Ward, "A program for calculating photonic band structures and transmission coefficients of complex structures," *Comput. Phys. Commun.* **85**, 306–322 (1995).
92. K. Yee, "Numerical solution of initial boundary value problems involving Maxwell's equations in isotropic media," *IEEE Trans. Antennas Propagat.* **14**, 302–307 (1966).
93. A. Taflove, *Computational Electrodynamics: The Finite-Difference Time-Domain Method*, Artech House, Boston (1995).
94. K. S. Kunz and R. J. Luebbers, *The Finite Difference Time Domain Method for Electromagnetics*, CRC Press, Boca Raton, FL (1993).
95. K. Sakoda, "Numerical study on localized defect modes in 2D triangular photonic crystals," *J. Appl. Phys.* **84**, 1210–1214 (1998).
96. P. R. Villeneuve, S. Fan, and J. D. Joannopoulos, "Microcavities in photonic crystals: mode symmetry, tunability, and coupling efficiency," *Phys. Rev. B* **54**, 7837–7842 (1996).
97. O. Painter, J. Vuckovic, and A. Scherer, "Defect modes of a 2D photonic crystal in an optically thin dielectric slab," *J. Opt. Soc. Am. B* **16**, 275–285 (1999).
98. N. Kawai, M. Wada, and K. Sakoda, "Numerical analysis of localized defect modes in a photonic crystal: 2D triangular lattice with square rods," *Jpn. J. Appl. Phys.* **37**, 4644–4647 (1998).

99. N. Stojic, J. Glimm, Y. Deng, and J. W. Haus, "Transverse magnetic defect modes in 2D triangular-lattice photonic crystals," *Phys. Rev. E* **64**, 056614 (2001).
100. K. Sakoda and H. Shiroma, "Numerical method for localized defect modes in photonic lattices," *Phys. Rev. B* **56**, 4830–4835 (1997).
101. E. Yablonovitch, T. J. Gmitter, R. D. Meade, A. M. Rappe, K. D. Brommer, and J. D. Joannopoulos, "Donor and acceptor modes in photonic band structure," *Phys. Rev. Lett.* **67**, 3380–3383 (1991).
102. S.-Y. Lin, V. M. Hietala, S. K. Lyo, and A. Zaslavsky, "Numerical method for localized defect modes in photonic lattices," *Appl. Phys. Lett.* **56**, 3233–3235 (1996).
103. T. Ueta, K. Ohtaka, N. Kawai, and K. Sakoda, "Limits on quality factors of localized defect modes in photonic crystals due to dielectric loss," *J. Appl. Phys.* **84**, 6299–6304 (1998).
104. P. Paddon and J. F. Young, "Two-dimensional vector-coupled-mode theory for textured planar waveguides," *Phys. Rev. B* **61**, 2090–2101 (2000).
105. T. Ochiai and K. Sakoda, "Nearly free-photon approximation for 2D photonic crystal slabs," *Phys. Rev. B* **64**, 045108 (2001).
106. J. Martorell, R. Vilaseca, and R. Corbalan, "Second harmonic generation in a photonic crystal," *Appl. Phys. Lett.* **70**, 702–704 (1997).

List of symbols

A_m	amplitude in the mth layer for waves traveling in the positive z direction
$\mathbf{B}$	B field vector
c	speed of light in vacuum
C_m	amplitude in the mth layer for waves traveling in the negative z direction
$\mathbf{D}$	electric displacement vector
$\mathbf{E}$	electric field vector
$\mathbf{E}_{n\mathbf{k}}$	plane wave Fourier component of the electric field vector for the nth band and wave vector $\mathbf{k}$
f	volume fraction of low-permittivity medium in a two-medium structure
G	magnitude of $\mathbf{G}$
$\mathbf{G}$	reciprocal lattice vector
$\mathbf{H}$	H field vector
$\mathbf{H}_{n\mathbf{k}}$	plane wave Fourier component of the H field vector for the nth band and wave vector $\mathbf{k}$
$\mathbf{I}$	2×2 unit matrix
k	Magnitude of Bloch wave vector for 1D PBG at normal incidence
$\mathbf{k}$	wave vector
k_a, k_b	wave numbers in media a and b
k_m	wave number in the mth layer for waves traveling in the positive z direction
L	thickness of a periodic structure
$\mathbf{M}$	2×2 transfer matrix
n_m	index of refraction of the mth medium
n_{eff}	effective index of refraction for the PBG
P	medium polarization
r_1	complex-valued reflection amplitude for a unit cell
r_N	complex-valued reflection amplitude for an N-period PBG
t	time
t_1	complex-valued transmission amplitude for a unit cell
t_N	complex-valued transmission amplitude for an N-period PBG
T_N	transmission for an N-period PBG
v_g	group velocity
$\mathbf{x} \equiv (x, y, z)$	position vector
α	angle of planewave incidence
α_m	angle of refraction in the mth layer
$\delta(\cdot)$	Dirac delta function
$\delta_{\mathbf{G},\mathbf{G}'}, \delta_{m,m'}$	Kronecker delta functions
ϵ	relative permittivity
ϵ_0	permittivity of vacuum

ϵ_a, ϵ_b	relative permittivities of media a and b
$\epsilon_{\mathbf{GG}'}$	relative permittivity matrix in the plane wave basis
η	inverse relative permittivity
$\eta_{\mathbf{GG}'}$	inverse relative permittivity matrix in the plane wave basis
θ_0	complex angle representation for eigenvalues of the transfer matrix $\mathbf{M}$
Θ_m	Heaviside function for the mth medium
λ	eigenvalue of matrix $\mathbf{M}$
λ_0	wavelength in vacuum
μ	relative permeability
μ_0	permeability of vacuum
μ_a, μ_b	relative permeabilities of media a and b
ρ_N	density of modes for an N-period PBG
χ	linear dielectric susceptibility
ω	angular frequency

Joseph W. Haus has been the director of the Electro-Optics Program at the University of Dayton for the past 4 years. He received BS and MS degrees in physics in 1971 and 1972 from John Carroll University and a PhD degree in physics from Catholic University in 1975. After spending almost 7 years in Germany, he was appointed a National Research Council (NRC) visiting scientist at the U.S. Army Missile Command for the period 1983 to 1985 and a professor at Rensselaer Polytechnic Institute for 15 years. He hold two patents and has two more pending. He is a fellow of the OSA and a member of SPIE, APS, and IEEE-LEOS. He held the Hitachi Ltd. Quantum Materials Chair at the University of Tokyo in 1991 and 1992. Dr. Haus has published more than 200 research papers. His research is devoted to nonlinear optics in heterogeneous materials including photonic band gap structures and metal-semiconductor heterostructured nanoparticles.

Chapter 4

Quantum Dots: Phenomenology, Photonic and Electronic Properties, Modeling and Technology

Fredrik Boxberg and Jukka Tulkki

4.1. Introduction 109
- 4.1.1. What are they? 109
- 4.1.2. History 111

4.2. Fabrication 112
- 4.2.1. Nanocrystals 114
- 4.2.2. Lithographically defined quantum dots 115
- 4.2.3. Field-effect quantum dots 116
- 4.2.4. Self-assembled quantum dots 117

4.3. QD spectroscopy 119
- 4.3.1. Microphotoluminescence 119
- 4.3.2. Scanning near-field optical spectroscopy 121

4.4. Physics of quantum dots 122
- 4.4.1. Quantum dot eigenstates 123
- 4.4.2. Electromagnetic fields 124
- 4.4.3. Photonic properties 126
- 4.4.4. Carrier transport 127
- 4.4.5. Carrier dynamics 129
- 4.4.6. Dephasing 130

4.5. Modeling of atomic and electronic structure 131
- 4.5.1. Atomic structure calculations 131
- 4.5.2. Quantum confinement 132

4.6. QD technology and perspectives 133
- 4.6.1. Vertical-cavity surface-emitting QD laser 134
- 4.6.2. Biological labels 134

4.6.3. Electron pump 135
4.6.4. Applications you should be aware of 136
References 137
List of symbols 142

4.1 Introduction

4.1.1 What are they?

The research of microelectronic materials is driven by the need to tailor electronic and optical properties for specific component applications. Progress in epitaxial growth and advances in patterning and other processing techniques have made it possible to fabricate "artificial" dedicated materials for microelectronics.[1] In these materials, the electronic structure is tailored by changing the local material composition and by confining the electrons in nanometer-size foils or grains. Due to quantization of electron energies, these systems are often called quantum structures. If the electrons are confined by a potential barrier in all three directions, the nanocrystals are called quantum dots (QDs). This review of quantum dots begins with discussion of the physical principles and first experiments and concludes with the first expected commercial applications: single-electron pumps, biomolecule markers, and QD lasers.

In nanocrystals, the crystal size dependency of the energy and the spacing of discrete electron levels are so large that they can be observed experimentally and utilized in technological applications. QDs are often also called mesoscopic atoms or artificial atoms to indicate that the scale of electron states in QDs is larger than the lattice constant of a crystal. However, there is no rigorous lower limit to the size of a QD, and therefore even macromolecules and single impurity atoms in a crystal can be called QDs.

The quantization of electron energies in nanometer-size crystals leads to dramatic changes in transport and optical properties. As an example, Fig. 4.1 shows the dependence of the fluorescence wavelength on the dimensions and material composition of the nanocrystals. The large wavelength differences between the blue, green, and red emissions result here from using materials having different band gaps: CdSe (blue), InP (green), and InAs (red). The fine-tuning of the fluorescence emission within each color is controlled by the size of the QDs.

The color change of the fluorescence is governed by the "electron in a potential box" effect familiar from elementary text books of modern physics.[3] A simple potential box model explaining the shift of the luminescence wavelength is shown in the inset of Fig. 4.1. The quantization of electron states exists also in larger crystals, where it gives rise to the valence and conduction bands separated by the band gap. In bulk crystals, each electron band consists of a continuum of electron states. However, the energy spacing of electron states increases with decreasing QD size, and therefore the energy spectrum of an electron band approaches a set of discrete lines in nanocrystals.

As shown in Sec. 4.4, another critical parameter is the thermal activation energy characterized by $k_B T$. For the quantum effects to work properly in the actual devices, the spacing of energy levels must be large in comparison to $k_B T$, where k_B is Boltzmann's constant, and T the absolute temperature. For room-temperature operation, this implies that the diameter of the potential box must be at most a few nanometers.

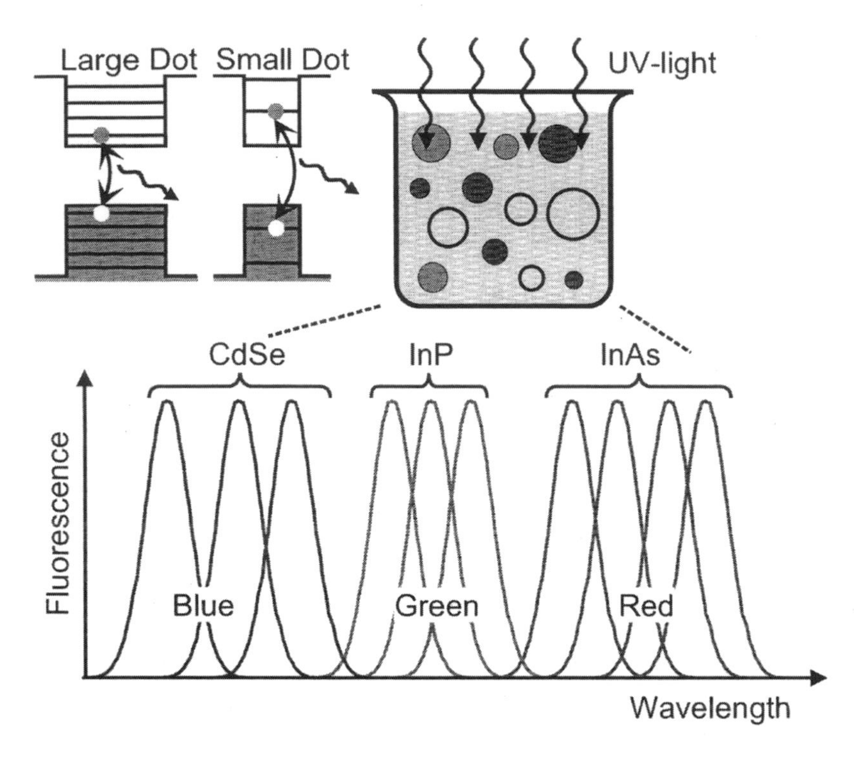

Figure 4.1 Nanocrystal quantum dots (NCQD) illuminated by UV-light emit light at a wavelength that depends both on the material composition and the size of the NCQDS. Large differences in the fluorescence wavelength result from different band gaps of the materials. Within each color (blue, green, and red) the wavelength is defined by the different sizes of the NCQDs.

In quantum physics, the electronic structure is often analyzed in the terms of the density of electron states (DOS). The prominent transformation from the continuum of states in a bulk crystal to the set of discrete electron levels in a QD is depicted in Fig. 4.2. In a bulk semiconductor [Fig. 4.2(a)], the DOS is proportional to the square root of the electron energy. In quantum wells (QWs) [see Fig. 4.2(b)], the electrons are restricted into a foil that is just a few nanometers thick. The QW DOS consists of a staircase, and the edge of the band (lowest electron states) is shifted to higher energies. However, in QDs the energy levels are discrete [Fig. 4.2(c)] and the DOS consists of a series of sharp (delta-function-like) peaks corresponding to the discrete eigenenergies of the electrons. Due to the finite life time of electronic states, the peaks are broadened and the DOS is a sum of Lorenzian functions. Figure 4.2(c) also depicts another subtle feature of QDs: In an experimental sample not all QDs are of the same size. Different sizes mean different eigenenergies; and the peaks in the DOS are accordingly distributed around some average energies corresponding to the average QD size. In many applications, the active device material contains a large ensemble of QDs. Their joint density of states then includes a statistical broadening characterized by a Gaussian function.[4] This broadening is

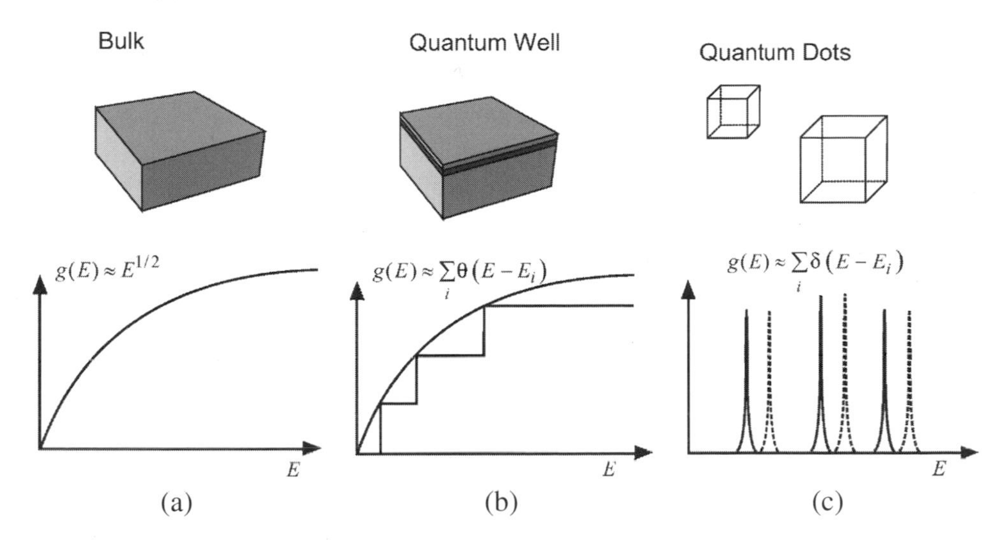

Figure 4.2 The density of electron states (DOS) in selected semiconductor crystals. The DOS of (a) bulk semiconductor, (b) quantum well, and (c) quantum dots.

often called inhomogeneous in distinction to the lifetime broadening, often called homogeneous broadening.[4]

4.1.2 History

Fabrication of QDs became possible because of the development of epitaxial growth techniques for semiconductor heterostructures. The prehistory of QDs began in the early 1970s with nanometer-thick foils called quantum wells. In QWs charge carriers (electrons and holes) become trapped in a few-nanometers-thick layer of wells, where the band gap is smaller than in the surrounding barrier layers. The variation of the band gap is achieved by changing the material composition of the compound semiconductor.[5]

The energy quantization in the optical absorption of a QW was first reported by Dingle et al.[6] in 1974. The photon absorption spectrum exhibits a staircase of discrete exciton resonances, whereas in the photon absorption of a bulk semiconductor only one exciton peak and the associated continuum is found. The transport properties of QW superlattices (periodic system of several QWs) were studied in the early 1970s by Esaki and Tsu.[7] The resonance tunneling effect and the related negative differential resistance was reported by Chang et al.[8] in 1974. These works began the exponential growth of the field during the 1970s; for a more complete list of references, see Bimberg et al.[4]

The experimental findings of Dingle et al.[6] were explained by the envelope wave function model that Luttinger and Kohn[9] developed for defect states in semiconductor single crystals. Resonant tunneling of electrons was explained in terms of quantum-mechanical transmission probabilities and Fermi distributions at source and drain contacts. Both phenomena were explained by the mesoscopic behavior of the electronic wave function,[9] which governs the eigenstates at the scale of several tens of lattice constants.

By the end of 1970, nanostructures could be fabricated in such a way that the mesoscopic variation of the material composition gave rise to the desired electronic potentials, eigenenergies, tunneling probabilities, and optical absorption. The quantum engineering of microelectronic materials was promoted by the Nobel prizes awarded in 1973 to L. Esaki for the discovery of the tunneling in semiconductors and in 1985 to K. von Klitzing et al.[10] for the discovery of the quantum Hall effect. Rapid progress was made in the development of epitaxial growth techniques: Molecular beam epitaxy[11] (MBE) and chemical vapor deposition[12] (CVD) made it possible to grow semiconductor crystals at one-monolayer accuracy.

In the processing of zero-dimensional (0D) and 1D structures, the development of electron beam lithography made it possible to scale down to dimensions of a few nanometers. Furthermore, the development of transmission electron microscopy (TEM), scanning tunneling microscopy (STM), and atomic force microscopy (AFM) made it possible to obtain atomic-level information of the nanostructures.

Figure 4.3 presents the discovery of level quantization in QDs reported by Ekimov and Onushenko[13] in 1984. The resonance structures are directly related to the energy quantization. One of the first measurements[14] of transport through a QD is shown in Fig. 4.4. In this case, the conductance resonance can be related to discrete charging effects that block the current unless appropriate QD eigenstates are accessible for electronic transport.

4.2 Fabrication

In the following, we limit our discussion to selected promising QD technologies including semiconductor nanocrystal QDs (NCQD), lithographically made QDs

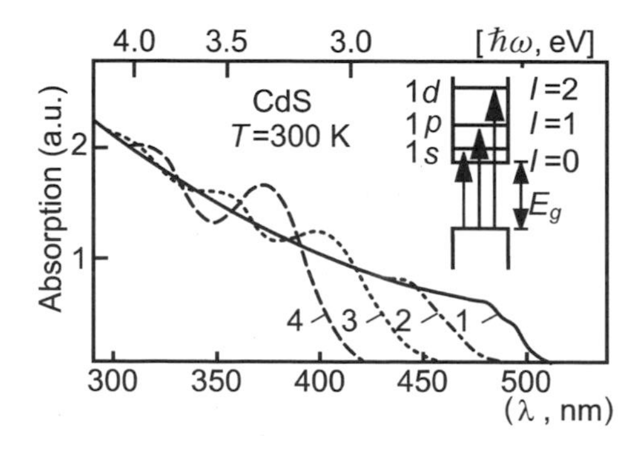

Figure 4.3 Photoabsorption by a set of CdS nanocrystals having different average radii as follows: (1) 38 nm, (2) 3.2 nm, (3) 1.9 nm, and (4) 1.4 nm. The inset marks the dipole transitions that are seen as resonances in absorption. The threshold of the absorption is blue-shifted when the size of the QD becomes smaller.[13]

(LGQDs), field-effect QDs (FEQDs), and self-assembled QDs (SAQDs). The main emphasis is on semiconductor QDs. Selected material parameters of the nanostructures are listed in Table 4.1.

Table 4.1 Selected room-temperature properties for the previously discussed materials.[32]

Material	Band gap (eV)	Electron mass (m_0)	Hole mass (m_0)	Permittivity (ε_0)
CdS	2.482(d)	0.165^a	$m^A_{p\perp} = 0.7$	$\varepsilon^{\perp}(0) = 8.28$
			$m^A_{p\parallel} = 5$	$\varepsilon^{\perp}(\infty) = 5.23$
				$\varepsilon^{\parallel}(0) = 8.73$
				$\varepsilon^{\parallel}(\infty) = 5.29$
CdSe	1.738(d)	0.112^a	$m^A_{p\perp} = 0.45$	$\varepsilon^{\perp}(0) = 9.29$
			$m^A_{p\parallel} \geq 1$	$\varepsilon^{\parallel}(0) = 10.16$
			$m^B_{p\perp} = 0.92$	$\varepsilon(\infty) = 5.8$
GaAs	1.5192(d)	0.0635	$m_{hh[100]} = 0.33$	$\varepsilon(0) = 12.80$
			$m_{hh[111]} = 0.33$	$\varepsilon(\infty) = 10.86$
			$m_{lh[100]} = 0.090$	
			$m_{lh[111]} = 0.077$	
InAs	0.4180(d)	0.023	$m_{hh} = 0.57$	$\varepsilon(0) = 14.5$
			$m_{[100]} = 0.35$	$\varepsilon(\infty) = 11.6$
			$m_{[111]} = 0.85$	
InP	1.344(d)	0.073	$m_{hh} = 0.65$	$\varepsilon(0) = 12.61$
			$m_{lh} = 0.12$	$\varepsilon(\infty) = 9.61$
Si	1.13(i)	$m_{\perp} = 0.1905^I$	$m_{hh} = 0.537^I$	$\varepsilon = 11.9$
		$m_{\parallel} = 0.9163^I$	$m_{lh} = 0.153^I$	

(^{I}The effective masses of Si are low temperature data $T = 4.2$ K). The low- and high-frequency permittivities are denoted by $\varepsilon(0)$ and $\varepsilon(\infty)$, respectively. The superscripts $\perp$ and $\parallel$ correspond to permittivities for the electric field perpendicular ($\overline{E} \perp c$) and parallel ($\overline{E} \parallel c$) with the c axis of the crystal.

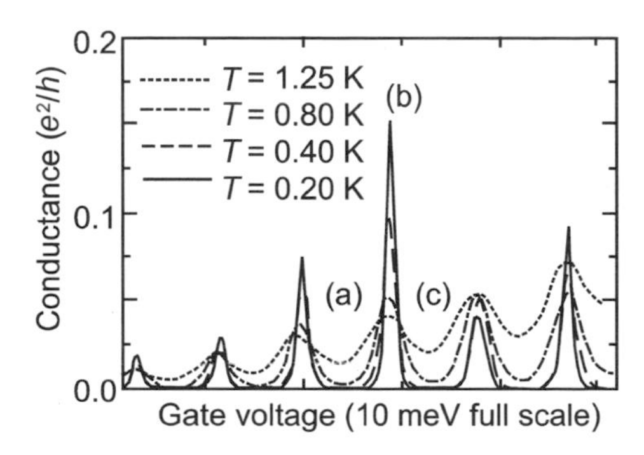

Figure 4.4 Conductance through a QD as a function of the gate voltage. Regions (a) and (c) indicate blocking of current by the Coulomb charging effect. In (b) electrons can tunnel from source to drain through empty electron states of the QD, thereby leading to a peak in the conductance. Note the rapid smearing of the resonance as the temperature increases.[14] (Reprinted with permission from Ref. 14, © 1991 The American Physical Society.)

4.2.1 Nanocrystals

A nanocrystal (NC) is a single crystal having a diameter of a few nanometers. A NCQD is a nanocrystal that has a smaller band gap than the surrounding material. The easiest way to produce NCQDs is to mechanically grind a macroscopic crystal. Currently NCQDs are very attractive for optical applications because their color is directly determined by their dimensions (see Fig. 4.1). The size of the NCQDs can be selected by filtering a larger collection of NCQDs or by tuning the parameters of a chemical fabrication process.

4.2.1.1 CdSe nanocrystals

Cadmium selenide (CdSe) and zinc selenide (ZnSe) NCQDs are approximately spherical crystalites with either wurtzite or zinc-blend structure. The diameter ranges usually between 10 and 100 Å. CdSe NCQDs are prepared by standard processing methods.[15] A typical fabrication procedure for CdSe NCQDs is described in Ref. 16. $Cd(CH_3)_2$ is added to a stock solution of selenium (Se) powder dissolved in tributylphosphine (TBP). This stock solution is prepared under N_2 in a refrigerator, while tri-*n*-octylphosphine oxide (TOPO) is heated in a reaction flask to 360°C under argon (Ar) flow. The stock solution is then quickly injected into the hot TOPO, and the reaction flask is cooled when the NCQDs of the desired size is achieved. The final powder is obtained after precipitating the NCQDs with methanol, centrifugation, and drying under nitrogen flow. The room-temperature quantum yield and photostability can be improved by covering the CdSe NCQDs with, e.g., cadmium sulphide (CdS).

By further covering the CdSe NCQDs by CdS, for example, the room-temperature quantum yield and photostability can be increased. The almost ideal crystal structure of a NCQD can be seen very clearly in the TEMs shown in Fig. 4.5.

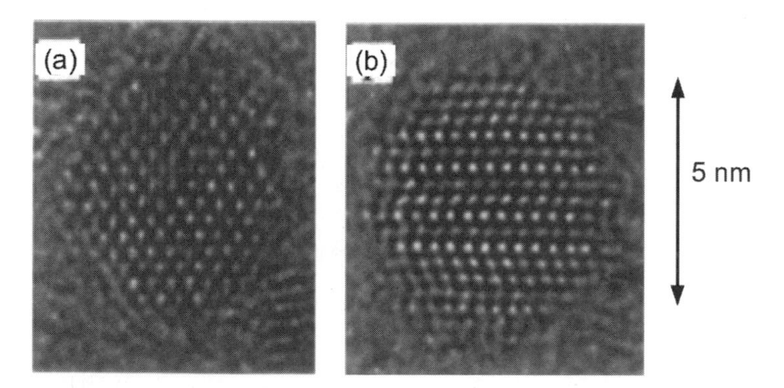

Figure 4.5 TEM images of CdSe/CdS core/shell NCQDs on a carbon substrate in (a) [001] projection and (b) [100] projection. Dark areas correspond to atom positions. The length bar at the right indicates 50 Å. (Reprinted with permission from Ref. 16, © 1997 American Chemical Society.)

Electron confinement in CdSe NCQDs is due to the interface between CdSe and the surrounding material. The potential barrier is very steep and at most equal to the electron affinity of CdSe. Even if the growth technique is fairly easy, it is very difficult to integrate single NCQDs into semiconductor chips in a controlled way, whereas the possibility to use them as biological labels or markers is more promising.[2]

4.2.1.2 Silicon nanocrystals

Silicon/silicon dioxide (Si/SiO_2) NCQDs are Si clusters completely embedded in insulating SiO_2.[17] They are fabricated by ion-implanting Si atoms into either ultra-pure quartz or thermally grown SiO_2. The NCs are then formed from the implanted atoms under thermal annealing. The exact structure of the resulting NCQDs is not known. Pavesi et al.[17] reported successful fabrication of NCQDs with a diameter around 3 nm and a NCQD density of 2×10^{19} cm^{-3}. The high-density results[17] in an even higher light wave amplification (100 cm^{-1}) than for seven stacks of InAs QDs (70 to 85 cm^{-1}). The main photoluminescence peak was measured at $\lambda = 800$ nm. The radiative recombination in these QDs is not very well understood, but Pavesi et al. [17] suggested that the radiative recombinations take place through interface states. Despite the very high modal gain, it is very difficult to fabricate an electrically pumped laser structure of Si NCQD due to the insulating SiO_2.

4.2.2 Lithographically defined quantum dots

4.2.2.1 Vertical quantum dots

A vertical quantum dot (VQD) is formed by either etching out a pillar from a QW or a double barrier heterostructure[18,19] (DBH). Figure 4.6 shows the main steps in the fabrication process of a VQD. The AlGaAs/InGaAs/AlGaAs DBH was grown epitaxially, after which a cylindrical pillar was etched through the DBH. Finally, metallic contacts were made for electrical control[19] of the QD. Typical QD dimensions are a diameter of about 500 nm and a thickness of about 50 nm. The confinement potential due to the AlGaAs barriers is about 200 meV. The optical quality of VQDs is usually fairly poor due to the etched boundaries. However,

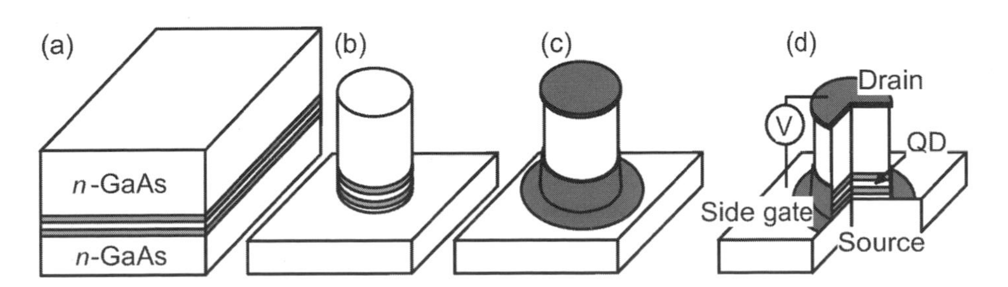

Figure 4.6 Fabrication process of a VQD consists mainly of (a) epitaxial growth of a BDH, (b) etching of a pillar through the DBH, and (c) the metallization (following Ref. 19). The QD of the device is defined by the DBH and the side gate. (d) The final device.

VQDs are attractive for electrical devices because of the well-controlled geometry and the well-defined electrical contacts.

4.2.2.2 Si quantum dots

Si QDs discussed here are lithographically defined Si islands either completely isolated by SiO_2 or connected to the environment through narrow Si channels. Si QDs can be fabricated using conventional CMOS technology on a silicon-on-insulator (SOI) wafer. The SOI wafer enables complete electrical isolation from the substrate. Figure 4.7 shows schematically the fabrication process[20] of Si QDs. A narrow wire is etched using electron beam lithography from the top Si layer. The QD is then formed in the wire by thermal oxidation. The oxidation rate is sensitive to the local O_2 influx and the local strain field. Both depend strongly on the geometry and, as a result, the center of the Si wire is oxidized very slowly compared to the rest. Therefore, the oxidation process gives rise to constrictions pinning off the wire from the leads, resulting in a Si QD in the center of the wire. This technique has been developed[21] further to fabricate double QDs and even memory and logical gate devices.[22] The main advantage of this technique is the easy integration to CMOS circuits. Si QDs do also have the potential to operate at room temperature due to very high carrier confinement ($V_C \approx 3$ eV) and small size. However, these Si QDs cannot be used for optical applications due to the low quantum efficiency of Si.

4.2.3 Field-effect quantum dots

In a FEQD, the charge carriers are confined into a 2D electron gas (2DEG) by a modulation-doped heterojunction. Within the 2DEG plane, the charges are electrostatically confined by external gates. Figure 4.8(a) shows schematically a typical device geometry, whereas Fig. 4.8(b) represents a more sophisticated double QD system. The ohmic contacts in Fig. 4.8(a) represent any kind of electric contacts to the QD. The effective potential of a FEQD is very smooth and, within the plane

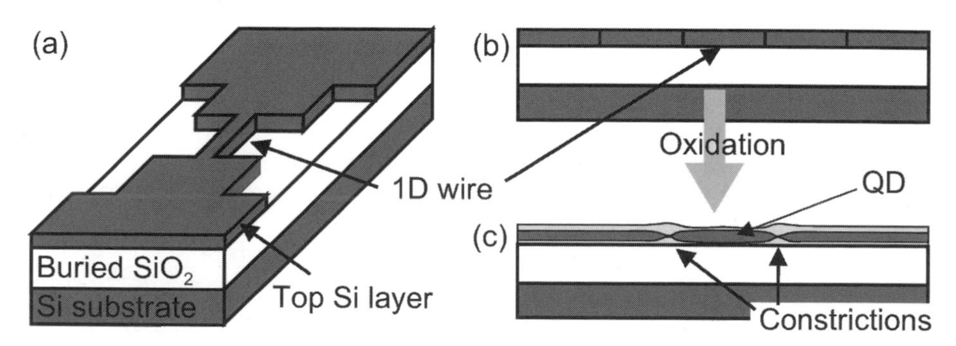

Figure 4.7 Fabrication process of a Si QD: (a) bird's eye perspective of a narrow Si wire, etched from the top Si layer of a SOI wafer; (b) cross section along the center of the wire; and (c) during thermal oxidation of the structure, the center of the wire is pinned off from the top Si layer. The result is a QD in the wire.

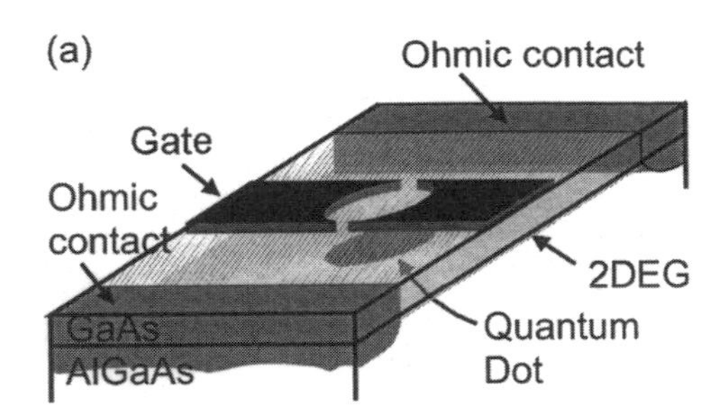

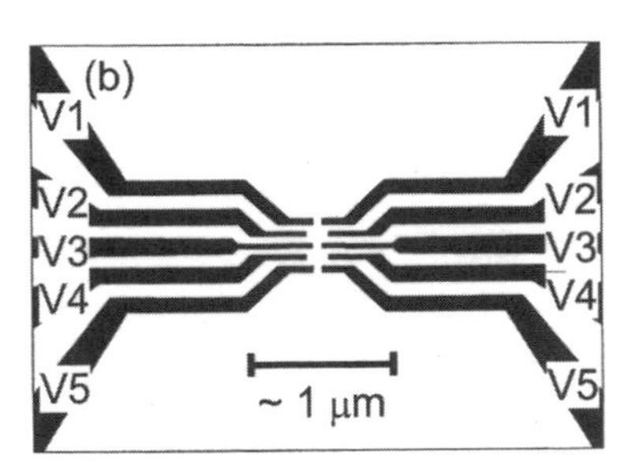

Figure 4.8 Field-induced quantum dots. (a) A schematic drawing of a FEQD in a 2DEG at the material interface between AlGaAs and GaAs. The ohmic contacts represents any electric contacts to the QD. (b) Schematic drawing of top gates of a double QD device. By using several gates, one can set the tunneling barriers (V1 and V5), the interdot tunneling coupling (V3) of multiple QDs, the number of electrons and energy levels in each QD (V4). (Reprinted with permission from Ref. 23, © 2001 The American Association for the Advancement of Science.)

of the 2DEG, its shape is close to a parabola depending on the gates. For a FEQD having a diameter around 200 nm, the spacing of the energy level is typically[23,24] tens of micro eV. These types of QDs are not expected to operate at room temperature because of the shallow potential profile. However, FEQDs are attractive for low-temperature infrared light detectors because of a very smooth gate-induced potential and high-quality heterostructure interfaces.

4.2.4 Self-assembled quantum dots

In self-assembly of QDs, one makes use of an island formation in epitaxial growth. The effect is similar to the formation of water droplets on a well-polished surface. The islands can either be QDs themselves or induce QDs in a nearby QW. The major self-assembly growth techniques are vapor phase epitaxy (VPE) and MBE.

Generally, the epitaxial growth proceeds in atomic layer-by-layer mode. However, islands are formed if there is a large lattice mismatch between the materials and/or if the surface energy of the deposited material is different from that of the substrate. The deposited material minimizes its potential energy by forming islands on the substrate. In the Stranski-Krastanow (S-K) mode, the growth starts in layer-by-layer mode and proceeds into the island mode after exceeding a critical thickness (see Fig. 4.9). Dislocation-free S-K growth has been observed in, e.g., InAs on GaAs[25] and InP on GaAs.[26] Typical island densities are 10^9 to 10^{12} cm^{-2}, depending on the growth conditions. Self-organized growth of III–V semiconductors is currently the most promising fabrication technique of optically active QDs.

4.2.4.1 Quantum dot island

The self-assembled island is a QD itself if the island is embedded in a material with a larger band gap than that of the island material. An example is provided by InAs islands in GaAs. Figure 4.10 shows QD islands schematically and a high-resolution scanning tunneling micrograph of a true InAs island. Very promising

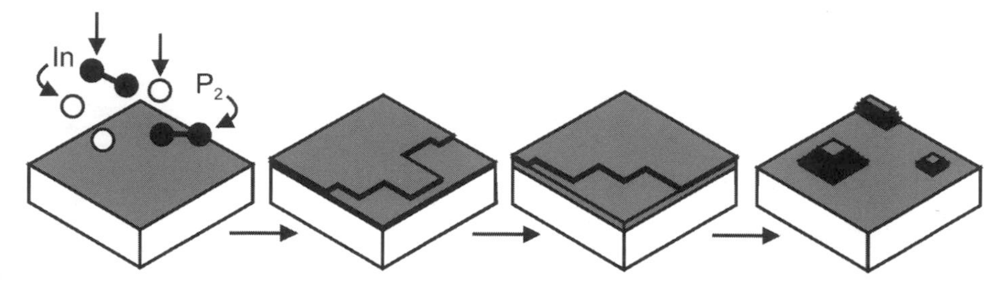

Figure 4.9 In the S-K mode, the growth of QDs starts in atomic layer-by-layer growth, but when the thickness of the overgrowth layer exceeds a critical thickness, islands begin to form.

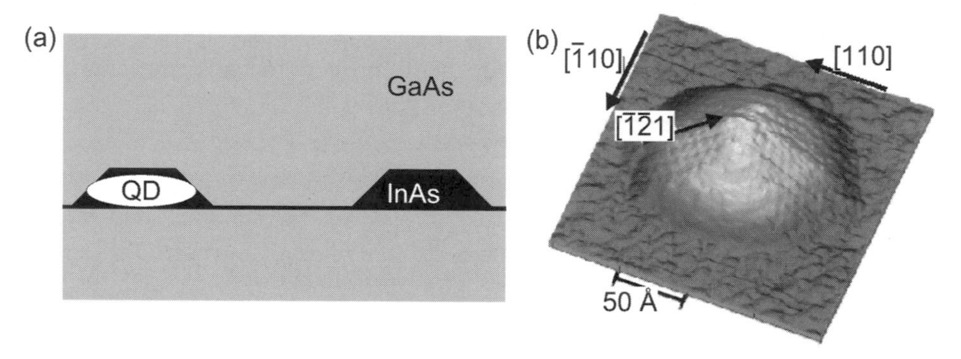

Figure 4.10 (a) Schematic image of an InAs QD island embedded in GaAs and (b) *in situ* STM image, from Ref. 29, of an InAs island.

laser structures have been fabricated using these types of quantum dots by stacking several island layers on top of each other.[27] Typical QD heights range from 5 to 15 nm and widths range from 15 to 25 nm. This means that there are very few electrons and holes per QD. The total charge confinement is a combination of strain, piezoelectric fields, and material interface effects. For a dot of 13.6 nm height, the calculated confinement energy of the electron ground state is about 180 meV.[28]

4.2.4.2 Strain-induced quantum dots

Strain is always present in self-assembled islands as well as in the substrate close to the island. The magnitude of the strain depends on the lattice constants and elastic moduli of the materials. If there is a QW close to the quantum dot, the strain field penetrates it also and affects its energy bands. The island can therefore induce a lateral carrier confinement in the QW. This results[26] in a total QD confinement in the QW. Typical stressor island heights range from 12 to 18 nm and the QW thickness is around 10 nm.[30] The lateral strain-induced confinement is very smooth and has the shape of a parabola. The strain-induced electron confinement is about 70 meV deep.[31] The resulting QD is pretty large and contains in general tens of electron-hole pairs. Figure 4.11(a) shows schematically a strain-induced QD and Fig. 4.11(b) shows a transmission electron micrograph (TEM) of self-assembled InP islands on GaAs.

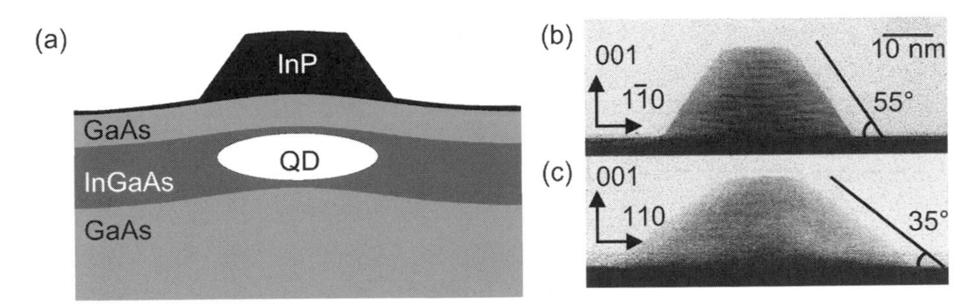

Figure 4.11 (a) A QD induced in an InGaAs QW by an InP island and (b) TEM images of an InP island on a (001) GaAs substrate (after Georgsson et al.[30]).

4.3 QD spectroscopy

Even if QDs are called artificial atoms, the rich structure of the atomic spectra is not easily accessible in QD spectroscopy. Unlike atoms, the QDs are not all identical, which gives rise to inhomogeneous broadening. When several millions of QDs are probed simultaneously, the spectral lines become 10 to 100 times broader than the natural linewidth. The details of optical spectra or electronic states can be seen only in single QD measurements.

Two microscopic methods developed for single QD spectroscopy are discussed in this section. In addition, an experimental setup that combines interband (transition *between* the valence and conduction bands) optical excitation and intraband (transitions *within* the conduction or valence bands) far infrared (FIR) excitation of QDs is described briefly.

4.3.1 Microphotoluminescence

In a microphotoluminescence (μPL) experiment, the sample is photoexcited with a laser beam focused by a microscope. The same microscope can also be used to collect the luminescence in the far field mode, see Fig. 4.12(a). The resolution of μPL is limited by conventional ray optics to a few microns. Measurements can be made both by continuous wave (cw) and pulse excitation. In the latter case, either photon-counting electronics or a streak camera is used for photon detection. μPL often requires appropriate preprocessing of the QD sample. Etching form pillars (mesas) with a diameter smaller than the resolution of the microscope reduces the number of photoexcited QDs so that eventually only a single QD is probed.

A typical μPL of InGaAs self-assembled QDs (SAQDs) is shown in Fig. 4.13. In this particular measurement, a high resolution is obtained more easily by preprocessing the mesa than by focusing the excitation or detection. Figure 4.13(a) shows μPL spectra recorded from mesas of different sizes. In the large mesas, there are several QD of different sizes, which gives rise to different luminescence energies. Figure 4.13(b) shows PL from a mesa having only one QD. The intensity of the excitation increases from bottom to top. The bottom panel shows PL (feature X)

from a single exciton (a complex of electron and hole coupled by the Coulomb interaction) confined in the ground state of the QD. This is called a ground state exciton, which means that the exciton wave function is governed by the single-electron and single-hole ground state orbitals. At higher excitation intensities, a biexciton line (feature X_2) appears. It comes from the decay of a bound exciton-exciton pair into a photon and an exciton. The energy difference between the exciton and biexciton lines is the biexciton binding energy $\Delta X_2 = 3.1$ meV. The biexciton binding energy in bulk GaAs is 0.13 meV. The order-of-magnitude increase of the binding energy comes from enhanced correlation effects as the many-particle system is squeezed by the confinement potential.

At still higher excitation intensities another feature X_2^* is found. It is related to the higher exciton state in which the excited single-particle orbitals dominate the

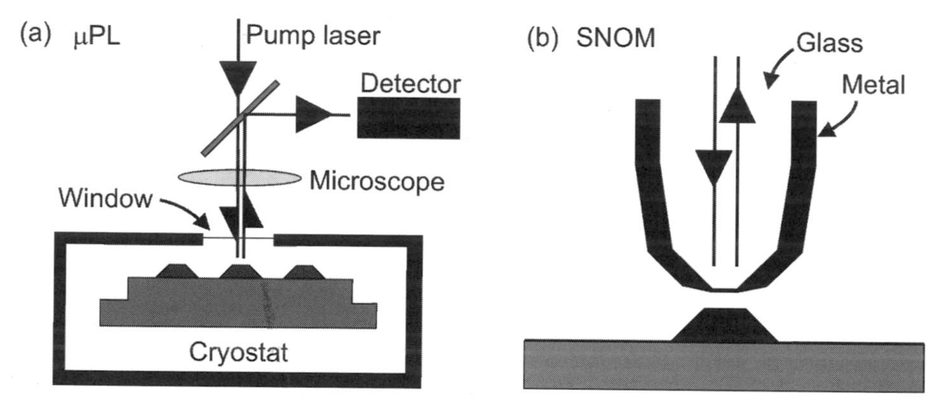

Figure 4.12 Experimental setup used in the (a) μPL and (b) scanning near-field optical microscope (SNOM) measurements. Both setups can be used in cw and time-resolved modes. Generally, either wide-area excitation or detection is used in the measurements. The resolution of the μPL is ~1 μm and the resolution of the SNOM ~ 100 nm.

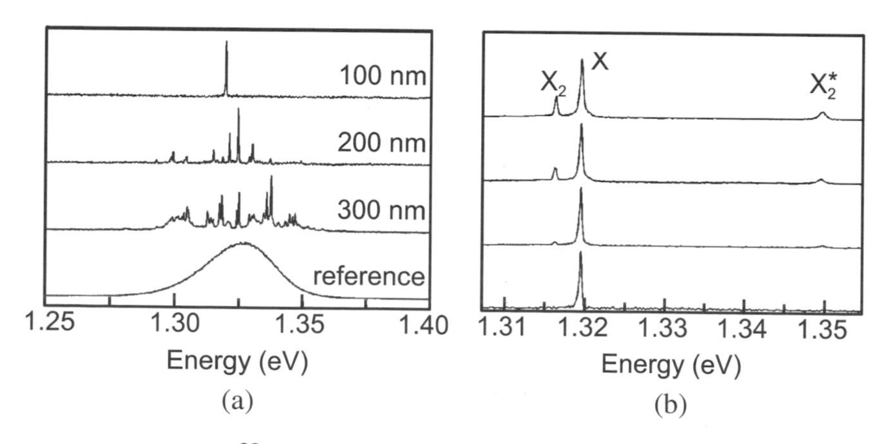

Figure 4.13 μPL spectra[33] of InGaAs SAQDs. (a) Spectra measured from mesas having diameters from 100 to 300 nm. In the larger mesas (diameters 200 and 300 nm), the spectrum is a superposition of emission coming from several QDs. (b) μPL from a mesa having only one QD. The intensity of excitation increases from bottom to the top panels.

exciton wave function. In general, the appearance of higher energy lines is related to a phenomenon called Pauli blocking[26] common to all fermion systems.

4.3.2 Scanning near-field optical spectroscopy

Another optical microprobe, called the scanning near-field optical microscope (SNOM) is depicted in Fig. 4.12(b). Note that the fiber aperture is much smaller than the excitation or emission wavelength. Excitation takes place by an evanescent electromagnetic wave. The resolution enables single QD scanning of unpatterned samples. Figure 4.14(a) shows a surface scan (energy integrated PL intensity) of an InGaAs QD sample. Emission from individual QDs is clearly visible. In Fig. 4.14(b), the energy spectrum of the surface scan has been analyzed and various excited states recognized for each dot.

As a case study, we finally discuss PL excitaton simultaneously with FIR radiation from a free electron laser (FEL) and with an argon ion laser. The experiment was done at the University of California Santa Barbara[35] (UCSB). SAQDs induced by self-organized InP islands on top of a near surface GaAs/InGaAs QW are studied in this experiment. Several millions of QDs were probed simultaneously. The experimental setup is shown in Fig. 4.15. The sample is pumped simultaneously with FIR and blue light. The wavelength of the FIR light from the FEL can be tuned to resonance with the intraband transition of electrons or holes. This gives rise to intraband resonance absorption or emission of FIR radiation. The recorded luminescence spectra are shown in Fig. 4.16. The emission from the ground state is enhanced and the emission from the higher excited states quenched when the FEL is turned on. The results are in contrast to similar measurements of QW photoluminescence where the luminescence is blue-shifted when FEL is turned on. The experimental result still lacks a theoretical explanation.

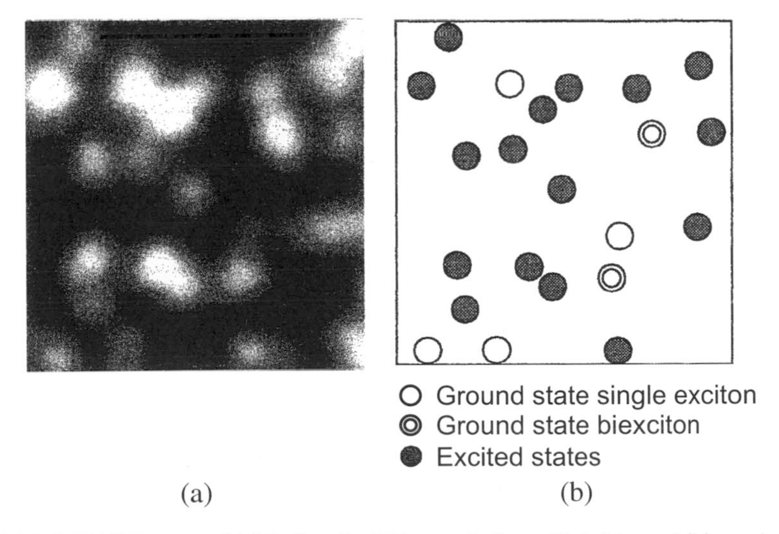

Figure 4.14 A SNOM scan of (a) InGaqAs QD sample from Ref. 34, and (b) analysis of the energy spectrum used to identify the various excited states in individual QDs.

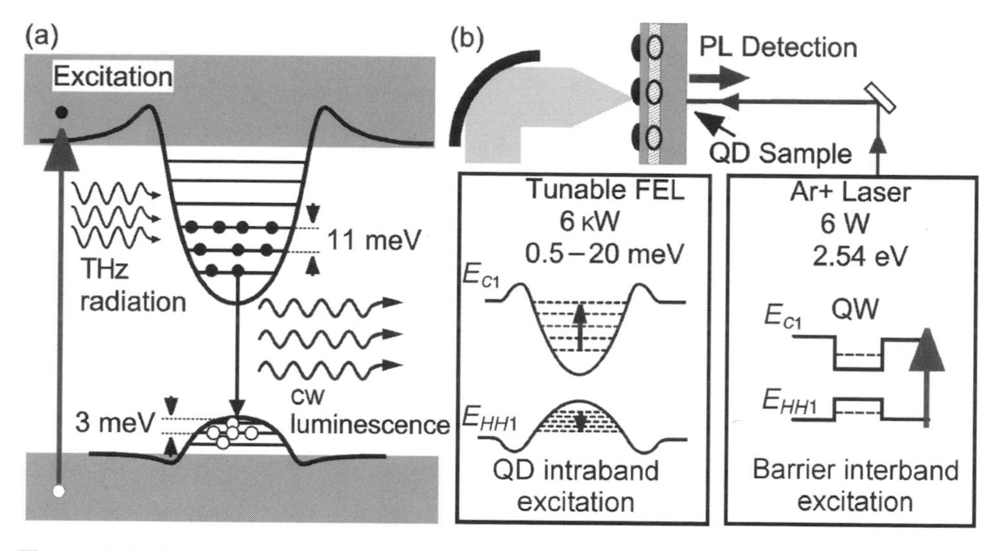

Figure 4.15 Photoluminescence excitation by a FIR FEL (tetrahertz radiation) and an Ar^+-ion laser.[35] The FIR frequence can be tuned to intraband resonance of electrons and holes. The PL spectrum is measured with and without FIR pumping.

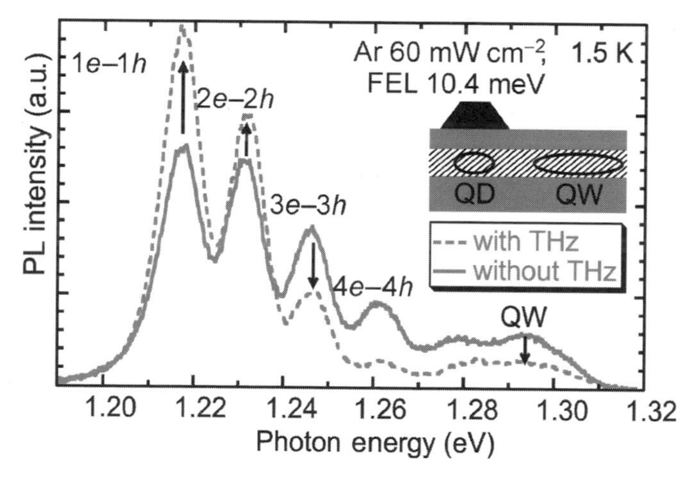

Figure 4.16 Luminescence spectrum of strain-induced QDs measured with and without tetrahertz (FIR) radiation.[35]

4.4 Physics of quantum dots

In a QD, the charge carriers occupy discrete states, just like the electrons of single atoms. Therefore, QDs are also referred to as *artificial atoms*. The atomic features result from 3D confinement. The most important physical principles and technologically relevant features of QDs include electron states, transitions between these, the influence of external electromagnetic fields, charge transport, and dynamics.

4.4.1 Quantum dot eigenstates

The carriers confined in QDs interact strongly with the surrounding material. This interaction depends on the surrounding material and it is distinctively different for metals and semiconductors. Hence, it is very difficult to formulate a general many-electron theory of QDs. In electronic QD devices such as single electron transistors (SETs), the electron addition spectrum is clearly dominated by the Coulomb effect.[36] However, in optical devices, charges are added in charge-neutral electron-hole pairs (excitons), and their energy spectrum is more complex.

Neglecting the coupling with the surrounding material, the electron-hole many-body Hamiltonian for a QD[37] can be written very generally:

$$\begin{aligned} H_{mb} = & \sum_i E_i^e \hat{c}_i^\dagger \hat{c}_i + \sum_i E_i^h \hat{h}_i^\dagger \hat{h}_i - \sum_{ijkl} \langle ij \hat{V}_{eh} kl \rangle \hat{c}_i^\dagger \hat{h}_j^\dagger \hat{c}_k \hat{h}_l \\ & + \frac{1}{2} \sum_{ijkl} \langle ij \hat{V}_{ee} kl \rangle \hat{c}_i^\dagger \hat{c}_j^\dagger \hat{c}_k \hat{c}_l + \frac{1}{2} \sum_{ijkl} \langle ij \hat{V}_{hh} kl \rangle \hat{h}_i^\dagger \hat{h}_j^\dagger \hat{h}_k \hat{h}_l . \end{aligned} \tag{4.1}$$

The operators $\hat{c}_i^\dagger$ ($\hat{c}_i$) and $\hat{h}_i^\dagger$ ($\hat{h}_i$) are the electron and hole creation (annihilation) operators of the state $|i\rangle$, respectively. The total energy is, hence, a sum of the single particle energies E_i^e and E_i^h (electron and hole) and the two-body Coulomb interactions: the electron-hole (eh), the electron-electron (ee), and the hole-hole (hh) interactions.

As an example, consider an axial symmetric QD disk with a harmonic lateral potential. The general single-particle Hamiltonian is

$$H^{(e/h)} = \frac{\mathbf{p}^2}{2m^*_{(e/h)}} + \frac{1}{2} m^*_{(e/h)} \omega^2 r^2 + H_E + H_{DM} + H_Z + H_\sigma , \tag{4.2}$$

where $m^*_{(e/h)}$ is the effective mass of an electron or hole, H_E is the electric field term, and $H_{DM} + H_Z + H_\sigma$ are the magnetic terms. The height of the QD is much smaller than its radius, and we can therefore neglect the vertical dimension in the lowest energy states. The zero-field single-particle energies are $E_i^{(e/h)} = \hbar\omega_+^{(e/h)}(n + 1/2) + \hbar\omega_-^{(e/h)}(m + 1/2)$ and the angular momenta $L_i^{(e/h)} = m - n$. The eigenstates of Eq. (4.2) are then characterized by the quantum numbers $n \in \{0, 1, 2, \ldots\}$, $m \in \{0, 1, 2, \ldots\}$, and the spin $\sigma \in \{\uparrow, \downarrow\}$. The ground state ($s$-band) of the Hamiltonian is doubly degenerate due to the spin while the first excited state (p-band) is fourfold degenerate. As a consequence, the conduction (valence) s-band can contain at maximum two electrons (holes) with opposite spin configuration. These analytic results show that the density of states of a QD differs remarkably from the parabolic DOS of a bulk semiconductor. Figure 4.2 shows schematically the DOS diagrams of a bulk semiconductor, a QW, and a QD. The band gap is present in all three DOS diagrams. However, the QD DOS consists only of sharp peaks determined by Eq. (4.1).

4.4.2 Electromagnetic fields

In the presence of an electromagnetic field defined by the vector potential $\mathbf{A}$ and the scalar potential ϕ, the following replacements are required in the Hamiltonian:

$$\begin{cases} \mathbf{p} \rightarrow \mathbf{p} - q\mathbf{A} \\ H \rightarrow H + q\phi, \end{cases} \tag{4.3}$$

where

$$\begin{cases} \mathbf{E} = \dfrac{\partial \mathbf{A}}{\partial t} - \nabla\phi \\ \mathbf{B} = \nabla \times \mathbf{A}. \end{cases} \tag{4.4}$$

The definition of the electromagnetic field is, however, ambiguous and the exact gauge can be chosen according to the prevailing situation.[38] Furthermore, the spin splitting is included by the following term:

$$H_\sigma = g\mu_B \boldsymbol{\sigma} \cdot \mathbf{B}, \tag{4.5}$$

where g is called the g factor ($g = 2$ for free electrons) and $\mu_B = e\hbar/(2m_0c)$ is the Bohr magneton. Generally in a semiconductor, the hole states are Luttinger spinor states and the exact effect of the magnetic field on them is, in general, more complicated than on the electron band spin states.

4.4.2.1 Quantum-confined Stark effect

The effect of an electric field on a quantum-confined state is called the quantum-confined Stark effect. An electric field is accounted for in the Hamiltonian through the term

$$H_E = -q\mathbf{r} \cdot \mathbf{E}. \tag{4.6}$$

This does not change the degeneracy of the s-states; however, the p-states are split into two different energy levels. The Stark effect on confined QD states is shown schematically in Fig. 4.17(a). The effective band gap and the exciton binding energy are reduced by the electric field. The electric field can either separate the electron and hole wave functions or bring them closer together. Therefore, the recombination rate W_{if} is affected, since it is proportional to the wave function overlap. The quantum-confined Stark effect on the exciton binding energy has been very nicely demonstrated by several groups.[39,40] Findeis et al.[39] measured the binding energy as a function of the field strength by studying a single self-assembled InGaAs QD; see Fig. 4.17(b).

4.4.2.2 Magnetic field

In the single-particle Hamiltonian, the magnetic field is included through Eq. (4.3). In general, a constant magnetic field ($\mathbf{B} \parallel z$) can be accounted for through three

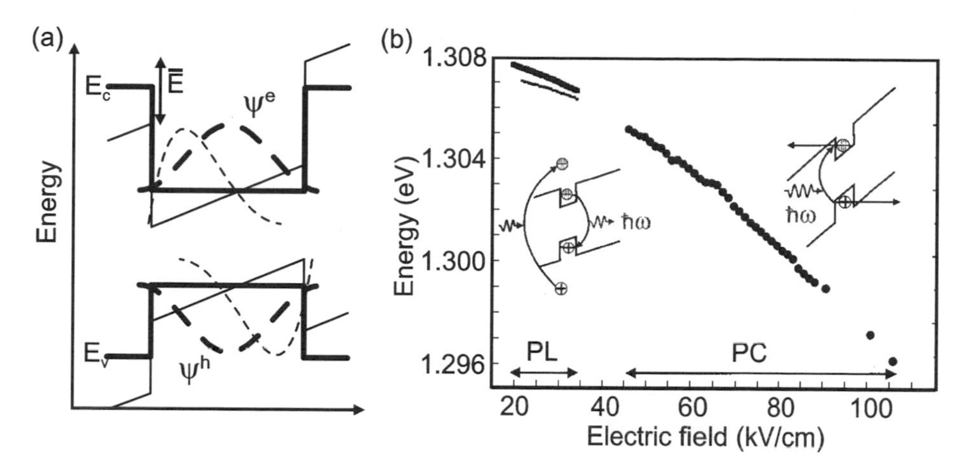

Figure 4.17 (a) The effect of a electric field on excitons of a QD. The black lines correspond to $E=0$ and the gray lines to $E\neq 0$. (b) Line position of the neutral QD ground state as function of the electric field measured in photoluminescence (PL) and photocurrent (PC) spectroscopy from Ref. 39.

different terms. These are the diamagnetic shift

$$H_{DM}^{(e/h)} = -\frac{\overline{\omega}^2 r^2}{2m^*_{(e/h)}}, \tag{4.7}$$

the Zeeman shift

$$H_Z^{(e/h)} = -\frac{eB\hat{L}_z}{2m^*_{(e/h)}}, \tag{4.8}$$

and the spin-splitting H_σ of Eq. (4.5). In Eqs. (4.7) and (4.8), the frequency $\overline{\omega} = eB/2$, L_z is the z component of the angular momentum ($L_z = m\hbar$), and $m_r^{(e/h)}$ is the relative effective mass of an electron (hole). The Zeeman term and the spin-splitting do not influence the single-particle orbitals, and can be added directly to the many-body Hamiltonian.

In a magnetic field perpendicular to the cylindrical QD, the single-particle energies (omitting the spin-splitting and the band gap energy E_g) are

$$E_i^{(e/h)}(\mathbf{B}) = \hbar\Omega_+^{(e/h)}\left(n+\frac{1}{2}\right) + \hbar\Omega_-^{(e/h)}\left(m+\frac{1}{2}\right), \tag{4.9}$$

with the frequencies $\Omega_\pm^{(e/h)} = 1/2(\sqrt{\omega_c^2 + 4\omega_0^2} \pm \omega_c)$, where $\omega_c = qB/m^*_{(e/h)}$ is the cyclotron frequency. Not only the single-particle energies are affected but also the wave functions. Many of the most important magnetic effects, such as the spin-splitting, Zeeman effect, diamagnetic shift, and the formation of Landau levels in high magnetic fields predicted by the single-particle theory have been confirmed experimentally.[33,41]

4.4.3 Photonic properties

The photonic properties of a QD depend completely on the exciton states, which reflect the geometry and symmetry of the QD. Electrons can lose or gain only energy quanta, equal to the difference between two exciton states; i.e., only transitions between the discrete eigenstates are allowed due to energy conservation. This effect is seen in the form of peaks in the QD photoluminescence [see Fig. 4.18(a)]. The peaks are broadened by inhomogeneous broadening, which is due to the size distribution of a large QD ensemble. Figure 4.18(a) shows the PL of SAQDs after photoexcitation at different intensities. We see that when the excitation intensity is small, only the ground states become populated. However, for higher excitation intensity, the lowest states become filled (Pauli blocking), the PL from these peaks saturates. As a result, higher excitation intensity leads to occupation of higher states and accordingly, more PL peaks appear above the ground state energy.

Figure 4.18(b) shows schematically the lowest optical transitions in a single QD. In the neutral ground state of an intrinsic and unexcited QD, all states below the band gap are filled with electrons and all states above the band gap are empty [lower right of Fig. 4.18(b)]. By absorbing a photon, an electron can be excited across the band gap, thereby leaving behind a hole. We call the electron-hole pair an exciton X_1 [left part of Fig. 4.18(b)]. If there are two electron-hole pairs, we call it a biexciton X_2 [upper right of Fig. 4.18(b)]. Here X_1 is fourfold degenerate, but only two configurations are optically addressable, because a photon transition from X_0 to the other two states would not conserve the total angular momentum. The optically inactive states are called dark states. The polarization (σ^+ or σ^-) and energy ($\hbar\omega$) of the absorbed photon defines which of the degenerate states will be created, i.e., for the transition $X_1[\uparrow\downarrow] \rightarrow X_2[\uparrow\downarrow\downarrow\uparrow]$ a σ^+ polarized photon

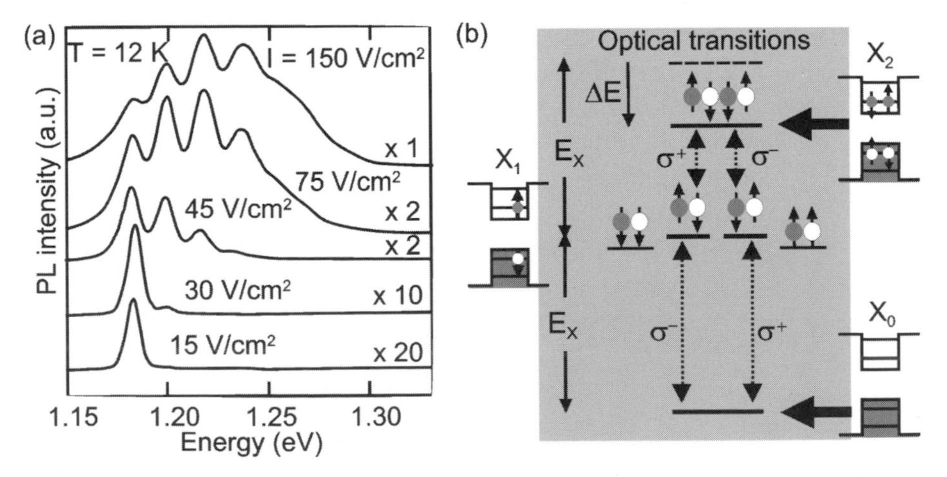

Figure 4.18 (a) Photoluminescence of an ensemble of strain-induced quantum dots at $T = 12$ K (Ref. 26). When increasing the excitation intensity, more and more states become populated starting with the states lowest in energy for low excitation. (b) Optical transitions in a single QD between the ground state X_0, the one exciton state X_1, and the biexciton state X_2. (Reprinted with permission from Ref. 26, © 1995 The American Physical Society.)

with the energy $\hbar\omega = E_X - \Delta E$ will be required [the upper left dotted arrow of Fig. 4.18(b)]. If the QD is doped or there is an electrically injected electron or hole in the QD, we call it a charged ground state X_0^- or X_0^+. The excited state of X_0^- (X_0^+) will also be charged, since a photoexcitation does not charge the QD.

The radiative transition rate or the probability that a photon is either emitted or absorbed by an exciton can be calculated by perturbation theory.[42] Fermi's golden rule gives the transition probability per time which is equal to the reciprocal of the radiative life time; thus

$$\frac{1}{\tau_r} = \frac{2\pi}{\hbar^2} |\langle f | H_{\text{ED}} | i \rangle|^2 \delta(E_i - E_f \pm h\omega), \tag{4.10}$$

where + and − correspond to photon absorption and emission, respectively; and $h\omega$ is the photon energy, which has to be equal to the difference between the energy of the initial $|i\rangle$ and the final excitonic state $|f\rangle$. The first-order electrodynamic perturbation is

$$H_{\text{ED}} - \frac{q}{m_0} \mathbf{A} \cdot \mathbf{p}, \tag{4.11}$$

where $\mathbf{A}$ is the vector potential of the radiation field, and $q = -e$ is the electron charge.

In the envelope wave function picture, the overlap integral in Eq. (4.10) becomes a sum of two terms, one term proportional to the Kane's optical transition matrix element for bulk material and one term the vector product of the envelope function. The former term dominates in interband transitions and the latter in intraband transitions.

4.4.4 Carrier transport

A rich variety of QD transport phenomena has been studied experimentally and theoretically. Here we discuss two limiting cases: ballistic transport and the Coulomb blockade regime. Ballistic transport is observed when the phase of the electron is preserved while it moves in a conducting channel trough a QD. In this regime, the conductance is described in the low-bias-voltage limit by the Landauer-Büttiker formula[43] and the conductance is quantized in units of $2e^2/h$. In ballistic transport the QD confinement potential gives rise to potential resonances in the transmission. When QD states are occupied by one or more electrons, interference between the elastic and inelastic scattering channels gives rise to interference effects known as Fano profiles.[44] Both of these resonance phenomena are well known from atomic collision physics.

Future applications of QDs in electronic devices are likely to utilize transport in the Coulomb blockade regime. The energy needed to place an additional electron on a QD (addition energy) is analogous to the electron affinity of atoms. The electron addition spectrum is mainly characterized by the Coulomb energy, which is much greater than the lowest excitation energies of the electrons on the QD. The addition energy can be measured with a single electron transistor[36,45] (SET).

Figure 4.19 shows the operation principles of an SET. The QD is separated from two leads by two high-potential barriers, and the energy levels within the QD can be shifted by a gate electrode. When the SET is biased with a voltage V_B, current starts to flow through the QD if and only if there is an electron state in between the chemical potentials μ_L and μ_R of the left and the right leads. The electrons tunneling into the dot have energy $E = \mu_L$ and can therefore only occupy states below μ_L. Analogously, the tunneling electron at the QD with energy $E = \mu_{N+1}$ can tunnel to the right lead only if $\mu_{N+1} > \mu_R$. This phenomenon is known as the Coulomb blockade. If a small bias V_B is applied to the SET and V_G is varied, clear current peaks appear as a function of V_G. Moreover, the peak separation is directly proportional to the addition energy[45]

$$\mu_{N+1} - \mu_N = e\frac{C_G}{C}\left(V_G^{N+1} - V_g^N\right), \tag{4.12}$$

where C_G is the gate capacitance; and $C = C_G + C_L + C_R$ is the total QD capacitance to the gate (C_G) and the left (C_L) and right leads (C_R). Room-temperature SETs have been fabricated.[20] However, the width of the current peaks tends to increase with temperature and QD size (see also Fig. 4.4).

Two QDs separated by potential barriers between the leads constitute a single electron pump. This device can be operated like a sluice gate in a water channel. With a single electron pump, one can control the current very accurately by cycling

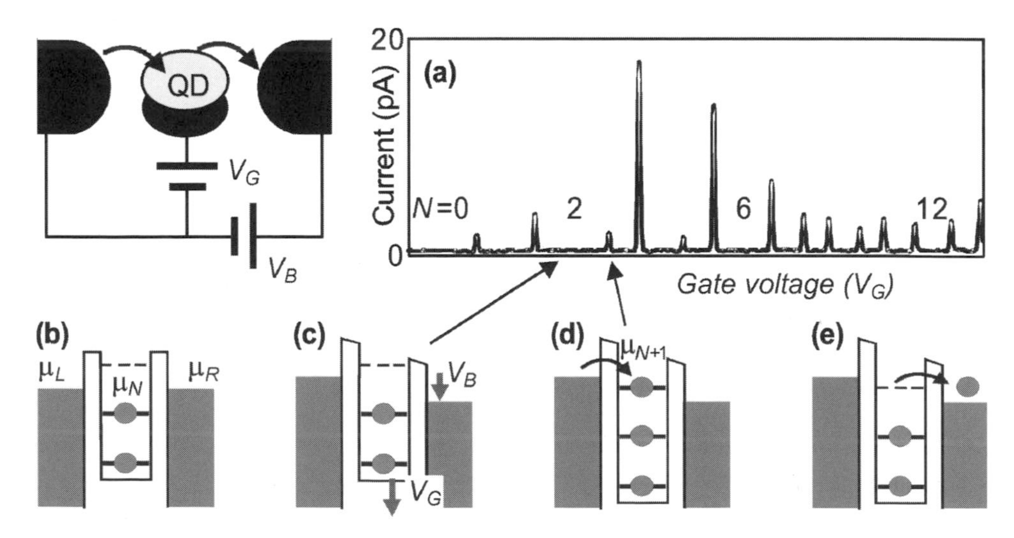

Figure 4.19 (a) Current though a QD SET as a function of the gate voltage from Ref. 36. The current peaks correspond to conducting QD states between the chemical potentials μ_L and μ_R of the left and right conductors, respectively. The inset shows the electrical scheme of the SET. The energy diagrams for are (b) a nonbiased SET, (c) an SET in the Coulomb blockade, and (d)–(e) conducting SETs. (Reprinted with permission from Ref. 36, © 1996 The American Physical Society.)

the gate voltages of the two QDs. In the ideal case, electrons are moved from one lead to the other, one by one. For more details on double QDs see Ref. 46. Technological applications of SETs are also discussed in Sec. 4.6.3.

4.4.5 Carrier dynamics

Intraband and interband relaxation of carriers in semiconductor QDs has been studied using femtosecond-pulse lasers and fast photon-detection electronics.[47,48] Figure 4.20 shows time-resolved PL spectra of a strain-induced SAQD sample[47] (see also Secs. 4.2.4.2 and 4.3.1). The main PL lines originate from transitions from the ith conduction level to the jth valence level, where $i = j$ for allowed dipole transitions. The 1-1 transition has the longest lifetime, while the luminescence from the QW and from higher excited states fade out much faster. The carriers relax quickly from the QW into the QD by Coulomb (carrier-carrier) scattering. Within the QD, carriers relax to lower energy levels by carrier-carrier and carrier-phonon interactions. For high carrier densities, the carrier-carrier scattering and Auger transitions always provide a very fast relaxation mechanism. For low carrier densities, the phonon-assisted relaxation dominates.

There are two phonon relaxation mechanisms; longitudinal optical (LO) and longitudinal acoustic (LA) phonon scattering. For the electrons the LA relaxation is a lot slower ($\tau^e_{\mathrm{LA}} \approx 10^3$ ns) than for holes ($\tau^h_{\mathrm{LA}} \approx 30$ ps), whereas the LO relaxation is very slow for both electrons and holes. The LO relaxation is slow because of the discrete QD DOS since a simultaneous energy and momentum conservation is impossible for LO relaxation. It is therefore theoretically predicted that the intraband relaxation, for low carrier densities, is much slower in QDs than in the corresponding bulk semiconductor. This phenomenon is called *phonon bottleneck*[49] and it was verified experimentally by Braskén et al.[50]

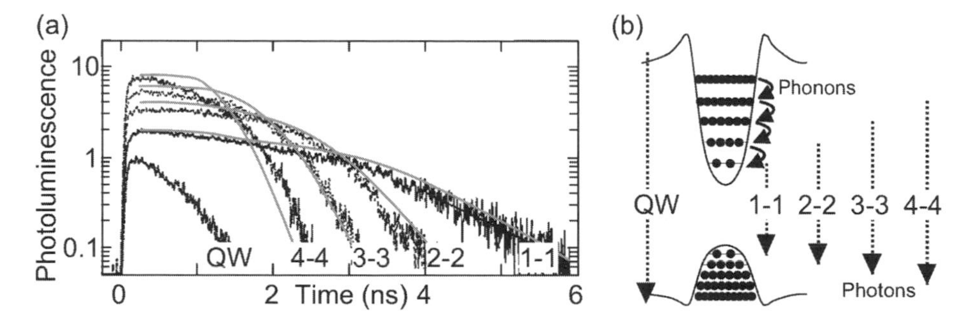

Figure 4.20 (a) Time-resolved PL spectra of strain-induced SAQDs. The black lines are measurements, while the gray lines are modeled energy level populations. The labels correspond to the initial conduction and final valence states. Reprinted with permission from Ref. 47. (b) Schematic energy diagram. The dotted arrows show the radiative transitions of (a) and the solid arrows represents the phonon-related intraband transitions. (Reprinted with permission from Ref. 47, © 1997 The American Physical Society.)

4.4.6 Dephasing

The phase of a quantum state evolves with time according to the time-dependent Schrödinger equation, however, several interactions can destroy the phase information of a state. The loss of phase information is called dephasing. The dephasing time τ_ϕ is influenced by both elastic and inelastic collisions. In bulk and QW semiconductors, τ_ϕ is usually taken to be the intraband relaxation time.

The dephasing time is equal to the polarization relaxation time, which is inversely proportional to the homogeneous linewidth Γ of an optical transition; i.e.,

$$\tau_\phi = \frac{2}{\Gamma}. \tag{4.13}$$

The dephasing time defines the time scale on which coherent interaction of light with medium takes place. Therefore it gives the ultimate time scale for realization of coherent control in a quantum system.

Elastic collisions can disturb the phase of the carrier wave functions in QDs without changing the populations of the carrier energy levels. Accordingly, the spectral lines are broadened. At room temperature, τ_ϕ can be of the order of hundreds of femtoseconds for typical QDs. The value is of the same order as for bulk and QWs. At low temperatures ($T < 50$ K) dephasing and spectral broadening is usually attributed to acoustic phonons. LO phonons can via a second-order process change the phase of the QD carriers without changing the carrier energies.[51]

Htoon et al. studied the relation between the dephasing time and the relaxation energy $E_{\rm rel}$ in an ensemble of QDs of different sizes.[52] The relaxation energy is defined as the energy difference between the initial excited state and the final state after the relaxation. For relaxation from the first excited states with $E_{\rm rel} \approx E_{\rm LO}$ the relaxation was very efficient ($\tau_\phi < 7$ ps). However, from the first excited states with relaxation energies $15 < E_{\rm rel} < 20$ meV, the relaxation times were very long (40 ps $< \tau_\phi <$ 90 ps). The relaxation time was still shorter than the radiative relaxation time and therefore, a quenching of the ground state PL intensity was not observed in the experiment. Moreover, higher excited QD states were found to have relaxation times of $\tau_\phi \leq 7$ ps. The wave functions of higher excited states overlap energetically and spatially continuum states of other QDs and the wetting layer.[52] Therefore, these states have also more accessible final states, which gives rise to a short relaxation time.

When the energy separation of the QD carrier states is much different from the phonon energies, the dephasing time increases. This is due to a smaller accessible phase space of elastic phonon scattering and the fact that the inelastic scattering becomes forbidden by energy conservation. Borri et al. report for InAs SAQDs dephasing times[53] as long as $\tau_\phi = 630$ ps. For quantum information applications (see Sec. 4.6.4) of QDs, the dephasing time has to be at least a few hundred picoseconds.

4.5 Modeling of atomic and electronic structure

Theory and computational modeling of QDs is required for interpretation of experimental data and for the development of predictive models of materials and devices based on QDs. Theories developed for calculation of the electronic and optical properties of QDs have been very successful in explaining most of the experimental data. Theoretical models describe QDs either at mesoscopic or atomic scale.

Mesoscopic-scale models account for changes in the material composition and material interfaces in terms of effective masses and other material parameters obtained for *bulk* semiconductors.[5] The multiband $\mathbf{k} \cdot \mathbf{p}$ method is an example of this approach. The particular advantage of mesoscopic models is that they relate the modification of the conduction and valence band edges in the QD directly to the changes in the wave function and thereby to changes in the electronic and optical properties as the materials or dimensions of a QD are varied.

Atomistic theories such as the pseudopotential method[54] and the tight binding approximation[55] also make use of parametric representation of the effective potential or parametrization of other atomic level electronic quantities. These methods are computationally more intensive and therefore limited to smaller systems.

4.5.1 Atomic structure calculations

Most nanometer-size semiconductor and metallic structures include strain fields that are due to processing of materials (e.g., oxidation of Si) or due to the lattice mismatch of overgrown materials. The strain induces a modification of the band edge through the displacement of ion cores and through the piezoelectric effect (in compound semiconductors). Depending on the origin of the strain, different theoretical methods are used in the modeling.

Strain fields created during in the oxidation of Si should be calculated from the general dynamic equations of visco-elastic fluids. The theoretical basis is well understood, but the values of the necessary material parameters describing the flow of SiO_2 at high oxidation temperatures are not known accurately.[56] The strain-induced deformation of the band edges is expected[56] to be very important in Si-based SET.

The modelling of strain due to lattice mismatch is more straightforward as long as the lattice remains coherent (strain is not relaxed by dislocation). In this regime, strain calculations can be made using either continuum elasticity (CE) approximation or by atomic elasticity (AE) such as the valence force field method.[57] Full-scale device models can be used in the CE approximation. These calculations are usually made using commercial finite-element software.[58] The atomic level models are computationally more intensive and limited to smaller systems. A comparison of CE and AE calculations of strain fields in InAs QDs is shown in Fig. 4.21. The results agree well except at the edges of the structure. The differences in the

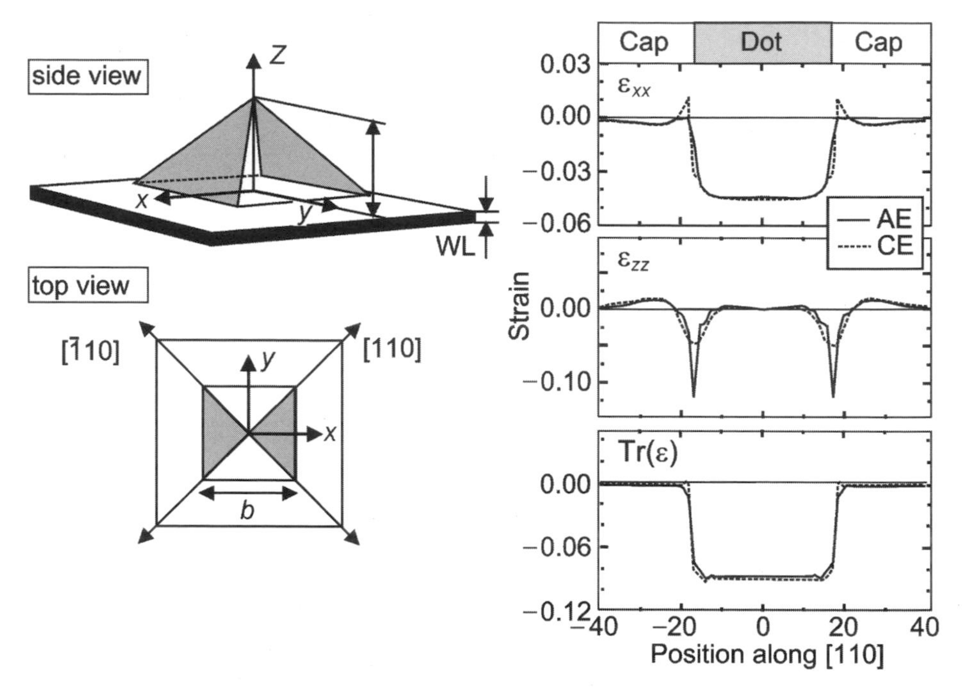

Figure 4.21 Comparison of strain distributions for an InAs QD calculated by AE and CE approximations: (a) the model and crystal orientation and (b) the diagonal strain components along the [110] direction. The solid line corresponds to the AE method and the dashed line to the CE method.[57]

calculated band structure modification are also small. The accuracy of the calculation depends more critically on the reliability of the deformation potentials that are required in calculation of the band edge deformation.

4.5.2 Quantum confinement

The theoretical methods used in electron structure calculation can be classified as follows. The *single-band effective mass model*, accounts for the solid only via the effective mass. This allows for a simple inclusion of the many body effects.[59] The effective mass approximation is reasonably accurate for metallic and compound semiconductor 2D electron gas structures. Correlation effects have been studied intensively using this method. Figure 4.22 shows[60] as an example the behavior of a many-electron system in a strong magnetic field. Coulomb interaction leads to reorganization of the electron density into a Wigner crystal. Theory has also predicted shell structures in QDs similar to those found in ordinary atoms. The existence of shell structures have been confirmed by experiments.[45]

The two-band effective mass model has been recently used by Braskén et al. to analyze electron-hole pair correlation in strain-induced[61] QDs. A strong pair correlation was found in systems including up to four electron-hole pairs. They found strong pair correlation up to systems including four electron-hole pairs. Unfortunately the *ab initia* direct diagonalization method used in Ref. 61 becomes

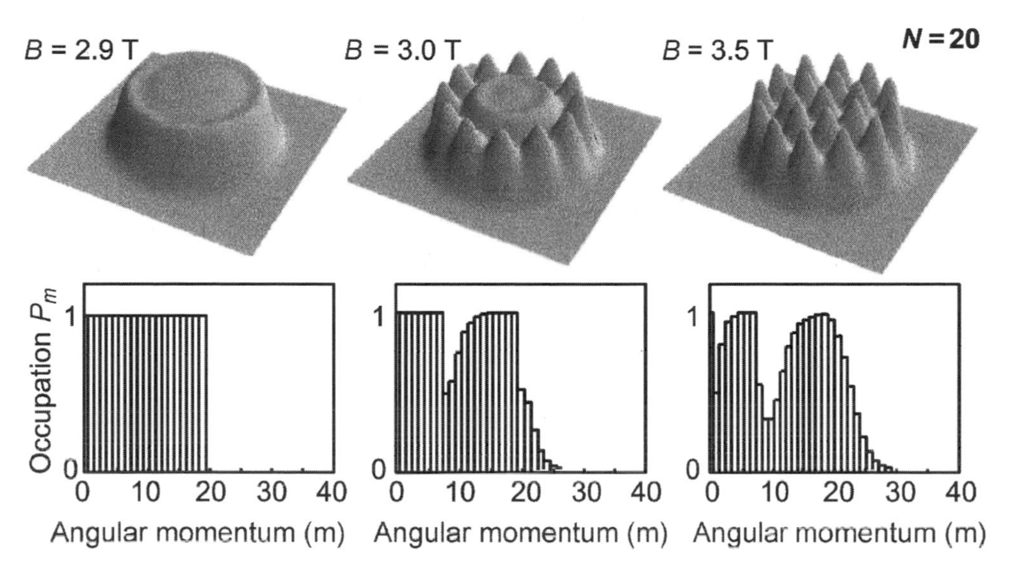

Figure 4.22 Self-consistent charge densities for a 20-electron GaAs QD for different values of the magnetic field; $B = 2.9$ T, $B = 3.0$ T, and $B = 3.5$ T. Below: Angular momentum occupations P_m. (Reprinted with permission from Ref. 60, © 1999 The American Physical Society.)

computationally very heavy for larger number of particles. Therefore, it is still unclear how much of the electron-hole quasi-particle (exciton picture) is left when a QD confines several tens of electrons and holes.

The *multiband methods* such as the *eight-band* $\mathbf{k} \cdot \mathbf{p}$ *method*, are required in the modelling of valence bands of compound semiconductors. However, many-body calculations based on this approach become computationally very demanding.[62] The *pseudopotential method* starts from the atomic-level effective potentials and enables calculation of systems including up to a few millions of atoms.[54] The *tight-binding approximation* can easily be generalized for electronic structure calculations of QDs. It is computationally heavy, since the calculation includes every single atom of the structure.[55] The pseudopotential method and the tight-binding method can be used to model the electronic structure down to atomic dimensions. These models are not limited to the near-band-edge regime as the effective mass and the related multiband models are. In compound semiconductors, one should include the Coulomb interaction between electrons and holes (excitonic effect) to obtain a reasonably accurate description of electron states. This part of the correlation energy can be included also in the pseudopotential and tight-binding calculations.

4.6 QD technology and perspectives

QDs are promising materials for nanotechnology devices. Thus far, biological markers made of NCQDs are the only commercial QD products available in the nanotechnology superstore. However, many more products are coming soon, including QD lasers and SETs.

4.6.1 Vertical-cavity surface-emitting QD laser

The vertical-cavity surface-emitting QD laser (QDVCSEL) is being studied actively in several research laboratories[63] and it could be the first commercial QD device. The room-temperature continuous wave operation has been demonstrated under laboratory conditions by several groups and there is a strong activity to commercialize the QDVCSEL. Figure 4.23 depicts a typical QDVCSEL. The active material contains layers of SAQDs sandwiched between super lattices of heterojunctions working as distributed Bragg-reflecting (DBR) mirrors. The laser field is confined in the microcavity, where the stimulated emission takes place in a single-cavity mode.

QDs are very promising as an active material of a laser[64] due to the single-atom-like discrete energy spectrum. In an ideal QD laser, the narrow photon emission peaks of the QDs are tailored to match the cavity mode. The QD laser is mainly motivated by a small threshold current and low power consumption. In addition, QD lasers can be made nearly chirp-free by appropriate fine-tuning[4] of the DOS. Furthermore, the spectral bandwidth of the output light is small due to the QD DOS.

4.6.2 Biological labels

A biological label is a marker that can be attached to a biological molecule, e.g., a virus or a protein. The molecule can then be traced by radioactive or optical detection. The aim of QD markers is to replace radioactive markers by optical ones. Previous optical markers, such as rhodamine 6G (R6G), suffer from very low luminescence intensity and photobleaching.

Chan and Nie[65] have fabricated QD markers, using ZnS covered CdSe QDs (see also Sec. 4.2.1.1). A polar carboxyl acid group was used to render the coated QDs water soluble and to couple the QDs covalently to various biomolecules. Figure 4.24(a) shows schematically the QD marker attached to a protein. Figure 4.24(b) shows a fluorescence image of QD immunoglobulin-G (IgG)

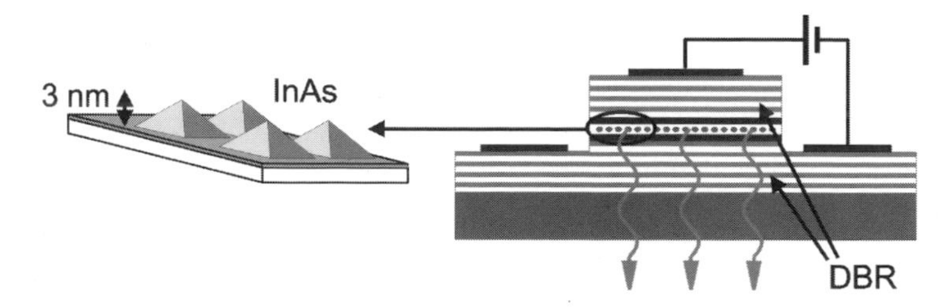

Figure 4.23 Typical QDVCEL. The active material contains layers of SAQDs sandwiched between (DBR) mirrors, which are doped to facilitate the injection of carriers to the InAs QDs where they recombine radiatively.

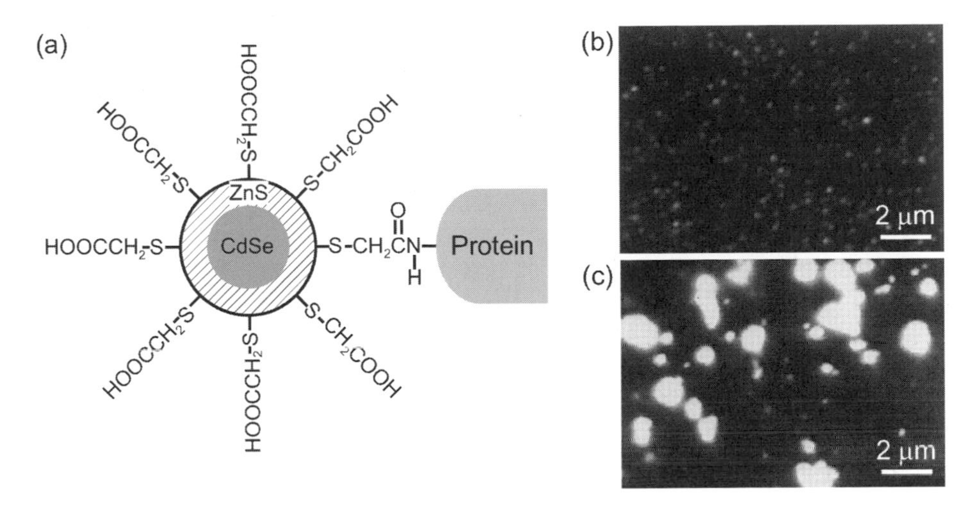

Figure 4.24 (a) Schematic of a QD marker. Right: fluorescence images of IgG labeled with (b) QDs and (c) antibody-induced agglutination of QDs labeled with IgG. (Reprinted with permission from Ref. 65, © 1998 American Association for the Advancement of Science.)

conjugates that were incubated with bovine serum albumin (BSA), whereas in Fig. 4.24(c) the conjugates are, in addition, subjected to a specific polyclonal antibody. It is clearly seen that the antibodies recognize the IgG and aggregate the QDs.

Chan and Nie showed the potential of QD markers for labeling biomolecules. The fluorescence intensity of a single QD is as strong as that of ~20 R6G molecules and the color of the QD marker can be tuned from blue to red. Moreover, the QD emission ($t_{1/2} = 960$ s) is nearly 100 times as stable as R6G ($t_{1/2} = 10$ s) against photobleaching. However, it is still a major technological challenge to decrease the QD size variation. The QD size must be very accurately defined for detection and distinguishing between several biomolecules simultaneously.

4.6.3 Electron pump

An electron pump (EP) is made of two QDs separated by a thin potential barrier and connected to one lead each through thick barriers as shown in Fig. 4.25. The EP consists, thus, of two SETs (see Sec. 4.4.4) in series. Due to a very transparent interdot barrier, the energy levels and populations of the QDs are strongly coupled.

Figure 4.25 shows the operation principles of an EP. The current through the QDs is controlled by the gate voltages V_{G1} and V_{G2}. The left panel of Fig. 4.25 shows the QD population diagram as a function of the gate voltages, where (i, j) correspond to i electrons in QD_1 (= left QD) and j electrons in QD_2 (= right QD). Pumping one electron from the left lead to the right lead through the QDs corresponds to one loop in the population diagram. Figures 4.25(a) through 4.25(d) show the energy band configuration at six different stages of a pumping loop. At 4.25(a)

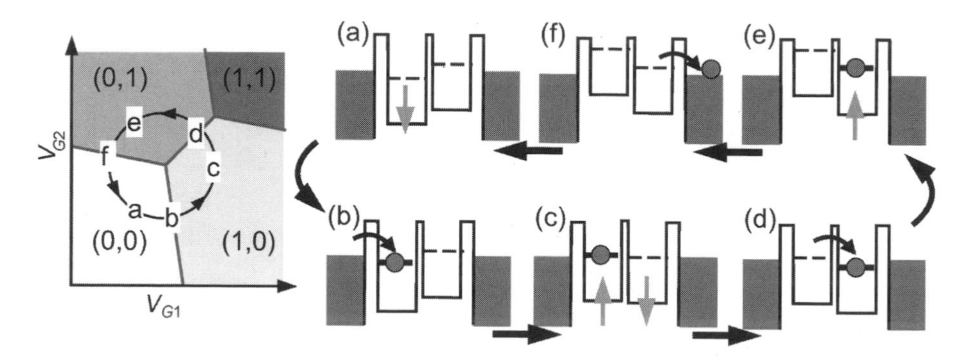

Figure 4.25 Operation principles of an electron pump. Initially both QDs are empty. By increasing V_{G1} (a), an electron is enabled to tunnel into QD_1 (b). Then V_{G2} is increased and V_{G1} decreased (c), and the electron tunnels farther to QD_2 (d). Finally, by increasing V_{G2} (e), the electron is made to tunnel into the right lead (f).

both QDs are empty. The electron energy levels of QD_1 are lowered by increasing V_{G1} until an electron tunnels [Fig. 4.25(b)] into QD_1 from the left lead. By raising V_{G2} and lowering V_{G1} [Fig. 4.25(c)] the electron tunnels farther into QD_2 [Fig. 4.25(d)]. Finally, V_{G2} is decreased [Fig. 4.25(e)] until the electron tunnels to the right lead [Fig. 4.25(f)]. One loop is completed. For more detailed information on double QD systems, see Ref. 46 and the references therein.

The current of an EP is directly related to the frequency of the gate voltages and there will be a current from left to right even at a zero bias voltage between the leads when the gate voltages are cycled appropriately. The EP enables extremely accurate current manipulation by pumping electrons one by one. The device may enable a new metrological standard for either electrical current or capacitance. The capacitance standard is based on pumping a known number of electrons on a capacitor. Keller et al. showed that this is possible with an error in the electron number[66] of 15 ppb.

4.6.4 Applications you should be aware of

QDs are very promising in many optical applications due to the possibility to tailor the DOS. Therefore it is also possible to fabricate, e.g., infrared (IR) detectors for low-energy photons. Lee et al. have proposed to use InAs QDs in a GaAs QW for photoconductivity detection of IR light.[67] The measured peak response was as high as 4.7 A per 1 W of incidence radiation in the photon energy range of 100 to 300 meV.

A QD single-photon source (SPS) is a photon emitter from which photons are emitted one by one on demand. The driving force of the development of SPS is quantum computation and cryptography-like applications.[68,69] To use photons as quantum bits (qubits), it is necessary to have a source for single photons with a predefined energy and polarization. Michler et al.[70] have demonstrated a SPS[70] using InAs QDs in a GaAs microdisk.

The QDs could also themselves serve as either classical (storing charge) or qubit (storing the quantum mechanical state) memories. One possibility of a qubit memory is to use a double quantum dot (DQD) molecule where the value of the qubit is based on the state of an exciton in the system.[71] For example, if the electron is in the first QD and the hole in the second QD the qubit value is $|0, 1\rangle$. Moreover, if both particles are in the first QD, the value is $|0, 0\rangle$ etc. The qubit value could also be stored using the electron spin or the state of a single particle in a QD.

Quantum dots could in principle also be used in cellular automata. Amlani et al.[72] presented logical AND and OR gates consisting of a cell, composed of four QDs connected in a ring by tunnel junctions, and two single-QD electrometers. In this system, the digital data is encoded in the position of two electrons.

References

1. M. A. Kastner, "Artifical atoms," *Phys. Today* **46**(1), 24–31 (1993).
2. M. Bruchez, Jr., M. Moronne, P. Gin, S. Weiss, and A. P. Alivisatos, "Semiconductor nanocrystals as fluorescent biological labels," *Science* **281**, 2013–2016 (1998).
3. R. W. Robinett, *Quantum Mechanics*, Oxford University Press, New York (1997).
4. D. Bimberg, M. Grundmann, and N. N. Ledentsov, *Quantum Dot Heterostructures*, John Wiley & Sons, West Sussex, UK (1999).
5. J. Singh, *Physics of Semiconductors and Their Heterostructures*, McGraw-Hill, New York (1993).
6. R. Dingle, R. W. Wiegmann, and C. H. Henry, "Quantum states of confined carriers in very thin $Al_xGa_{1-x}As$-GaAs-$Al_xGa_{1-x}As$ heterostructures," *Phys. Rev. Lett.* **33**, 827–830 (1974).
7. L. Esaki and R. Tsu, "Superlattice and negative differential conductivity in semiconductors," *IBM J. Res. Dev.* **14**, 61–65 (1970).
8. L. Chang, L. Esaki, and R. Tsu, "Resonant tunneling in semiconductor double barriers," *Appl. Phys. Lett.* **24**, 593–595 (1974).
9. J. M. Luttinger and W. Kohn, "Motion of electrons and holes in perturbed periodic fields," *Phys. Rev.* **97**, 869–883 (1955).
10. K. von Klitzing, G. Dorda, and M. Pepper, "New method for high-accuracy determination of the fine-structure constant based on quantized hall resistance," *Phys. Rev. Lett.* **45**, 494–497 (1980).
11. M. A. Herman and H. Sitter, *Molecular Beam Epitaxy: Fundamentals and Current Status*, 2nd ed., Springer, Berlin (1996).
12. M. L. Hitchman and K. F. Jensen, Eds., *Chemical Vapor Deposition Principles and Applications*, Academic, London (1993).
13. A. I. Ekimov and A. A. Onushenko, "Size quantization of the electron energy spectrum in a microscopic semiconductor crystal," *JETP Lett.* **40**, 1136–1138 (1984).

14. Y. Meir, N. S. Wingreen, and P. A. Lee, "Transport through a strongly interacting electron system: theory of periodic conductance oscillations," *Phys. Rev. Lett.* **66**, 3048–3051 (1991).
15. C. B. Murray, D. J. Morris, and M. G. Bawendi, "Synthesis and characterization of nearly monodisperse CdE (E = S, Se, Te) semiconductor nanocrystallites," *J. Am. Chem. Soc.* **115**, 8706–8715 (1993).
16. X. Peng, M. C. Schlamp, A. V. Kadavanich, and A. P. Alivisatos, "Epitaxial growth of highly luminescent CdSe/CdS core/shell nanocrystals with photostability and electronic accessibility," *J. Am. Chem. Soc.* **119**, 7019–7029 (1997).
17. L. Pavesi, L. D. Negro, C. Mazzoleni, G. Franzò, and F. Priolo, "Optical gain in silicon nanocrystals," *Nature* **408**, 440–444 (2000).
18. R. Steffen, T. Koch, J. Oshinowo, F. Faller, and A. Forchel, "Photoluminescence study of deep etched InGaAs/GaAs quantum wires and dots defined by low-voltage electron beam lithography," *Appl. Phys. Lett.* **68**, 223–225 (1996).
19. S. Tarucha, D. G. Austing, and T. Honda, "Resonant tunneling single electron transistors," *Superlat. Microstruct.* **18**, 121–130 (1995).
20. Y. Takahashi, M. Nagase, H. Namatsu, K. Kurihara, K. Iwdate, Y. Nakajima, S. Horiguchi, K. Murase, and M. Tabe, "Fabrication technique for Si single-electron transistor operating at room remperature," *Electron. Lett.* **31**, 136–137 (1995).
21. Y. Ono, Y. Takahashi, K. Yamazaki, M. Nagase, H. Namatsu, K. Kurihara, and K. Murase, "Fabrication method for IC-oriented Si single-electron transistors," *IEEE Trans. Electron Devices* **47**, 147–153 (2000).
22. Y. Takahashi, A. Fujiwara, Y. Ono, and K. Murase, "Silicon single-electron devices and their applications," in *Proceedings in 30th International Symposium on Multiple-Valued Logic (ISMVL 2000)*, IEEE Computer Society, 411–420 (2000).
23. H. Jeong, A. M. Chang, and M. R. Melloch, "The Kondo effect in an artificial quantum dot molecule," *Science* **293**, 2221–2223 (2001).
24. C. Livermore, C. H. Crouch, R. M. Westervelt, K. L. Campman, and A. C. Gossard, "The coulomb blockade in coupled quantum dots," *Science* **274**, 1332–1335 (1996).
25. J. M. Moison, F. Houzay, F. Barthe, L. Leprince, E. André, and O. Vatel, "Self-organized growth of regular nanometer-scale InAs dots on GaAs," *Appl. Phys. Lett.* **64**, 196–198 (1994).
26. H. Lipsanen, M. Sopanen, and J. Ahopelto, "Luminescence from excited states in strain-induced $In_xGa_{1-x}As$ quantum dots," *Phys. Rev. B* **51**, 13868–13871 (1995).
27. H. Saito, K. Nishi, and S. Sugou, "Ground-state lasing at room temperature in long-wavelength InAs quantum-dot lasers on InP(311)B substrates," *Appl. Phys. Lett.* **78**, 267–269 (2001).
28. M. Grundmann, O. Stier, and D. Bimberg, "InAs/GaAs pyramidal quantum dots: strain distribution, optical phonons, and electronic structure," *Phys. Rev. B* **52**, 11969–11981 (1995).

29. K. J. J. Marquez, L. Geelhaar, and K. Jacobi, "Atomically resolved structure of InAs quantum dots," *Appl. Phys. Lett.* **78**, 2309–2311 (2001).
30. K. Gergsson, N. Carlsson, L. Samuelson, W. Seifert, and L. R. Wallenberg, "Transmission electron microscopy investigation of the morphology of InP Stranski-Krastanow island grown by metalorganic chemical vapor deposition," *Appl. Phys. Lett.* **67**, 2981–2982 (1995).
31. J. Tulkki and A. Heinämäki, "Confinement effect in a quantum well dot induced by an InP stressor," *Phys. Rev. B* **52**, 8239–8243 (1995).
32. Landolt-Börnstein, *Group III:Condensed Matter*, Springer, Berlin (2002).
33. A. Kuther, M. Bayer, A. Forchel, A. Gorbunov, V. B. Timofeev, F. Schäfer, and J. P. Reithmaier, "Zeeman splitting of excitons and biexcitons in single $In_{0.60}Ga_{0.40}As$/GaAs self-assembled quantum dots," *Phys. Rev. B* **58**, 7508–7511 (1998).
34. T. Saiki, K. Nishi, and M. Ohtsu, "Low temperature near-field photoluminescence spectroscopy of InGaAs single quantum dots," *Jpn. J. Appl. Phys.* **37**, 1638–1642 (1998).
35. G. Yusa, "Modulation of zero-dimensional carrier distributions in quantum dots by DC-fields, THz-fields, and photons," PhD Dissertation, The University of Tokyo (Feb. 1999).
36. S. Tarucha, D. G. Austing, T. Honda, R. J. v. Hage, and L. P. Kouwenhoven, "Shell filling and spin effects in a few electron quantum dot," *Phys. Rev. Lett.* **77**, 3613–3616 (1996).
37. P. Hawrylak, "Excitonic artifical atoms: engineering optical properties of quantum dots," *Phys. Rev. B* **60**, 5597–5608 (1999).
38. V. B. Beretskii, E. M. Lifshitz, and L. P. Pitaevskii, *Relativistic Quantum Theory*, 1st ed., Addison-Wesley, New York (1971).
39. F. Findeis, M. Baier, E. Beham, A. Zrenner, and G. Abstreiter, "Photocurrent and photoluminscence of a single self-assembled quantum dot in electric fields," *Appl. Phys. Lett.* **78**, 2958–2960 (2001).
40. S. A. Empedocles and M. G. Bawendi, "Quantum-confined stark effect in single CdSe nanocrystallite quantum dots," *Science* **278**, 2114–2117 (1997).
41. R. Cingolani, R. Rinaldi, H. Lipsanen, M. Sopanen, R. Virkkala, K. Maijala, J. Tulkki, J. Ahopelto, K. Uchida, N. Miura, and Y. Arakawa, "Electron-hole correlation in quantum dots under a high magnetic field (up to 45T)," *Phys. Rev. Lett.* **83**, 4832–4835 (1999).
42. B. K. Ridley, *Quantum Processes in Semiconductors*, Oxford University Press, New York (2002).
43. D. K. Ferry and S. M. Goodnick, *Transport in Nanostructures*, Cambridge University Press, New York (1997).
44. K. Kobayashi, H. Aikawa, S. Katsumoto, and Y. Iye, "Tuning of the fano effect through a quantum dot in an Aharonov-Bohm interferometer," *Phys. Rev. Lett.* **88**, 256806 (2002).
45. L. P. Kouwenhoven, D. G. Austing, and S. Tarucha, "Few-electron quantum dots," *Rep. Prog. Phys.* **64**, 701–736 (2001).

46. W. G. v. d. Wiel, S. D. Franceschi, J. M. Elzerman, T. Fujisawa, S. Tarucha, and L. P. Kouwenhoven, "Electron transport through double quantum dots," *Rev. Mod. Phys.* **75**, 1–22 (2003).
47. S. Grosse, J. H. H. Sandmann, G. von Plessen, J. Feldmann, H. Lipsanen, M. Sopanen, J. Tulkki, and J. Ahopelto, "Carrier relaxation dynamics in quantum dots: scattering mechanisms and state-filling effects," *Phys. Rev. B* **55**, 4473–4476 (1997).
48. M. Grundmann and D. Bimberg, "Theory of random population for quantum dots," *Phys. Rev. B* **55**, 9740–9745 (1997).
49. H. Benisty, C. M. Sotomayor-Torrès, and C. Weisbuch, "Intrinsic mechanism for the poor luminescence properties of quantum-box systems," *Phys. Rev. B* **44**, 10945–10948 (1991).
50. M. Braskén, M. Lindberg, M. Sopanen, H. Lipsanen, and J. Tulkki, "Temperature dependence of carrier relaxation in strain-induced quantum dots," *Phys. Rev. B* **58**, 15993–15996 (1998).
51. A. V. Uskov, A.-P. Jauho, B. Tromborg, J. Mørk, and R. Lang, Dephasing times in quantum dots due to elastic LO phonon-carrier collisions," *Phys. Rev. Lett.* **85**, 1516–1519 (2000).
52. H. Htoon, D. Kulik, O. Baklenov, A. L. Holmes, Jr., T. Takagahara, and C. K. Shih, "Carrier relaxation and quantum decoherence of excited states in self-assembled quantum dots," *Phys. Rev. B* **63**, 241303 (2001).
53. P. Borri, W. Langbein, S. Schneider, U. Woggon, R. L. Sellin, D. Ouyang, and D. Bimberg, "Ultralong dephasing time in InGaAs quantum dots," *Phys. Rev. Lett.* **87**, 157401 (2001).
54. A. J. Williamson and A. Zunger, "Pseudopotential study of electron-hole excitations in colloidal free-standing InAs quantum dots," *Phys. Rev. B* **61**, 1978–1991 (2000).
55. R.-H. Xie, G. W. Bryant, S. Lee, and W. Jaskolski, "Electron-hole correlations and optical excitonic gaps in quantum-dot quantum wells: tight-binding approach," *Phys. Rev. B* **65**, 235306 (2002).
56. D.-B. Kao, J. P. McVittie, W. D. Nix, and K. C. Saraswat, "Two-dimensional thermal oxidation of silicon. II. Modeling stress effects in wet oxides," *IEEE Trans. Electron Devices* **35**, 25–37 (1988).
57. C. Pryor, J. Kim, L. W. Wang, A. J. Williamson, and A. Zunger, "Comparison of two methods for describing the strain profiles in quantum dots," *J. Appl. Phys.* **83**, 2548–2554 (1998).
58. "Ansys," Swanson Analysis Systems Inc., Houston, PA.
59. S. M. Reimann and M. Manninen, "Electronic structure of quantum dots," *Rev. Mod. Phys.* **74**, 1283–1342 (2002).
60. S. M. Reimann, M. Koskinen, M. Manninen, and B. R. Mottelson, "Quantum dots in magnetic fields: phase diagram and broken symmetry at the maximum-density-droplet edge," *Phys. Rev. Lett.* **83**, 3270–3273 (1999).
61. M. Braskén, M. Lindberg, D. Sundholm, and J. Olsen, "Spatial carrier-carrier correlations in strain-induced quantum dots," *Phys. Rev. B* **64**, 035312 (2001).

62. O. Stier, "Electronic and optical properties of quantum dots and wires," Wissenschatt und Technik Verlag, Berlin (2001).
63. H. Saito, K. Nishi, I. Ogura, S. Sugou, and Y. Sugimoto, "Room-temperature lasing operation of a quantum-dot vertical-cavity surface-emitting laser," *Appl. Phys. Lett.* **69**, 3140–3142 (1996).
64. V. I. Klimov, A. A. Mikhailovsky, S. Xu, A. Malko, J. A. Hollingsworth, C. A. Leatherdale, H.-J. Eisler, and M. G. Bawendi, "Optical gain and stimulated emission in nanocrystal quantum dots," *Science*, **290**, 314–317 (2000).
65. W. C. Chan and S. Nie, "Quantum dot bioconjugates for ultrasensitive nonisotopic detection," *Science* **281**, 2016–2018 (1998).
66. M. W. Keller, J. M. Martinis, N. M. Zimmerman, and A. H. Steinbach, "Accuracy of electron counting using a 7-junction electron pump," *Appl. Phys. Lett.* **69**, 1804–1806 (1996).
67. S.-W. Lee, K. Hirakawa, and Y. Shimada, "Bound-to-continuum intersubband photoconductivity of self-assembled InAs quantum dots in modulation-doped heterostructures," *Appl. Phys. Lett.* **75**, 1428–1430 (1999).
68. C. Macchiavello, G. Palma, A. Zeilinger, and C. Macchi, Eds., *Quantum Computation and Quantum Information Theory*, World Scientific, London (2000).
69. A. K. Ekert, *The Physics of Quantum Information*, D. Bouwmeester, A. Ekert, A. Zeilinger, and D. Bouwmeester, Eds., Springer, Berlin (2000).
70. P. Michler, A. Kiraz, C. Becher, W. V. Schoenfeld, P. M. Petroff, L. Zhang, E. Hu, and A. Imamoglu, "A quantum dot single-photon turnstile device," *Science* **290**, 2282–2285 (2000).
71. M. Bayer, P. Hawrylak, K. Hinzer, S. Fafard, M. Korkusinski, Z. R. Wasilewski, O. Stern, and A. Forchel, "Coupling and entangling of quantum states in quantum dot molecules," *Science* **291**, 451–453 (2001).
72. I. Amlani, A. O. Orlov, G. Toth, G. H. Bernstein, C. S. Lent, and G. L. Snider, "Digital logic gate using quantum-dot cellular automata," *Science* **284**, 289–291 (1999).

List of symbols

ε	permittivity
ε_0	permittivity of vacuum
$\varepsilon^{\perp}(0)$, $\varepsilon^{\parallel}(0)$, $\varepsilon(0)$	low-frequency permittivities
$\varepsilon^{\perp}(\infty)$, $\varepsilon^{\parallel}(\infty)$, $\varepsilon(\infty)$	high-frequency permittivities
λ	wavelength
μ_B	Bohr magneton
μ_L, μ_R, μ_N	chemical potentials
σ	z projection of spin
$\boldsymbol{\sigma}$	spin vector
σ^+, σ^-	light polarization
ϕ	scalar potential
ω, ω_+, ω_-, $\overline{\omega}$, Ω_+, Ω_-	angular frequencies
ω_c	cyclotron frequency
ΔX_2	energy difference between two exciton states
Ψ^e, Ψ^h	electron and hole wave functions
$\hbar$	Planck constant
c	speed of light in vacuum
$\hat{c}_i^{\dagger}$, $\hat{c}_i$	electron creation and annihilation operators
e	electron charge
$\langle f\|E_{\mathrm{ED}}\|i\rangle$	electric dipole matrix element of the radiative transition from state $\|i\rangle$ to state $\|f\rangle$
g	g factor
$\hat{h}_i^{\dagger}$, $\hat{h}_i$	hole creation and annihilation operators
$\langle ij\|\hat{V}_{ee}\|kl\rangle$	matrix element of the electron-electron Coulomb interaction
$\langle ij\|\hat{V}_{eh}\|kl\rangle$	matrix element of the electron-hole Coulomb interaction
$\langle ij\|\hat{V}_{hh}\|kl\rangle$	matrix element of the hole-hole Coulomb interaction
k_B	Boltzmann constant
m_0	rest mass of a free electron
m_{lh}, $m_{lh[100]}$, $m_{lh[111]}$	effective light hole masses
m_{hh}, $m_{hh[100]}$, $m_{hh[111]}$	effective heavy hole masses
$m_{\perp}$, $m_{\parallel}$, m_e^*	effective electron masses
$m_{p\perp}^A$, $m_{p\parallel}^A$, $m_{p\parallel}^A$, $m_{[100]}$, $m_{[111]}$, m_h^*	effective hole masses
m_r^e, m_r^h	relative effective masses of an electron and a hole
$\mathbf{p}$	momentum
q	charge
$\mathbf{r}$	position vector
$\mathbf{A}$	vector potential
$\mathbf{B}$	magnetic field

C, C_G, C_L, C_R	capacitances
$\mathbf{E}$	electric field
E_C, E_{C1}	conduction bandedge
E_G	energy band gap
E_V, E_{HH1}	valence bandedge
E_i^e, E_i^h	single-particle electron and hole energies
E_X	exciton binding energy
H	Hamiltonian
H_{mb}	many-body Hamiltonian
$H_E, H_{\mathrm{DM}}, H_Z, H_\sigma$	Hamiltonian terms related to an electromagnetic field
H^e, H^h	single-electron and single-hole Hamiltonians
L_i	angular momentum
T	temperature
V_B	bias voltage
V_C	height of the confining potential barrier
V_C, V_{C1}, V_{C2}	gate voltages
W_{if}	recombination rate between states i and f
X, X_0, X_1, X_2, X_2^*	exciton states
X^+, X^-	charged exciton states

Fredrik Boxberg received his MSc degree from the Helsinki University of Technology (HUT), Espoo, Finland, in 2000, and joined the Laboratory of Computational Engineering at HUT, where he is currently pursuing doctoral research on the electronic and optical properties of mesoscopic systems. He is being supervised by Prof. J. Tulkki. He has published two journal papers.

Jukka Tulkki received his doctorate in physics from the Helsinki University of Technology (HUT), Espoo, Finland, in 1986. His thesis addressed inelastic resonance scattering of x rays. From 1985 to 1987, he was an acting associate professor of physics at HUT. He held research appointments at the University of Helsinki (1988), the Academy of Finland in 1990 to 1993, and at HUT in 1994 to 1998. Since 1998, he has been a professor of computational engineering at HUT. He is the author or coauthor of more than 60 journal papers and numerous conference presentations. His current research interests include x-ray physics, atomic collision physics, semiconductors, electronic and optical properties of mesoscopic systems and microsystems.

Chapter 5

Nanoelectromagnetics of Low-Dimensional Structures

Sergey A. Maksimenko and Gregory Ya. Slepyan

5.1. Introduction 146
5.2. Electron transport in carbon nanotube 148
  5.2.1. Dispersion properties of π-electrons 148
  5.2.2. Bloch equation for π-electrons 151
5.3. Linear electrodynamics of carbon nanotubes 153
  5.3.1. Dynamic conductivity 153
  5.3.2. Effective boundary conditions 156
  5.3.3. Surface electromagnetic waves 157
  5.3.4. Edge effects 159
5.4. Nonlinear processes in carbon nanotubes 162
  5.4.1. Current density spectrum in an isolated CN 163
  5.4.2. Negative differential conductivity in an isolated CN 167
5.5. Quantum electrodynamics of carbon nanotubes 170
  5.5.1. Maxwell equations for electromagnetic field operators 170
  5.5.2. Spontaneous decay of an excited atom in a CN 172
5.6. Semiconductor quantum dot in a classical electromagnetic field 177
  5.6.1. Model Hamiltonian 178
  5.6.2. Equations of motion 182
  5.6.3. QD polarization 183
5.7. Interaction of QD with quantum light 184
  5.7.1. Model Hamiltonian 184
  5.7.2. Equations of motion 186
  5.7.3. Interaction with single-photon states 187
  5.7.4. Scattering of electromagnetic Fock qubits 189
  5.7.5. Observability of depolarization 192
5.8. Concluding remarks 194
Acknowledgments 194
References 194
List of symbols 203

5.1 Introduction

The ongoing rapid progress in the synthesis of a variety of different kinds of nanostructures with fascinating physical properties irreducible to properties of bulk media symbolizes a fundamental breakthrough in the physics and chemistry of condensed matter, significantly extending our knowledge of the nature of solids and our capabilities to control their properties. Solid state nanostructures are constitutive and geometric nanononhomogeneities in semiconductor and dielectric mediums. Fullerenes and nanotubes,[1–4] semiconductor structures with reduced dimensionality—quantum wells, wires and dots,[5–7] and sculptured thin films[8] can be mentioned as examples. Despite their different physical natures, these objects share the common property of having extremely small dimensions in one or more directions. These dimensions are about one or two orders of magnitude bigger than the characteristic interatomic distance, so that (1) *spatial confinement of charge carriers* is fully developed, thereby providing a discrete spectrum of energy states in one or several directions. Apart from that, the intrinsic spatial nonhomogeneity of nanostructures dictates (2) *nanoscale nonhomogeneity of electromagnetic fields* in them. Whereas the first factor lies in the focus of current research activity in nanosciences, the role of the second factor is often underestimated. This chapter stresses complementary characters of these two key factors whose interplay drastically modifies the electronic and optical properties of nanostructures as compared to bulk media.

Conventionally, condensed-matter physics is completely associated with homogeneous media, which are characterized by corresponding dispersion equations for coupled states of the electromagnetic field and material particles. The solutions of a dispersion equation describe the eigenwaves of the media—the so-called quasi-particles—which differ from usual (free) particles by the complex behavior of their dispersion characteristics (energy versus quasi-momentum). The embedding of nanoscale nonhomogeneities in a homogeneous media creates conditions for diffraction and scattering of quasi-particles and for their mutual transformation, in the same way as in irregular waveguides.

An important role is played by the resonant interactions between different modes and the corresponding matching conditions. The first step in the incorporation of resonant interactions of quasi-particles was made in the theory of quantum semiconductor superlattices.[9] Their high-frequency and optical properties turned out to be very unusual: negative differential conductivity, propagation of longitudinal (plasma) waves, and so on. The interaction of different modes in nanostructures appears to be even more complex due to the greater variety in interacting modes and the complex 3D geometry of the nonhomogeneities. It is no wonder that the electronic and electromagnetic properties of nanomaterials appear to be richer and more diverse. In particular, quantization of the charge-carrier motion and the pronounced nonhomogeneity of the electromagnetic field inside and in the vicinity of a nano-object often lead to spatially nonlocal electromagnetic response, provide peculiar manifestations of instabilities and nonlinearity, and make nano-objects attractive for use in quantum networks to store and process quantum information.[10,11] Thus, a new branch of the physics of nanostructures—*nanoelectromagnetics*—is currently emerging. It incorporates and modifies traditional electrodynamical methods and approaches, and it introduces new methods for new problems.

Milestones in the development of electrodynamics have always been related to practical problems arising from new ideas relating to the transmission and processing of electromagnetic signals. Advances in quantum electronics led to the development of the theory of open quasi-optical resonators.[12] The synthesis of high-quality optical fibers made fiber optic communication feasible, which led to the development of the theory of open dielectric waveguides (including irregular and nonlinear waveguides).[13,14] Progress in microwave microelectronics stimulated research on the electrodynamics of microstrips and other planar structures.[15] Modern electromagnetic theory is characterized by the development of highly efficient numerical methods simulating diffraction from lossy objects of arbitrary spatial configurations.[16] Undoubtedly, electromagnetic simulation of nanostructures is one of the main research directions for modern electrodynamics.

Among a variety of nanostructures, research on the properties of carbon nanotubes (CNs)—quasi-1D carbon macromolecules—has continued to grow unabated for more than a decade.[1–4] In particular, the modern quantum theory of quasi-1D conductors predicts monomolecular electronic devices whose operation relies on quantum charge-transport processes.[17,18] CN-based transistors,[19] tunneling diodes based on doped CN junctions,[20] and Schottky diodes in CNs heterojunctions[21,22] are actively studied. One more important attribute of CNs is pronounced field-electron emission, a property that makes CNs attractive for cathodes in electronic devices.[4]

Recent progress in the synthesis of sheets of nanoscale 3D confined narrow-gap insertions in a host semiconductor—the so-called quantum dots (QDs)—enables realization of the idea[23] of using structures with size quantization of charge carriers as active media for double-heterostructure lasers. It was predicted about two decades ago that lasers based on QDs would show radically changed characteristics as compared to conventional quantum-well lasers.[24,25] A large body of results on physical properties of QDs and their utilization for QD laser design is now available.[6,26] Another important class of problems attracting much attention in the semiconductor community concerns the electromagnetics of microcavities exposed to classical or quantum light; see Ref. 27 and the references therein. In that connection, the applications of semiconductor QDs in cavity quantum electrodynamics[28–30] (QED) and as potential quantum-light emitters[31–34] are being actively discussed.

This chapter focuses on some problems of the electromagnetics of isolated CNs and QDs, thus introducing the reader to nanoelectromagnetics of low-dimensional nanostructures. Both microscopic and macroscopic models can be utilized to study electromagnetic response properties of nanostructures. The macroscopic approach implies their phenomenological description by means of elctrodynamical constitutive relations. In that case, well-developed traditional methods and approaches originating from the microwave and antenna theory[12–16,35–38] can easily be extended to boundary-value problems of nanostructures. In contrast, microscopic approaches do not use *a priori* constitutive relations. For instance, electrodynamics is supplemented with quantum-mechanical modeling of charge carriers transport on the basis of reasonable field approximations. Such a microscopic approach is essentially more complicated but more consistent in comparison with the macroscopic phenomenological description.

The remainder of this chapter is based on a series of works that cover problems of linear electrodynamics of CNs,[39–44] nonlinear transport in and nonlinear optics of CNs,[44–49] QED of CNs,[44,50] and classical and quantum optics of QDs with the local fields accounted for.[51–59] Only isolated CNs and QDs are considered, with collective effects inherent in macroscopic ensembles of such particles being well beyond this chapter's scope. Note that the electromagnetic response theory of individual nano-objects supplemented with the traditional homogenization techniques[60] can be successfully applied to nanocomposite materials. Also, the material presented here can be extended to cover the constitutive modeling required for nanoelectromagnetics of microcavities.

5.2 Electron transport in carbon nanotube

5.2.1 Dispersion properties of π-electrons

Surface carbon structures, i.e., fullerenes and nanotubes, appear as results of certain deformations of a planar monoatomic graphite layer (graphene), whose crystalline structure is illustrated in Fig. 5.1. In fullerenes, discovered[61] in 1985, the graphene plane is transformed into a closed sphere or spheroid containing regular hexagons (their number depends on the fullerene dimension) and 12 regular pentagons. In a CN, originally synthesized[62] in 1991, the graphite surface is transformed into an extended hollow cylindrical structure; see Fig. 5.2. Thus, carbon atoms in CNs are

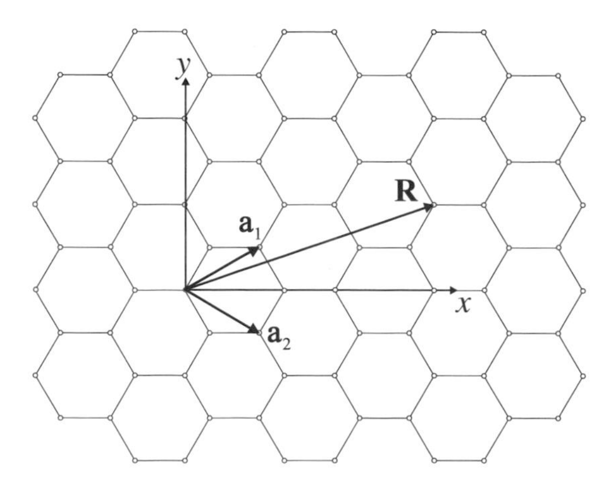

Figure 5.1 Configuration of the graphene crystalline lattice $\mathbf{R} = m\mathbf{a}_1 + n\mathbf{a}_2$.

situated regularly on a helical line with a certain wrapping angle (geometric chiral angle).

Let $\mathbf{R}$ be the relative position vector between two sites on the honeycomb lattice of the graphene plane, as shown in Fig. 5.1. In terms of the lattice basic vectors $\mathbf{a}_1$ and $\mathbf{a}_2$, $\mathbf{R} = m\mathbf{a}_1 + n\mathbf{a}_2$, where m and n are integers. Thus, the geometric configuration of CNs can be classified by the dual index (m, n)—with $(m, 0)$ for *zigzag* CNs, (m, m) for *armchair* CNs, and $0 < n \neq m$ for *chiral* CNs. The cross-sectional radius of a CN and its geometric chiral angle are given by[4]

$$R_{\mathrm{CN}} = \frac{\sqrt{3}}{2\pi} b\sqrt{m^2 + mn + n^2}, \qquad \theta_{\mathrm{CN}} = \tan^{-1}\left(\frac{\sqrt{3}n}{2m + n}\right), \tag{5.1}$$

where $b = 0.142$ nm is the C—C bond length in graphene. Typically, CNs are 0.1 to 10 μm in length and their cross-sectional radius varies within the range 1 to 10 nm, while $0 \leq \theta_{\mathrm{CN}} \leq 30$ deg. Recently, the synthesis of CNs of extremely small radius of ~ 0.4 nm has been reported.[63] In this chapter, a 2D Cartesian coordinate system (x, y) is used for graphene and the circular cylindrical coordinate system (ρ, φ, z) for any CN, with the CN axis parallel to the z axis. The x axis is oriented along a hexagonal side. The transition from graphene to a zigzag CN is established by the substitution $\{x \to z,\ y \to \phi\}$, while the transition from graphene to an armchair CN requires the substitution $\{y \to z,\ x \to \phi\}$. Gaussian units are used throughout, in conformity with CN literature.

Both single-wall and multiwall nanotubes have been synthesized.[4] The multilayer nanotubes have the form of several coaxial cylinders (the distance between the layers is 0.34 nm, while the number of cylinders is ordinarily 10 to 12). Although the theory presented here has been developed for a single-wall CN, it should be noted that a multiwall CN can be treated as an ensemble of single-wall nanotubes with a broad diameter distribution.[41] Along with CNs, nanotubes doped with nitrogen and boron are also known.[4]

The electromagnetic processes in any media essentially depend on its electronic properties. The properties of electrons in CNs and electron transfer processes in them have been studied in detail, both theoretically and experimentally. The theoretical analysis is usually confined to dynamics of π-electrons within the tight-binding approximation,[64,65] which allows for interaction between only three ad-

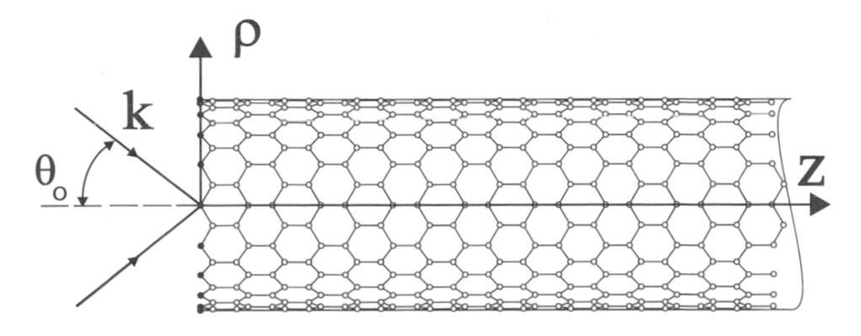

Figure 5.2 Model of an open-ended carbon nanotube. The wavevector **k** shows the direction of propagation of the cylindrical wave in the analysis of edge effects in nanotubes.

jacent atoms of the hexagonal structure. In the framework of this model, electron properties of graphene are described by the well-known dispersion law[66]

$$\mathcal{E}_{c,v}(\mathbf{p}) = \pm\gamma_0\sqrt{1 + 4\cos(ap_x)\cos\left(\frac{a}{\sqrt{3}}p_y\right) + 4\cos^2\left(\frac{a}{\sqrt{3}}p_y\right)}, \tag{5.2}$$

where $\gamma_0 \approx 2.7$ eV is the overlapping integral, $a = 3b/2\hbar$, $\hbar$ is the Planck constant, and $p_{x,y}$ are the projections of the quasi-momentum. The upper and lower signs in Eq. (5.2) refer to the conduction and valence bands, marked by the indices c and v, respectively. The range of definition of the quasi-momentum $\mathbf{p}$ (the first Brillouin zone) spans the hexagons shown in Fig. 5.3. The vertices are the Fermi points where $\mathcal{E} = 0$, which is indicative of the absence of the forbidden zone for π-electrons in graphene. Note that graphene is a semimetal: it lacks a band gap, but the density of states at the Fermi level is zero.

The dispersion properties of nanotubes essentially differ from the dispersion properties of graphene because of the difference in topology. In the cylindrical structure, electrons residing at the origin and at the point $\mathbf{R} = m\mathbf{a}_1 + n\mathbf{a}_2$ are identical, which quantizes the transverse quasi-momentum component:

$$p_\varphi = \hbar s/R_{\text{CN}}, \quad s = 1, 2, \ldots, m. \tag{5.3}$$

The axial component of quasi-momentum p_z remains continuous. The relationships in Eqs. (5.2) and (5.3) and the substitution $\{p_x \to p_z,\ p_y \to p_\varphi\}$ yield the dispersion law for the zigzag nanotubes as follows:

$$\mathcal{E}_{c,v}(p_z, s) = \pm\gamma_0\sqrt{1 + 4\cos(ap_z)\cos\left(\frac{\pi s}{m}\right) + 4\cos^2\left(\frac{\pi s}{m}\right)}. \tag{5.4}$$

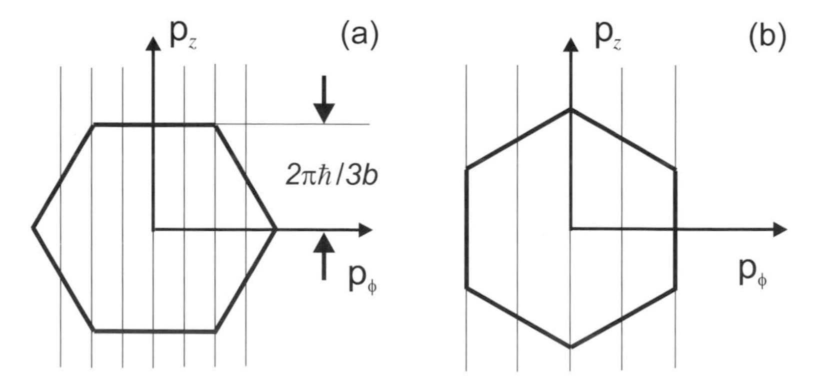

Figure 5.3 Configuration of the first Brillouin zone for (a) zigzag and (b) armchair CNs.

To evaluate the electron dispersion relation for armchair CNs from (5.2), the substitution $\{p_x \to p_\varphi,\ p_y \to p_z\}$ must be carried out; accordingly,

$$\mathcal{E}_{c,v}(p_z, s) = \pm\gamma_0\sqrt{1 + 4\cos\left(\frac{\pi s}{m}\right)\cos\left(\frac{a}{\sqrt{3}}p_z\right) + 4\cos^2\left(\frac{a}{\sqrt{3}}p_z\right)}. \tag{5.5}$$

As follows from Eqs. (5.4) and (5.5), the first Brillouin zone in a CN is not a hexagon. Rather, it a set of 1D zones: rectilinear segments inside the hexagon. Depending on whether or not these lines pass through the hexagon's vertexes (Fermi points), the band gap in the electron spectrum either disappears or appears. Unlike graphene, the density of states at the Fermi level in 1D zones is nonzero. Accordingly, a nanotube is either metallic or semiconducting. As can be seen from Fig. 5.3, armchair CNs exhibit metallic conductivity at any m; whereas zigzag CNs behave as a metal only for $m = 3q$, where q is an integer. For a metallic CN of a small radius and for a CN of a very large radius ($m \to \infty$), the approximate dispersion law for π-electrons, $\mathcal{E}_{c,v}(\mathbf{p}) = \pm v_F|\mathbf{p} - \mathbf{p}_F|$, has been proposed;[67] here, $v_F = a\gamma_0$ is the velocity of π-electrons at the Fermi level and $\mathbf{p}_F$ is a constant vector defined as the quasi-momentum at the Fermi level. In both cases, the foregoing approximate dispersion law is applicable, because the regions near the Fermi points give the maximum contribution to the conductivity.

When a CN is placed in either an axial[68] or a transverse[69] magnetostatic field, the type of its conductivity changes. Due to this feature, the conductivity can be controlled over a wide range by varying the magnetization vector. An important property of chiral nanotubes is that a voltage applied across the ends produces an azimuthal current component.[45,70–72] As a result, the trajectory of the current in the CN is helical, although graphene has isotropic conductivity.

5.2.2 Bloch equation for π-electrons

The theory of optical properties of CNs applied in the present chapter involves a direct solution of the quantum-mechanical equations of motion for π-electrons. Consider an infinitely long rectilinear single-wall CN oriented along the z axis and excited by the component of electromagnetic field polarized along this axis: $\mathbf{E}(\mathbf{r}, t) = \mathbf{e}_z E_z(\mathbf{r}, t)$. Let the field be incident normally to the CN axis. In the tight-binding approximation, the motion of electrons in the CN crystalline lattice potential $W(\mathbf{r})$ is described by the Schrödinger equation

$$i\hbar\frac{\partial\Psi}{\partial t} = -\frac{\hbar^2}{2m_0}\Delta\Psi + \left[W(\mathbf{r}) - e(\mathbf{E}\mathbf{r})\right]\Psi, \tag{5.6}$$

where e and m_0 are the electron charge and mass, respectively. The solution can be represented by the Bloch wave expansion

$$\Psi = \sum_q C_q \Psi_q(\mathbf{p}, \mathbf{r}), \tag{5.7}$$

where the index q stands for the collection of quantum numbers characterizing states of π-electrons with a given quasi-momentum. In the framework of the two-band model, the index takes the values either v or c. Amplitudes $u_q(\mathbf{r})$ of the Bloch functions

$$\Psi_q(\mathbf{p}, \mathbf{r}) = \hbar^{-1/2} \exp(i\mathbf{pr}/\hbar) u_q(\mathbf{r}) \tag{5.8}$$

are periodic with respect to an arbitrary lattice vector $\mathbf{R}$. The expansion of Eq. (5.7) does not contain states of the continuous spectrum; consequently, consideration is restricted to the effects below the ionization threshold. The coefficients C_q satisfy the equation[73]

$$i\hbar \frac{\partial C_q}{\partial t} = \mathcal{E}_q C_q - i\hbar e E_z \frac{\partial C_q}{\partial p_z} - eE_z \sum_{q'} C_{q'} R_{qq'}, \tag{5.9}$$

where

$$R_{qq'} = \frac{i\hbar}{2} \int_0^\infty r dr \int_{S_{\mathrm{uc}}} \left(u_q^* \frac{\partial u_{q'}}{\partial p_z} - \frac{\partial u_q^*}{\partial p_z} u_{q'} \right) dS, \tag{5.10}$$

and S_{uc} is the honeycomb cell on the CN surface. After using the standard representation of the density matrix elements $\rho_{qq'} = C_q C_{q'}^*$, Eq. (5.9) can be transformed to the following system of equations:[49]

$$\begin{aligned}
&\frac{\partial \rho_{vv}}{\partial t} + eE_z \frac{\partial \rho_{vv}}{\partial p_z} = -\frac{i}{\hbar} eE_z \big(R_{vc}^* \rho_{vc} - R_{vc} \rho_{cv} \big), \\
&\frac{\partial \rho_{vc}}{\partial t} + eE_z \frac{\partial \rho_{vc}}{\partial p_z} = -\frac{i}{\hbar} eE_z \big[R_{vc}(2\rho_{vv} - 1) - \Delta R \rho_{vc} \big] - i\omega_{vc} \rho_{vc}, \\
&\rho_{vv} + \rho_{cc} = 1.
\end{aligned} \tag{5.11}$$

Here, $\omega_{vc} = (\mathcal{E}_c - \mathcal{E}_v)/\hbar$ is the frequency of the transition. The transition frequency as well as the matrix elements R_{vc} and $\Delta R = R_{vv} - R_{cc}$ are evaluated in the tight-binding approximation, taking into account transverse quantization of the charge carriers' motion and the hexagonal structure of the CN crystalline lattice. For zigzag CNs, the matrix element

$$R_{vc}(p_z, s) = -\frac{b\gamma_0^2}{2\mathcal{E}_c^2(p_z, s)} \left[1 + \cos(ap_z) \cos\left(\frac{\pi s}{m} \right) - 2\cos^2\left(\frac{\pi s}{m} \right) \right]. \tag{5.12}$$

The analogous expression for armchair (m, m) CNs is as follows:

$$R_{vc}(p_z, s) = -\frac{\sqrt{3}b\gamma_0^2}{2\mathcal{E}_c^2(p_z, s)}\sin\left(\frac{a}{\sqrt{3}}p_z\right)\sin\left(\frac{\pi s}{m}\right). \tag{5.13}$$

For CNs of both types, the condition $\Delta R/R_{vc} \ll 1$ can easily be derived, which enables us to reject the term ΔR in Eq. (5.11). The solution of Eq. (5.11) enables evaluation of the surface density of the induced axial current as follows:

$$j_z(t) = \frac{2e\gamma_0}{(\pi\hbar)^2 R_{\mathrm{CN}}}\sum_{s=1}^{m}\int\left[\frac{\partial\mathcal{E}_c}{\partial p_z}\rho_{vv} + \mathcal{E}_c R_{vc}\mathrm{Im}[\rho_{vv}]\right]dp_z. \tag{5.14}$$

Equations (5.11), supplemented with Eqs. (5.4), (5.5), (5.12), and (5.13), constitute a basic system for the analysis of the optical properties of CNs. Generally, its solution is nonlinear with respect to E_z. In the linear regime, the system of Eq. (5.11) can be linearized. In the case of weak nonlinearity, optical properties of CNs are derived from Eq. (5.11) using a polynomial expansion in E_z; so that the nonlinear properties of CNs are characterized by nonlinear optical susceptibilities of different orders. Of course, for the high-intensity external fields of subpicosecond optical pulses ($>10^{10}$ W/cm^2), the formalism of nonlinear susceptibilities becomes inefficient. In Ref. 49, the system of Eq. (5.11) was solved numerically in the time domain. The method of characteristics[74] was used for integration of the system. Initial distribution of electrons in zones was specified by the Fermi equilibrium distribution function at room temperature and the periodicity on the Brillouin zone boundaries was exploited.

Generally, relaxation terms describing inelastic scattering of π-electrons propagating in CNs should be introduced in Eq. (5.11). This can be done either phenomenologically[44] (e.g., in the framework of the relaxation-time approximation[9]) or on the basis of microscopic theory of electron-phonon interactions.[69,75] An alternative approach is to solve Eq. (5.11) without relaxation terms and then introduce corresponding corrections into the final results.

5.3 Linear electrodynamics of carbon nanotubes

5.3.1 Dynamic conductivity

As pointed out in the previous section, to obtain the linear optical response of a CN, the linearized Eqs. (5.11) must be solved, and the surface current density Eq. (5.14) must be evaluated. In the linear regime, the Bloch equations can be solved in the weak-field limit by the Fourier transform method. If we neglect spatial dispersion, the optical response proves to be spatially local. In that case, for Fourier amplitudes of the axial current and field, the relation[44]

$$j_z(\omega) = \sigma_{zz}(\omega)E_z(\omega) \tag{5.15}$$

is obtained, where

$$\sigma_{zz}(\omega) = -\frac{ie^2\omega}{\pi^2\hbar R_{\text{CN}}}\left\{\frac{1}{(\omega+i0)^2}\sum_{s=1}^{m}\int_{1\text{stBZ}}\frac{\partial F_c}{\partial p_z}\frac{\partial \mathcal{E}_c}{\partial p_z}dp_z - 2\sum_{s=1}^{m}\int_{1\text{stBZ}}\mathcal{E}_c|R_{vc}|^2\frac{F_c - F_v}{\hbar^2(\omega+i0)^2 - 4\mathcal{E}_c^2}dp_z\right\} \tag{5.16}$$

is the axial conductivity of the CN. In this expression, the integration is performed over the first Brillouin zone (BZ),

$$F_{c,v}(p_z, s) = \frac{1}{1+\exp[\mathcal{E}_{c,v}(p_z, s) - \mu_{\text{ch}}/k_B T]} \tag{5.17}$$

is the equilibrium Fermi distribution function, T is the temperature, and k_B is the Boltzmann constant. The chemical potential is denoted by μ_{ch}; in graphite and undoped CNs, $\mu_{\text{ch}} = 0$.

The CN conductivity law [Eq. (5.15)] is analogous to constitutive relations for 3D conducting media in classical electrodynamics. However, there is a significant distinction: as in classical electrodynamics, the derivation of Eq. (5.16) employed macroscopic spatial averaging, but a surface element was used instead of an infinitesimally small volume. Thus, $j_z(\omega)$ is the surface current density.

The relaxation effect is phenomenologically incorporated in Eq. (5.16) by substituting $(\omega + i0)^2 \to \omega(\omega + i/\tau)$. The mean time of the electronic free pass in nanotubes (relaxation time) τ is estimated[76] by $\tau = 3 \times 10^{-13}$ s. The first term on the right side of Eq. (5.16) describes the intraband motion of π-electrons, and corresponds to the first term in Eq. (5.14). The second term on the right side of Eq. (5.16) describes direct transitions between the valence and the conductivity bands, and corresponds to the second term in Eq. (5.14). Note that the contribution of interband transitions is negligible in the frequency region determined by the condition

$$\omega < \omega_\ell = \begin{cases} 2v_F/R_{\text{CN}}, & \text{for metallic CNs,} \\ 2v_F/3R_{\text{CN}}, & \text{for semiconducting CNs.} \end{cases} \tag{5.18}$$

For typical nanotubes, the low-frequency edge of the optical transition band ω_l falls in the infrared regime. Figure 5.4 illustrates the behavior of the axial conductivity at frequencies of optical transitions.

Figure 5.5 shows $\sigma_{zz}(\omega)$ for zigzag CNs as a function of the radius (index m). For armchair CNs, this function is monotonic, because those CNs always exhibit metallic conductivity. Irrespective of the nanotube type, its conductivity tends[66] to the same limit equal to the graphene conductivity as $m \to \infty$.

The foregoing results demonstrate that a rigorous microscopic transport theory must be utilized for elaboration of the electrodynamics of CNs: constitutive

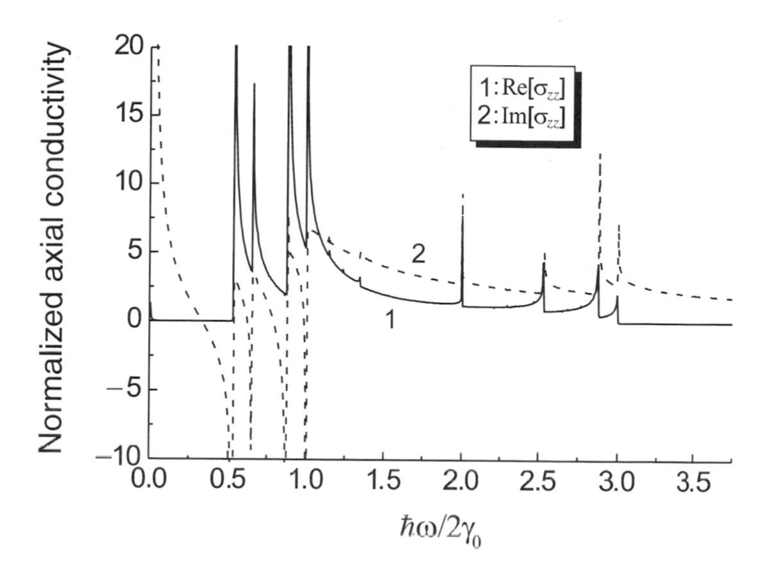

Figure 5.4 Frequency dependence of the total axial conductivity $\sigma_{zz}(\omega)$ of the (9, 0) metallic zigzag CN; $\tau = 3 \times 10^{-12}$ s and $T = 295$ K. The axial conductivity is normalized by $e^2/2\pi\hbar$. (Reprinted with permission from Ref. 41, © 1999 The American Physical Society.)

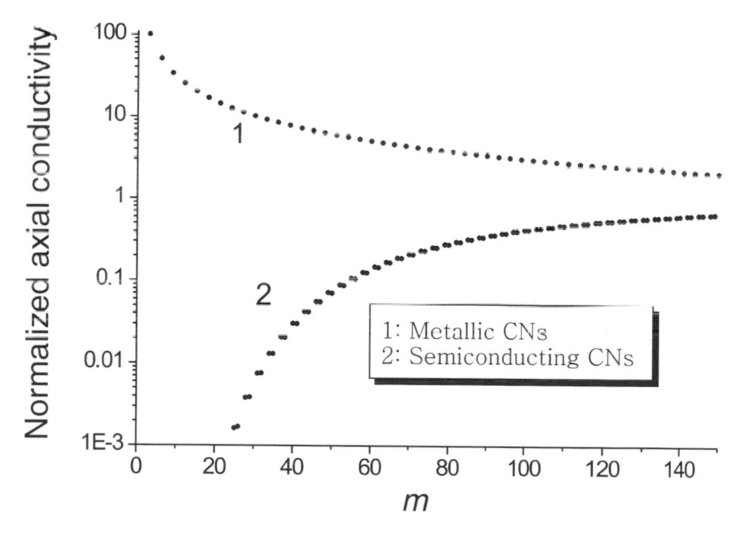

Figure 5.5 Normalized semiclassical conductivity $\sigma_{zz}(\omega)/\sigma_\infty$ for zigzag CNs as a function of m (and therefore of the cross-sectional radius R_{CN}); $\sigma_\infty = \lim_{m\to\infty} \sigma_{zz}$, $\tau = 3 \times 10^{-12}$ s, and $T = 264$ K. (Reprinted with permission from Ref. 4, © 1999 The American Physical Society.)

relations for CNs cannot be properly introduced without such a theory. In particular, phenomenological models proposed in Refs. 77 to 79 prove to be unsatisfactory since they assume the CN conductivity to be identical to the graphene conductivity. Figure 5.4 shows that such an approximation is adequate only for large-radius CNs ($m > 100$), where specific properties of CNs as low-dimensional structures do not manifest themselves since the role of transverse quantization becomes negligible.

5.3.2 Effective boundary conditions

Using the model of axial conductivity introduced in the previous section, we can now impose effective boundary conditions (EBCs) on the nanotube surface. This approach provides a general method for solving a wide range of problems of nanotube electrodynamics. The basic idea is to replace the periodic structure by a smooth one-sided surface on which appropriate EBCs for the electromagnetic field are imposed. These EBCs are chosen in such a manner that the spatial structure of the electromagnetic field induced by the effective current that flows on the smooth homogeneous surface and the spatial structure of the electromagnetic field generated by the real current in the lattice are identical at a certain distance from the surface. The lattice parameters are included in the so-called EBC coefficients.

The EBCs are obtained as a result of the spatial averaging of macroscopic fields over a physically infinitesimal element of the cylindrical surface. The condition that the tangential electric field component and the axial component of the magnetic field be continuous on the CN surface yields

$$E_{\varphi,z}|_{\rho=R_{\mathrm{CN}}+0} - E_{\varphi,z}|_{\rho=R_{\mathrm{CN}}-0} = 0, \quad H_z|_{\rho=R_{\mathrm{CN}}+0} - H_z|_{\rho=R_{\mathrm{CN}}-0} = 0. \tag{5.19}$$

The next condition follows from the equation for the CN axial conductivity.[41,42] Its derivation utilizes the relation between the surface current density $j_z(\omega)$ and the discontinuity of the magnetic field component H_φ at the CN surface; i.e.,

$$H_\varphi|_{\rho=R_{\mathrm{CN}}+0} - H_\varphi|_{\rho=R_{\mathrm{CN}}-0} = \frac{4\pi}{c}\sigma_{zz}(\omega)E_z|_{\rho=R_{\mathrm{CN}}}, \tag{5.20}$$

where c is the speed of light in the vacuum.

In the regime of optical transitions, the electromagnetic response of a CN is significantly influenced by the spatial dispersion of π-electrons provided, in particular by the Coulomb screening effect. A theory of this effect in quasi-1D structures is available.[80]

Spatial dispersion results in the CN conductivity σ_{zz} becoming a 1D integral operator. As an example, consider the propagation in the CN along its axis of a traveling wave with nonzero z directed component of the electric field: $E_z(\mathbf{r}, t) = \mathrm{Re}\{E_z^0 \exp[i(hz - \omega t)]\}$. The plane wave propagation considered in Sec. 5.2.2 corresponds to the particular case $h = 0$. For such a traveling wave, the conductivity acquires a dependence on the wave number h; i.e., $\sigma_{zz} = \sigma_{zz}(h, \omega)$. A concise expression for $\sigma_{zz}(h, \omega)$ is available.[42] Spatial dispersion is incorporated into EBCs by the change $\sigma_{zz}(\omega) \rightarrow \sigma_{zz}(\omega)[1 + \gamma(\omega)\partial^2/\partial z^2]^{-1}$, where $\gamma(\omega) = l_0/[k(1 + i/\omega\tau)]^2$, $k = \omega/c$ is the free-space wave number, and the coefficient

$$l_0 = \frac{k^2}{2\sigma_{zz}(0,\omega)} \frac{\partial^2 \sigma_{zz}(h,\omega)}{\partial h^2}\bigg|_{h=0} \left(1 + \frac{i}{\omega\tau}\right)^2 \tag{5.21}$$

characterizes the contribution of the spatial field nonhomogeneity.[40–42] After taking the foregoing into account, the EBC of Eq. (5.20) changes to

$$\Delta(\omega)\big(H_{\varphi}|_{\rho=R_{\rm CN}+0} - H_{\varphi}|_{\rho=R_{\rm CN}-0}\big) = \frac{4\pi}{c}\sigma_{zz}(\omega)E_z|_{\rho=R_{\rm CN}} \tag{5.22}$$

while Eq. (5.19) remain valid; here $\Delta(\omega) = 1 + \gamma(\omega)\partial^2/\partial z^2$. The estimates $l_0 \approx (v_F/c)^2$ for metallic nanotubes of a small radius, and $l_0 \approx 3(v_F/c)^2/4$ for any nanotube of a large radius, are available.[41,42] For metallic CNs, calculations yield $l_0 \sim 10^{-5}$.

The conditions of Eqs. (5.19) and (5.20), or Eqs. (5.19) and (5.22), constitute a complete system of EBCs for the electromagnetic field on the CN surface. They are analogous to the Weinstein–Sivov boundary condition[38] for grid structures and small-period grids in the microwave literature.

5.3.3 Surface electromagnetic waves

To exemplify the EBC method, let us examine the propagation of surface waves along an isolated infinite CN in free space, assuming that the nanotube exhibits axial conductivity. The eigenwaves under study satisfy the homogeneous Maxwell equations, boundary conditions Eqs. (5.19) and (5.20), and the condition that there are no exterior current sources at infinity. The problem formulated thus is similar to the eigenwave problem for microwave slow-wave helical structures and can be solved by the field-matching technique.[14,16,36]

The entire space is divided into two cylindrical partial domains—the domains inside and outside the tube. The electromagnetic field is represented by the scalar Hertz potential[35] Π_e. Using the Maxwell equations and the radiation conditions in the limit $\rho \to \infty$, we obtain

$$\Pi_e = A\exp(ihz + il\varphi)\begin{cases} I_l(\kappa\rho)K_l(\kappa R_{\rm CN}), & \rho < R_{\rm CN}, \\ I_l(\kappa R_{\rm CN})K_l(\kappa\rho), & \rho > R_{\rm CN}, \end{cases} \tag{5.23}$$

where A is an arbitrary constant; $\kappa = \sqrt{h^2 - k^2}$; while I_l and K_l are the modified cylindrical Bessel functions of the first and second kinds, respectively. The representation of Eq. (5.23) directly satisfies the EBCs of Eq. (5.19). Using an expression for the Wronskian of the modified Bessel functions, the dispersion relation for the surface wave in a CN is obtained:

$$\left(\frac{\kappa}{k}\right)^2 I_l(\kappa R_{\rm CN})K_l(\kappa R_{\rm CN}) = \frac{ic}{4\pi k R_{\rm CN}\sigma_{zz}}\left[1 - \frac{1+(\kappa/k)^2}{(1+i/\omega\tau)^2}l_0\right]. \tag{5.24}$$

Figure 5.6 shows the complex-valued slow-wave coefficient $\beta = k/h$ for the axially symmetric ($l = 0$) surface wave in the metallic (9, 0) CN obtained numerically from Eq. (5.24). Axially asymmetric modes are discussed elsewhere.[41] At low frequencies ($\omega < 1/\tau$), when $kb < 10^{-7}$ (where b is the C$-$C bond length),

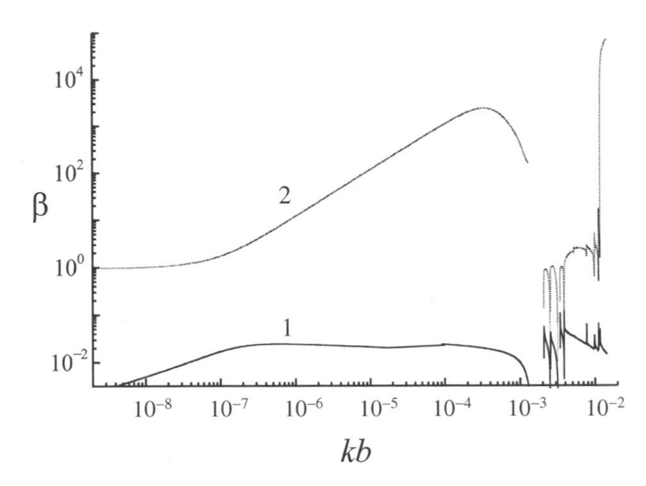

Figure 5.6 Frequency dependence of the complex-valued slow-wave coefficient β for the axially symmetric surface wave in a (9, 0) metallic zigzag CN. Input parameters are the same as for Fig. 5.4; 1, $\mathrm{Re}[\beta]$; and 2, $-\mathrm{Re}[\beta]/\mathrm{Im}[\beta]$. (Reprinted with permission from Ref. 41, © 1999 The American Physical Society.)

the nanotube demonstrates strong attenuation: $\mathrm{Im}[\beta] \sim \mathrm{Re}[\beta]$. One can thus conclude that the nanotubes are of no interest as surface waveguides at low frequencies. It is important that, for nanotubes of typical lengths of $l_{\mathrm{CN}} \sim 1\ \mu\mathrm{m}$, $l_{\mathrm{CN}}\mathrm{Re}[h] \ll 1$. This means that the CNs transmit low-frequency electric signals similar to electric circuits without wave effects. Unlike the low-frequency limit, in the infrared regime ($10^{-5} < kb < 10^{-3}$ or $3 \times 10^{12}\ \mathrm{s}^{-1} < \omega/2\pi < 3 \times 10^{14}\ \mathrm{s}^{-1}$), nanotubes permit the propagation of slowly decaying surface waves. Analysis has shown that, in the infrared regime, the slow-wave coefficient $\mathrm{Re}[\beta]$ of semiconducting CNs is $1/10$ that of the metallic CNs, while the respective values of $\mathrm{Im}[\beta]$ are comparable. Therefore, attenuation in semiconducting CNs is significantly higher than in the metallic CNs. Moreover, as semiconducting CNs are characterized by high slow-wave coefficients ($2 \times 10^{-3} < \mathrm{Re}[\beta] < 2 \times 10^{-2}$), the electromagnetic field in such CNs is tightly localized near the surface.

Note also that the slow-wave coefficient $\mathrm{Re}[\beta]$ and the phase velocity $v_{\mathrm{ph}} = \mathrm{Re}[\omega/h]$ are almost frequency independent. Therefore, a wave packet will propagate in the nanotube without significant distortions, which is very important for possible application in nanoelectronics. This demonstrates that CNs can serve in the infrared regime as dispersionless surface-wave nanowaveguides, which may become high-efficiency nanoelectronic elements.

Practical application of CNs as waveguiding structures and antenna elements requires the generation of different types of irregularities in CNs like those that are formed in ordinary macroscopic waveguides in the microwave range. There are several types of irregularities observed experimentally, e.g., junction of two CNs with different diameters,[81] T junctions formed by fusing two CNs of different diameters and chiralities perpendicular to each other,[82] differently configured Y junctions,[82–84] crossed CN junctions,[85] and setup of two CNs contacted to a gapped superconductor.[86] Note that the embedding of irregularities in a macroscopic waveguide does not change its electronic properties, but manifests itself in the scattering of electromagnetic waves and mode transformation. In contrast, irregularities in a nanowaveguide may drastically change the conductivity's character—which can be taken into account by means of corresponding transformation of EBCs [Eqs. (5.19) and (5.20) or Eqs. (5.19) and (5.22), equivalently] in the region adjoining the irregularity.

5.3.4 Edge effects

There are two alternative mechanisms for the manifestation of edge effects in CNs: electronic and electromagnetic. The first mechanism is provided by the modification of the CN electronic structure entailed by an edge; in particular, new electronic states are localized in the vicinity of an edge.[87] The second mechanism is related to electromagnetic wave diffraction at a CN edge. Here consideration is focused on the electromagnetic diffraction edge effects, which are similar to those that occur in wire antennas.[35] Indeed, at optical frequencies, the length and radius of real CNs satisfy the conditions $kR_{CN} \ll 1$ and $kl_{CN} \sim 1$. These are the same conditions that characterize microwave wire antennas. This analogy not only indicates the importance of studying resonances associated with the finite length of nanotubes, but also indicates the analytical method: the problem of surface wave diffraction by an open end of the semi-infinite nanotube is solved by the Wiener–Hopf technique.[35] A finite nanotube can be analyzed with the help of the modified factorization method or by using the approximate solutions of integral equations for induced current as is done in antenna theory.

Consider the diffraction of an E-polarized cylindrical electromagnetic wave by an open end of a nanotube. Let the wave travel at the angle θ_0 to the tubule axis (see Fig. 5.2). The scalar Hertz potential of this wave is given by

$$\Pi_e^{(\mathrm{inc})} = -\frac{i}{k\sin^2\theta_0} H_l^{(\mu)}(k\rho\sin\theta_0)\exp(ikz\cos\theta_0 + il\varphi), \tag{5.25}$$

where $H_l^{(\mu)}$ are the cylindrical Hankel functions of the first and second kinds ($\mu = 1$ or 2). The scalar Hertz potential of the scattered field satisfies the Helmholtz equation and is related to the electromagnetic field in the standard manner.[35] This enables the use of EBC in Eqs. (5.19) and (5.22) to derive boundary conditions for the potential.[43,44] These conditions should be supplemented by the radiation conditions as well as the edge condition (which requires that no source is present on a sharp edge). The edge condition implies that the field energy in any finite spatial region containing the edge is finite.

The boundary-value problem formulated can be solved by the Wiener–Hopf technique.[43] For convenience, the space is assumed to be filled by a lossy media in which $k = k' + ik''$, and the limit $k'' \to 0$ is taken in the final expressions. Application of the Jones approach[36] leads to a functional equation of the Wiener–Hopf type for two unknown functions $J_+(\alpha)$ and $Y_-(R_{CN},\alpha)$ in the band enclosing the real axis in the complex α plane; i.e.,

$$J_+(\alpha)G(\alpha)R_{CN} = Y_-(R_{CN},\alpha) - \frac{\widetilde{\Phi}(\alpha)}{\xi(\omega)\kappa^2}, \tag{5.26}$$

where $\kappa = \sqrt{\alpha^2 - k^2}$. The subscripts $\pm$ stand for functions that are analytical in the upper and lower half planes, respectively, while

$$G(\alpha) = K_l(\kappa R_{CN})I_l(\kappa R_{CN}) - \frac{1 - l_0\alpha^2/k^2}{R_{CN}\xi(\omega)\kappa^2}, \qquad \widetilde{\Phi}(\alpha) = \frac{kH_l^{(\mu)}(kR_{CN}\sin\theta_0)}{\alpha + k\cos\theta_0}. \tag{5.27}$$

The function $\widetilde{\Phi}(\alpha)$ is the Fourier transform of the function $\Phi(\varphi,z)\exp(-il\varphi)$, where

$$\Phi(\varphi,z) = \xi(\omega)k^2\sin^2\theta_0\Pi_e^{(\mathrm{inc})}(R_{CN},\varphi,z), \qquad \xi(\omega) = -4\pi i\sigma_{zz}(\omega)/ck. \tag{5.28}$$

The main idea of the technique used to solve Eq. (5.26) consists of factorization and decomposition of known functions that enter Eq. (5.26) to obtain[35] two independent expressions for $J_+(\alpha)$ and $Y_-(R_{CN},\alpha)$. As a result, the general solution for the diffracted field is expressed by quadratures. The function $G(\alpha)$ can be factorized as[36]

$$\ln\left[G_\pm(\alpha)\sqrt{\alpha \pm k}\right] = \frac{1}{2\pi i}\int_{-\infty\pm\alpha_0}^{+\infty\pm\alpha_0} \ln\left[\sqrt{\alpha'^2 - k^2}\,G(\alpha')\right]\frac{d\alpha'}{\alpha' \pm \alpha}, \tag{5.29}$$

where α_0 is a real number such that $0 < \alpha_0 < \mathrm{Im}(k)$. The factor $\sqrt{\alpha'^2 - k^2}$ provides the asymptotic behavior of the integrand necessary for the convergence of the integral. The integral in Eq. (5.29) cannot be analytically evaluated; therefore, the factorization has to be performed numerically. By following the standard Wiener–Hopf procedure,[36] we arrive at the following formula for the z component of the field outside the tubule:

$$E_z(\rho,\varphi,z) = \frac{\eta_l(\theta_0)}{2\pi ik}\exp(il\varphi)\int_C \exp(-i\alpha z)\frac{(\alpha - k)K_l(\kappa\rho)I_l(\kappa R_{CN})}{(\alpha + k\cos\theta_0)G_+(\alpha)}\,d\alpha. \tag{5.30}$$

Here,

$$\eta_l(\theta_0) = \frac{H_l^{(\mu)}(kR_{CN}\sin\theta_0)}{G_+(k\cos\theta_0)(1+\cos\theta_0)}, \tag{5.31}$$

and the integration path C is shown in Fig. 5.7 with the solid line. The field inside the tubule is obtained by interchanging ρ and R_{CN} in Eq. (5.30). The remaining components of **E** and **H** can be obtained in a similar manner.

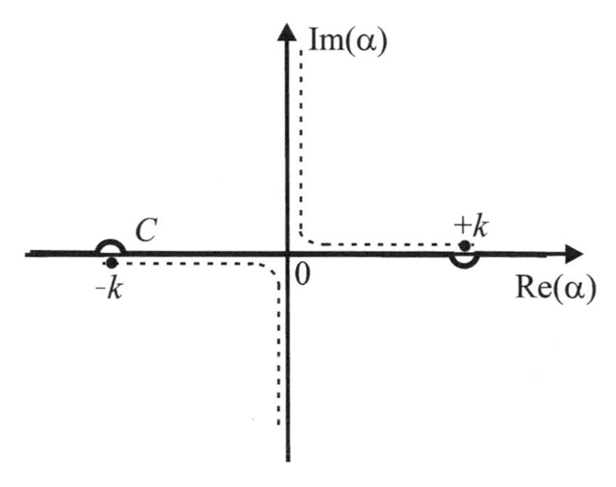

Figure 5.7 Contour C in the complex α-plane. The dashed lines show the branch cuts.[43]

Equation (5.30) is an exact analytical expression for the field scattered by a semi-infinite CN; and it holds true both near and far from the nanotube. In the near zone, the convergence of integrals in Eq. (5.30) is very slow and one should be careful when integrating numerically.

In the far zone, the integrals in Eq. (5.30) can be estimated asymptotically by the saddle-point method. The standard procedure results in

$$E_z \sim F(\theta, \theta_0) \sin\theta \frac{\exp\{ik\sqrt{\rho^2 + z^2}\}}{k\sqrt{\rho^2 + z^2}}, \tag{5.32}$$

where

$$F(\theta, \theta_0) = \eta_l(\theta_0) \frac{J_l(kR_{\mathrm{CN}} \sin\theta)}{G_-(k\cos\theta)(\cos\theta + \cos\theta_0)} \cot\left(\frac{\theta}{2}\right) \exp\left(-i\frac{\pi}{4}\right), \tag{5.33}$$

and $\theta = \pi/2 + \arctan(z/\rho)$. The function $F(\theta, \theta_0)$ is conventionally referred to as the edge scattering pattern. The total scattering pattern also contains components associated with surface polaritons.[43]

To illustrate the foregoing results, the far-zone scattered power density $P_l(\theta) \sim |F_l(\theta, \theta_0)|^2$ was calculated. The assumption $l = 0$ was made, because this term dominates for realistic incident fields. Figure 5.8 shows the scattered power density versus frequency and angle for the (9, 0) metallic CN. In this figure, the frequency dependence at a fixed angle θ exhibits strong oscillations: the scattered field significantly increases at frequencies that correspond to the optical transitions. The resonance scattering maximums are higher for metallic CNs than for semiconducting CNs. The figure also shows that a relatively small detuning from the exact resonance frequencies significantly reduces the intensity of the scattered field. Thus, one can conclude that, physically, the intense field scattering in CNs is related to the induction in the CN of a plasmon (which propagates from the CN edge along its axis) by the incident field. Therefore, the solution of Eqs. (5.32) and (5.33) shows

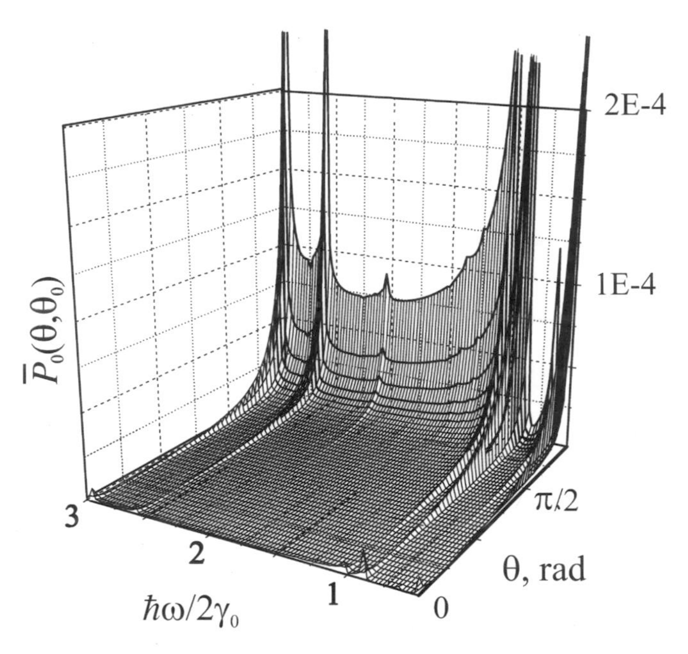

Figure 5.8 Density of the scattered power (normalized by the incident power density) $\overline{P}_0(\theta,\theta_0) = P_0(\theta)/[4\pi/c(1+\cos\theta_0)^2]$, for the metallic (9, 0) nanotube at frequencies of interband transitions when $\theta_0 = \pi/4$.[43]

that edge resonances play a significant role in the scattering process. This solution is the basis for solving the problem of electromagnetic scattering by a finite-length CN.

5.4 Nonlinear processes in carbon nanotubes

Nanostructures, and CNs in particular, exhibit a strong spatial nonhomogeneity and a large number of elementary resonances. A sufficiently strong dynamic nonlinearity is also typical of the nanostructures in a wide frequency range from the microwave to the ultraviolet regimes. This nonlinearity can manifest itself in various electromagnetic processes, such as solitonic propagation, optical instability, dynamical chaos, and the generation of high-order harmonics. These processes are of interest from two points of view. First, they can be used for the diagnostics of nanostructures. Second, these processes open new unique possibilities for controlling electromagnetic radiation, which is very promising for many optical and nanoelectronic applications.

The spatial nonhomogeneity of nanostructures hampers the description of nonlinear electromagnetic effects observed in them, because it involves nonlinear diffraction. Therefore, special simplifications are required in any particular case to reduce the original problem to a mathematical model that can be studied analytically or numerically. Two approaches should be mentioned that are most promising as applied to nanostructures. The first approach singles out the contribution of resonances by expanding the field in a set of specially chosen eigenmodes.[88] This approach was developed to solve nonlinear problems of macroscopic electromagnetics and was used, in particular, to analyze bistability in nonlinear diffraction lattices in the vicinity of the so-called Rayleigh–Wood anomalies.[88] An alternative approach—the electrodynamics of nonlinear composites—is based on the macroscopic averaging of the electromagnetic field in an ensemble of a large number of nonlinear scatterers whose sizes and distances are much smaller than the wavelength.[89–91]

As an example of a nonlinear problem, consider the generation within a CN of high-order harmonics of the incident field.[46–49] Interest in the generation of the high-order harmonics is caused primarily by searching for ways to create coherent far-ultraviolet and soft x-ray sources. Gases[92,93] and solid surfaces[94] have been studied as possible nonlinear media. Generation of high-order odd harmonics in gases is caused by the tunneling of electrons from atomic orbitals to the continuous-spectrum states and back, under the effect of a strong oscillating pumping field. In solid surfaces, harmonics (both even and odd) are generated by transitions of electrons through solid vacuum interfaces at high (relativistic) velocities. Pumping is provided by subpicosecond pulses of a titanium-sapphire laser with power density $\sim 10^{14}$ W/cm^2. The harmonic spectrum has a very characteristic shape in both cases:[92–94] it falls off for the first few harmonics, then exhibits a plateau when all the harmonics have approximately the same intensity, and ends with a sharp cutoff. The pumping wave-to-high-harmonics power conversion factor, which is between 10^{-6} and 10^{-7} in the plateau region. In the next section, high-order harmonic generation by conduction electrons confined at the cylindrical surface of a CN is considered.[46,48,49]

5.4.1 Current density spectrum in an isolated CN

Let a CN interact with an intense laser pulse whose electric field is polarized along the CN axis. Assume that the pumping frequency ω_1 satisfies inequalities (5.18), so that the contribution of interband transitions to the π-electrons motion can be neglected. This means that the motion of π-electrons is quasi-classical. After expanding $\mathcal{E}_c(p_z, s)/\gamma_0$ and $F_c(p_z, s)$ of Eq. (5.17) as Fourier series in p_z with coefficients $\mathcal{E}_{\rm sq}$ and $F_{\rm sq}$, respectively, the surface current density can be represented by[46]

$$j_z(t) = \sum_{M=0}^{\infty} j_z^{(2M+1)}(\omega_1) \sin\big[(2M+1)\omega_1 t\big], \tag{5.34}$$

where the coefficients

$$j_z^{(2M+1)}(\omega_1) = j_0 \sum_{s=1}^{m} \sum_{q=1}^{\infty} q \mathcal{E}_{\rm sq} F_{\rm sq} J_{2M+1}(\Lambda q) \tag{5.35}$$

involve the cylindrical Bessel functions $J_N(x)$ and

$$F_{sq} = \frac{a}{2\pi}\int_0^{2\pi/a} F_c(p_z, s)\exp(-iaqp_z)\,dp_z,$$
$$\mathcal{E}_{sq} = \frac{a}{2\pi\gamma_0}\int_0^{2\pi/a} \mathcal{E}_c(p_z, s)\exp(-iaqp_z)\,dp_z. \quad (5.36)$$

In these expressions, $j_0 = 8e\gamma_0/\pi\hbar R_{CN}$ and $\Lambda = \Omega_{st}/\omega_1$, where Ω_{st} is the angular Stark frequency; $\Omega_{st} = aeE_z/\sqrt{3}$ and $\Omega_{st} = aeE_z$ for armchair and zigzag CNs, respectively.

Figure 5.9 displays typical spectrums of the surface current density for metallic and semiconducting CNs at various pumping field intensities. If a titanium-sapphire laser with $\lambda = 0.8$ μm is used for pumping, $\Lambda = 1$ corresponds to the field $E_1 = 7 \times 10^9$ V/m, or the intensity $I_1 = 1.3 \times 10^{13}$ W/cm^2. The most important feature of the spectrums shown is the absence of cutoff frequencies.[92–94] This feature is due to the dispersion law for π-electrons in conducting CNs. Figure 5.9 also shows that a harmonic's spectrum falls much faster with the harmonic's number in semiconducting CNs than in metallic CNs.

Figure 5.10 shows the light intensity generated in the spectral range $300 < \lambda < 750$ nm, around the third harmonic (TH) of the Cr:forsterite laser[49] at 417 nm. The spectrums represent a continuous background superimposed on a narrow spectral line corresponding to the TH of the pump frequency. The TH generated by all samples of CNs in measurements*[49] is indeed emerging from a broad background, as illustrated in Fig. 5.10 for a sample of aligned multiwall CNs. The relative intensity of this background is, however, much higher than in the theoretical prediction. The

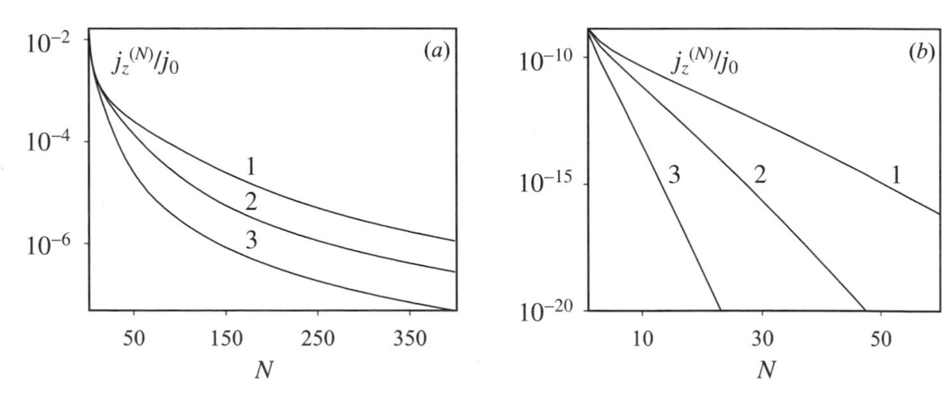

Figure 5.9 Envelope of the spectrum of high-order harmonics of the nonlinear current induced in (a) metallic (12, 0) and (b) semiconducting (11, 0) zigzag nanotubes by pumping pulses of different intensities: (1) $\Lambda = 1.0$, (2) 0.5, and (3) 0.2. The normalization factor j_0 for the metallic and semiconducting nanotubes is 2.6×10^6 A/m and 2.8×10^6 A/m, respectively; and $N = 2M + 1$ is the harmonic's number, $M = 0, 1, 2 \ldots$. (Reprinted with permission from Ref. 48, © 2001 The American Physical Society.)

*Experiments were carried out at Max Born Institute (Berlin, Germany) and Gothenburg University & Chalmers University of Technology (Gothenburg, Sweden).

formation of a local plasma through the emission of free electrons from the CNs could be the reason for this discrepancy.

TH generation with nonresonant excitation in bulk crystals or gases can be described, in general, by the third-order polarization $P^{(3)}(3\omega_1) = \chi^{(3)}(3\omega_1)E^3(\omega_1)$, even for rather high intensities below the optical damage threshold.[95] One would also expect a similar behavior for the TH yield of CNs. Figure 5.11(a) shows the theoretical dependence of the TH yield on the driving field intensity for various types of CNs. Surprisingly, for pump intensities as low as 10^{10} W/cm^2, the expected power law for the intensity dependence is broken; thus,

$$j_z^{(3)}(\omega_1) \sim E_z^p(\omega_1), \tag{5.37}$$

with the exponent p lying between 2.04 and 2.58 for the considered types of CNs.

The theory also predicts that p depends not only on the type of the CN and its diameter, but also on the pumping frequency. The experimental dependencies measured for samples of nonaligned multiwall CNs in Fig. 5.11(b) show good agreement with this theoretical prediction. Physically, this fact indicates that the interaction of CNs with an intense laser pulse can not be described by a perturba-

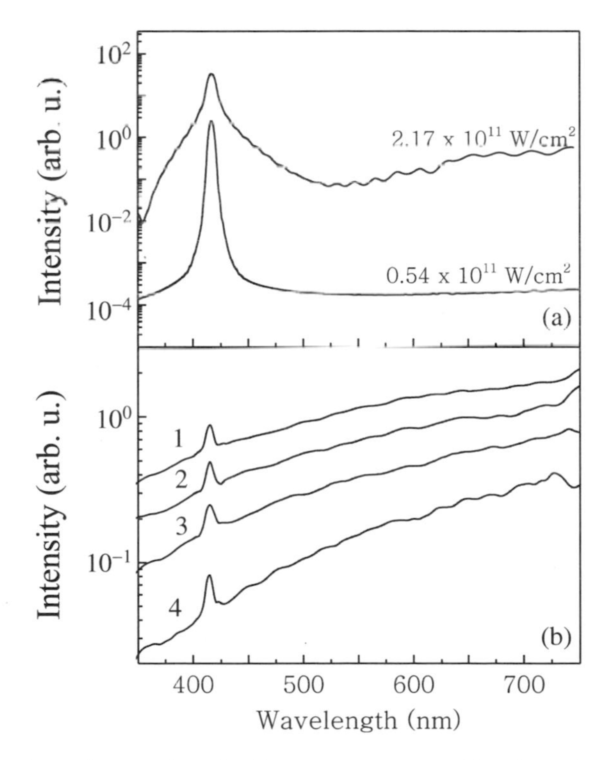

Figure 5.10 Broad background and TH signal generated by the interaction of intense laser radiation with aligned multiwall CNs: (a) theory and (b) experiment. Input intensities: 1, 2.3×10^{11} W/cm^2; 2, 1.7×10^{11} W/cm^2; 3, 1.3×10^{11} W/cm^2; and 4, 0.8×10^{11} W/cm^2. The experimental curves are corrected for the efficiencies of the monochromator, the photomultiplier and the transmission of the KG5 filters in the detection system.[49]

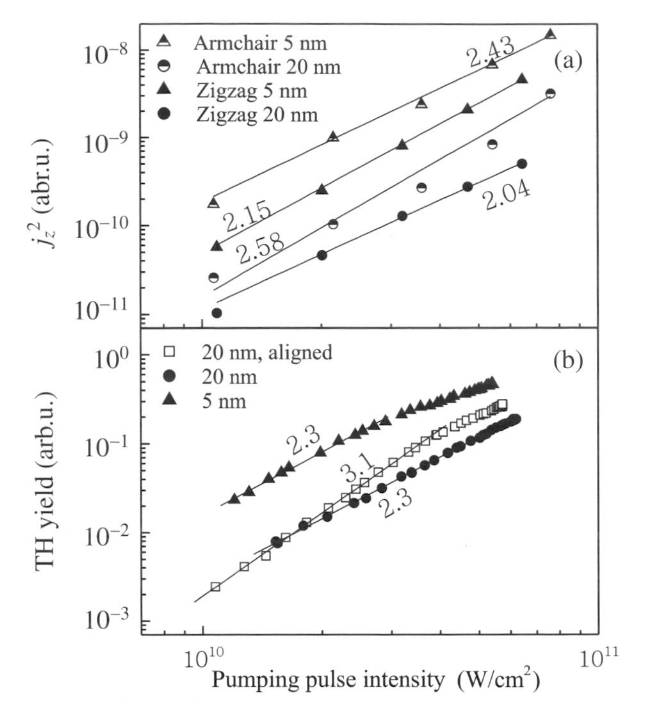

Figure 5.11 Dependence of the TH generation efficiency on the intensity of the pumping pulse: (a) theory and (b) experiment.[49]

tional approach, even for relatively low pump intensities. Note that the theoretical and experimental values of p agree remarkably well, except for the case of the orthogonally aligned array of CNs.

An interesting experimental result, shown in Fig. 5.11(b), is the observed decrease in slope of the TH intensity at a pump laser pulse intensity of $\sim 3 \times 10^{10}$ W/cm^2. Such saturation of the TH signal is also predicted by the theory, however at incident intensities about two orders of magnitude higher.

A similar situation, where the power expansion of polarization does not work, occurs for the fifth harmonic of the current density j_z at intensities $\sim 10^{10}$ to 10^{11} W/cm^2. Analysis shows that both theoretical and experimental values of the exponent p for the fifth harmonic differ from 5 and, at the same time, are close to each other (4.0 and 4.26 for experiment and theory, respectively).

In conclusion, the interaction of strong laser fields with samples of CNs can not be described by a power expansion of the polarization. This results in the violation of the general expressions $j_z^{(3)}(\omega_1) \sim E_z^3(\omega_1)$ and $j_z^{(5)}(\omega_1) \sim E_z^5(\omega_1)$ for the dependence of the third and fifth harmonics yields on the input laser field, even for intensities as low as 10^{10} to 10^{11} W/cm^2. The results from a fully quantum theoretical model show good agreement with experimental findings.

In this section, the high-order harmonics of the current density in a single CN have been studied. The next step is the study of high-order harmonics in arrays of aligned CNs with allowance for dispersion. Such an array is effectively an anisotropic birefringent media.[40] A consistent analysis of the problem stated with allowance for phase matching is given elsewhere.[48]

5.4.2 Negative differential conductivity in an isolated CN

In the quasi-static regime, nonlinear properties of the charge carriers in CNs also exhibit themselves as portions with the negative differential conductivity (NDC), $dI/dV < 0$, in the current-voltage (I-V) characteristics.[20,47,96] In a CN interacting simultaneously with dc and ac fields in the vicinity of a particular operating point in the I-V characteristics, instability evolves. This makes CNs attractive as potential nanoscale amplifying diodes similar to the macroscopic tunneling ones.

The I-V characteristics for tunneling electrons in individual single-wall CNs at low temperatures have been measured.[76,97] At temperatures such that $k_B T \ll \epsilon_c$ and $k_B T \ll \Delta\epsilon$, conduction occurs through well-separated discrete electron states; here, ϵ_c is the charging energy, and $\Delta\epsilon = \pi\hbar v_F / l_{CN}$ is the energy level spacing. It is reported[76] that $\Delta\epsilon \simeq 0.6$ meV for CNs with $l_{CN} \simeq 3$ μm; and the estimate $\epsilon_c \simeq 1.4e^2 \ln(l_{CN}/R_{CN})/l_{CN}$ has also been made.[98] Thus, $\epsilon_c \simeq 2.5$ meV for CNs of radius $R_{CN} \simeq 0.7$ nm, which is in good agreement with experimental data.[76] Under these conditions, current is produced by the electrons tunneling through a CN in the presence of the Coulomb blockade induced by the long-range (unscreened) Coulomb interaction. Due to this mechanism, the observed I-V characteristics are analogous to those obtained via scanning tunneling microscopy. As a result, the normalized differential conductivity $(V/I)(dI/dV)$ proves to be proportional to the local density of states. Therefore, the I-V characteristics of Refs. 76 and 97 carry important information on the nanotube electron structure. On the other hand, tunneling in macromolecules (in nanotubes, in particular) can serve as a basis for monomolecular transistors.[19]

In this section, the I-V characteristics of CNs at room temperature, when $k_B T \ll \epsilon_c$ and $k_B T \ll \Delta\epsilon$, are theoretically analyzed. Consider a single-wall zigzag nanotube exposed to a homogeneous axial dc field E_z. We apply the semi-classical approximation, considering the motion of π-electrons as the classical motion of free quasi-particles in the field of the crystalline lattice with dispersion law Eq. (5.4) extracted from quantum theory. The motion of quasi-particles in an external axial dc electric field is described by the Boltzmann kinetic equation wherein $\partial/\partial t = \partial/\partial z = 0$ is assumed and the collision integral is taken in the relaxation-time approximation. Depending on the relaxation time, the relaxation term can describe electron-phonon scattering, electron-electron collisions, etc.

The surface current density is determined by Eq. (5.14). Expansions of $F_c(p_z, s)$ and $\mathcal{E}_c(p_z, s)/\gamma_0$ into Fourier series in p_z, carried out by analogy with the previous section, lead us to the equation[47]

$$j_z(E_z) = \frac{j_0}{2} \sum_{q=1}^{\infty} \frac{q^2 \Omega_{st} \tau}{1 + (q\Omega_{st}\tau)^2} \sum_{s=1}^{m} F_{sq} \mathcal{E}_{sq}, \tag{5.38}$$

with F_{sq} and $\mathcal{E}_{sq}$ defined by Eq. (5.36). Equation (5.38) is the basis for evaluation of the I-V characteristics of CNs.

Direct numerical integration in Eq. (5.36) for the coefficients F_{sq} and $\mathcal{E}_{sq}$ is technically difficult because the integrands are rapidly oscillating functions. Therefore, the following technique is suggested. The change of variable $z = \exp(iap_z)$ transforms the original integrals into integrals over the closed path $|z| = 1$ in the complex plane. The integrands have two pairs of branch points in the z plane. The integrand for F_{sq} also has an infinite number of first-order poles inside the unit circle. According to the Cauchy residue theorem, the integrals can be written in terms of integrals over banks of the branch cuts plus series of residues (for F_{sq}). The integrals over the cut banks do not contain oscillating functions and can easily be calculated numerically. The residue series converge rapidly and can also be summed numerically.

Let us estimate constraints that follow from this theoretical model. As has been stated previously, the model describes motion of the quasi-particles by the classical Boltzmann kinetic equation. Thus, both interband transitions and quantum-mechanical corrections to the intraband motion are not accounted for in this model. The first of these approximations is valid when the inequality $\Omega_{st} \leq \omega_l$ holds true, where ω_l is given by Eq. (5.18). The second assumption requires that Ω_{st} does not exceed the allowed band width, which is of the order of γ_0. This estimate and inequality (5.18) reduce both constraints imposed on the Stark frequency to the limitation on the intensity of the external electric field $|E_z| < \gamma_0/2eR_{CN}$.

The adopted theoretical model also neglects the Coulomb interaction between electrons. The role of this interaction in CNs has been addressed.[98–100] It has been found that the short-range electron-electron interaction typical for CN arrays does not significantly contribute at high temperatures. Since the Coulomb interaction in an isolated CN is unscreened, it exhibits itself in a different manner to provide an observable effect over a wide temperature range. Therefore, the results obtained from the adopted model are applicable primarily to CN arrays. For a single CN, this model should be modified to allow for the long-range Coulomb interaction. A change in the temperature dependence of the relaxation time τ is expected as the only result of the Coulomb interaction.[99]

Figure 5.12 shows the I-V characteristic of undoped (with zero chemical potential) metallic zigzag nanotubes. When the strength of the imposed electric field is low, j_z is a linear function of E_z, corresponding to ohmic conductivity. By increasing the imposed electric field strength, $\partial j_z/\partial E_z$ decreases until the current density reaches its maximum value j_z^{max} at $E_z = E_z^{max}$. Increasing the intensity of the applied electric field will further decrease j_z. Thus, the negative differential conductivity $\partial j_z/\partial E_z < 0$ is predicted.

The imposed field strength $E_z^{max} \approx 3.2 \times 10^3$ V/cm at which the NDC begins to be found to be unexpectedly weak. Indeed, nonlinearity in these structures is determined by the quantity aE_z. In quantum superlattices the spacing is about 10^{-6} cm,[9] which is much greater than the C—C bond length b in graphene. Nevertheless, the NDC is observed in them almost at the same strength of the imposed

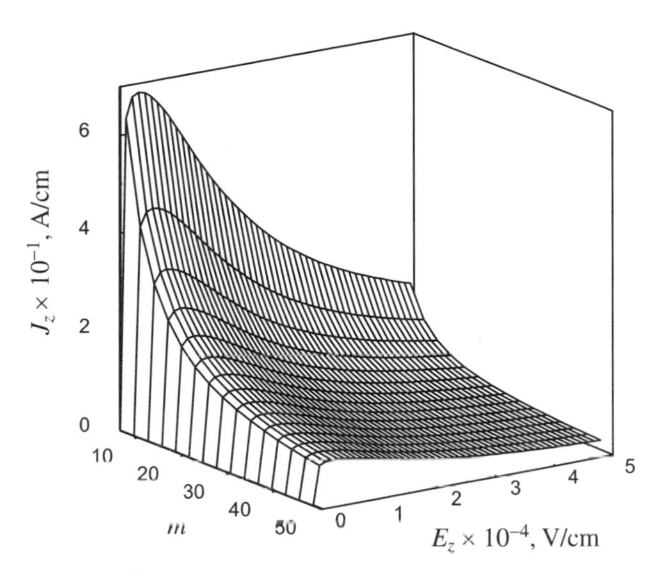

Figure 5.12 I-V characteristics of metallic zigzag nanotubes at $T = 287.5$ K and $\tau = 3 \times 10^{-12}$ s. (Reprinted with permission from Ref. 47, © 2000 The American Physical Society.)

field. Therefore, nonlinearity in CNs is much stronger than in quantum superlattices.

To explain this phenomenon, let us compare the nonlinear conductivity mechanisms in CNs and superlattices. In quantum superlattices, the dispersion law is $\mathcal{E}_{v,c}(p_z) = \pm\gamma_0'[1 - \cos(2ap_z/3)]$, where γ_0' is the overlapping integral. Applying the method described previously to this dispersion law, the expression $j_z(E_z) = \sigma_{zz}E_z/(1 + i\tau\Omega_{\rm st})$ is obtained instead of Eq. (5.38), where $\sigma_{zz} = \lim_{E_z\to 0}(\partial j_z/\partial E_z)$ is the linear conductivity. The comparison of these two expressions for the current density shows that a specific feature of CNs is the production of high-order Stark harmonics. Calculations show that the number of significant Stark harmonics is within 70 to 150 for metallic CNs and within 200 to 300 for the semiconducting CNs. As a result, the high-order Stark components play a significant role in CNs, and the integral nonlinearity in CNs is much stronger than in superlattices. Impurities and defects in the lattice provide an additional mechanism for carrier scattering, which can be described quantitatively by the substitution $\tau \to \tau' = \tau\tau_1(\tau + \tau_1)^{-1}$, where the relaxation time τ_1 is determined by the impurities and defects. Since $\tau' < \tau$, doping increases $E_z^{\max}$ and decreases $\partial j_z/\partial E_z$ in the NDC regime.

The predicted NDC effect in CNs is expected to be observable in sufficiently long CNs at room temperatures. As was emphasized before, the NDC causes the current instability. One can expect that simultaneously applied dc and ac fields will result in dynamic electron localization (which is the nonlinear phase of the instability) and in the 2D analog of the self-induced transparency. The effects mentioned are responsible for the absolute negative conductivity, which is thus predicted in CNs. Due to this phenomenon, regions must appear where nanotubes exhibit absolute negative conductivity and active properties, which hints at the possibility of developing microwave and infrared oscillator nanodiodes in single CNs as well as in CN arrays.

The predicted NDC mechanism is not alone in creating the NDC effect. Another mechanism observed in nonhomogeneous nanotubes is caused by tunneling of π-electrons through the potential barrier near the nonhomogeneity.[20,96]

5.5 Quantum electrodynamics of carbon nanotubes

5.5.1 Maxwell equations for electromagnetic field operators

In most cases, electromagnetic modeling of nanostructures assumes the number of photons involved in the process to be large enough to describe the electromagnetic field by classical equations. At the same time, peculiarities of traditional QED effects—such as spontaneous emission and electromagnetic fluctuations—as well as recently raised ideas to use nanostructures for storage and processing of quantum information, provide a growing interest for developing the QED of nanostructures and, in particular, CNs. The quantum nature of the electromagnetic field in CNs should then be taken into account. Since the nano-object (i.e., the CN) is an nonhomogeneity much smaller than the photon wavelength, this issue appears to be significantly more complex than QED problems in homogeneous media. This section is focused on the problem of spontaneous emission of an atom located inside or in the vicinity of a CN.

Standard schemes of the electromagnetic field quantization are based on modal representations: in free space, these modes are plane waves; in cavities, they are eigenmodes. The quantum description of the electromagnetic field replaces coefficients of such modal representations by operators of creation and annihilation of photons associated with a particular mode.

Since nanostructures are strongly nonhomogeneous open systems, it is usually difficult to find an appropriate system of eigenmodes. Therefore, an alternative approach developed recently for lossy dispersive media[101] appears to be more convenient for the QED of nanostructures. This approach rejects the modal representation and allows for quantization in the Maxwell equations: the vectors **E** and **H** are replaced by corresponding operators that satisfy the appropriate commutation relations and define observable quantities as mean values of these operators.

Let us therefore introduce the electric field operator $\widehat{\mathbf{E}}(\mathbf{r}) = \widehat{\mathbf{E}}^{(+)}(\mathbf{r}) + \widehat{\mathbf{E}}^{(-)}(\mathbf{r})$, where

$$\widehat{\mathbf{E}}^{(+)}(\mathbf{r}) = \int_0^\infty \underline{\widehat{\mathbf{E}}}(\mathbf{r}, \omega)\, d\omega, \quad \widehat{\mathbf{E}}^{(-)}(\mathbf{r}) = \left[\widehat{\mathbf{E}}^{(+)}(\mathbf{r})\right]^\dagger, \tag{5.39}$$

and † indicates the Hermitian conjugate. The magnetic field operator $\widehat{\mathbf{H}}(\mathbf{r})$ is defined in the same manner.

Operators $\underline{\widehat{\mathbf{E}}}$ and $\underline{\widehat{\mathbf{H}}}$ are subject to radiation conditions at infinity. They satisfy the Maxwell equations

$$\nabla \times \underline{\widehat{\mathbf{E}}} = ik\,\underline{\widehat{\mathbf{H}}}, \qquad \nabla \times \underline{\widehat{\mathbf{H}}} = -ik\,\underline{\widehat{\mathbf{E}}} + \frac{4\pi}{c}\widehat{\mathbf{J}}^{\text{ext}}, \tag{5.40}$$

where $\widehat{\mathbf{J}}^{\text{ext}}$ is the external current operator. The effective boundary conditions of Eqs. (5.19) and (5.20) for Eq. (5.40) are rewritten as

$$\begin{aligned} &\mathbf{n} \times \left(\underline{\widehat{\mathbf{E}}}|_{\rho=R_{\text{CN}}+0} - \underline{\widehat{\mathbf{E}}}|_{\rho=R_{\text{CN}}-0}\right) = \mathbf{0}, \\ &\mathbf{n} \times \left(\underline{\widehat{\mathbf{H}}}|_{\rho=R_{\text{CN}}+0} - \underline{\widehat{\mathbf{H}}}|_{\rho=R_{\text{CN}}-0}\right) + \tfrac{4\pi}{c}\widehat{J}_z^{ns}\mathbf{e}_z = \tfrac{4\pi}{c}\sigma_{zz}(\omega)\widehat{E}_z\mathbf{e}_z, \end{aligned} \tag{5.41}$$

where $\mathbf{n}$ is the unit vector along the exterior normal to the CN surface, $\widehat{J}_z^{ns}$ is the operator of an axial noise current, and the axial dynamical conductivity of CN $\sigma_{zz}(\omega)$ is given by Eq. (5.16). The axial noise current is expressed as $\widehat{J}_z^{ns} = \{\hbar\omega\text{Re}[\sigma_{zz}(\omega)]/\pi\}^{-1}\,\hat{f}(\mathbf{R},\omega)$ in terms of 2D scalar field operator $\hat{f}(\mathbf{R},\omega)$ satisfying standard bosonic commutation relations

$$\begin{aligned} &\left[\hat{f}(\mathbf{R},\omega), \hat{f}^{\dagger}(\mathbf{R}',\omega')\right] = \delta(\mathbf{R}-\mathbf{R}')\delta(\omega-\omega'), \\ &\left[\hat{f}(\mathbf{R},\omega), \hat{f}(\mathbf{R}',\omega')\right] - \left[\hat{f}^{\dagger}(\mathbf{R},\omega), \hat{f}^{\dagger}(\mathbf{R}',\omega')\right] = 0, \end{aligned} \tag{5.42}$$

where $\delta(\cdot)$ is the Dirac delta function, $(\hat{f}_1, \hat{f}_2) = \hat{f}_1\hat{f}_2 - \hat{f}_2\hat{f}_1$, and $\mathbf{R}$ lies on the CN surface. The axial noise current is responsible[101] for the correct commutation relations of the operators $\underline{\widehat{\mathbf{E}}}$ and $\underline{\widehat{\mathbf{H}}}$. The homogeneous Maxwell Eqs. (5.40) along with the boundary conditions of Eq. (5.41) describe the QED of CNs.

One of the most important applications[102,103] of this quantization scheme is the dynamics of an excited two-level atom located inside (or near) a dielectric object with relative permittivity $\varepsilon(\mathbf{r},\omega)$. This problem considers an electric dipole transition, characterized by the dipole moment $\boldsymbol{\mu}$ and frequency ω_A, in an electrically neutral atom located at position $\mathbf{r} = \mathbf{r}_A$. The general expression for the spontaneous radiation time τ_{sp} for this system in Markovian approximation is[101]

$$\Gamma_{\text{sp}} = \frac{1}{\tau_{\text{sp}}} = \frac{8\pi}{\hbar}k_A^2\mu_\alpha\mu_\beta\text{Im}\left[G_{\alpha\beta}(\mathbf{r}_A,\mathbf{r}_A,\omega_A)\right], \tag{5.43}$$

where $k_A = \omega_A/c$, and $G_{\alpha\beta}$ are the components of the classical dyadic Green's function that accounts for the dielectric object. This notation implies summation over repetitive indexes.

Expression (5.43) can be interpreted physically as follows. Spontaneous emission is the process of interaction between an excited atom and the vacuum states of the electromagnetic field. The vacuum states are diffracted by the dielectric object similar to the diffraction of conventional electromagnetic fields. This effect is taken into account[103] by the difference between the dyadic $G_{\alpha\beta}$ and the free-space Green's dyadic $G^{(0)}_{\alpha\beta}$. Since $\mathrm{Im}[G^{(0)}_{\alpha\beta}(\mathbf{r}_A, \mathbf{r}_A, \omega_A)] = \omega_A \delta_{\alpha\beta}/6\pi c$, Eq. (5.43) simplifies to the formula[104]

$$\Gamma^{(0)}_{\mathrm{sp}} = \frac{1}{\tau^{(0)}_{\mathrm{sp}}} = \frac{4\omega_A^3}{3\hbar c^3}|\boldsymbol{\mu}|^2 \tag{5.44}$$

for the spontaneous decay time in free space. Note also that the Lamb shift in the transition frequency due to the presence of the dielectric object, which differs from that in free space, has been reported.[103]

Equation (5.43) was used to analyze various physical situations: for example, to calculate the spontaneous decay time of an excited atom in a spherical microcavity,[103] and to study the influence of the local field effects on the spontaneous emission in optically dense gases and solid dielectrics.[102] Next, the foregoing quantization scheme is used to study the spontaneous emission process in CNs.

5.5.2 Spontaneous decay of an excited atom in a CN

Consider the spontaneous decay of an excited atom located inside a CN at a distance ρ_0 from its axis.[50] The dipole moment of the atom is assumed to be aligned with the z axis. Note that application of the EBC method to the problem of the spontaneous decay of an atom inside a nanotube has already yielded Eq. (5.43).

Since the dipole moment of the atom is parallel to the CN axis, only the longitudinal component G_{zz} of the dyadic Green's function is of physical interest. Let us represent this component in terms of the scalar Green's function $\overline{G}$ of the atom in the CN as follows:

$$G_{zz} = \frac{1}{k^2}\left(\frac{\partial^2 \overline{G}}{\partial z^2} + k^2 \overline{G}\right). \tag{5.45}$$

In turn, $\overline{G}$ can be represented as

$$\overline{G} = \begin{cases} \widetilde{G}^{+}, & \rho > R_{\mathrm{CN}} \\ G_0 + \widetilde{G}^{-}, & \rho < R_{\mathrm{CN}}, \end{cases} \tag{5.46}$$

where $G_0 = \exp(ik\rho)/4\pi\rho$ is the free-space scalar Green's function. The unknown functions $\widetilde{G}^{\pm}$ satisfy the homogeneous Helmholtz equation and boundary conditions on the CN surface, which follow from the EBCs of Eqs. (5.19) and (5.20) as

$$\widetilde{G}^{+}\big|_{\rho=R_{\mathrm{CN}}} = (G_0 + \widetilde{G}^{-})\big|_{\rho=R_{\mathrm{CN}}},$$

$$\left[\Delta(\omega)\frac{\partial}{\partial\rho}(\widetilde{G}^{+} - \widetilde{G}^{-}) - \xi(\omega)\left(\frac{\partial^2 \overline{G}}{\partial z^2} + k^2\overline{G}\right)\widetilde{G}^{+}\right]\Bigg|_{\rho=R_{\mathrm{CN}}} = \Delta(\omega)\frac{\partial G_0}{\partial\rho}\bigg|_{\rho=R_{\mathrm{CN}}}.$$

The function $\xi(\omega)$ is related by Eq. (5.28) to the axial conductivity $\sigma_{zz}(\omega)$ of Eq. (5.16).

Let us seek $\widetilde{G}^{\pm}$ as expansions in terms of cylindrical functions. Using Eq. (5.43), we arrive at the expression

$$\Gamma_{\text{sp}} = \zeta(\omega_A)\Gamma_{\text{sp}}^{(0)}, \tag{5.47}$$

where

$$\zeta(\omega_A) = 1 + \frac{3\pi R_{\text{CN}}}{16k_A^3} \sum_{p=-\infty}^{\infty} \text{Im} \int_C \frac{\beta_A \kappa_A^4 I_p^2(\kappa_A \rho_0) K_p^2(\kappa_A R_{\text{CN}})}{1 - \beta_A R_{\text{CN}} \kappa_A^2 I_p(\kappa_A \rho_0) K_p(\kappa_A R_{\text{CN}})}\, dh \tag{5.48}$$

for the spontaneous decay rate of an atom in an isolated CN. Here, $\kappa_A = \sqrt{h^2 - k_A^2}$ and $\beta_A = -\xi(\omega_A)/[1 - \gamma(\omega_A)h^2]$. The integration path C in the complex plane is shown in Fig. 5.7. The quantity ζ_0 directly characterizes the effect of diffraction of the vacuum states on the spontaneous decay rate of an atom in the nanotube. Note that the integral in Eq. (5.48) can not be reduced to an integral with finite limits as was done elsewhere[105] for a perfectly conducting cylinder.** This is due to the contribution to the spontaneous decay of surface waves propagating in the CN. By analogy with the classical diffraction theory, one can expect this contribution to be significant.

For the inner region ($r_A < R_{\text{CN}}$), Eq. (5.48) is modified by the simple interchange of r_A and R_{CN} in the numerator of the integrand. Note the divergence of the integral in Eq. (5.48) at $r_A = R_{\text{CN}}$, i.e., when the atom is located directly on the CN surface. This divergence originates from the averaging procedure over a physically infinitely small volume when describing the optical properties of a CN. Such an averaging does not assume any additional atoms on the CN surface; to take them into consideration the procedure must be modified. Thus, the domain of applicability of the presented model is restricted by the condition $|r_A - R_{\text{CN}}| > b$.

The decay of the excited atom interacting with media may proceed both via real photon emission (radiative decay) and via virtual photon emission with subsequent excitation in the media of quasi-particles (nonradiative decay). Both of these decay channels are present in the atomic spontaneous decay rate Γ_{sp} described by Eqs. (5.47) and (5.48).

The partition of the total Γ_{sp} into radiative and nonradiative contributions is not a trivial problem. For an atom near a microsphere, the radiative contribution Γ_r has been estimated by using the Poynting vector.[103] The radiative contribution has also been estimated for an atom inside an optical fiber.[106] Following this approach, let us estimate the spontaneous emission intensity $I(\mathbf{r}, t)$ at large distances $|\mathbf{r}| \to \infty$. In a spherical coordinate system, $(|\mathbf{r}|, \phi, \theta)$, with its origin fixed on the atom, we

**Modeling as a perfectly conducting cylinder is inadequate[42] for CNs.

obtain

$$I(\mathbf{r},t) \simeq \frac{1}{|\mathbf{r}|^2} k_A^4 |\mu|^2 \sin^2\theta \left| \sum_{p=-\infty}^{\infty} \Xi_p(-ik_A \sin\theta) e^{ip\phi} \right|^2 \exp(-\Gamma_{\mathrm{sp}} t), \quad (5.49)$$

with

$$\Xi_p(x) = \begin{cases} \dfrac{I_p(xr_A)}{1 + R_{\mathrm{CN}}\beta_A x^2 I_p(xR_{\mathrm{CN}}) K_p(xR_{\mathrm{CN}})}, & r_A < R_{\mathrm{CN}}, \\ I_p(xr_A) - \dfrac{R_{\mathrm{CN}}\beta_A x^2 I_p^2(xR_{\mathrm{CN}}) K_p(xr_A)}{1 + R_{\mathrm{CN}}\beta_A x^2 I_p(xR_{\mathrm{CN}}) K_p(xR_{\mathrm{CN}})}, & r_A > R_{\mathrm{CN}}. \end{cases} \quad (5.50)$$

Then the relative contribution of the radiative channel is given by

$$\begin{aligned} \frac{\Gamma_r}{\Gamma_{\mathrm{sp}}} &= \frac{c}{2\pi\hbar\omega_A} \lim_{|\mathbf{r}|\to\infty} \int_0^\infty dt \int_0^{2\pi} d\phi \int_0^\pi |\mathbf{r}|^2 I(\mathbf{r},t) \sin\theta \, d\theta \\ &= \frac{3}{4\zeta(\omega_A)} \sum_{p=-\infty}^{\infty} \int_0^\pi \left|\Xi_p(-ik_A \sin\theta)\right|^2 \sin^3\theta \, d\theta. \end{aligned} \quad (5.51)$$

Figure 5.13 shows the values of $\zeta(\omega_A)$ calculated according to Eq. (5.48) for metallic and semiconducting zigzag CNs. The atom is supposedly located on the CN axis. The frequency range $0.305 < \hbar\omega_A/2\gamma_0 < 0.574$ corresponds to visible light. Lower frequencies $\hbar\omega_A/2\gamma_0 < 0.305$ correspond to infrared waves emitted

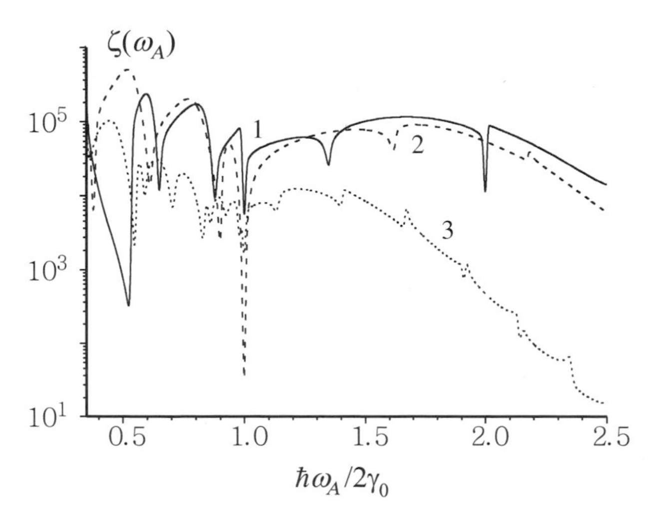

Figure 5.13 Graph of $\zeta(\omega_A)$ calculated from Eq. (5.48) for an atom located on the axis of a zigzag CN of order $(n, 0)$: 1, (9, 0); 2, (10, 0); and 3, (23, 0). Surface axial conductivity σ_{zz} appearing in (5.48) was calculated in the relaxation-time approximation[9] with $\tau = 3 \times 10^{-12}$ s (Reprinted with permission from Ref. 50, © 2002 The American Physical Society.)

by highly excited Rydberg atomic states. A large difference (of three to four orders of magnitude) is seen in the values of $\zeta(\omega_A)$ for metallic and semiconducting CNs. The difference is caused by the Drude-type conductivity (intraband electronic transitions) dominating at infrared and visible frequencies, the relative contribution of the intraband transitions to the total CN conductivity being larger in metallic than in semiconducting CNs.[41,42,107]

As the frequency increases, interband transitions start manifesting themselves and $\zeta(\omega_A)$ becomes irregular. At high frequencies, there is no significant difference between metallic and semiconducting CNs of approximately equal radius. The function $\zeta(\omega_A)$ has dips when ω_A equals the interband transition frequencies; in particular, there is a dip at $\hbar\omega_A = 2\gamma_0$ for all CNs considered. It is essential that $\zeta(\omega_A) \gg 1$ throughout the entire frequency range considered. This enables us to formulate the central result of the present analysis: the spontaneous decay probability of an atom in the vicinity of a CN is larger by a few orders of magnitude than that of the same atom in free space. In other words, the Purcell effect[108] is extraordinarily strong in CNs. This is physically explained by the photon vacuum renormalization: the density of photonic states (and, as a consequence, the atomic decay rate) near a CN effectively increases as per $\zeta(\omega)\omega^2/\pi c^3$, since, along with ordinary free photons, photonic states coupled with CN electronic quasi-particle excitations appear. The presence of a CN is seen to drastically accelerate the spontaneous decay process of an excited atomic state.

The possible existence of slow surface electromagnetic waves in CNs has been demonstrated.[41,42] Such waves are responsible for the strong Purcell effect for an atom in a spherical microcavity,[103] which conclusion is in qualitative agreement with the results of the present analysis. However, there is the risk of going beyond the applicability limits of the two-level model and Markovian approximation.[101] Indeed, considering the spontaneous radiation of the atom in the near-surface regime, one gets

$$\zeta(\omega_A) \approx \frac{3\varepsilon''(\omega_A)}{8\mid \varepsilon(\omega_A)+1\mid^2} \frac{1}{(k_A|r_A - R_{\rm CN}|)^3} + O\big(|r_A - R_{\rm CN}|^{-1}\big), \tag{5.52}$$

for the tangential atomic dipole orientation;[103] here, $\varepsilon(\omega)$ is the relative permittivity of the subsurface media, and $\varepsilon''(\omega) = \mathrm{Im}[\varepsilon(\omega)]$. Seemingly, approaching the surface, one obtains arbitrary large Γ. However, in doing so one has to remain within the applicability domain of the macroscopic approximation.

Equation (5.52) was derived under the condition that $|r_A - R_{\rm CN}|$ is much smaller than all other parameters, or, more physically, when the atom is placed so close to the surface that it *sees* a quasi-plane and the surface curvature is irrelevant. For CNs of small enough radius ($m \approx 10$ to 30), this condition contradicts the inequality $|r_A - R_{\rm CN}| > b$ determining the applicability limits of the macroscopic approximation for this particular task. As a consequence, the CN surface curvature turns out to be essential and Eq. (5.48) can not be, in principle, reduced to any equation similar to Eq. (5.52). Thus, the large Purcell effect in CNs has nothing to do with the near-surface regime.

Figure 5.14 shows $\zeta(\omega_A)$ for an atom located outside a CN at different distances outside the CN surface. The qualitative behavior of $\zeta(\omega_A)$ is similar to that in Fig. 5.13 for an atom inside a CN. It is seen that $\zeta(\omega_A)$ rapidly decreases with increasing distance—as it should be, in view of the fact that photonic states coupled with CN electronic excitations are spatially localized on the CN surface, and their coupling strength with the excited atom decreases with increasing distance of the atom from CN. Figure 5.15 shows the ratio $\Gamma_r/\Gamma_{\text{sp}}$ calculated according to Eq. (5.51) for an atom located on the axis of a CN.

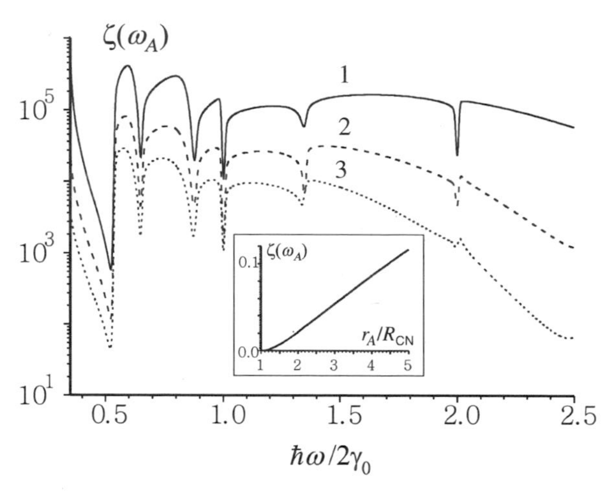

Figure 5.14 Plot of $\zeta(\omega_A)$ for an atom at different distances outside a zigzag (9, 0) CN: 1, $r_A = 1.5R_{\text{CN}}$; 2, $2.0R_{\text{CN}}$; and 3, $2.5R_{\text{CN}}$. Inset: $\xi(\omega_A)$ at $\omega_A = 3\gamma_0/\hbar$ as a function of r_A/R_{CN} for an atom near a (9, 0) CN modeled as a perfectly conducting cylinder. (Reprinted with permission from Ref. 50, © 2002 The American Physical Society.)

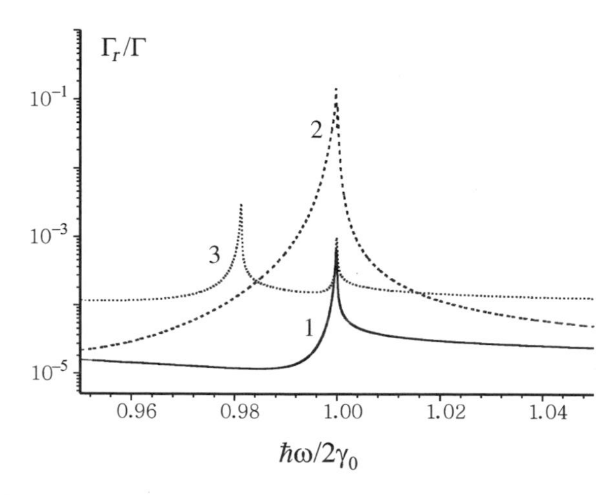

Figure 5.15 Ratio $\Gamma_r/\Gamma_{\text{sp}}$ calculated from Eq. (5.51) for an atom located at the axis of a zigzag CN of order $(n, 0)$: 1, (9, 0); 2, (10, 0); and 3, (23, 0). (Reprinted with permission from Ref. 50, © 2002 The American Physical Society.)

Note that $\Gamma_r/\Gamma_{\text{sp}} = W_s(\omega_A)/\hbar\omega_A$, with $W_s(\omega_A)$ being the total power of atomic spontaneous radiation far from the CN. The ratio is very small, indicating that nonradiative decay dominates over radiative decay. However, the radiative decay is seen to essentially contribute in the vicinity of the interband transition frequencies. Therefore, the frequency dependence of $W_s(\omega_A)$—which quantity, in principle, can be measured experimentally—reproduces the specific features of the CN electronic structure. The main conclusion one can draw from Fig. 5.15 is that the Purcell effect in CNs, along with the increase of the atomic spontaneous decay rate, manifests itself by decreasing the power of spontaneous radiation.

The presented model of atomic spontaneous decay in the presence of a CN allows, as a limiting case, to consider the CN as a perfectly conducting cylinder.[105] The inset in Fig. 5.14 shows $\zeta(\omega_A)$ at $\omega_A = 3\gamma_0/\hbar$ ($k_A R_{\text{CN}} \simeq 0.01$) as a function of r_A/R_{CN} for this case. The dependence is similar to that for a z-oriented[105] dipole at $k_A R_{\text{CN}} = 1$. For the atom inside the CN, Eq. (5.48) yields $\zeta(\omega_A) \to 0$ as $\sigma_{zz} \to \infty$. The result is natural since, in this case, only one electromagnetic eigenmode can propagate in the CN; this mode is essentially transverse and, consequently, is not coupled with the axially oriented atomic dipole moment. However, the actual $\zeta(\omega_A)$ behavior is quite different from that predicted by the perfectly conducting cylinder model, since the latter does not account for CN electronic quasi-particle excitations responsible for the nonradiative atomic decay dominating the total spontaneous decay process.

The theory may be generalized to cover the transverse atomic electric dipole orientation, electric quadrupole and magnetic dipole atomic transitions, the properties[109] of organic molecules inside and/or outside CNs. The mechanism that was revealed of the photon vacuum renormalization is likely to manifest itself in other phenomena in CNs such as Casimir forces or electromagnetic fluctuations.

The presented results can be tested by methods of atomic fluorescent spectroscopy and may have various physical consequences. In particular, the effect of the drastic increase of the atomic spontaneous decay rate may turn out to be of practical importance in problems of the laser control of atomic motion,[110] increasing the ponderomotive force acting on a atom moving in the vicinity of a CN in a laser field. One might expect the Purcell effect peculiarities predicted for CNs to manifest themselves in macroscopic anisotropically conducting waveguides with strong wave deceleration (for example, in microwave spiral or collar waveguides with highly excited Rydberg atoms inside).

5.6 Semiconductor quantum dot in a classical electromagnetic field

An exquisite description of quantum dots has been provided in this book by Boxberg and Tulkki.[111] The remainder of this chapter therefore deals only with the nanoelectromagnetics of QDs.

5.6.1 Model Hamiltonian

Let an isolated QD imbedded in a host semiconductor be exposed to a classical electromagnetic field. Further consideration is restricted to a two-level model, which treats the QD as a set of electron-hole pairs that are strongly confined in space.[112] The electron wave function is stated as

$$\Psi_{q\nu}(\mathbf{r}) = F_q(\mathbf{r}) u_\nu(\mathbf{r}), \tag{5.53}$$

where the index ν takes the values e and g, which correspond to excited and ground bands of the electron, respectively, and $u_\nu(\mathbf{r})$ are the Bloch function amplitudes as per Eq. (5.8). The function $F_q(\mathbf{r})$ varies slowly on the atomic scale envelope satisfying the Schrödinger equation. For a spherical QD of radius R_{QD},

$$F_q(\mathbf{r}) \equiv F_{nlm} = C_{nl} Y_{lm}(\vartheta, \varphi) J_{l+1/2}(\kappa_{nl} r / R_{\mathrm{QD}}) / \sqrt{r}, \tag{5.54}$$

where $Y_{lm}(\vartheta, \varphi)$ are the spherical harmonics,[104] κ_{nl} is the nth root of the Bessel function $J_{l+1/2}(x)$, (r, ϑ, φ) is the triad of spherical coordinates, and the indices n and l identify a particular mode in the electron-hole pair's spectrum. The coefficients $C_{nl} = \sqrt{2}[R_{\mathrm{QD}} J_{l+3/2}(\kappa_{nl})]^{-1}$ orthonormalize F_{nlm}. The function $F_q(\mathbf{r})$ must be found numerically for QDs of more complicated shapes.

In terms of field operators

$$\widehat{\Psi}_\nu^\dagger(\mathbf{r}, t) = \sum_q a_{q\nu}^\dagger(t) \Psi_{q\nu}^*(\mathbf{r}), \quad \widehat{\Psi}_\nu(\mathbf{r}, t) = \sum_q a_{q\nu}(t) \Psi_{q\nu}(\mathbf{r}), \tag{5.55}$$

the polarization single-particle operator is expressed by

$$\widehat{\mathbf{P}}(\mathbf{r}, t) = e\mathbf{r} \sum_{\nu, \nu'} \widehat{\Psi}_\nu^\dagger(\mathbf{r}, t) \widehat{\Psi}_{\nu'}(\mathbf{r}, t), \tag{5.56}$$

where $a_{q\nu}^\dagger$ and $a_{q\nu}$ stand for the electron creation and annihilation operators, respectively. These operators satisfy the anticommutative relations usual for fermions.[104] After taking the periodicity of the Bloch functions $u_\nu(\mathbf{r})$ into account and considering the envelope function $F_q(\mathbf{r})$ to be constant over the unit cell of the QD crystalline lattice, the polarization operator averaged over the unit cell's volume V_{uc} is obtained as

$$\widehat{\mathbf{P}}(\mathbf{r}) = e\mathbf{r}\big(\widehat{F}_{\mathrm{e}}^\dagger \widehat{F}_{\mathrm{e}} + \widehat{F}_{\mathrm{g}}^\dagger \widehat{F}_{\mathrm{g}}\big) + \big(\boldsymbol{\mu} \widehat{F}_{\mathrm{e}}^\dagger \widehat{F}_{\mathrm{g}} + \boldsymbol{\mu}^* \widehat{F}_{\mathrm{g}}^\dagger \widehat{F}_{\mathrm{e}}\big), \tag{5.57}$$

where

$$\widehat{F}_\nu(\mathbf{r}, t) = \sum_q a_{q\nu}(t) F_q(\mathbf{r}), \tag{5.58}$$

and

$$\mu = \frac{e}{V_{\rm uc}} \int_{V_{\rm uc}} \mathbf{r} u_{\rm e}^{*}(\mathbf{r}) u_{\rm g}(\mathbf{r})\, d^{3}\mathbf{r} \tag{5.59}$$

is the dipole moment of the electron-hole pair. Thus, the averaged polarization operator is expressed in terms of slow-varying envelopes; and the Bloch functions define only the electron-hole pair's dipole moment, which is further considered as an external input parameter. Note that the first term on the right side of Eq. (5.57) describes intraband motion, while the second term corresponds to the interband transitions.

Further analysis is restricted to the two-level model, which allows the neglect of all terms in the sum of Eq. (5.58) except one (whose index is omitted from here onwards). As a result, Eq. (5.57) is reduced to

$$\widehat{\mathbf{P}}(\mathbf{r}) = |F(\mathbf{r})|^{2}\big[e\mathbf{r}\big(a_{\rm e}^{\dagger} a_{\rm e} + a_{\rm g}^{\dagger} a_{\rm g}\big) + \big(-\boldsymbol{\mu}\hat{b}^{\dagger} + \boldsymbol{\mu}^{*}\hat{b}\big)\big], \tag{5.60}$$

where $\hat{b}^{\dagger} = a_{g}a_{e}^{\dagger}$ and $\hat{b} = a_{g}^{\dagger}a_{e}$ are the creation and annihilation operators for electron-hole pairs.

Any QD is essentially a multilevel system. However, the joint contribution of all transitions lying far away from a given resonance can be approximated by a nonresonant relative permittivity ε_{h}. The host semiconductor relative permittivity is also assumed to be equal to ε_{h}. For analytical tractability, let ε_{h} be frequency independent and real-valued. This enables us to put $\varepsilon_{h} = 1$ without loss of generality. The substitutions

$$\left\{c \to c/\sqrt{\varepsilon_{h}},\ \boldsymbol{\mu} \to \boldsymbol{\mu}/\sqrt{\varepsilon_{h}}\right\} \tag{5.61}$$

in the final expressions will restore the case $\varepsilon_{h} \neq 1$.

In the strong confinement regime, the Coulomb interaction is assumed to be negligible, so that electrons and holes in a QD are independently mobile and spatial quantization is entailed by the interaction of the particles with the QD boundary. The Hamiltonian formalism describes the system "QD + electromagnetic field" and takes the role of the QD boundary into account. Apparently, the most sequential and rigorous approach is based on the concept of spatially varying interaction coefficients.[113,114] However, the use of this approach for systems with stepwise interaction coefficients entails that the Hamilton equations are inapplicable at the discontinuity. The same problem is found in the macroscopic electrodynamics of stratified media.[16] By analogously introducing a transient layer and reducing its thickness, one can obtain boundary conditions complementary to the Hamilton equations for the system under analysis. However, this approach is much too complicated and has been implemented only for the only simplest configuration: the interaction of a material layer with normally incident light.[113,114] Note that even in this simplest case, the local field effects are left unconsidered.

A constructive approach for QDs assumes them to be electrically small.[57] This assumption neglects the retardation of the electromagnetic field inside the QD. The spatial averaging of the electric field over the QD volume is thereby introduced.

In this approximate framework, the "QD + electromagnetic field" system is described by the Hamiltonian $\mathcal{H} = \mathcal{H}_0 + \mathcal{H}_{IL}$, where $\mathcal{H}_0 = \epsilon_e a_e^\dagger a_e + \epsilon_g a_g^\dagger a_g$ is the Hamiltonian of carrier motion, while $\epsilon_{g,e}$ are the energy eigenvalues. The term

$$\mathcal{H}_{IL} = -\int_{V_{\text{QD}}} \widehat{\mathbf{P}}(\mathbf{r}, t)\mathbf{E}_L(\mathbf{r})\, d^3\mathbf{r} \tag{5.62}$$

describes interaction with the electromagnetic field, where the polarization operator $\widehat{\mathbf{P}}$ is given by Eq. (5.60), $\mathbf{E}_L$ is the field inside the QD, and V_{QD} is the QD volume. Thus, the light-matter interaction Hamiltonian is defined in the dipole approximation,[101,110] i.e., a negligibly small term proportional to **AA** is rejected. Such an approximation is valid, at least, in the vicinity of the exciton resonance.[115,116] Note that the model can also describe higher excitonic modes; then, operators $\hat{b}^\dagger$ and $\hat{b}$, respectively, move up the exciton into the next energy level and return it back.

The field inside the QD—involved in Eq. (5.62)—is different from the external exciting field $\mathbf{E}_0$. Further analysis is aimed to express Hamiltonian Eq. (5.62) in terms of the field $\mathbf{E}_0$ assuming the QD to be electrically small. A time-domain integral relation follows from the Maxwell equations to yield[117,118]

$$\mathbf{E}_L(\mathbf{r}, t) = \mathbf{E}_0(\mathbf{r}, t) + \nabla\nabla \cdot \int_{V_{\text{QD}}} \mathbf{P}(\mathbf{r}', t)\, \frac{d^3\mathbf{r}'}{|\mathbf{r} - \mathbf{r}'|} \tag{5.63}$$

in the small-QD approximation.

Substitution of this equation into Eq. (5.62) enables us to present the interaction Hamiltonian by

$$\mathcal{H}_{IL} = \mathcal{H}_{0L} + \Delta\mathcal{H}, \tag{5.64}$$

where

$$\mathcal{H}_{I0} = -\int_{V_{\text{QD}}} \widehat{\mathbf{P}}(\mathbf{r}, t)\mathbf{E}_0\, d^3\mathbf{r}, \tag{5.65}$$

and

$$\Delta\mathcal{H} = \int_{V_{\text{QD}}}\int_{V_{\text{QD}}} \widehat{\mathbf{P}}(\mathbf{r}, t)\nabla\nabla \cdot \mathbf{P}(\mathbf{r}', t)\frac{d^3\mathbf{r}\, d^3\mathbf{r}'}{|\mathbf{r} - \mathbf{r}'|}. \tag{5.66}$$

The term $\Delta\mathcal{H}$ is the local-field correction to the interaction Hamiltonian. Since the QD is assumed to be electrically small (and, in consequence, the field inside QD is uniform), $\mathcal{H}_{0L}$ can be expressed in terms of the average over the QD volume polarization operator $\widehat{\mathbf{P}}_v$ as

$$\mathcal{H}_{I0} = -V_{\mathrm{QD}}\mathbf{E}_0\widehat{\mathbf{P}}_v, \tag{5.67}$$

where $\widehat{\mathbf{P}}_v = V_{\mathrm{QD}}^{-1}(-\boldsymbol{\mu}\hat{b}^\dagger + \boldsymbol{\mu}^*\hat{b})$.

The described procedure removes the intraband-motion contribution from the polarization operator. After assuming this contribution to be negligible because of the nonresonant nature of intraband transitions, thc substitution $\widehat{\mathbf{P}}(\mathbf{r},t) = V_{\mathrm{QD}}\widehat{\mathbf{P}}_v(t)|F(\mathbf{r})|^2$ in Eq. (5.66) results in

$$\Delta\mathcal{H} = 4\pi\, V_{\mathrm{QD}}\widetilde{N}_{\alpha\beta}\widehat{\mathbf{P}}_{\mathrm{v}\beta}\langle\widehat{\mathbf{P}}_{\mathrm{v}}\rangle_\alpha, \tag{5.68}$$

where

$$\widetilde{N}_{\alpha\beta} - -\frac{V_{\mathrm{QD}}}{4\pi}\int_{V_{\mathrm{QD}}}\int_{V_{\mathrm{QD}}}\frac{\partial^2}{\partial x_\alpha \partial x_\beta}\frac{|F(\mathbf{r})F(\mathbf{r}')|^2}{|\mathbf{r}-\mathbf{r}'|}d^3\mathbf{r}\,d^3\mathbf{r}'.$$

The Hamiltonian given by Eqs. (5.67) and (5.68) implies the relation

$$\mathbf{E}_L = \mathbf{E}_0 - 4\pi\, V_{\mathrm{QD}}|F(\mathbf{r})|^2\underline{\mathbf{N}}\langle\widehat{\mathbf{P}}_v\rangle \tag{5.69}$$

between the local and the exciting fields, with the components of the depolarization dyadic $\underline{\mathbf{N}}$ bcing[119]

$$N_{\alpha\beta} = -\frac{V_{\mathrm{QD}}}{4\pi}\frac{\partial^2}{\partial x_\alpha \partial x_\beta}\int_{V_{\mathrm{QD}}}\frac{|F(\mathbf{r}')|^2}{|\mathbf{r}-\mathbf{r}'|}d^3\mathbf{r}'. \tag{5.70}$$

The second term on the right side of Eq. (5.69) is the depolarization field in the QD. Generally, owing to the term $|F(\mathbf{r})|^2$, this field is nonlocal with respect to the macroscopic polarization and, consequently, is nonhomogeneous. Neglecting nonlocality in the strong confinement regime permits the approximation $|F(\mathbf{r})|^2 \simeq 1/V_{\mathrm{QD}}$, which leads to the model of a QD as a dielectric particle.[57,112]

Let the electron-hole pair's dipole moment be directed along the unit vector $\mathbf{e}_x$ in a Cartesian coordinate system related to the QD; i.e., $\boldsymbol{\mu} = \mu\mathbf{e}_x$. Then the total Hamiltonian is represented by

$$\mathcal{H} = \mathcal{H}_0 + \mathcal{H}_{I0} + \Delta\mathcal{H}, \tag{5.71}$$

where

$$\mathcal{H}_{I0} = -V E_{0x}\widehat{P}_{\mathrm{v}x}, \tag{5.72}$$

$$\Delta\mathcal{H} = 4\pi\widetilde{N}_x(\mu^*\hat{b} - \mu\hat{b}^\dagger)(\mu^*\langle\hat{b}\rangle - \mu\langle\hat{b}^\dagger\rangle) \tag{5.73}$$

and

$$\widetilde{N}_x = \boldsymbol{\mu}(\underline{\widetilde{\mathbf{N}}}\boldsymbol{\mu})/|\mu|^2 \equiv \mathbf{e}_x(\underline{\widetilde{\mathbf{N}}}\mathbf{e}_x). \tag{5.74}$$

Thus, in the total Hamiltonian we have separated the contribution $\mathcal{H}_{I0}$ of the interaction of electron-hole pairs with the exciting field from the contribution $\Delta\mathcal{H}$ of depolarization. This enables consideration of the local field effects without an explicit solution of the electrodynamic boundary-value problem. This is of special importance for the quantization of the electromagnetic field: as $\Delta\mathcal{H}$ is expressed in terms of dynamic variables of particle motion, $\widetilde{N}_x$ contains complete information about electromagnetic interaction.[‡]

5.6.2 Equations of motion

Let $|\tilde{\psi}(t)\rangle$ be the wave function of the "QD + classical electromagnetic field" system. In the interaction representation, this system is described by the Schrödinger equation

$$i\hbar\frac{\partial|\psi\rangle}{\partial t} = \mathcal{H}_{\text{int}}|\psi\rangle, \tag{5.75}$$

where

$$\mathcal{H}_{\text{int}} = \exp(i\mathcal{H}_0 t/\hbar)(\mathcal{H}_{I0} + \Delta\mathcal{H})\exp(-i\mathcal{H}_0 t/\hbar) \tag{5.76}$$

and

$$|\psi\rangle = \exp(i\mathcal{H}_0 t/\hbar)|\tilde{\psi}\rangle. \tag{5.77}$$

The function $|\psi\rangle$ can be represented by the sum

$$|\psi\rangle = A(t)|e\rangle + B(t)|g\rangle,$$

where $A(t)$ and $B(t)$ are coefficients to be found; while $|g\rangle$ and $|e\rangle$ are the wave functions of QD in the ground and excited states, respectively. Then, the macroscopic polarization is determined by

$$P_{\text{v}x} = \langle\tilde{\psi}|\widehat{P}_{\text{v}x}|\tilde{\psi}\rangle = \text{Re}\left[\frac{2}{V_{\text{QD}}}\mu^* A(t)B^*(t)e^{-i\omega_0 t}\right], \tag{5.78}$$

and the asterisk stands for the complex conjugate.

Within the confines of the slowly varying amplitude approximation, the exciting field is given by $E_{0x} = \text{Re}[\overline{E}(t)\exp(-i\omega t)]$, with $\overline{E}(t)$ as the slowly varying amplitude. Then, after taking Eq. (5.78) as well as the identities $\hat{b}^\dagger|e\rangle = \hat{b}|g\rangle = 0$, $\hat{b}|e\rangle = |g\rangle$ and $\hat{b}^\dagger|g\rangle = -|e\rangle$ into account, and neglecting rapidly oscillating terms,

[‡]The situation with $\Delta\mathcal{H}$ is, to a certain extent, analogous to the situation with $\mathbf{A}\cdot\mathbf{A}$. The latter term is expressed in terms of a field dynamical variable (the vector potential), but it contains information about the location of the particle because the vector potential is taken at the location of the particle.

the equations of motion[44,56]

$$i\hbar\frac{\partial A}{\partial t} = \hbar(\Delta\omega)A|B|^2 - \frac{1}{2}\overline{E}(t)\mu B e^{i(\omega_0-\omega)t},$$
$$i\hbar\frac{\partial B}{\partial t} = \hbar(\Delta\omega)B|A|^2 - \frac{1}{2}\overline{E}^*(t)\mu^* A e^{-i(\omega_0-\omega)t}, \tag{5.79}$$

emerge, where

$$\Delta\omega = \frac{4\pi}{\hbar V_{\mathrm{QD}}}\widetilde{N}_x|\mu|^2. \tag{5.80}$$

These equations constitute a basic self-consistent system describing the interaction of a QD with the electromagnetic field. The consistency is provided by the depolarization-induced first terms on the right side of the equations. Physically, the system of Eq. (5.79) is analogous to Bloch equations for optically dense media.[120] Relaxation can easily be included in Eq. (5.79) either by introduction of the phenomenological transverse and longitudinal relaxation times[120] or by a suitable modification of the initial Hamiltonian of Eq. (5.71).

5.6.3 QD polarization

An excited QD can be analyzed by supplementing Eq. (5.79) with the initial conditions $A(0) = 1$ and $B(0) = 0$. In the linear approximation with respect to the electromagnetic field, we can set $A(t) \approx 1$. Physically, this restricts the analysis to temporal intervals essentially less than the relaxation time of the given resonant state. Then, the equations of motion reduce to

$$i\hbar\frac{\partial B}{\partial t} = \hbar(\Delta\omega)B - \frac{1}{2}\overline{E}^*(t)\mu^* e^{i(\omega_0-\omega)t}. \tag{5.81}$$

For time-harmonic excitation, i.e., for $\overline{E}(t) = \overline{E} = \text{const}$, this equation is exactly integrable; thus,

$$B(t) \approx -\frac{\overline{E}^*\mu^*}{2\hbar(\omega_0 - \Delta\omega - \omega)}\left[e^{-i(\omega_0-\omega)t} - e^{-i\Delta\omega t}\right], \tag{5.82}$$

with $\Delta\omega$ determined by Eq. (5.80). Therefore, depolarization leads to the shift $\Delta\omega$ of the resonant frequency. This shift has been predicted on the basis of several different phenomenological models.[53,54,112,121] It has been predicted and experimentally verified that this shift in nonspherical QDs provides polarization splitting of the gain band.[51,52] Note also that the depolarization effect has been proposed by Gammon et al.[122] as a hypothesis to explain the experimentally observed polarization-dependent splitting of the photoluminescence spectrum of a single nonspherical QD. Finally, Eq. (5.80) has been obtained by other means too.[51,52]

The spin-degeneracy of electron-hole pairs results in duplication of $\Delta\omega$ because the total polarization of the system is provided by superposition of two partial polarizations corresponding to two spin components. Then, expressing the macroscopic polarization in terms of $B(t)$, we find

$$P_{vx} = \mathrm{Re}\left\{ \frac{|\mu|^2}{\hbar V_{QD}(\omega + \Delta\omega - \omega_0)} \overline{E}\left[e^{-i\omega t} - e^{-i(-\Delta\omega+\omega_0)t}\right]\right\}. \tag{5.83}$$

For a ground-state QD, the initial conditions are $A(0) = 0$ and $B(0) = 1$. Accordingly,

$$A(t) \approx \frac{\overline{E}\mu}{2\hbar(\omega_0 + \Delta\omega - \omega)}\left[e^{i(\omega_0-\omega)t} - e^{-i\Delta\omega t}\right]. \tag{5.84}$$

Thus, in the ground state, the local field effects manifest themselves in the same shift $\Delta\omega$ of the resonance, but with the opposite sign. If we introduce a finite radiation linewidth, the interaction of a ground-state QD with the electromagnetic field corresponds to absorption, while interaction of an excited QD corresponds to stimulated emission. In other words, the optical absorption and gain of an isolated QD could be distinguished owing to the depolarization shift—blue in the former case and red in the latter.

5.7 Interaction of QD with quantum light

5.7.1 Model Hamiltonian

At the first glance, Eq. (5.69) remains valid for nonclassical fields, if one inserts operators instead of the corresponding fields. However, such is not the case; and the relation between the exciting and the local fields in QED requires different handling. The time-domain integral equation

$$\widehat{\mathbf{E}}_L(\mathbf{r},t) = \widehat{\mathbf{E}}_0(\mathbf{r},t) + \left(\nabla\nabla\cdot - \frac{1}{c^2}\frac{\partial^2}{\partial t^2}\right) \times 4\pi \int_{-\infty}^{t}\int_{V_{QD}} G^{(\mathrm{ret})}(\mathbf{r}-\mathbf{r}', t-t')\widehat{\mathbf{P}}(\mathbf{r}',t')\, d^3\mathbf{r}'\, dt', \tag{5.85}$$

must now be used, where the retarded Green's function is given by[114]

$$G^{(\mathrm{ret})}(\mathbf{r},t) = \lim_{\varpi\to 0+} \frac{1}{(2\pi)^4} \iint \frac{\exp[i(\mathbf{kr}-\omega t)]}{\mathbf{k}^2 - (\omega + i\varpi)^2/c^2}\, d^3\mathbf{k}\, d\omega, \tag{5.86}$$

and the polarization operator is given by

$$\widehat{\mathbf{P}}(\mathbf{r},t) = |F(\mathbf{r})|^2\left(-\boldsymbol{\mu}\hat{b}^\dagger + \boldsymbol{\mu}^*\hat{b}\right) = |F(\mathbf{r})|^2(\boldsymbol{\mu}|\mathrm{e}\rangle\langle\mathrm{g}| + \boldsymbol{\mu}^*|\mathrm{g}\rangle\langle\mathrm{e}|).$$

To derive a relation for the field and polarization operators, we first construct the interaction Hamiltonian

$$\mathcal{H}_{IL} = -\frac{1}{2}\int_{V_{\mathrm{QD}}} \left(\widehat{\mathbf{P}}\widehat{\mathbf{E}}_L + \widehat{\mathbf{E}}_L\widehat{\mathbf{P}}\right) d^3\mathbf{r}. \tag{5.87}$$

The operators $\widehat{\mathbf{P}}$ and $\widehat{\mathbf{E}}_L$ are generally noncommutative, since the field operator $\widehat{\mathbf{E}}_L$ is not transverse.§ Next, we substitute Eq. (5.85) into Eq. (5.87) and separate out the Hamiltonian component corresponding to the depolarization field as

$$\Delta\mathcal{H} = -2\pi \int_{-\infty}^{t}\int_{V_{\mathrm{QD}}}\int_{V_{\mathrm{QD}}} \left(\frac{\partial^2}{\partial x_\alpha \partial x_\beta} - \delta_{\alpha\beta}\frac{1}{c^2}\frac{\partial^2}{\partial t^2}\right) G^{(\mathrm{ret})}(\mathbf{r}-\mathbf{r}', t-t')$$
$$\times \left[\widehat{P}_\alpha(\mathbf{r}',t')\widehat{P}_\beta(\mathbf{r},t) + \widehat{P}_\beta(\mathbf{r},t)\widehat{P}_\alpha(\mathbf{r}',t')\right] d^3\mathbf{r}\, d^3\mathbf{r}'\, dt'. \tag{5.88}$$

Equation (5.88) can be drastically simplified by applying the mean-field approximation. In accordance with that approximation, the replacement $\widehat{P}_\alpha(\mathbf{r}',t') \to \langle \widehat{P}_\alpha(\mathbf{r}',t')\rangle \widehat{I}$ is implemented in Eq. (5.88), and the retardation inside the QD is neglected because of it being electrically small. Therefore, $G^{(\mathrm{ret})}(\mathbf{r},t) \sim \delta(t)/|\mathbf{r}|$ and the $O(\partial^2/\partial t^2)$ terms in Eq. (5.88) are omitted. That equation then reduces to Eq. (5.68). Analogous approximations being applied to Eq. (5.85) lead to the formula

$$\widehat{\mathbf{E}}_L = \widehat{\mathbf{E}}_0 - 4\pi V_{\mathrm{QD}} |F(\mathbf{r})|^2 \underline{\mathbf{N}} \langle \widehat{\mathbf{P}}_v \rangle \widehat{I}, \tag{5.89}$$

which is the nonclassical alternative to Eq. (5.69).

In order to obtain the total Hamiltonian of the "QD + quantum electromagnetic field" system, the right side of Eq. (5.71) has to be augmented by the term $\mathcal{H}_F$ corresponding to the free-space field, and the replacement $E_{0x} \to \widehat{E}_{0x}$ must be implemented in the term $\mathcal{H}_{I0}$. In the quantum optics of nonhomogeneous media, the problem of representing the electromagnetic field operator exists, since the local fields are nonhomogeneous. Unlike conventional approaches, the proposed[57] scheme of electromagnetic field quantization does not encounter this problem, since the interaction Hamiltonian is represented in terms of the exciting field but not the local field; thus, the usual plane wave expansion is applicable to the operator $\widehat{E}_{0x}$, and the role of the QD boundary is taken into account by the term $\Delta\mathcal{H}$ in Eq. (5.73). Thus, the Hamiltonian for the quantum electromagnetic field is

$$\mathcal{H} = \mathcal{H}_0 + \Delta\mathcal{H} + \mathcal{H}_{I0} + \mathcal{H}_F, \tag{5.90}$$

where $\mathcal{H}_{I0} = -V_{\mathrm{QD}}\widehat{P}_{vx}\widehat{E}_{0x}$ and

$$\widehat{E}_{0x} = i\sum_k \sqrt{\frac{2\pi\hbar\omega_k}{\Omega}}\left(c_k e^{i\mathbf{kr}} - c_k^\dagger e^{-i\mathbf{kr}}\right). \tag{5.91}$$

§The second term on the right side of Eq. (5.85) contains a longitudinal component.

In these equations, Ω is the normalization volume, whereas $c_k^\dagger$ and c_k are the photon creation and annihilation operators, respectively. In accordance with QED literature, the index k introduces summation over different photonic modes $\mathbf{k}$; $k = |\mathbf{k}|$ and $\omega_k = c|\mathbf{k}|$. The definition in Eq. (5.91) of the electric field operator restricts consideration to states of quantum light, which are superpositions of photons with a given polarization—the so-called factorized states of light.[101] Equation (5.91) yields

$$\mathcal{H}_F = \hbar \sum_k \omega_k \left(c_k^\dagger c_k + \frac{1}{2} \right), \tag{5.92}$$

and

$$\mathcal{H}_{I0} = -\hbar \sum_k \left(g_k b^\dagger c_k - g_k^* b c_k^\dagger \right), \tag{5.93}$$

where $g_k = -i\mu\sqrt{2\pi\omega_k/\hbar\Omega}\exp(i\mathbf{k}\mathbf{r}_c)$, and $\mathbf{r}_c$ is the radius vector of the QD geometrical center.

The Hamiltonian of Eq. (5.90) conforms to the use of Eq. (5.89) for field operators in lieu of Eq. (5.69) for classical fields. The term $\widehat{\mathbf{E}}_0$ in Eq. (5.89) represented by a superposition of photons[123] is an auxiliary field that can be interpreted as an incident field only in the classical limit. Such a simple interpretation is inapplicable for quantum light. Indeed, the operator $\widehat{\mathbf{E}}_0$, in general, is not identical to the field either inside or outside the QD; moreover, this term can arise even in the absence of any external sources (for example, in spontaneous transitions). Note also that $\widehat{\mathbf{E}}_0$ is transverse, and can be represented as a superposition of "genuine" photons.[123] However, the total field inside the QD is not transverse[118,124] due to the second term on the right side of Eq. (5.85).

5.7.2 Equations of motion

In the interaction representation, the "QD + quantum electromagnetic field" system with the Hamiltonian of Eq. (5.90) is described by Eq. (5.75), after the substitution $\mathcal{H}_0 \to \mathcal{H}_0 + \mathcal{H}_F$ has been performed therein. Then the wave function of the system can be written as

$$|\psi(t)\rangle = \sum_{k,n_k=0} \left[A_k^{n_k}(t)|\mathrm{e}\rangle + B_k^{n_k}(t)|\mathrm{g}\rangle \right] |n_k\rangle, \tag{5.94}$$

where $A_k^{n_k}(t)$ and $B_k^{n_k}(t)$ are functions to be found, $|n_k\rangle$ denotes the field states where there are n photons in mode $\mathbf{k}$ and no photons in all other modes, and $|0\rangle$ is the wave function of the electromagnetic field in the vacuum state. In view of Eq. (5.94), Eq. (5.78) for macroscopic polarization is transformed into

$$P_{\mathrm{v}x} = \mathrm{Re}\left\{ \frac{2}{V_{\mathrm{QD}}} \mu^* e^{-i\omega_0 t} \sum_{k,n_k=0} A_k^{n_k}(t) \left[B_k^{n_k}(t) \right]^* \right\}. \tag{5.95}$$

Then, after some standard manipulations with the Schrödinger Eq. (5.75), the infinite chain

$$i\frac{dA_k^{n_k}}{dt} = \Delta\omega B_k^{n_k} \sum_{\varsigma,m_\varsigma} A_\varsigma^{m_\varsigma}\left(B_\varsigma^{m_\varsigma}\right)^* + g_k\sqrt{n_k+1}B_k^{n_k+1}e^{-i(\omega_k-\omega_0)t}$$

$$+\,\delta_{n_k,0}\sum_\varsigma (1-\delta_{\varsigma k})g_\varsigma B_\varsigma^{1_\varsigma}e^{-i(\omega_\varsigma-\omega_0)t}, \tag{5.96}$$

$$i\frac{dB_k^{n_k}}{dt} = \Delta\omega A_k^{n_k} \sum_{\varsigma,m_\varsigma} \left(A_\varsigma^{m_\varsigma}\right)^* B_\varsigma^{m_\varsigma} + g_k^*\sqrt{n_k}A_k^{n_k-1}e^{i(\omega_k-\omega_0)t},$$

of coupled nonlinear differential equations for slowly varying amplitudes emerges for any n_k. This system of equations serves as a basis for further analysis with different initial conditions. Note that accounting for the depolarization field is a specific property of this system that makes it nonlinear and couples all quantum states of the electromagnetic field. These properties distinguish this system from conventional equations of quantum electrodynamics. Equations (5.96) satisfy the conservation law

$$\sum_{k,n_k=0}\left[\left|A_k^{n_k}(t)\right|^2 + \left|B_k^{n_k}(t)\right|^2\right] = 1, \tag{5.97}$$

which dictates the normalization of the functions involved.

In the limit $\widetilde{N}_x \to 0$, the system of Eq. (5.96) splits into recurrent sets of linear equations coupling only the $|n_k\rangle$ and $|n_k+1\rangle$ states. Then the system becomes equivalent to the ordinary equations of motion of a two-level atom exposed to a quantum electromagnetic field.[110]

5.7.3 Interaction with single-photon states

The phenomenon of spontaneous emission from a QD can be treated as the interaction of an excited QD with two states of the electromagnetic field, $|0\rangle$ and $|1_k\rangle$. The initial conditions

$$A_k^0(0) = 1, \quad B_k^0(0) = B_k^{1_k}(0) = A_k^{1_k}(0) = 0 \tag{5.98}$$

for spontaneous emission describe an excited state of the electron-hole pair with zero photons at the initial instant. After neglecting all other states, Eqs. (5.96) reduce to following form:

$$\begin{aligned}\frac{dA_k^0}{dt} &= -i\sum_\varsigma g_\varsigma B_\varsigma^{1_\varsigma}e^{-i(\omega_\varsigma-\omega_0)t},\\ \frac{dB_k^{1_k}}{dt} &= -ig_k^* A_k^0 e^{i(\omega_k-\omega_0)t}.\end{aligned} \tag{5.99}$$

When investigating this new system of equations, we should take the natural width of the resonant transition into account. Hence, we can not assume $A_k^0(t) \approx 1$, as was done to derive Eq. (5.82).

The integration of the second of Eqs. (5.99) with respect to time and substitution of the result in the first of Eqs. (5.99) leads to the Volterra integrodifferential equation

$$\frac{dA_k^0}{dt} = \int_0^t K\big(t - t'\big) A_k^0(t')\, dt' \tag{5.100}$$

with the kernel

$$K(t) = -\sum_k |g_k|^2 e^{-i(\omega_k - \omega_0)t}. \tag{5.101}$$

By means of the replacement

$$\sum_k [\cdot] \rightarrow \frac{\Omega}{(2\pi)^3} \int_0^{2\pi} d\varphi \int_0^{\pi} \sin\theta\, d\theta \int_0^{\infty} k^2 [\cdot]\, dk,$$

which corresponds to the limit $\Omega \rightarrow \infty$, and subsequent integration,[125] Eq. (5.101) is reduced to the simple result

$$K(t) = -\Gamma_{\rm sp}^{(0)} \delta(t)/2, \tag{5.102}$$

where $\Gamma_{\rm sp}^{(0)}$ is given by Eq. (5.44).

Note that nonresonant transitions can be accounted for by means of Eq. (5.61), and the result is the same as reported by Thränhardt et al.[126] An analogous result has been obtained for the spontaneous emission of an excited atom imbedded in a lossy dispersive dielectric media.[127]

Equation (5.100) with the Dirac delta function as the kernel has the solution

$$A_k^0(t) = \exp\big[-\Gamma_{\rm sp}^{(0)} t/2\big]. \tag{5.103}$$

In the frequency domain, this solution defines the Lorentz shape $[\omega - \omega_0 + i\Gamma_{\rm sp}^{(0)}/2]^{-1}$ for the spontaneous emission line.

Unlike absorption and stimulated emission, the spontaneous emission line does not experience any depolarization shift. The depolarization also does not influence the resonance linewidth. A similar situation appears in the interaction of QD with any pure state of electromagnetic field.[57] To clarify this conclusion, let us consider the mean value

$$\langle \widehat{E}_{0x} \rangle = \langle \tilde{\psi} | \widehat{E}_{0x} | \tilde{\psi} \rangle = -2\,\mathrm{Im} \sum_{k,n_k} e^{i(\mathbf{k}\mathbf{r}_c - \omega_k t)} \sqrt{2\pi\hbar\omega_k/\Omega}$$
$$\times \sqrt{n_k + 1}\big[(A_k^{n_k})^* A_k^{n_k+1} + (B_k^{n_k})^* B_k^{n_k+1}\big] \tag{5.104}$$

of the electric field operator $\widehat{E}_{0x}$ of Eq. (5.91) for the wave function Eq. (5.94). It follows from this expression that $\langle E_{0x}\rangle = 0$ for any state of the electromagnetic field with a fixed number of photons. Thus, if the initial state is a pure state (as for spontaneous emission), its mean value is equal to zero and it does not induce the depolarization field.

Absorption of a single photon with the wave number $\varsigma = |\varsigma|$ is determined by the initial conditions $B_k^{1_k}(0) = \delta_{k\varsigma}$ and $A_k^0(0) = 0$. Then the solution of the system of Eqs. (5.96) is given by

$$A_\varsigma^0(t) = \frac{g_\varsigma}{\omega - \omega_0 + i\Gamma_{\mathrm{sp}}^{(0)}/2}\left[e^{i(\omega_0-\omega)t} - e^{-\Gamma_{\mathrm{sp}}^{(0)}t/2}\right]. \tag{5.105}$$

Thus, differing from the emission and absorption of classical electromagnetic waves—see Eqs. (5.82) and (5.84), respectively—the spontaneous emission and absorption of a single photon are characterized by the same resonant frequency and the same radiative linewidth. In other words, single-photon processes are insensitive to the depolarization field. This is because the mean electric field of a single photon is equal to zero, as per Eq. (5.104).

5.7.4 Scattering of electromagnetic Fock qubits

A superposition of two arbitrary quantum field states is referred to as a qubit. While a Fock qubit is the superposition of two arbitrary Fock states that are eigenfunctions of the Hamiltonian $\mathcal{H}_F$ of Eq. (5.92), Fock states are those states that have a fixed number of photons.

Let a ground-state QD interact with the electromagnetic field in the Fock qubit state of the mode ς: $\beta_{N_\varsigma}|N_\varsigma\rangle + \beta_{N_\varsigma+1}|N_\varsigma + 1\rangle$. Here β_{N_ς} and $\beta_{N_\varsigma+1}$ are the complex-valued quantities such that $|\beta_{N_\varsigma}|^2 + |\beta_{N_\varsigma+1}|^2 = 1$. The physical principles behind the generation of arbitrary quantum states of light and, particularly, electromagnetic qubits have been described in detail elsewhere.[128,129] Explicit expressions for wave functions can easily be found, allowing analytical treatment.

Suppose that $N_\varsigma \geq 2$.*** The dynamical properties of the system are described by Eq. (5.96) supplemented by the initial conditions

$$B_k^{n_k}(0) = (\delta_{N_k,n_k}\beta_{N_k} + \delta_{N_k+1,n_k}\beta_{N_k+1})\delta_{k\varsigma}, \quad A_k^{n_k}(0) = A_k^0(0) = B_k^0(0) = 0$$

with $n_k \geq 1$. As analysis is confined to a specific photon mode here, the index ς in N_ς is further omitted; also, analysis is restricted to temporal intervals that are small in comparison to the radiation lifetime. Then the approximate relations $B_\varsigma^N(t) \approx \beta_N = \text{const}$ and $B_\varsigma^{N+1}(t) \approx \beta_{N+1} = \text{const}$ hold true. As a result, the

***The cases $N_\varsigma = 0$ and $N_\varsigma = 1$ can be considered by analogy but lead to mathematically different results.[57]

amplitudes A_ς^N and A_ς^{N+1} satisfy the coupled differential equation

$$\frac{d}{dt}\begin{pmatrix} A_\varsigma^N \\ A_\varsigma^{N+1} \end{pmatrix} = -i\Delta\omega \begin{pmatrix} |\beta_N|^2 & \beta_N\beta_{N+1}^* \\ \beta_N^*\beta_{N+1} & |\beta_{N+1}|^2 \end{pmatrix} \begin{pmatrix} A_\varsigma^N \\ A_\varsigma^{N+1} \end{pmatrix} + \begin{pmatrix} f_\varsigma(t) \\ 0 \end{pmatrix}, \quad (5.106)$$

while the amplitude A_ς^{N-1} satisfies the differential equation

$$\frac{dA_\varsigma^{N-1}}{dt} = -ig_\varsigma\sqrt{N}\beta_N e^{-i(\omega_\varsigma-\omega_0)t}, \quad (5.107)$$

where $f_\varsigma(t) = -ig_\varsigma\sqrt{N+1}\beta_{N+1}\exp[-i(\omega_\varsigma-\omega_0)t]$.

If $f_\varsigma(t) = 0$ in Eq. (5.106), the partial solutions of the type $A_\varsigma^{N+1,N} \sim \exp(-idt)$ satisfy the characteristic equation $d^2 - d\Delta\omega = 0$, which has two roots, $d_1 = 0$ and $d_2 = \Delta\omega$, with $\Delta\omega$ defined by Eq. (5.80). Thus, the eigenstate spectrum of system Eq. (5.106) contains states with resonant frequency both unshifted and shifted due to depolarization; and these eigenstates become degenerate as $\widetilde{N}_x \to 0$. The gap between the resonances significantly exceeds the linewidth: $\Delta\omega \gg \Gamma_{\rm sp}^{(0)}/2$.

The general solution of Eq. (5.106) is the superposition of the two eigenstates as follows:

$$\begin{aligned} A_\varsigma^N(t) &= c_{1\varsigma}(t) + c_{2\varsigma}(t)e^{-i\Delta\omega t}, \\ A_\varsigma^{N+1}(t) &= -\frac{\beta_N^*}{\beta_{N+1}^*}c_{1\varsigma}(t) + \frac{\beta_{N+1}}{\beta_N}c_{2\varsigma}(t)e^{-i\Delta\omega t}. \end{aligned} \quad (5.108)$$

These equations employ the time-varying coefficients

$$\begin{aligned} c_{1\varsigma}(t) &= -g_\varsigma\sqrt{N+1}\beta_{N+1}|\beta_{N+1}|^2\frac{e^{i(\omega_0-\omega)t}-1}{\omega_0-\omega}, \\ c_{2\varsigma}(t) &= -g_\varsigma\sqrt{N+1}\beta_{N+1}|\beta_N|^2\frac{e^{i(\omega_0-\omega+\Delta\omega)t}-1}{\omega_0-\omega+\Delta\omega}, \end{aligned} \quad (5.109)$$

while Eq. (5.107) gives

$$A_\varsigma^{N-1}(t) = -g_\varsigma\sqrt{N}\beta_N\frac{[e^{i(\omega_0-\omega)t}-1]}{\omega_0-\omega}. \quad (5.110)$$

Equations (5.108) and (5.110) enable us to derive an explicit expression for the transition probability as

$$w(t) = \frac{d}{dt}\Big[\big|A_\varsigma^{N-1}(t)\big|^2 + \big|A_\varsigma^N(t)\big|^2 + \big|A_\varsigma^{N+1}(t)\big|^2\Big].$$

The QD effective scattering cross section is proportional to the quantity $w(\infty) = \lim_{t\to\infty} w(t)$. Certain elementary manipulations and the substitution $\sin(\alpha t)/\pi\alpha \to \delta(\alpha)$ then yield

$$w(\infty) = 2\pi |g_\varsigma|^2 \{[N|\beta_N|^2 + (N+1)|\beta_{N+1}|^4]\delta(\omega_0 - \omega) + (N+1)|\beta_N|^2|\beta_{N+1}|^2\delta(\omega_0 + \Delta\omega - \omega)\}. \tag{5.111}$$

The substitution $\Delta\omega \to -\Delta\omega$ should be performed in Eq. (5.111) for stimulated emission. Figure 5.16 schematically represents the QD optical response defined by Eq. (5.111) for the absorption of a Fock qubit.

On neglecting depolarization—i.e., in the limit $\Delta\omega \to 0$—Eq. (5.111) reduces to

$$w(\infty) = 2\pi |g_\varsigma|^2 [N + |\beta_{N+1}|^2]\delta(\omega_0 - \omega), \tag{5.112}$$

thus, the resonance line is not shifted.

When the incident field contains the only photon state, the substitutions $\beta_{N+1} \to 0$ and $\beta_N \to 1$ (or $\beta_{N+1} \to 1$ and $\beta_N \to 0$) must be carried out in Eq. (5.111). The former case gives $w(\infty) \sim N\delta(\omega_0 - \omega)$, whereas the latter case leads to the identical expression with $N \to N+1$ substituted. Thus, single-photon states are characterized by unshifted resonances just as when depolarization is neglected; however, the resonance amplitudes are quite different.

The foregoing analysis demonstrates that, in general, two spectral lines are present in the effective scattering cross section. One of these lines has the fre-

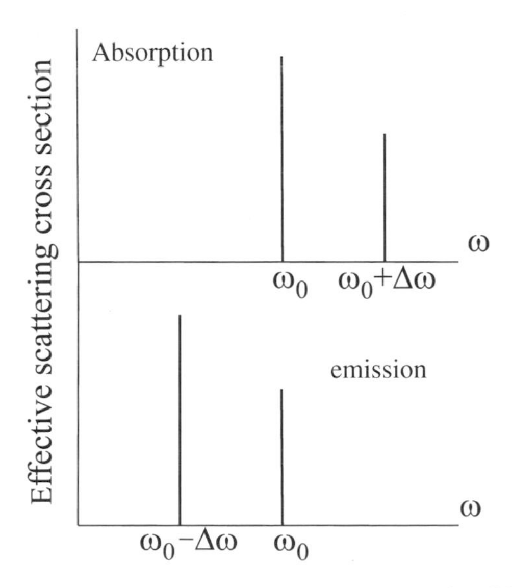

Figure 5.16 Fine structure of the electromagnetic response of a QD illuminated by quantum light. For the depicted cases of absorption and emission, the weighting coefficients β_N and β_{N+1} were chosen differently. (Reprinted with permission from Ref. 57, © 2002 The American Physical Society.)

quency of the exciton transition, while the other is shifted owing to the induced depolarization of the QD.

The shifted line is due to macroscopic polarization of the QD. This conclusion follows from using Eq. (5.108) in Eq. (5.95) to get

$$P_{\mathrm{v}x} = \mathrm{Re}\left\{ \frac{|\mu|^2}{\hbar V_{\mathrm{QD}}} \langle \widehat{\overline{E}} \rangle \frac{[e^{-i\omega t} - e^{-i(\omega_0+\Delta\omega)t}]}{\omega_0 - \omega + \Delta\omega} \right\}, \tag{5.113}$$

where $\langle \widehat{\overline{E}} \rangle = -2\hbar\beta_N^* \beta_{N+1} g_\varsigma \sqrt{N+1}/\mu$ in accordance with Eq. (5.104). As Eq. (5.113) is analogous to Eq. (5.83), one can conclude that the shifted line is related only with the classical polarization. In contrast, the unshifted line is due to the quantum nature of the electromagnetic field. Indeed, the classical approach implies that the scattering cross section is completely determined by the QD macroscopic polarization, as shown in Sec. 5.6.3.

Since $w(t) = d|A|^2/dt$ for classical light, Eq. (5.84) gives $w(\infty) \sim \delta(\omega_0 + \Delta\omega - \omega)$. Thus, the quantum nature of the electromagnetic field gives rise to an electromagnetic response that is not related to the media polarization, but is conditioned by the field eigenstates with a fixed number of photons. Spontaneous emission is an example. The key result of this section is that electromagnetic field states with fixed and fluctuating numbers of photons react differently to the local fields. The former states do not "feel" the local fields, while the latter ones demonstrate a shift of resonant frequency.

5.7.5 Observability of depolarization

The basic physical result in this section thus far has been the prediction of a fine structure of the absorption (emission) line in a QD interacting with quantum light. Instead of a single line with a frequency corresponding to the exciton transition ω_0, a doublet appears with one component blue-shifted or red-shifted by $\Delta\omega$ as per Eq. (5.80). This fine structure is due to depolarization of a QD and has no analog in classical electrodynamics. The value of the shift depends only on the geometrical properties of a QD, while the intensities of components are completely determined by statistics of the quantum light. In the limiting cases of classical light and single-photon states, the doublet reduces to a singlet, which is shifted in the former case and unshifted in the latter. A physical interpretation of the depolarization effect can be given by analogy with the $\mathbf{k} \cdot \mathbf{p}$ theory of bulk crystals utilizing the concept of the electron-hole effective mass.[57]

The shift can be estimated using well-known data for QDs. To incorporate a host media, Eqs. (5.80) and (5.44) are subjected to the substitutions of Eq. (5.61); i.e.,

$$\Delta\omega = \frac{\pi \widetilde{N}_x}{\tau V_{\mathrm{QD}}} \left(\frac{c}{\sqrt{\varepsilon_h}\omega_0} \right)^3. \tag{5.114}$$

For a spherical InGaAs/GaAs QD ($\widetilde{N}_x = 1/3$) of radius $R_{\rm QD} \simeq 3$ nm, when $\varepsilon_h = 12$ and radiation lifetime[6] $\tau_{\rm sp}^{(0)} \simeq 1$ ns, Eq. (5.114) yields $\hbar\Delta\omega \simeq 1$ meV at the wavelength $\lambda = 1.3\,\mu$m. This value correlates well with a theoretical estimate,[130] and is of the same order of magnitude as polarization-dependent splitting.[51,52] Note that the Bohr radius for such QDs is about 10 nm, so that the strong confinement approximation is valid.

A recent low-temperature measurement of the QD dipole moment[131] gives $\tau_{\rm sp}^{(0)} \simeq 0.05$ to 0.15 ns; however, the lateral size of that QD is much larger than its thickness and the Bohr radius. Since $N_x \to 0$ in this case, we do not predict an observable depolarization shift for such QDs.

For experimental detection of the predicted fine structure, the value $\hbar\Delta\omega$ must exceed the linewidths of the doublet components; i.e., $\Delta\omega \gg \Gamma_{\rm sp}^{(0)}/2$ and $\Delta\omega \gg \Gamma_{\rm hom}/2$, where $\Gamma_{\rm hom}$ is the homogeneous broadening of the spectral line due to dephasing. As follows from Eq. (5.114), the first inequality is fulfilled at $N_x \gg (2\pi)^2 V_{\rm QD}/\lambda^3$—i.e., for any realistic QDs of arbitrary shapes. Analysis shows that the exciton-phonon interaction determines the $\Gamma_{\rm hom}$ magnitude. Recent low-temperature ($T = 20$ to 40 K) measurements give[33,132,133] $\hbar\Gamma_{\rm hom} \sim 1$ to 20 μeV, while a similar estimate follows[134] from calculations at $T = 77$ K. Thus, at low temperatures, the predicted value of the shift turns out to be sufficiently large to be measurable.

At room temperatures, the quantity $\hbar\Gamma_{\rm hom}$ grows to between 0.2 and 1 meV.[6,133,134] Then, line broadening may result in overlapping of the doublet components. Even so, local-field effects are important for adequate prediction of the spectral lineshape of a QD illuminated by quantum light.

As stated earlier, the depolarization shift has opposite signs for absorptive and inverted exciton levels. This property of QDs exposed to classical light has been elucidated on the basis of classical electrodynamics.[51,121] Results obtained for classical light are often extended to quantum light by using the concept of Einstein coefficients.[123] Such a transformation applied to single-photon states, however, leads to a paradox: the energies of absorbed and emitted photons differ by $2\hbar\Delta\omega$. On the contrary, in accordance with Sec. 5.7.3, the single-photon processes are insensitive to the depolarization field, so that the spontaneous emission and absorption of a single photon occur at the same resonant frequency ω_0. The depolarization shift occurs only in QDs exposed to light with a fluctuating number of photons (classical light is the limiting case of such states of electromagnetic field). In that situation, the energy defect $2\hbar\Delta\omega$ can physically be interpreted as follows: the defect $2\hbar\Delta\omega$ is stipulated in the total nonclassical Hamiltonian of Eq. (5.90) by the term $\Delta\mathcal{H}$ in Eq. (5.68). This equation describes the electromagnetic interaction of an oscillating electron-hole pair. In QED, that interaction is transferred by a virtual photon with energy $\hbar\Delta\omega$, which is extracted from the external field and returns randomly. Obviously, such an interaction mechanism is impossible in external fields with a fixed number of photons, such as the Fock states; therefore, the depolarization field is not excited in QDs exposed to Fock states and, consequently, the depolarization shift does not then exist.

5.8 Concluding remarks

This chapter ranged over several linear and nonlinear electromagnetic problems and associated issues of electron transport through carbon nanotubes, which are quasi-1D nanostructures. QED as applied to these nanostructures was also formulated and used for consideration of atomic spontaneous emission near a CN and of local-field contribution in the quantum optics of QDs.

The choice of problems and methodology presented here were dictated by the following reasoning: first, stress was laid on the close connection between traditional problems of classical electrodynamics of microwaves and new problems arising from technological progress in synthesis and application of nanostructures. Such a connection enables us to extend to nanostructures the rich experience and mathematical approaches that are known well in the classical electrodynamics community. Second, the chapter demonstrated the peculiarities of electromagnetic problems in nanostructures irreducible to problems in classical electrodynamics due to the complex conductivity law and pronounced field nonhomogeneity.

While CNs and QDs are nice examples to demonstrate the correctness of both ideas, the chosen methodology can be applied to all kinds of nanostructures. Certainly, this chapter touched on only a restricted set of problems, a set that is far from being complete. It shows, however, that the range of problems is very wide and that the methods and techniques of traditional electrodynamics (linear, nonlinear, and quantum) can be successfully adapted to nanostructures.

Acknowledgments

The authors are grateful to Profs. D. Bimberg and N. N. Ledentsov, Drs. I. Herrmann, A. Hoffmann, and O. M. Yevtushenko for long-term collaboration, and to Dr. I. Bondarev for helpful discussions and fruitful cooperation in the investigation of the spontaneous emission process in carbon nanotubes. The research is partially supported through the NATO Science for Peace Program under project SfP-972614, the BMBF (Germany) under Project No. BEL-001-01, and the Belarus Foundation for Fundamental Research under Project Nos. F02-176 and F02 R-047.

References

1. M. S. Dresselhaus, G. Dresselhaus, and P. C. Eklund, *Science of Fullerenes and Carbon Nanotubes*, Academic Press, New York (1996).
2. T. W. Ebbesen, Ed., *Carbon Nanotubes: Preparation and Properties*, CRC Press, Boca Raton, FL (1997).
3. R. Saito, G. Dresselhaus, and M. S. Dresselhaus, *Physical Properties of Carbon Nanotubes*, Imperial College Press, London (1998).
4. M. S. Dresselhaus, G. Dresselhaus, and Ph. Avouris, *Carbon Nanotubes*, Springer, Berlin (2001).

5. L. Banyai and S. W. Koch, *Semiconductor Quantum Dots*, World Scientific, Singapore (1993).
6. D. Bimberg, M. Grundmann, and N. N. Ledentsov, *Quantum Dot Heterostructures*, Wiley, Chichester (1999).
7. W. W. Chow and S. W. Koch, *Semiconductor Laser Fundumentals: Physics of the Gain Materials*, Springer, Berlin (1999).
8. A. Lakhtakia and R. Messier, "The past, the present, and the future of sculptured thin films," in *Introduction to Complex Mediums for Optics and Electromagnetics*, W. S. Weiglhofer and A. Lakhtakia, Eds., SPIE Press, Bellingham, WA (2003).
9. F. G. Bass and A. A. Bulgakov, *Kinetic and Electrodynamic Phenomena in Classical and Quantum Semiconductor Superlattices*, Nova Science, New York (1997).
10. S. Ya. Kilin, "Quantum information," *Phys. Usp.* **42**, 435–452 (1999).
11. E. Biolotti, I. D. Amico, P. Zanardi, and F. Rossi, "Electrooptical properties of quantum dots: application to quantum information processing," *Phys. Rev. B* **65**, 075306 (2002).
12. L. A. Weinstein, *Open Resonators and Open Waveguides*, Golem, New York (1969).
13. V. V. Shevchenko, *Tapers in Open Waveguides*, Golem, Boulder, CO (1971).
14. M. J. Adams, *An Introduction to Optical Waveguides*, Wiley, New York (1981).
15. K. C. Gupta, R. Garg, and R. Chadha, *Computer Aided Design of Microwave Circuits*, Artech House, Boston (1981).
16. A. S. Ilyinsky, G. Ya. Slepyan, and A. Ya. Slepyan, *Propagation, Scattering and Dissipation of Electromagnetic Waves*, Peter Peregrinus, London (1993).
17. Ph. Avouris, T. Hertel, R. Martel, T. Schmidt, H. R. Shea, and R. E. Walkup, "Carbon nanotubes: nanomechanics, manipulation, and electronic devices," *Appl. Surf. Sci.* **141**, 201–209 (1999).
18. P. G. Collins and Ph. Avouris, "Nanotubes for electronics," *Sci. Am.* **283**(6), 62–69 (2000).
19. S. J. Tans, A. R. M. Verschueren, and C. Dekker, "Room-temperature transistor based on a single carbon nanotube," *Nature* **393**, 49–52 (1998).
20. A. A. Farajian, K. Estarjani, and Y. Kawazoe, "Nonlinear coherent transport through doped nanotube junctions," *Phys. Rev. Lett.* **82**, 5084–5087 (1999).
21. F. Leonard and J. Tersoff, "Role of Fermi-level pinning in nanotube Schottky diodes," *Phys. Rev. Lett.* **84**, 4693–4696 (2000).
22. A. A. Odintsov, "Schottky barriers in carbon nanotube heterojunctions," *Phys. Rev. Lett.* **85**, 150–153 (2000).
23. R. Dingle and C. H. Henry, "Quantum effects in heterostructure lasers," U.S. Patent No. 3,982,207 (1976).
24. Y. Arakawa and H. Sakaki, "Multidimensional quantum well lasers and temperature dependence of its threshold current," *Appl. Phys. Lett.* **40**, 939–941 (1982).

25. M. Asada, M. Miyamoto, and Y. Suematsu, "Gain and the threshold of three dimensional quantum dot lasers," *IEEE J. Quantum Electron.* **22**, 1915–1933 (1986).
26. N. N. Ledentsov, "Long-wavelength quantum-dot lasers on GaAs substrates: from media to device concepts," *IEEE J. Sel. Top. Quantum Electron.* **8**, 1015–1024 (2002).
27. G. Khitrova, H. M. Gibbs, F. Jahnke, and S. W. Koch, "Nonlinear optics of normal-mode-coupling semiconductor microcavities," *Rev. Mod. Phys.* **71**, 1591–1639 (1999).
28. M. Pelton and Y. Yamamoto, "Ultralow threshold laser using a single quantum dot and a microsphere cavity," *Phys. Rev. A* **59**, 2418–2421 (1999).
29. M. V. Artemyev, U. Woggon, R. Wannemacher, H. Jaschinski, and W. Langbein, "Light trapped in a photonic dot: microspheres act as a cavity for quantum dot emission," *Nano. Lett.* **1**, 309–314 (2001).
30. J. M. Gérard and B. Gayral, "InAs quantum dots: artificial atoms for solid-state cayity-quantum electrodynamics," *Phys. E* **9**, 131–139 (2001).
31. P. Michler, A. Imamoglu, M. D. Mason, P. J. Carson, G. F. Strouse, and S. K. Buratto, "Quantum correlation among photons from a single quantum dot at room temperature," *Nature* **406**, 968–970 (2000).
32. C. Santori, M. Pelton, G. Solomon, Y. Dale, and Y. Yamamoto, "Triggered single photons from a quantum dot," *Phys. Rev. Lett.* **86**, 1502–1505 (2001).
33. D. V. Regelman, U. Mizrahi, D. Gershoni, E. Ehrenfreund, W. V. Schoenfeld, and P. M. Petroff, "Semiconductor quantum dot: a quantum light source of multicolor photons with tunable statistics," *Phys. Rev. Lett.* **87**, 257401 (2001).
34. E. Moreau, I. Robert, L. Manin, V. Thierry-Mieg, J. M. Gérard, and I. Abram, "Quantum cascade of photons in semiconductor quantum dots," *Phys. Rev. Lett.* **87**, 183601 (2001).
35. L. A. Weinstein, *The Theory of Diffraction and the Factorization Method*, Golem, New York (1969).
36. R. Mittra and S. W. Lee, *Analytical Techniques in the Theory of Guided Waves*, Macmillan, New York (1971).
37. N. A. Khiznjak, *Integral Equations of Macroscopic Electrodynamics* (in Russian), Naukova Dumka, Kiev (1986).
38. D. J. Hoppe and Y. Rahmat–Samii, *Impedance Boundary Conditions in Electromagnetics*, Taylor & Francis, Washington, DC (1995).
39. G. Ya. Slepyan, S. A. Maksimenko, A. Lakhtakia, O. M. Yevtushenko, and A. V. Gusakov, "Electronic and electromagnetic properties of nanotubes," *Phys. Rev. B* **57**, 9485–9497 (1998).
40. A. Lakhtakia, G. Ya. Slepyan, S. A. Maksimenko, O. M. Yevtushenko, and A. V. Gusakov, "Effective media theory of the microwave and the infrared properties of composites with carbon nanotube inclusions," *Carbon* **36**, 1833–1838 (1998).

41. G. Ya. Slepyan, S. A. Maksimenko, A. Lakhtakia, O. M. Yevtushenko, and A. V. Gusakov, "Electrodynamics of carbon nanotubes: dynamic conductivity, impedance boundary conditions and surface wave propagation," *Phys. Rev. B* **60**, 17136–17149 (1999).
42. S. A. Maksimenko and G. Ya. Slepyan, "Electrodynamic properties of carbon nanotubes," in *Electromagnetic Fields in Unconventional Materials and Structures*, O. N. Singh and A. Lakhtakia, Eds., 217–255, Wiley, New York (2000).
43. G. Ya. Slepyan, N. A. Krapivin, S. A. Maksimenko, A. Lakhtakia, and O. M. Yevtushenko, "Scattering of electromagnetic waves by a semi-infinite carbon nanotube," *Arch. Elektron. Übertrag* **55**, 273–280 (2001).
44. S. A. Maksimenko and G. Ya. Slepyan, "Electrodynamics of carbon nanotubes," *J. Commun. Technol. Electron.* **47**, 235–252 (2002).
45. O. M. Yevtushenko, G. Ya. Slepyan, S. A. Maksimenko, A. Lakhtakia, and D. A. Romanov, "Nonlinear electron transport effects in a chiral carbon nanotube," *Phys. Rev. Lett.* **79**, 1102–1105 (1997).
46. G. Ya. Slepyan, S. A. Maksimenko, V. P. Kalosha, J. Herrmann, E. E. B. Campbell, and I. V. Hertel, "Highly efficient high harmonic generation by metallic carbon nanotubes," *Phys. Rev. A* **61**, R777–R780 (1999).
47. A. S. Maksimenko and G. Ya. Slepyan, "Negative differential conductivity in carbon nanotubes," *Phys. Rev. Lett.* **84**, 362–365 (2000).
48. G. Ya. Slepyan, S. A. Maksimenko, V. P. Kalosha, A. V. Gusakov, and J. Herrmann, "High-order harmonic generation by conduction electrons in carbon nanotube rope," *Phys. Rev. A* **63**, 053808 (2001).
49. C. Stanciu, R. Ehlich, V. Petrov, O. Steinkellner, J. Herrmann, I. V. Hertel, G. Ya. Slepyan, A. A. Khrutchinski, S. A. Maksimenko, F. Rotermund, E. E. B. Campbell, and F. Rohmund, "Experimental and theoretical study of third-order harmonic generation in carbon nanotubes," *Appl. Phys. Lett.* **81**, 4064–4066 (2002).
50. I. V. Bondarev, G. Ya. Slepyan, and S. A. Maksimenko, "Spontaneous decay of excited atomic states near a carbon nanotube," *Phys. Rev. Lett.* **89**, 115504 (2002).
51. G. Ya. Slepyan, S. A. Maksimenko, V. P. Kalosha, J. Herrmann, N. N. Ledentsov, I. L. Krestnikov, Zh. I. Alferov, and D. Bimberg, "Polarization splitting of the gain band in quantum wire and quantum dot arrays," *Phys. Rev. B* **59**, 1275–1278 (1999).
52. S. A. Maksimenko, G. Ya. Slepyan, V. P. Kalosha, S. V. Maly, N. N. Ledentsov, J. Herrmann, A. Hoffmann, D. Bimberg, and Zh. I. Alferov, "Electromagnetic response of 3D arrays of quantum dots," *J. Electron. Mater.* **29**, 494–503 (2000).
53. S. A. Maksimenko, G. Ya. Slepyan, N. N. Ledentsov, V. P. Kalosha, A. Hoffmann, and D. Bimberg, "Light confinement in a quantum dot," *Semicond. Sci. Technol.* **15**, 491–496 (2000).

54. S. A. Maksimenko, G. Ya. Slepyan, V. P. Kalosha, N. N. Ledentsov, A. Hoffmann, and D. Bimberg, "Size and shape effects in electromagnetic response of quantum dots and quan-tum dot arrays," *Mater. Sci. Eng. B* **82**, 215–217 (2001).
55. G. Ya. Slepyan, S. A. Maksimenko, V. P. Kalosha, A. Hoffmann, and D. Bimberg, "Effective boundary conditions for planar quantum-dot structures," *Phys. Rev. B* **64**, 125326 (2001).
56. S. A. Maksimenko, G. Ya. Slepyan, A. Hoffmann, and D. Bimberg, "Local field effect in an isolated quantum dot: self-consistent microscopic approach," *Phys. Status Solidi (a)* **190**, 555–559 (2002).
57. G. Ya. Slepyan, S. A. Maksimenko, A. Hoffmann, and D. Bimberg, "Quantum optics of a quantum dot: local-field effects," *Phys. Rev. A* **66**, 063804 (2002).
58. G. Ya. Slepyan, S. A. Maksimenko, A. Hoffmann, and D. Bimberg, "Excitonic composites," in *Advances in Electromagnetics of Complex Media and Metamaterials*, S. Zouhdi, A. Sihvola, and M. Arsalane, Eds., 385–402, Kluwer, Dordrecht (2003).
59. S. A. Maksimenko and G. Ya. Slepyan, "Quantum dot array: electromagnetic properties of," in *Encyclopedia of Nanoscience and Nanotechnology*, J. A. Schwarz, C. Contescu, and K. Putyera, Eds., Marcel Dekker, New York (2004).
60. A. Lakhtakia, Ed., *Selected Papers on Linear Optical Composite Materials*, SPIE Press, Bellingham, WA (1996).
61. H. W. Kroto, J. R. Heath, S. C. O'Brien, R. F. Curl, and R. E. Smalley, "C^{60}: Buckminsterfullerene," *Nature* **318**, 162–165 (1985).
62. S. Iijima, "Helical microtubules of graphitic carbon," *Nature* **354**, 56–58 (1991).
63. L.-Ch. Qin, X. Zhao, K. Hirahara, Y. Miyamoto, Y. Ando, and S. Iijima, "The smallest carbon nanotube," *Nature* **408**, 50 (2000).
64. R. Saito, M. Fujita, G. Dresselhaus, and M. S. Dresselhaus, "Electronic structure of graphene tubules based on C_{60}," *Phys. Rev. B* **46**, 1804–1811 (1992).
65. M. F. Lin and K. W.-K. Shung, "Plasmons and optical properties of carbon nanotubes," *Phys. Rev. B* **50**, 17744–17747 (1994).
66. P. R. Wallace, "The band theory of graphite," *Phys. Rev.* **71**, 622–634 (1947).
67. J. W. Mintmire and C. T. White, "Universal density of states for carbon nanotubes," *Phys. Rev. Lett.* **81**, 2506–2509 (1998).
68. W. Tian and S. Datta, "Aharonov–Bohm-type effect in graphene tubules: a Landauer approach," *Phys. Rev. B* **49**, 5097–5100 (1994).
69. H. Suzuura and T. Ando, "Phonons and electron-phonon scattering in carbon nanotubes," *Phys. Rev. B* **65**, 235412 (2002).
70. Y. Miyamoto, S. G. Louie, and M. L. Cohen, "Chiral conductivities of nanotubes," *Phys. Rev. Lett.* **76**, 2121–2124 (1996).
71. P. Kral, E. J. Mele, and D. Tomanek, "Photogalvanic effects in heteropolar nanotubes," *Phys. Rev. Lett.* **85**, 1512–1515 (2000).

72. O. V. Kibis, "Electronic phenomena in chiral carbon nanotubes in the presence of a magnetic field," *Physica* **12**, 741–744 (2002).
73. Yu. A. Il'inskii and L. V. Keldysh, *Electromagnetic Response of Material Media*, Plenum, New York (1994).
74. R. Courant and D. Hilbert, *Methods of Mathematical Physics*, Vol. 1, Interscience, New York (1962).
75. L. M. Woods and G. D. Mahan, "Electron-phonon effects in graphene and armchair (10,10) single-wall carbon nanotubes," *Phys. Rev. B* **61**, 10651–10663 (2000).
76. S. J. Tans, M. H. Devoret, H. Dai, A. Thess, R. E. Smalley, L. J. Geerligs, and C. Dekker, "Individual single-wall carbon nanotubes as quantum wires," *Nature* **386**, 474–477 (1997).
77. F. J. García-Vidal, J. M. Pitarke, and J. B. Pendry, "Effective media theory of the optical properties of aligned carbon nanotubes," *Phys. Rev. Lett.* **78**, 4289–4292 (1997).
78. W. Lü, J. Dong, and Z.-Ya. Li, "Optical properties of aligned carbon nanotube systems studied by the effective-media approximation method," *Phys. Rev. B* **63**, 033401 (2000).
79. J. M. Pitarke and F. J. García-Vidal, "Electronic response of aligned multishell carbon nanotubes," *Phys. Rev. B* **64**, 073404 (2001).
80. M. F. Lin and D. S. Chuu, "π-plasmons in carbon nanotube bundles," *Phys. Rev. B* **57**, 10183–10187 (1998).
81. R. Tamura and M. Tsukada, "Analysis of quantum conductance of carbon nanotube junctions by the effective-mass approximation," *Phys. Rev. B* **58**, 8120–8124 (1998).
82. M. Menon, D. Srivastava, and S. Saini, "Fullerene-derived molecular electronic devices," *Semicond. Sci. Technol.* **13**, A51–A54 (1998).
83. A. N. Andriotis, M. Menon, D. Srivastava, and L. Chernozatouskii, "Ballistic switching and rectification in single wall carbon nanotube Y-junctions," *Appl. Phys. Lett.* **79**, 266–268 (2001).
84. A. N. Andriotis, M. Menon, D. Srivastava, and L. Chernozatonskii, "Rectification properties of carbon nanotube Y-junctions," *Phys. Rev. Lett.* **87**, 066802 (2001).
85. Y.-G. Yoon, M. S. C. Mazzoni, H. J. Choi, J. Ihm, and S. G. Louie, "Structural deformation and intertube conductance of crossed carbon nanotube junctions," *Phys. Rev. Lett.* **86,** 688–691 (2001).
86. C. Bena, S. Vishveshwara, L. Balents, and M. P. A. Fisher, "Quantum entanglement in carbon nanotubes," *Phys. Rev. Lett.* **89**, 037901 (2002).
87. A. Rochefort, D. R. Salahub, and Ph. Avouris, "Effects of finite length on the electronic structure of carbon nanotubes," *J. Phys. Chem. B* **103**, 641–646 (1999).
88. F. G. Bass, A. Ya. Slepyan, and G. Ya. Slepyan, "Resonant oscillations of diffraction structures with weak nonlinearity," *Microwave Opt. Technol. Lett.* **19**, 203–208 (1998).

89. G. Ya. Slepyan, S. A. Maksimenko, F. G. Bass, and A. Lakhtakia, "Nonlinear electromagnetics in chiral media: self-action of waves," *Phys. Rev. E* **52**, 1049–1058 (1995).
90. V. M. Shalaev, *Nonlinear Optics of Random Media: Fractal Composites and Metal-Dielectric Films*, Springer, Berlin (2000).
91. T. G. Mackay, "Homogenization of linear and nonlinear complex composite materials," in *Introduction to Complex Mediums for Optics and Electromagnetics*, W. S. Weiglhofer and A. Lakhtakia, Eds., SPIE Press, Bellingham, WA (2003).
92. A. L'Huillier, K. Schafer, and K. Kulander, "Higher-order harmonic generation in xenon at 1064 nm: the role of phase matching," *Phys. Rev. Lett.* **66**, 2200–2203 (1991).
93. J. Krause, K. Schafer, and K. Kulander, "High-order harmonic generation from atoms and ions in the high intensity regime," *Phys. Rev. Lett.* **68**, 3535–3538 (1992).
94. D. von der Linde, T. Engers, G. Jenke, P. Agostini, G. Grillon, E. Nibbering, A. Mysyrowicz, and A. Antonetti, "Generation of high-order harmonics from solid surfaces by intense femtosecond laser pulses," *Phys. Rev. A* **52**, R25–R27 (1995).
95. Y. R. Shen, *The Principles of Nonlinear Optics*, Wiley, New York (1984).
96. F. Leonard and J. Tersoff, "Negative differential resistance in nanotube devices," *Phys. Rev. Lett.* **85**, 4767–4770 (2000).
97. A. Bezryadin, A. R. M. Verschueren, S. J. Tans, and C. Dekker, "Multiprobe transport experiments on individual single-wall carbon nanotubes," *Phys. Rev. Lett.* **80**, 4036–4039 (1998).
98. R. Eigger and A. O. Gogolin, "Effective low-energy theory for correlated carbon nanotubes," *Phys. Rev. Lett.* **79**, 5082–5085 (1997).
99. C. L. Kane, L. Balents, and M. P. A. Fisher, "Coulomb interactions and mesoscopic effects in carbon nanotubes," *Phys. Rev. Lett.* **79**, 5086–5089 (1997).
100. Yu. A. Krotov, D.-H. Lee, and S. G. Louie, "Low energy properties of (n, n) carbon nanotubes," *Phys. Rev. Lett.* **78**, 4245–4248 (1997).
101. W. Vogel, D.-G. Welsch, and S. Wallentowitz, *Quantum Optics: An Introduction*, Wiley, New York (2001).
102. M. Fleischhouer, "Spontaneous emission and level shifts in absorbing disordered dielectrics and dense atomic gases: a Green's-function approach," *Phys. Rev. A* **60**, 2534–2539 (1999).
103. H. T. Dung, L. Knöll, and D.-G. Welsch, "Decay of an excited atom near an absorbing microsphere," *Phys. Rev. A* **64**, 013804 (2001).
104. V. B. Berestetskii, E. M. Lifshitz, and L. P. Pitaevskii, *Quantum Electrodynamics*, Pergamon Press, Oxford (1982).
105. V. V. Klimov and M. Ducloy, "Allowed and forbidden transitions in an atom placed near an ideally conducting cylinder," *Phys. Rev. A* **62**, 043818 (2000).
106. T. Sondergaard and B. Tromborg, "General theory for spontaneous emission in active dielectric microstructures: example of a fiber amplifier," *Phys. Rev. A* **64**, 033812 (2001).

107. S. Tasaki, K. Maekawa and T. Yamabe, "π-band contribution to the optical properties of carbon nanotubes: effects of chirality," *Phys. Rev. B* **57**, 9301–9318 (1998).
108. E. M. Purcell, "Spontaneous emission probabilities at radio frequencies," *Phys. Rev.* **69**, 681 (1946).
109. E. P. Petrov, V. N. Bogomolov, I. I. Kalosha, and S. V. Gaponenko, "Spontaneous emission of organic molecules embedded in a photonic crystal," *Phys. Rev. Lett.* **81**, 77–80 (1998).
110. M. O. Scully and M. S. Zubairy, *Quantum Optics*, Cambridge University Press, Cambridge (2001).
111. F. Boxberg and J. Tulkki, "Quantum dots: phenomenology, photonic and electronics properties, modeling, and technology," in *Nanometer Structures: Theory, Modeling, and Simulation*, A. Lakhtakia, Ed., SPIE Press, Bellingham, WA (2004).
112. S. Schmitt-Rink, D. A. B. Miller, and D. S. Chemla, "Theory of the linear and nonlinear optical properties of semiconductor microcrystallites," *Phys. Rev. B* **35**, 8113–8125 (1987).
113. P. D. Drummond and M. Hillery, "Quantum theory of dispersive electromagnetic modes," *Phys. Rev. A* **59**, 691–707 (1999).
114. M. Hillery and P. D. Drummond, "Noise-free scattering of the quantized electromagnetic field from a dispersive linear dielectric," *Phys. Rev. A* **64**, 013815 (2001).
115. V. Savona, Z. Hradil, A. Quattropani, and P. Schwendimann, "Quantum theory of quantum-well polaritons in semiconductor microcavities," *Phys. Rev. B* **49**, 8774–8779 (1994).
116. S. Savasta and R. Girlanda, "Quantum description of the input and output electromagnetic fields in a polarizable confined system," *Phys. Rev. A* **53**, 2716–2726 (1996).
117. R. E. Kleinman, "Low frequency electromagnetic scattering," in *Electromagnetic Scattering*, P. L. E. Uslenghi, Ed., 1–28, Academic Press, New York (1978).
118. Y. Ohfuti and K. Cho, "Nonlocal optical response of assemblies of semiconductor spheres," *Phys. Rev. B* **51**, 14379–14394 (1995).
119. A. D. Yaghjian, "Electric dyadic Green's function in the source region," *Proc. IEEE* **68**, 248–263 (1980).
120. C. M. Bowden and J. P. Dowling, "Near-dipole-dipole effects in dense media: generalized Maxwell–Bloch equations," *Phys. Rev. A* **47**, 1247–1251 (1993).
121. B. Hanewinkel, A. Knorr, P. Thomas, and S. W. Koch, "Optical near-field response of semiconductor quantum dots," *Phys. Rev. B* **55**, 13715–13725 (1997).
122. D. Gammon, E. S. Snow, B. V. Shanabrook, D. S. Katzer, and D. Park, "Fine structure splitting in the optical spectra of single GaAs quantum dots," *Phys. Rev. Lett.* **76**, 3005–3008 (1996).

123. V. L. Ginzburg, *Theoretical Physics and Astrophysics*, Pergamon Press, Oxford (1979).
124. H. Ajiki and K. Cho, "Longitudinal and transverse components of excitons in a spherical quantum dot," *Phys. Rev. B* **62**, 7402–7412 (2000).
125. H. T. Dung and K. Ujihara, "Three-dimensional nonperturbative analysis of spontaneous emission in a Fabry–Perot microcavity," *Phys. Rev. A* **60**, 4067–4082 (1999).
126. A. Thränhardt, C. Ell, G. Khitrova, and H. M. Gibbs, "Relation between dipole moment and radiative lifetime in interface fluctuation quantum dots," *Phys. Rev. B* **65**, 035327 (2002).
127. M. E. Crenshaw and C. M. Bowden, "Lorentz local-field effects on spontaneous emission in dielectric media," *Phys. Rev. A* **63**, 013801 (2000).
128. A. S. Parkins, P. Marte, P. Zoller, and H. J. Kimble, "Synthesis of arbitrary quantum states via adiabatic transfer of Zeeman coherence," *Phys. Rev. Lett.* **71**, 3095–3098 (1993).
129. C. K. Law and J. H. Eberly, "Arbitrary control of a quantum electromagnetic field," *Phys. Rev. Lett.* **76**, 1055–1058 (1996).
130. O. Keller, "Local fields in the electrodynamics of mesoscopic media," *Phys. Rept.* **268**, 85–262 (1996).
131. J. R. Guest, T. H. Stievater, X. Li, J. Cheng, D. G. Steel, D. Gammon, D. S. Katzer, D. Park, C. Ell, A. Thränhardt, G. Khitrova, and H. M. Gibbs, "Measurement of optical absorption by a single quantum dot exiton," *Phys. Rev. B* **65**, 241310(R) (2002).
132. D. Birkedal, K. Leosson, and J. M. Hvam, "Long lived coherence in self-assembled quantum dots," *Phys. Rev. Lett.* **87**, 227401 (2001).
133. P. Borri, W. Langbein, S. Schneider, U. Woggon, R. L. Sellin, D. Ouyang, and D. Bimberg, "Exciton relaxation and dephasing in quantum-dot amplifiers from room to cryogenic temperature," *IEEE J. Sel. Top. Quantum Electron.* **8**, 984–991 (2002).
134. O. Verzelen, R. Ferreira, and G. Bastard, "Excitonic polarons in semiconductor quantum dots," *Phys. Rev. Lett.* **88**, 146803 (2002).

List of symbols

a	$3b/2\hbar$
$\mathbf{a}_{1,2}$	lattice basis vectors
$a_{qv}^{\dagger}$ and a_{qv}	electron creation and annihilation operators
$b = 0.142$ nm	interatomic distance in graphene
$\hat{b}^{\dagger}$ and $\hat{b}$	creation and annihilation operators for electron-hole pairs
c	speed of light in free space (i.e., vacuum)
$c_k^{\dagger}$ and c_k	photon creation and annihilation operators
e	electron charge
$\mathbf{e}_z$	unit vector along the CN axis
$\mathbf{E}$	electric field
$\widehat{\mathbf{E}}$	electric field operator
$F(\theta, \theta_0)$	edge-scattering pattern
$F_{c,v}(p_z, s)$	equilibrium Fermi distribution functions in conduction and valence bands
$G_{\alpha\beta}$	components of the classical dyadic Green's function
$G_{\alpha\beta}^{(0)}$	components of the classical free-space dyadic Green's function
$G^{(\mathrm{ret})}(\mathbf{r}, t)$	retarded Green's function
$H_l^{(1,2)}(\cdot)$	cylindrical Hankel functions of the first and second kinds
$\mathbf{H}$	magnetic field
$\widehat{\mathbf{H}}$	magnetic field operator
h	wave number of the surface wave
$\mathcal{H}_0$	Hamiltonian of carrier motion
i	$\sqrt{-1}$
$I_l(\cdot)$	modified cylindrical Bessel functions of the first kind
$I(\mathbf{r}, t)$	spontaneous emission intensity
$\widehat{\mathbf{J}}^{\mathrm{ext}}$	external current operator
$\hat{j}_z^{\mathrm{ns}}$	axial noise current operator
$J_N(\cdot)$	Bessel functions
j_z	surface axial current density
k	free-space wave number
k_B	Boltzmann constant
$K_l(\cdot)$	modified cylindrical Bessel functions of the second kind
l_{CN}	CN length
l_0	parameter to characterize spatial dispersion
(m, n)	dual index to characterize CNs
m_0	electron mass

$\mathbf{n}$	unit vector along the exterior normal to the CN surface
$\underline{\underline{\mathbf{N}}}$	depolarization diadic
$\widehat{\mathbf{P}}(\mathbf{r}, t)$	the polarization single-particle operator
$\mathbf{p}$	π-electron quasi-momentum
$\mathbf{p}_F$	π-electron quasi-momentum at the Fermi level
$P_l(\theta)$	far-zone scattered power density
$p_{x,y,z}$	projections of the quasi-momentum
$\mathbf{r}$	position vector
$\mathbf{r}_c$	position vector of QD geometrical center
$\mathbf{R}$	relative position vector between two sites on the honeycomb lattice
$R_{qq'}$	matrix element of the interband transition
R_{CN}	carbon nanotube radius
R_{QD}	quantum dot radius
(r, ϑ, φ)	spherical coordinates
t	time
T	temperature
$u_q(\mathbf{r})$	amplitude of a Bloch wave
v_F	velocity of π-electron at the Fermi level
v_{ph}	phase velocity
V	QD volume
V_{uc}	unit cell volume of crystalline lattice
$W(\mathbf{r})$	potential of the CN crystalline lattice
$w(t)$	transition probability
(x, y)	2D Cartesian coordinate system for graphene
$Y_{lm}(\vartheta, \varphi)$	spherical harmonics
$\mathcal{E}_{c,v}$	energy of π-electrons in the conduction and the valence bands
β	complex-valued slow-wave coefficient
β_N	complex-valued Fock-qubit coefficients
$\gamma_0 \approx 2.7$ eV	overlapping integral
Γ_{hom}	homogeneous broadening of the spectral line
Γ_r	radiative spontaneous decay rate
Γ_{sp}	spontaneous decay rate
$\Gamma_{\mathrm{sp}}^{(o)}$	spontaneous decay rate in free space
$\Delta\epsilon$	energy-level spacing
$\delta(x)$	Dirac delta function
$\delta_{\alpha\beta}$	Kronecker symbol
ϵ_{c}	charging energy
$\epsilon_{\mathrm{g,e}}$	energy eigenvalues of carrier-motion Hamiltonian
$\varepsilon(\mathbf{r}, \omega)$	relative permittivity
ε_h	relative permittivity of a host media

$\Lambda = \Omega_{st}/\omega_1$	dimensionless parameter to characterize pumping field strength
$\lambda = 2\pi/k$	free-space wavelength
μ_{ch}	chemical potential
$\boldsymbol{\mu}$	dipole moment
Π_e	scalar Hertz potential
τ	relaxation time
τ_{sp}	spontaneous radiation time
$\tau_{sp}^{(o)}$	spontaneous radiation time in free space
θ_0	angle of propagation of the cylindrical wave in a CN
θ_{CN}	geometric chiral angle of a CN
(ρ, φ, z)	circular cylindrical coordinate system for any CN with the CN axis parallel to the z axis
$\rho_{qq'}$	elements of the density matrix
$\sigma_{zz}(\omega)$	axial conductivity of the CN
Ψ	electron wave function in CN
Ψ_q	Bloch function
$\vert\psi(t)\rangle$	wave function of a QD exposed to the classical electromagnetic field
ω	angular frequency
ω_l	low-frequency edge of the optical transition band
ω_1	angular frequency of the pump field
ω_{vc}	angular transition frequency
Ω	normalization volume
Ω_{st}	angular Stark frequency
$\hbar = 1.05459 \times 10^{-34}$ J s	Planck constant

Sergey A. Maksimenko was born in Belarus in 1954. He received his MS degree in physics of heat and mass transfer from Belarus State University, Minsk, in 1976; his PhD degree in theoretical physics from Belarus State University in 1988; and his ScD degree in theoretical physics from the Institute of Physics, Minsk, in 1996. He is currently Deputy Vice-Rector of Belarus State University and heads the Laboratory at the Institute for Nuclear Problems at Belarus State University. He has authored or coauthored more than 80 conference and journal papers. Dr. Maksimenko was a member of the Scientific Advisory Committee of Bianisotropics '98. His current research interests are electromagnetic wave theory, diffraction by periodic media and structures, and electromagnetic processes in quasi-one- and zero-dimensional nanostructures in condensed matter.

Gregory Ya. Slepyan was born in Minsk, Belarus, in 1952. He received his MS degree in radioengineering from the Minsk Radioengineering Institute in 1974; his PhD degree in physics from the Belarus State University, Minsk, in 1979; and his ScD degree in physics from Kharkov State University, Kharkov, Ukraine, in 1988. He is currently a principal researcher at the Institute for Nuclear Problems at Belarus State University. He has authored or coauthored more than 100 theoretical conference and journal papers as well as two books. Dr. Slepyan is a member of the editorial board of *Electromagnetics* and he was a member of the Scientific Advisory Committee of the conferences Bianisotropics '97 and Bianisotropics 2000. His current research interests are diffraction theory, microwave and millimeter-wave circuits, nonlinear waves and oscillations, and nanostructures in condensed matter with applications to nanoelectronics.

Chapter 6

Atomistic Simulation Methods

Pierre A. Deymier, Vivek Kapila and Krishna Muralidharan

6.1. Introduction 208
6.2. Determininistic atomistic computer simulation methodologies 210
  6.2.1. Microcanonical molecular dynamics 210
  6.2.2. Canonical ensemble molecular dynamics 211
  6.2.3. Other ensembles 215
  6.2.4. Interatomic potentials 216
  6.2.5. Thermostating a buckyball: an illustrative example 217
6.3. Stochastic atomistic computer simulation methodologies 221
  6.3.1. Canonical Monte Carlo 221
  6.3.2. Grand canonical Monte Carlo 223
  6.3.3. Lattice Monte Carlo 225
  6.3.4. Self-assembly of surfactants 226
  6.3.5. Kinetic Monte Carlo 230
  6.3.6. Application of kinetic MC to self-assembly of protein subcellular nanostructures 230
6.4. Multiscale simulation schemes 233
  6.4.1. Coupling of MD and MC simulations 234
  6.4.2. Coupling of an atomistic system with a continuum 239
6.5. Concluding remarks 243
References 244
List of symbols 252

6.1 Introduction

From the time they were pioneered several decades ago, atomistic computer simulation methods such as molecular dynamics (MD) and Monte Carlo (MC) have led to great strides in the description of materials.[1] The limitation of atomistic methods to simulating systems containing a small number of particles is a pathological problem in spite of continuous progress in pushing the limit toward systems of ever increasing sizes.[2] System size is an especially critical issue when one desires a high degree of accuracy in modeling the interatomic forces between the atoms constituting the system with first-principle atomistic simulation approaches.[3] While small system size is an issue for atomistic simulations of bulk materials, the possibility of simulating small systems provides fresh opportunities for scientific advances in the field of nanomaterials. In contrast to modeling bulk materials, atomistic computer simulations could greatly speed up the development of materials at the nanoscale. Nanomaterials exhibit sizes intermediate between those of isolated atoms, molecules, and bulk materials with dimensions scaling from several to hundreds of nanometers. Such systems are ideal for computational studies using MD or MC methods, because these simulations can be done at the realistic size limit, imparting them with predictive capabilities. Therefore, nanomaterials offer a fertile ground for contributions from atomistic computer simulations.

There is already extensive literature on atomistic computer simulations of nanoscale systems; it is not our intention to present an exhaustive review of such studies. A few illustrative examples include: MD simulations of carbon nanotubes,[4–7] fullerenes,[8,9] nanoclusters of polymers,[10] and a plethora of nanostructures: nanorod, nanoindentation, nanomesa, and nanowire.[11–14] Self-assembly is regarded as an extremely powerful approach in the construction of nanoscale structures. Reviews on MC simulation studies of self-assembling processes in aqueous media have appeared in the literature recently.[15,16] MD and MC have also been extensively employed to simulate the formation of self-assembled monolayers on solid substrates.[17–20]

MD and MC methods find their origin in classical statistical mechanics.[21] Provided a model for the interactions between the atomic constituents of some system (for instance, in the form of interatomic or intermolecular potentials that describe the energy of the system as a function of its microscopic degrees of freedom) exists, one can sample deterministically (MD) or stochastically (MC) the microscopic states of the system. The microscopic degrees of freedom usually consist of the set of positions and momenta of the particles. The original intent of MD and MC is, once equilibrium is achieved, to use the concepts of temporal averaging[22] (MD) or statistical averaging[23] (MC) over the sampled microscopic states to calculate the properties of a macroscopic system. This calculation necessitates that the system studied satisfies two hypotheses; namely, the long-time limit and the thermodynamic limit.

The former requires that there must exist macroscopic states of the system for which the macroscopic state variables do not vary, although the microscopic degrees of freedom may undergo considerable variations. The long-time limit hypothesis thus implies that the system does not evolve macroscopically on a time scale large compared to the time scale of microscopic processes. This hypothesis leads into the concepts of ensembles. An ensemble consists of a large collection of macroscopically identical systems that are different in their microscopic states. Ensembles are therefore a construct of the mind that enables the calculation of statistical averages. The most common ensembles are:

- The microcanonical ensemble [constant energy (E), volume (V), and number of atoms (N) for a monoatomic system or number of atomic species (N_i) for multicomponent systems];
- Canonical ensemble [constant temperature (T), V and N];
- Isothermal-isobaric ensemble [constant T, pressure (P), N]; and
- Grand canonical ensemble [constant T, V, and chemical potential (μ)].

The thermodynamic limit hypothesis supposes that (1) the linear dimensions of the system are large compared to the scale of the constitutive elements (for instance, all spatial fluctuations must be included in the description of the system even if their length diverges) and (2) the edges or surface effects can be neglected.

A lot of effort was put into MD and MC methods to satisfy the two hypotheses, including the development of numerous thermostats to maintain the temperature of a system constant, and the application of appropriate boundary conditions such as periodic boundary conditions to mimic bulklike behavior. Moreover, atomistic computer simulation methods have also been used to study material systems beyond these limits. MD and MC simulations have demonstrated usefulness in unraveling the structure and properties of surfaces and interfaces.[24] Steady-state nonequilibrium molecular dynamics (NEMD) methods have enabled the calculation of nonlinear transport properties.[25] Other nonequilibrium processes using stochastic transition based on reaction/transition rates such as reaction kinetics,[26] nucleation and growth,[27] and growth and transport of biological nanostructures[28] have been simulated within the framework of kinetic Monte Carlo approach.

It is clear that most nanoscale systems (or most processes involving nanostructures) will not satisfy the thermodynamic and the long-time limits. Similar to experimental nanotechnology research and development that not only require manipulation and processing of nanoscale structures but also integration into larger systems, an atomistic computer simulation of a nanoscale system must address not only the simulation of the individual nanostructure but also its interactions with larger scale environments. A simple example of this integration is the coupling between a nanostructure and a thermostat in isothermal MD simulations. Models of physical, chemical, and biological systems at the nanoscale based on multiparticle simulations ought to address the issues related to their coupling to systems with spatial and temporal scales that exceed those of the nanostructure itself.

The aim of this chapter is, after presenting a brief review of several representative simulation methodologies, to illustrate with specific examples some of the issues relevant to atomistic computer simulation of nanoscale systems. Particular attention is paid to coupling systems with vastly different spatial scales and/or time scales. Several case studies are presented illustrating spatial and temporal scaling issues, namely, (1) interfacing an individual nanoscale system to "macroscopic" thermostats in MD simulations; (2) bypassing the hierarchy in relaxation times for the simulation of self-assembly of polymer surfactants with the MC method; (3) obtaining dynamical information from kinetic MC simulation of a coarse-grained biological nanostructure; (4) bridging simulation methodologies with different spatial scales (interfacing a small MD system to a larger MC system); and (5) coupling a small MD simulation to continuum mechanics.

6.2 Determininistic atomistic computer simulation methodologies

6.2.1 Microcanonical molecular dynamics

A molecular dynamics investigation consists of numerically solving the classical equations of motion of a set of interacting particles. The solution results in the trajectory of the system, that is, the temporal evolution of the positions and the momenta of all of the particles. The physical description of the system is made via a Hamiltonian that is written as the sum of kinetic energy and potential energy functions. The kinetic energy is typically a sum of quadratic functions of the particles' momenta. The potential energy is usually a function of the particles' positions. A simple formulation for the Hamiltonian of a system of N interacting identical particles is given by

$$H(\{\mathbf{p}\}, \{\mathbf{r}\}) = \sum_{i=1}^{N} \frac{\mathbf{p}_i^2}{2m} + V(\{\mathbf{r}\}), \tag{6.1}$$

where $\{\mathbf{p}\}$ and $\{\mathbf{r}\}$ stand for the momenta and positions of the N particles in some Cartesian coordinate system; i.e., $\{\mathbf{p}\} = \{\mathbf{p}_1, \mathbf{p}_2, \ldots, \mathbf{p}_N\}$ and $\{\mathbf{r}\} = \{\mathbf{r}_1, \mathbf{r}_2, \ldots, \mathbf{r}_N\}$. The function V can be derived from first principles or expressed in the form of semiempirical or empirical functions; some illustrative examples of such functions will be given in Sec. 6.2.4. By a "particle," one understands not only a physical object such as atoms or molecules, but also a pseudo-particle or any other nonphysical object that may be needed as part of the physical description of the system studied. For instance, to achieve isothermal conditions, artificial degrees of freedom may be added to a physical Hamiltonian (see Sec. 6.2.2.). In the simplest form of MD, the trajectories conserve the Hamiltonian (energy) and the number of particles.

Additional boundary conditions are often imposed on the simulated system. Depending on these conditions, the trajectories may conserve either volume or pressure/stress. Free-boundary conditions enable the system of particles to expand freely and therefore achieve constant pressure/stress conditions. One of the simplest forms of MD simulations is conducted at constant energy with a constant number of particles, and uses fixed periodic boundary conditions (PBCs) to maintain the volume constant. When imposing periodic boundary conditions, the potential energy function V includes the position of the particles in the periodic images of the simulation cell. The intent of PBCs is to mimic the behavior of a bulk material with the simulation of a computationally tractable small system. The artificial periodicity reduces free-surface effects that are inherent to small systems. With PBCs the local environment of particles at the edges of the simulation becomes bulklike. PBCs, however, impose unrealistic correlations in the simulated system for distances exceeding half of the shortest edge of the simulation cell. As noted previously, PBCs may not be suitable for the simulation of nanoscale systems where surface effects are an integral part of their behavior.

To solve numerically the equations of motion [Eq. (6.1)], time is discretized. The finite but small integration time step is typically a small fraction of the time necessary for one atomic vibration, typically on the order of 1 fs. Numerous numerical methods are used for the temporal integration of the coupled equations of motion. Some methods such as the leap-frog approach[29] offer the advantages of simplicity and low memory requirement, but entail the drawback of a small time step for accuracy. Other higher-order methods, such as the predictor-corrector method, allow for larger time steps but necessitate larger memory allocations as well. Issues that must be addressed in the choice of a numerical time integrator relate to the conservation of the total energy in a microcanonical ensemble MD. It is not the objective of this chapter to present these numerical methods in any detail, and we refer the reader to several books on the subject.[30–33]

6.2.2 Canonical ensemble molecular dynamics

Isolated systems conserve energy. Nanoscale systems are rarely isolated and are often in thermal contact with some environment that may act as a heat bath. Under such conditions, the nanosystem has to be modeled under isothermal conditions. The simulation of the system of interest including its surroundings, both at the atomic level, becomes quickly a terrifying problem due to the very large number of degrees of freedom required simply to model the surrounding environment.

The goal of isothermal MD (i.e., MD in the canonical ensemble) is reductionist. That is, canonical ensemble MD attempts to couple the atomistic degrees of freedom of the system of interest to a thermostat represented by a small number of variables only. Energy is a constant of motion in the microcanonical ensemble MD. In the canonical ensemble, the equipartition theorem establishes a relationship between the temperature of the system and its kinetic energy.[21] The total energy of the system fluctuates in an isothermal MD while the kinetic energy should become a constant of motion. Any isothermal scheme must satisfy the requirement that a time-averaged property computed along a trajectory from an isothermal MD must equal its canonical ensemble average.

In this section, we will focus on illustrating five different thermostats, namely, the Andersen thermostat,[34] the momentum rescaling method,[35] Hoover's constraint method,[36] the Nosé-Hoover thermostat,[36,37] and the chain of thermostats.[38]

Andersen's thermostat is stochastic. The system is coupled to a heat bath represented by stochastic forces that act on randomly selected particles. The collision effectively occurs by drawing new particle velocities from a Maxwell–Boltzmann distribution according to the desired temperature. Between collisions, the equations of motion are those of a constant energy MD. The Andersen thermostat produces the canonical distribution. The drawback of this method is that the dynamics is not continuous with well-defined quantities (such as energy and momentum).

The momentum rescaling method is an early primitive thermostat based on the equipartition relation. This relation states that the kinetic energy of a system of N particles is related to the temperature T through

$$\sum_{i=1}^{N} \frac{\mathbf{p}_i^2}{2m} = \frac{3}{2} N k_B T, \tag{6.2}$$

where k_B is the Boltzmann constant. The momenta of all of the particles are rescaled at any small interval of time by the factor $\sqrt{T_d/T_a}$, where T_d and T_a are the desired and actual temperatures, respectively. The momentum rescaling method does not reproduce the canonical distribution.

The constraint method is based on non-Newtonian dynamics. The equation of motion for the ith particle takes the form*

$$\ddot{\mathbf{r}}_i = \frac{\mathbf{F}_i}{m} + \alpha \dot{\mathbf{r}}_i, \quad i \in [1, N]. \tag{6.3}$$

From the constraint that the kinetic energy does not fluctuate (i.e., its time derivative is zero) follows the damping factor

$$\alpha = -\frac{\sum_{i=1}^{N} \dot{\mathbf{r}}_i . \mathbf{F}_i}{\sum_{i=1}^{N} m \dot{\mathbf{r}}_i^2}. \tag{6.4}$$

The equilibrium properties of this isothermal system have been shown to be those of the canonical ensemble.[39]

Deterministic isothermal molecular dynamics can be performed with both the Nosé-Hoover thermostat and Nosé-Hoover chain of thermostats. We first consider the Nosé-Hoover thermostat. For each component of the position and momentum

*Throughout the chapter a dot over a variable indicates differentiation with respect to time.

vectors, the following set of dynamical equations defines the Nosé-Hoover dynamics:

$$
\begin{aligned}
\dot{r}_i &= \frac{p_i}{m}, \quad i \in [1, N], \\
\dot{p}_i &= F_i - p_i \frac{p_\eta}{Q}, \quad i \in [1, N], \\
\dot{p}_\eta &= \sum_i \frac{p_i^2}{m} - N k_B T, \\
\dot{\eta} &= \frac{p_\eta}{Q}.
\end{aligned}
\tag{6.5}
$$

The dynamics of the thermostat degree of freedom η is driven by the imbalance between the actual kinetic energy and the desired kinetic energy (through the desired temperature). Here Q is a mass associated with the thermostat degree of freedom.

The Nosé-Hoover chain of thermostats couples the particles to a Nosé-Hoover thermostat which, in turn, is coupled to a second thermostat coupled to a third one, and so on, up to some nth thermostat. The n thermostats form the so-called chain. The dynamics of the Nosé-Hoover chain is driven by the following equations:

$$
\begin{aligned}
\dot{r}_i &= \frac{p_i}{m}, \quad i \in [1, N], \\
\dot{p}_i &= F_i - p_i \frac{p_{\eta_1}}{Q_1}, \quad i \in [1, N], \\
\dot{p}_{\eta_1} &= \left[\sum_i \frac{p_i^2}{m} - N k_B T \right] - p_{\eta_1} \frac{p_{\eta_2}}{Q_2}, \\
\dot{p}_{\eta_j} &= \left[\frac{p_{\eta_{j-1}}^2}{Q_{j-1}} - k_B T \right] - p_{\eta_j} \frac{p_{\eta_{j+1}}}{Q_{j+1}}, \\
\dot{p}_{\eta_n} &= \left[\frac{p_{\eta_{j-1}}^2}{Q_{n-1}} - k_B T \right], \\
\dot{\eta}_j &= \frac{p_{\eta_j}}{Q_j}.
\end{aligned}
\tag{6.6}
$$

The Nosé-Hoover and Nosé-Hoover chain of thermostats produce the proper canonical distribution under specific conditions. Some of these conditions may not be satisfied in the simulation of nanoscale systems. There are two issues to address. One relates to the ergodicity of the equations of motion. The other relates to the conditions for achieving the canonical distribution once ergodicity is achieved.

We first address the problem of ergodicity. In a microcanonical ensemble the trajectories of a system of particles $\{\mathbf{p}, \mathbf{r}\}$ (i.e., the temporal evolution of the system in the position and momentum space) must conserve energy. As a side note, the multidimensional space of positions and momenta of the particles is called the phase space. The condition of conservation of energy is written in the form $H(\{\mathbf{p}\}, \{\mathbf{r}\}) = E = \text{constant}$, and describes a hypersurface in phase space. The ergodic hypothesis as applied to that system implies that, given an infinite amount of time, the trajectory will cover the entire constant energy hypersurface. As a consequence, the temporal average of some quantity will equal its statistical ensemble average. For instance, the microcanonical trajectory of a 1D harmonic oscillator is ergodic and describes an ellipse in the 2D phase space. For a system of particles coupled to a Nosé-Hoover thermostat, the phase space includes the positions and momenta of the particles and of the thermostat. Provided that the Nosé-Hoover equations of motion are ergodic, the system constituted by the particles and thermostat should evolve on the constant generalized energy hyper surface:

$$H(\{\mathbf{p}\}, \{\mathbf{r}\}, \eta, p_\eta) = \sum_{i=1}^{N} \frac{\mathbf{p}_i^2}{2m} + V(\{\mathbf{r}\}) + \frac{p_\eta^2}{2Q} + dNk_BT\eta = \text{constant}, \tag{6.7}$$

where d is the dimensionality of the system.

The rationale for the coupling of the system to the thermostat is to allow the particles to explore a larger region of phase space. For instance, in the case of a harmonic oscillator, the intent is to explore phase space beyond its elliptic trajectory. The Nosé-Hoover equations of motion, however, do not guarantee ergodicity of the trajectory and therefore do not guarantee the canonical distribution. The harmonic oscillator is an example of such a pathological example.[36] This problem can be alleviated with the Nosé-Hoover chain of thermostats. This chain of thermostats may make the trajectory sufficiently chaotic to explore a larger region of phase space and therefore approach ergodicity.

Equation (6.6) conserves the total energy of the Nosé-Hoover extended system, namely,

$$H(\{\mathbf{p}\}, \{\mathbf{r}\}, \eta, p_\eta) = \sum_{i=1}^{N} \frac{\mathbf{p}_i^2}{2m} + V(\{\mathbf{r}\}) + \sum_{k=1}^{n} \frac{p_{\eta k}^2}{2Q_k} + dNk_BT\eta_1 + \sum_{k=2}^{n} k_BT\eta_k = \text{constant}. \tag{6.8}$$

The set of equations (6.6) cannot be derived from a Hamiltonian. Using the principles of non-Hamiltonian statistical mechanics, Tuckerman and Martyna[40] have shown that one requires conservation laws in addition to Eq. (6.8) to yield trajectories of the system of particles that reproduce the canonical distribution. If the system of particles is subjected to no external forces, i.e., $\sum_{i=1}^{N} \mathbf{F}_i = \mathbf{0}$, then there are d additional conservation laws (i.e., one can define d quantities whose time derivatives are zero).

The Nosé-Hoover equations of motion, Eqs. (6.5), also describe non-Hamiltonian dynamics. In that case, the canonical distribution is obtained if there is only one conservation law [i.e., Eq. (6.7)]. In absence of external forces, the Nosé-Hoover trajectories do not sample the phase space according to the canonical distribution unless one imposes that the net momentum be null: $\sum_{i=1}^{N} \mathbf{p}_i = \mathbf{0}$. We refer the reader to Frenkel and Smit[33] for a detailed discussion of these conditions.

In summary, the MD simulation of a nanoscale system, under isothermal conditions with the Nosé-Hoover thermostat or chain of thermostats, requires that special attention be paid to the presence or absence of external stimuli (forces) to generate the appropriate canonical distribution. Although stochastic thermostats such as Andersen's may be more forgiving, deterministic approaches are often preferred in that they are time-reversible and enable direct comparison of results generated by different investigators (if the same initial conditions are used). Additional care is also required to verify that the MD trajectories are ergodic. For small systems (e.g., 1D harmonic oscillator), the Nosé-Hoover trajectories are not ergodic. Other systems that are prone to nonergodic behavior include stiff systems. Systems of this type that are relevant to nanoscale science would be polymeric chains with stiff harmonic interactions[41] or quantum systems described classically by discrete path integrals.[42] The separation of time scale in such systems gives rise to nonergodic trajectories that do not sample the canonical phase space.[43] Some solutions to this problem involve for instance periodic refresh of the velocities[41] or decomposition into normal modes with multiple time scale integration techniques and thermostating with a Nosé-Hoover chain of thermostats.[44] To achieve the canonical distribution, some authors have pushed thermostating to the extreme limit of one Nosé-Hoover chain of thermostats per degree of freedom.[44,45]

6.2.3 Other ensembles

The conventional MD simulation using periodic boundary conditions is performed under constant volume conditions. As mentioned before, it is trivial to simulate a system of particles at constant pressure ($P = 0$) by employing free boundary conditions as may be true for numerous individual nanoscale systems. However, several constant-pressure MD schemes compatible with periodic boundary conditions have been proposed. At constant pressure, the volume of a system of N particles fluctuates. Andersen[34] replaced the atomic coordinates by scaled atomic coordinates. The scaling factor becomes an additional dynamical degree of freedom. Andersen interpreted the scaling factor as the volume of the system. The scaled coordinates are given as the ratio of the coordinates to a length given by the cubic root of the volume. A change in volume results in a homogeneous scaling of the particles' positions. The new degree of freedom possesses its own mass and is associated with a new momentum and kinetic energy. This mass is artificial. It is a measure of the inertia of the volume and controls its rate of change. The potential energy can be visualized as the mechanical work an external pressure would do on the volume. The dynamics of the volume is driven by the imbalance between an applied external pressure and the internal pressure. This latter quantity is related to the particles' positions and interatomic forces through the so-called Virial expression.[21] In absence of thermostats, the equations of motion of the particles and of the volume conserve a quantity closely related to the enthalpy of the system, $H = E + PV$. The trajectories sample the (N, P, H) ensemble.

Parrinello and Rahman[46] extended Andersen's scheme to nonuniform scaling of the simulation cell. Here changes in the orientation and length of the edges of the simulation cell are possible. This allows fluctuations not only in the volume but also in the shape of the cell, thereby enabling the study of crystal structural phase transformations. Nine additional degrees of freedom are necessary to describe the dynamical shape and size of a 3D simulation cell. Each extra degree of freedom possesses a momentum. The inertia of the borders of the simulation cell is therefore characterized by a second-order mass tensor. The potential energy associated with the borders of the cell is a measure of the elastic energy in the limit of linear elasticity. The dynamics of the cell are driven by the imbalance between an external stress tensor (with hydrostatic and nonhydrostatic components) and the internal stress tensor. Parrinello and Rahman's trajectories conserve a generalized enthalpy. The combination of isobaric and isothermal conditions was also undertaken by Andersen,[34] Parrinello and Rahman,[46] as well as, more recently, Tuckerman and Martyna.[40] Other attempts have been made to develop formalisms in other ensembles, including the grand canonical ensemble.[47]

6.2.4 Interatomic potentials

Atomistic computer simulation methods (MD and MC) require a description of the interparticle interactions to yield a microscopic model of the system. Accurate MD or MC results are contingent on the degree of realism of the microscopic description. The information about interparticle interactions is contained in the potential energy function V. For continuous potential functions, the force field acting on a given particle is simply equal to the negative of its gradient. A vast collection of microscopic models has been developed over many years. Early microscopic models range from discontinuous interactions (such as in the hard sphere model or square-well potential model) to pairwise additive continuous interatomic potentials of the Lennard-Jones, Buckingham and other variant forms.[48] Pair potentials were also developed for ionic crystals[49] and metals.[50] The development of microscopic models beyond pair potentials made it possible to describe more realistic systems.[51] Examples of early empirical many-body potentials for describing covalent bonds include the Stillinger-Weber potential for condensed phases of silicon,[52] Rahman-Stillinger potential for water,[53] and Tersoff's potential for carbon.[54] One particularly successful example of a Tersoffian potential for hydrocarbons is the reactive empirical bond-order potential developed by Brenner.[55,56] This type of potential can describe chemical reactivity, that is, chemical processes that involve bond breaking and bond forming.[57] Molecular mechanics nonreactive potential functions for organic substance based on harmonic descriptions of covalent bond stretching and bending are also available.[41] Charge transfer plays an important role in covalent bonding, especially near surfaces, interfaces and defects. Alavi et al. proposed a charge-transfer molecular dynamics that is conservative.[58] This model was applied to the study of silica containing bond-breaking ions.

Many-body potentials derived with the embedded-atom method (EAM) have been very successful at modeling the structure, properties and defects of metals.[59,60] EAM potential functions incorporate the energy associated with the action of embedding an atom within the electron cloud of neighboring atoms plus repulsive pair potential between the atoms. EAM potentials have had great success for face-centered cubic (FCC) metals. Angle-dependent forces are needed to explain the behavior of non-FCC metals. To that effect a modified EAM (MEAM) has been used with success.[61] MD simulations of metals have also been performed with semiempirical potentials based on a quantum mechanical tight-binding method.[62,63] *Ab initio* MD simulation is becoming a powerful alternative to atomistic simulations with empirical or semiempirical potential functions. This method requires no input potential model and solves simultaneously for the classical dynamics of atoms and the electronic structure. For instance, the method of Car and Parrinello unifies MD and density-functional theory.[64] The computational overhead due to the additional electronic degrees of freedom limits this kind of simulation to systems significantly smaller than those accessible with classical MD using empirical or semiempirical interatomic potential functions.

6.2.5 Thermostating a buckyball: an illustrative example

In this section, we illustrate the application of MD and, in particular, the effect of deterministic thermostats on the dynamical and structural behavior of a nanoscale system, namely, a fullerene buckyball. Specifically, the thermal decomposition of a C_{60} molecule is studied with the temperature of the molecule being controlled by an external thermostat. Kim and Tomanek examined the high-temperature behavior of fullerenes, which involves a consequent distortion, and the ultimate fragmentation of the molecule as the temperature is increased.[9] Their work included a detailed MD simulation study of the "melting" of the molecules of three prototype fullerenes, namely, C_{20}, C_{60}, and C_{240}. The force calculation was based on a linear combination of atomic orbital formalisms (involving the parameterization of *ab initio* local density functional results) for structures as different as C_2, carbon chains, graphite, and diamond.[9] On heating the molecule, many phases were identified at elevated temperatures. The system evolved from a buckyball to a "floppy-like" phase, then to a pretzel-like phase with 3D structure of connected carbon rings, and finally to carbon chain fragments. The temperature of the system was controlled using a Nosé-Hoover thermostat.

Following the same lines, we examine the thermal decomposition characteristics of a buckyball using the Tersoff's potential to represent the interatomic interactions.[54] The Tersoff's potential does not account for the different states of hybridization of a carbon atom explicitly which leads to results that are slightly different than the more realistic *ab initio* calculations of Kim and Tomanek. In the Tersoff's potential, the interatomic energy between any two neighboring atoms (i and j) is of the form: $V_{ij} = f_c(r_{ij})[f_R(r_{ij}) + b_{ij} f_A(r_{ij})]$. In this expression, V_{ij} is the bond energy, $f_R(r_{ij})$ represents a repulsive pair potential, $f_A(r_{ij})$ represents an attractive pair potential associated with bonding, and $f_c(r_{ij})$ is a smooth cut-off function. The Tersoff's potential includes a many-body environment-dependent bond order term b_{ij}. This term is associated with the attractive part (f_A) of V_{ij} and describes the modulation of the two-body potential due to the presence of other neighboring atoms—the k atoms. Also, b_{ij} is a measure of bond order and is a monotonically decreasing function of the number of neighbors j of atom i (i.e., the coordination number of i).

A more important part of this example is to compare and contrast the effect of the various thermostats on the thermal decomposition of the molecule. As pointed out earlier in previous sections, the various thermostats maintain the temperature of the system through different means; and thus their effect on an extremely small system could be dramatically different. On the other hand, the system could be perfectly oblivious to the types of the thermostats. We test four deterministic thermostats, namely, momentum rescaling, constraint method, Nosé-Hoover, and Nosé-Hoover chain on the C_{60} molecule as the molecule is heated. In this study, the MD simulations are in quasi-equilibrium, thus the effect of the thermostats on the evolution of thermodynamic quantities such as internal energies can be evaluated.

Following closely the procedure adopted by Kim and Tomanek, the system initially at rest at 0 K is heated up systematically, with the system temperature ramped up by 400 K for every 0.4 ps. The size of the time step used is 0.1 fs to ensure accurate integration of the equations of motion of the system. A simple one-step finite-difference Verlet method is used as the time integrator. Data is collected only during the final 0.2 ps at each temperature and for each thermostat. Five different runs (corresponding to different starting velocity distributions) are carried out to ensure better statistics. The variations of the internal energy, coordination number, and the atomic binding energy (ABE) are recorded as a function of temperature for the different thermostat runs.

The C_{60} molecule consists of sp^2 hybridized carbon atoms, each atom bonded to three other carbon atoms. As the molecule is heated, it constantly changes shape and finally fragments. Thus, in addition to the potential that governs the interatomic interactions, the role of the external thermostat becomes crucial in evaluating the fragmentation dynamics of the molecule.

Figure 6.1 represents the variation of the system temperature with time, for four different thermostats. As is evident from Fig. 6.1, the thermostats control the temperature efficiently, with no visible effect of the size of the system on any of the thermostat's ability to maintain the required temperature.

Next, we evaluate the average total energy of the system as a function of time (or equivalently the system temperature), as shown in Fig. 6.2. The variation in the energy for each case seems to follow very similar trends: a monotonous increase in energy with temperature, with a significant change in slope around 6000 K. We note that the velocity rescaling and the constraint methods differ significantly from each other. The dynamics of the C_{60} molecule is practically the same when using the Nosé-Hoover thermostat or the Nosé-Hoover chain of thermostats.

To follow the structural changes of the fullerene, the average coordination number of the atoms is also tabulated at every temperature. Figure 6.3 clearly shows

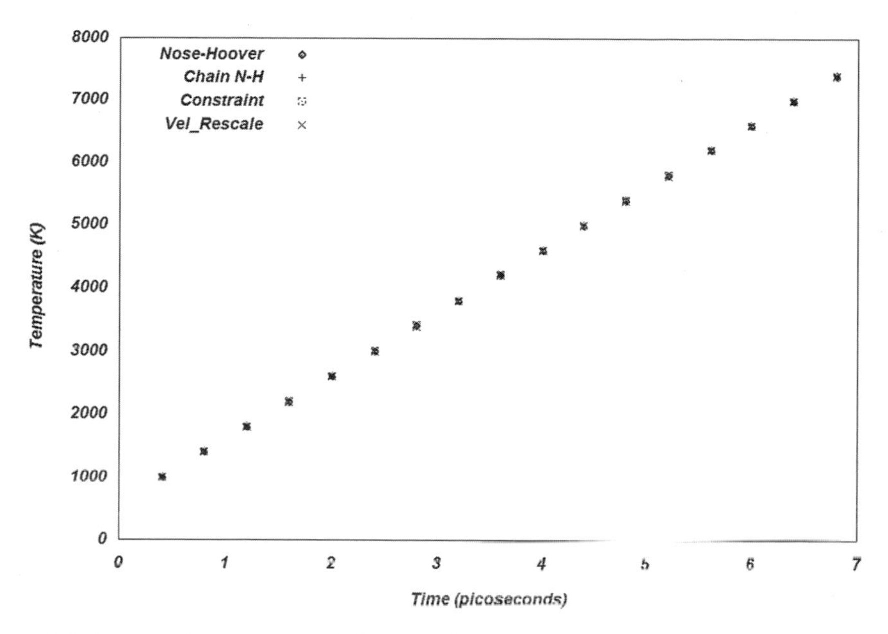

Figure 6.1 Average temperature of C_{60} as a function of time for the four thermostats studied.

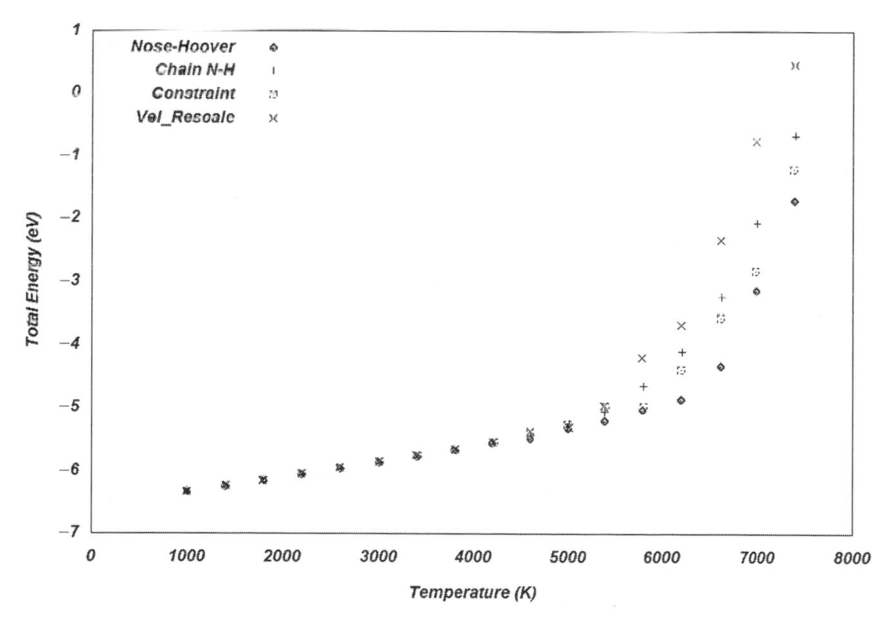

Figure 6.2 Average internal energy as a function of time for C_{60} thermalized with four different thermostats.

that the molecule starts to fragment once enough thermal energy is pumped into the system, as readily seen in the steady decrease in the average number of neighbors around 5500 K.

Also, from Figs. 6.2 and 6.3, it is obvious that the molecule completely fragments around 7000 K. The trends for all thermostats seem to be very similar. How-

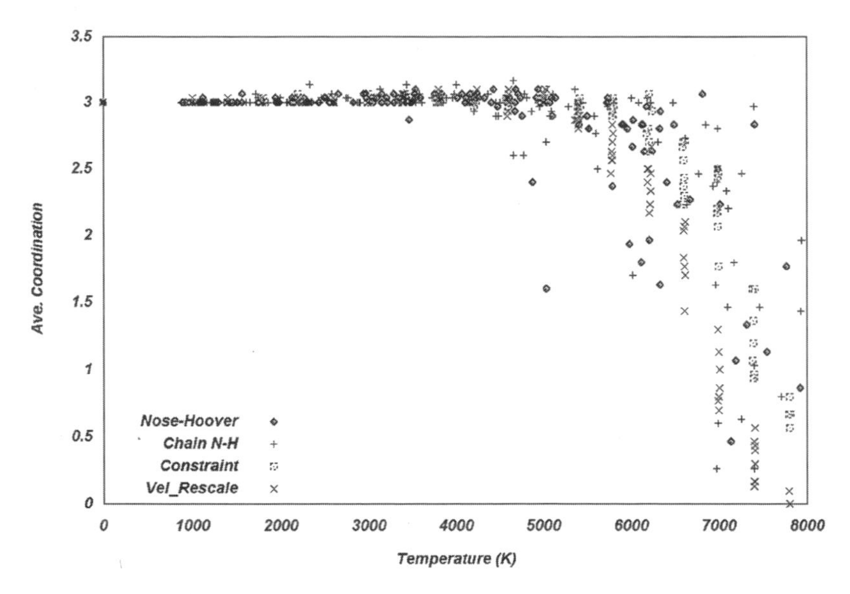

Figure 6.3 Average atomic coordination as a function of temperature.

ever, as expected, it appears that the Nosé-Hoover approaches (single thermostat or chain) and the constraint method lead to less fragmentation than the velocity rescaling method.

These observations differ from the work of Kim and Tomanek. In their work, they observed that the C_{60} molecule transformed initially into a "floppy" phase, then into a pretzel phase, and finally fragmented. But in the present study, the molecule started to uncoil around 5000 K and this continued until it fragmented at 7000 K. This was observed for all of the thermostats, leading to the conclusions that (1) uncoiling and subsequent fragmentation was dictated by the nature of the Tersoff's potential, and (2) the dynamics of the thermostats have a minimal effect on the high-temperature decomposition characteristics of the C_{60} molecule.

Finally, Fig. 6.4 represents the variation in the distribution of the coordination number of the atoms as a function of temperature for a representative case, namely, the Nosé-Hoover thermostat. All of the atoms are initially threefold coordinated, but with increasing temperature there is a slight decrease in the number of threefold coordinated atoms and a subsequent increase in fourfold coordinated atoms. This can be ascribed to the fact that the Tersoff's potential is parameterized for sp^3 carbon. At much higher temperatures, the uncoiling and the fragmentation of the molecule can readily be correlated to the increase in twofold and onefold coordinated carbon atoms.

Although the four thermostats used to thermalize a C_{60} molecule give results in qualitative agreement, the Nosé-Hoover thermostat and the Nosé-Hoover chain of thermostats provide consistency in the evolution of the internal energy and of the structure of the nanocluster. Velocity rescaling, known not to generate the canonical distribution, produces results differing significantly from the other three thermostats studied. These observations of the structural changes taking place during

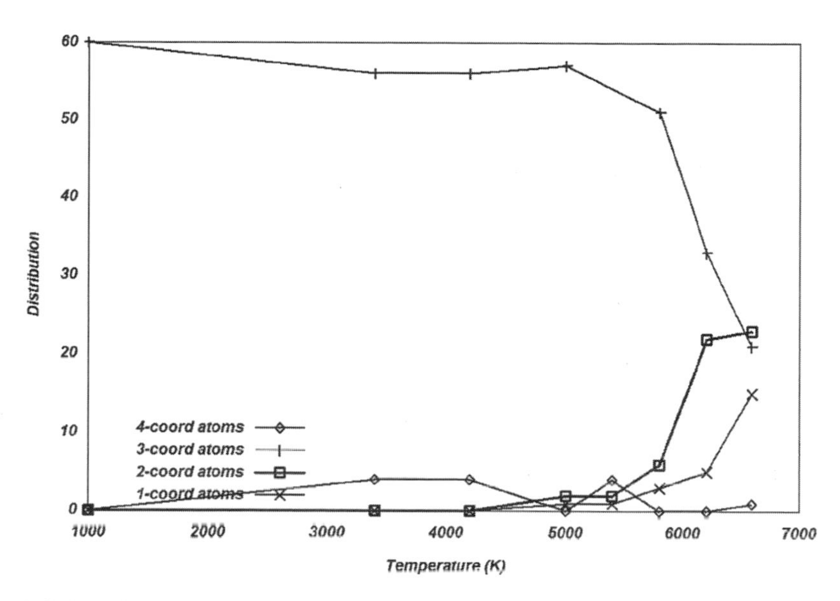

Figure 6.4 Atomic coordination in the C_{60} cluster as a function of temperature. Temperature is maintained with the Nosé-Hoover thermostat.

the heating of a fullerene differ from that of Kim and Tomanek.[9] This difference, however, may be assigned to the difference in interatomic potential. This example illustrates the importance of choosing the appropriate methodology for coupling an individual nanostructure to a "macroscopic" thermostat.

6.3 Stochastic atomistic computer simulation methodologies

6.3.1 Canonical Monte Carlo

MD simulations attempt to simulate the behavior of a system in real time by solving the equations of motion. While such an approach is required for determining the time-dependent properties such as diffusion, it may not be very well-suited *computationally* for determining time-independent properties. For example, the time step in an MD simulation is approximately one to two orders of magnitude smaller than the time for the fastest motion. In flexible molecules, such as hydrocarbon chain molecules, the highest frequency vibrations are due to bond stretching. A C−H bond vibrates with the repeat period of approximately 10 fs. An interesting problem in the nanoscale simulations is the self-assembly of chain molecules of surfactants and other polymers in solution and on surfaces. The timescales for the self-assembly of these molecules can range from several seconds to hours. Clearly, special efforts would be needed to develop integration schemes allowing for larger time steps for the investigation of self-assembly.

According to statistical mechanics, for an ergodic system, the time-averaged properties are equivalent to statistical averages in an appropriate ensemble.[21] An alternative approach for calculating the time-independent properties in cases such as surfactant self-assembly is to calculate the ensemble averages without attempting to simulate the real dynamics of the system. The MC simulations are stochastic techniques that generate a large number of states of the system and calculate the thermodynamic properties as statistical ensemble averages. Historically, the MC simulations were the first molecular simulations undertaken. Metropolis et al.[23] performed the first simulations of a liquid on the MANIAC computer at Los Alamos National Laboratory.

In molecular simulations, we are interested in calculating the thermodynamic properties of the system. As stated earlier, these properties are calculated as temporal averages in the MD simulations and as statistical averages in the MC simulations. The MC simulation samples a $3N$-dimensional space (particle positions) in contrast to a $6N$-dimensional space (particle positions and momenta) by an MD simulation. The momenta contribute only to an ideal gas term; and the deviations from the ideal gas in MC simulations are calculated by the potential energy term that depends only on the particle positions.[33,65]

According to statistical mechanics, any thermodynamic property A of a system can be evaluated as[30,33,65,66]

$$\langle A \rangle = \int A(\{\mathbf{r}\}) p(\{\mathbf{r}\})\, d\{\mathbf{r}\}, \tag{6.9}$$

where $p(\{\mathbf{r}\})$ is the probability of occurrence of the configuration $\{\mathbf{r}\}$. This probability depends on the potential energy of the configuration $V(\{\mathbf{r}\})$ and is given by

$$p(\{\mathbf{r}\}) = \frac{\exp[-\beta V(\{\mathbf{r}\})]}{\int \exp\{-\beta V[\{\mathbf{r}\}]\}\, d\{\mathbf{r}\}}, \tag{6.10}$$

where $\beta = 1/k_B T$. The integrals in Eqs. (6.9) and (6.10) are usually evaluated numerically. The simple techniques of evaluating these integrals, such as the trapezoidal rule or Simpson's rule,[67] are prohibitive due to the large number of calculations involved. These integrals can be evaluated more effectively by employing random sampling methods. In the simplest of these methods, the phase space is explored by generating a large number of states randomly and the integrals in the equation are replaced by the summations. Equation (6.9) then becomes

$$\langle A \rangle = \frac{\sum_{i=1}^{n} A_i(\{\mathbf{r}\}) \exp[-\beta V_i(\{\mathbf{r}\})]}{\sum_{i=1}^{n} \exp[-\beta V_i(\{\mathbf{r}\})]}, \tag{6.11}$$

where n is the number of randomly sampled states "i."

Simple random sampling often generates states that do not make significant contributions to the sample averages. Metropolis et al.[23] introduced the method of *importance sampling* that samples only the states having a Boltzmann factor with an appreciable value. In this method, the states are generated by following the Markov chain.[66] Each successive state in the Markov chain depends only on its immediate predecessor and has no memory of the previous states. This is important as it provides a clear distinction between the MD and MC approaches. The MD simulations follow the equations of motion and are connected in time, whereas the Markov chain enables unphysical moves in MC simulations and can relax the system much faster.

In practice, the importance sampling method is used frequently in the canonical ensemble MC (constant N, V, T). The implementation of the importance sampling method involves the generation of an initial random configuration of the system. The energy of the system in this initial configuration V_0 is calculated as a function of the positions of the particles. A new state is then generated by either a random displacement of a randomly selected particle or by a random displacement of all of the particles. The energy of the new state V_i is then calculated. The transition of the system from state $o \rightarrow i$ is always accepted if $\Delta V = V_i - V_0 < 0$. If the new state i results in $\Delta V > 0$, then the likelihood of the new state is based on a transition probability. The transition probability is calculated as $\exp(-\beta\Delta V)$ and a random number R is generated between 0 and 1. The new state is then accepted if the transition probability $\exp(-\beta\Delta V) > R$. Mathematically, the acceptance probability for the new configuration in the importance sampling method is expressed as

$$p = \min\left[1, \exp(-\beta\Delta V)\right]. \tag{6.12}$$

As with any other simulation technique, MC simulations with importance sampling also have efficiency-related issues. An algorithm is often considered to be efficient when approximately 50% of the moves are accepted. The acceptance rate of the MC moves invariably depends on the maximum displacement $dr_{\max}$ of the particles allowed in a step. If the maximum displacement is too small, it results in a large number of successful moves; however, the phase space is sampled very slowly. In contrast, if the maximum allowed displacement is too large, it results in high-energy overlaps and a large number of moves are rejected.

6.3.2 Grand canonical Monte Carlo

Grand canonical MC simulations are performed in an ensemble at constant (μ, V, T). These are particularly important in the studies of adsorption[68–70] as they enable the simulations of an open system (variable number of particles) at a constant chemical potential. The thermodynamic properties in the grand canonical ensemble are calculated by[30,33,65]

$$\langle A \rangle = \frac{\sum\limits_{N=0}^{\infty} \frac{\Lambda^{-3N}}{N!} \exp(\beta\mu N) \int\limits_{\Omega} A(\{\mathbf{r}\}) \exp[-\beta V(\{\mathbf{r}\})]\, d\{\mathbf{r}\}}{Z_{\mu VT}}, \tag{6.13}$$

where $Z_{\mu VT}$ is the grand canonical partition function given by

$$Z_{\mu VT} = \sum_{N=0}^{\infty} \frac{\Lambda^{-3N}}{N!} \exp(-\beta\mu N) \int_{\Omega} \exp[-\beta V(\{\mathbf{r}\})]\, d\{\mathbf{r}\}, \tag{6.14}$$

and

$$\Lambda = \frac{h}{(2\pi m k_B T)^{1/2}} \tag{6.15}$$

is the de Broglie thermal wavelength; Ω represents the phase space.

The trial moves for generating new configurations in the grand canonical ensemble consist of particle displacement (as done with the canonical MC), particle addition, and particle annihilation. In the particle addition move, a particle is inserted at a randomly selected position; and in the particle annihilation move, a randomly chosen particle is annihilated. The acceptance probabilities for particle insertion ($p\{N \to N+1\}$) and particle annihilation ($p\{N \to N-1\}$) are as follows:[30,33,65]

$$p(N \to N+1) = \min\left(1, \frac{1}{\Lambda^3(N+1)} \exp\{-\beta[\mu - V(N+1) + V(N)]\}\right), \tag{6.16a}$$

$$p(N \to N-1) = \min\left(1, \Lambda^3 N \exp\{-\beta[\mu + V(N-1) - V(N)]\}\right). \tag{6.16b}$$

In the grand canonical MC method, a random configuration of the system is generated initially. A particle is selected at random and then a move (particle displacement, creation, or destruction) is selected at random. The particle displacement follows the usual canonical MC method. That is, if the particle displacement results in a lower energy, then the new configuration is accepted. If the energy increases, then the move is accepted according to the transition probability given by Eq. (6.12). For the particle-creation and particle-annihilation moves, the energies for the new and old configurations are again calculated, and the new configurations are accepted according to the transition probabilities given by Eq. (6.16).

In the simple grand canonical MC methodology thus outlined, the probabilities of particle creation and annihilation can become very small when simulating dense systems. This is particularly true in the simulations of polyatomic molecules such as alkanes and surfactants. Particle creation becomes difficult in a dense system due to the high-energy overlaps with the neighboring particles, whereas the particle removal from dense systems results in the unfavorable high-energy configurations. Siepmann and Frenkel[71] introduced the concept of configurational bias sampling schemes based on the Rosenbluth–Rosenbluth[72] method to address this problem.

6.3.3 Lattice Monte Carlo

In the lattice MC method, the physical space is discretized on a 2D or a 3D lattice. The atoms or molecules occupy these lattice sites and interact with each other via nearest-neighbor pair potentials. Such lattice models greatly simplify the physical description of the system and prove very useful for rapid sampling of phase space. Although the lattice models are highly simplified, they still capture in many cases the essential physics of the processes occurring at the molecular level.

Lattice models are particularly useful in the examination of systems composed of long-chain polymer molecules. Wide ranges of time and length scales are required to adequately describe the behavior of polymers. The time scales range from approximately 10^{-14} s (i.e., the period of a bond vibration) through seconds, hours, or even longer, e.g., time for molecular diffusion and self-assembly. The size scales range from angstroms to nanometers to micrometers (e.g., length of the polymer to spatial extent of aggregates of molecules). The lattice models enable spatial coarse-graining of these features. The MC methods enable various moves without reference to their hierarchy of relaxation time. In the lattice MC method, therefore, many states can be generated rapidly and analyzed.

In the lattice MC method of polymers or surfactants, the chain molecule is first grown on either a 2D or a 3D lattice. The lattice model of polymer achieves the coarse graining of the physical space by employing a grid, and the coarse graining of the polymer molecule by using united atom models. The chemical groups (e.g., alkane groups) in the polymers are then represented as the vertices of the grids. The generation of an initial random configuration of the system consists of the random selection of a grid site, and a chemical group is placed on this site. The next bonded group is placed on a randomly chosen nearest-neighbor lattice site. The process is repeated until the entire chain has been grown. A self-avoiding random walk is used in growing the chain.

After generating the initial random configuration, the new states are generated by displacing the chain molecules on the grid. The lattice MC method provides opportunities to employ several multi-time-scale moves in the simulations of chain molecules. Examples of such moves include (in the order of increasing time scales) the flip,[73] reptation,[74] global chain translation, and cluster moves.[75] In the flip move, a group in a kink position is selected and moved to a diagonally opposite grid site. This small time-scale move results in a small change in the local conformation of the chain molecule. In the reptation move, one end of the chain molecule is selected at random and moved to a randomly selected empty nearest-neighbor site. The rest of the groups in the chain move in the direction of this end group and occupy the grid sites of their predecessors. The reptation move is therefore a long-time-scale move and capable of moving the entire chain in one attempt. In global chain translation, a chain is selected at random and moved to a different region in the grid while maintaining its original conformation. This is a diffusion-related move and has a very long relaxation time constant. In Fig. 6.5, the described moves are illustrated. The original configuration of the chain molecule is represented in

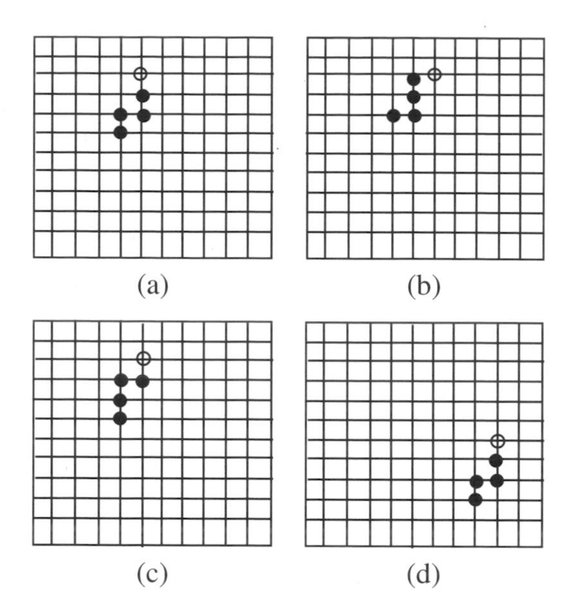

Figure 6.5 Lattice model of polymer: (a) original configuration, (b) reptation move, (c) flip move, and (d) global chain translation.

Fig. 6.5(a). The configurations shown in Figs. 6.5(b) through 6.5(d) result from the displacement of the chain molecules in the initial configuration of Fig. 6.5(a) using the reptation, flip, or chain translation moves, respectively.

Some special moves are required when simulating the self-assembly of surfactants. In the simulations of surfactants in which the chain molecules self-assemble to make aggregates of different shapes and sizes (e.g., micelles), the evolution of the system can become very slow if the simulation consists of solely the moves described in the previous paragraph. In such situations, it becomes necessary to employ moves such as cluster moves. A cluster can be defined as an assemblage of polymer chains that have at least one chemical group in a nearest-neighbor site to the group of a different chain. The cluster displacement move consists of random selection of a cluster and its relocation.

6.3.4 Self-assembly of surfactants

An important example of nanostructure is provided by the self-assembly of surfactant molecules in solutions or on surfaces. The fact that surfactant molecules, composed of hydrophobic tail groups and hydrophilic head groups, can aggregate or self-assemble in an aqueous environment has been exploited in many diverse areas of engineering and medical science in a variety of applications such as environmental, pharmaceutical, biological, and surface engineering.[76–79] Depending on their concentration and their geometry, surfactant molecules assemble in a spectrum of structures such as spherical micelles, cylindrical micelles, and membranes (e.g., bilayers).

MC simulation studies of surfactant solutions have been performed widely in the past 15 years. These techniques are generally based on lattice models in which a surfactant molecule is represented as a chain of chemical groups occupying certain grid sites on a 2D or 3D lattice. Extensive work has been done by Larson,[80–84] showing that surfactant self-assembly can be achieved by MC simulations without having to resort to any pre-assembled micellar structure or shape. Most of Larson's work has focused on three-component amphiphile-oil-water systems, and quantitative predictions of the phase behavior have been made by using a temperature integration method. By performing his simulations at different values of temperature and concentration, Larson has shown the ability of these models to predict self-assembly into lamellar, packed cylindrical, and spherical phases as well as bicontinuous structures.

More recently, lattice surfactant systems were studied by grand canonical MC (together with histogram-reweighting) techniques.[75] Both amphiphilic molecules of symmetric and asymmetric architectures were investigated. The osmotic pressure and chemical potential/volume relationships were determined with respect to temperature. The critical micelle concentration (CMC) was then determined as a function of temperature from the osmotic pressure curve. The CMC is that concentration above which addition of surfactant molecules results essentially in the formation of micelles.

Here we illustrate the lattice MC simulation of aqueous solutions of surfactants in a canonical ensemble.[85] The surfactant molecules are modeled as chains of connected grid sites on a 2D square lattice. The surfactant molecules contain 12 hydrophobic tail groups and one hydrophilic head group. A lattice site unoccupied by a head or tail group is assumed to be representing a solvent water molecule. For the calculation of the energy of the system, a discretized version of the potential energy function is used that captures the most essential features of the inter- and intramolecular interactions. In the context of the simulations of surfactants, this translates to the potential energy function of a form described by

$$V = (\varepsilon_{\mathrm{WW}} \cdot n_{\mathrm{WW}} + \varepsilon_{\mathrm{WT}} \cdot n_{\mathrm{WT}} + \varepsilon_{\mathrm{WH}} \cdot n_{\mathrm{WH}} + \varepsilon_{\mathrm{TH}} \cdot n_{\mathrm{TH}} + \varepsilon_{\mathrm{TT}} \cdot n_{\mathrm{TT}} + \varepsilon_{\mathrm{HH}} \cdot n_{\mathrm{HH}}), \tag{6.17}$$

where W, H, and T in the subscripts represent the solvent (water), the surfactant head group, and the surfactant tail group, respectively. The ε's represent the nearest-neighbor pair energies for the contacts, e.g., $\varepsilon_{\mathrm{WT}}$ is the energy of interaction for a tail group having a solvent group in a nearest neighbor site. The n's in Eq. (6.17) represent the number of nearest-neighbor pairs of groups in the subscripts. For convenience, the interaction energy for the solvent-solvent pair is set as the origin of energy. The sign and magnitude of the other interaction energies are expressed in reference to $\varepsilon_{\mathrm{WW}} = 0$. Due to the hydrophobic nature of the tail groups, the interaction energy for solvent-tail pair $\varepsilon_{\mathrm{WT}} > 0$. For hydrophilic head groups, the interaction energy for the head-solvent pair $\varepsilon_{\mathrm{WH}} < 0$. In addition to the short-range energies of Eq. (6.17), a model of a cationic surfactant (i.e., with charged head groups) would account for Coulomb interaction between head groups via a long-range repulsive term. A detailed discussion on the selection of the magnitude of the interaction energies (relative to the thermal energy $k_B T$) in the lattice MC simulations of the surfactants and the effect of the energy models used on the self-assembly behavior has been presented by Kapila et al.[85]

An initial configuration of the system is generated by growing N number of chains on the lattice. The simulation then follows the usual importance sampling MC algorithm of Eq. (6.12) using one or all of the moves described in Sec. 6.3.3. Important insights on the thermodynamic stability of aqueous surfactant solutions are gained in the self-assembly process by calculating the concentration of unaggregated surfactant molecules as a function of overall concentration of the surfactants. In addition to thermodynamics data, structural information can be obtained by, for instance, plotting the size distributions of surfactant aggregates. Such calculations permit the determination of important thermodynamic and structural quantities such as the CMC and the aggregation number. Several MC studies have been carried out for the measurement of the micellar properties: CMC, micellar size, micellar shape, aggregation number, and polydispersity. Brindle and Care,[86] Care,[87] and Desplat and Care[88] have studied both 2D and 3D lattice models of binary mixtures of water-surfactant systems in a canonical ensemble. The cluster size distribution has been determined as a function of temperature and concentration. Beyond the CMC, the cluster size distributions show a significant polydispersity; and a peak in these distributions is taken as indicative of micelle formation.

Figure 6.6 illustrates the concentrations of unaggregated surfactants (monomers) as functions of the overall surfactant concentration obtained from 2D simulations of an ionic surfactant.[85] The CMC of the system can be calculated from

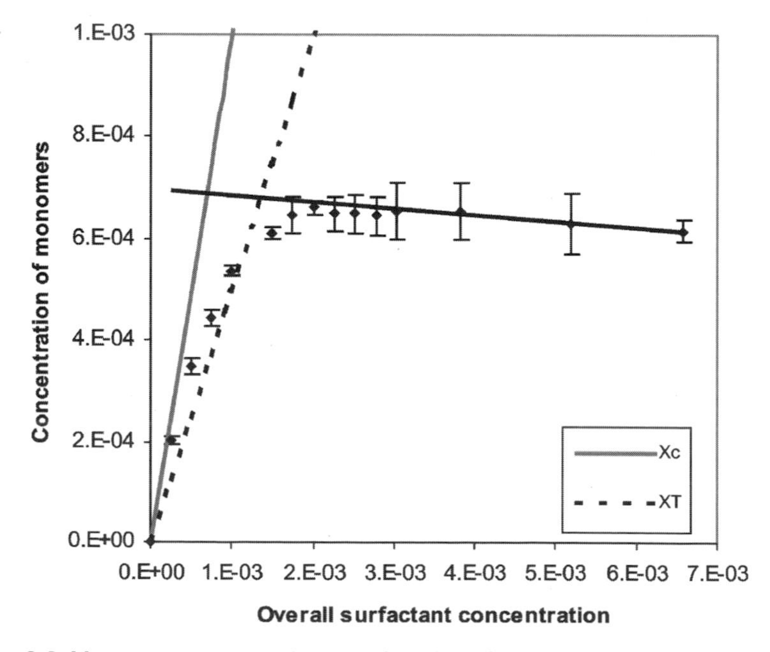

Figure 6.6 Monomer concentration as a function of overall surfactant concentration.

this plot as the concentration at which a line passing through the origin intersects a line fitting the high-concentration data. Several definitions for the low concentration line have been employed, including those by Care[86] (X_C) and Israelachvili[89] (X_T). The monomer concentration increases linearly with overall concentration below CMC; that is, the aqueous solution of surfactants is polydispersed with very few aggregates. At overall concentrations larger than the CMC, the monomer concentration plateaus and even decreases slightly. This indicates that, as monomers are added to the solution, they do not remain dispersed but participate in the formation of aggregates or micelles.

The detailed structure of a surfactant solution above the CMC is best illustrated in three dimensions. We present some results on the MC simulation of a model solution of cationic surfactants on a 3D cubic grid. The model and method are similar to that of the 2D solution reported previously. The 3D model differs from the 2D case in that the number of configurations available to a surfactant molecule is significantly larger.

Figure 6.7 shows a snapshot of the clusters of surfactant molecules, and Fig. 6.8 presents the size distribution of these clusters obtained from a 3D canonical MC simulation. The overall concentration exceeds the CMC. These figures show clearly the presence of micellar aggregates as well as a dispersion of the cluster sizes. Length scales in this solution range from the shortest one corresponding to the lattice spacing (or tail-tail or head-tail groups separation) to an individual surfactant molecule (i.e., several lattice spacings) to aggregates with radii extending over several tens of lattice spacings.

In an actual solution, each spatial scale has its own characteristic time such as that associated with the fast flip move, slower reptation, even slower individual surfactant diffusion, and the very slow diffusion of surfactant aggregates. The use of MC sampling emancipates us from this hierarchy of time scales and enables us to achieve equilibrium more efficiently. It is noteworthy that separation of time scales has also been achieved in some MD simulations.[44]

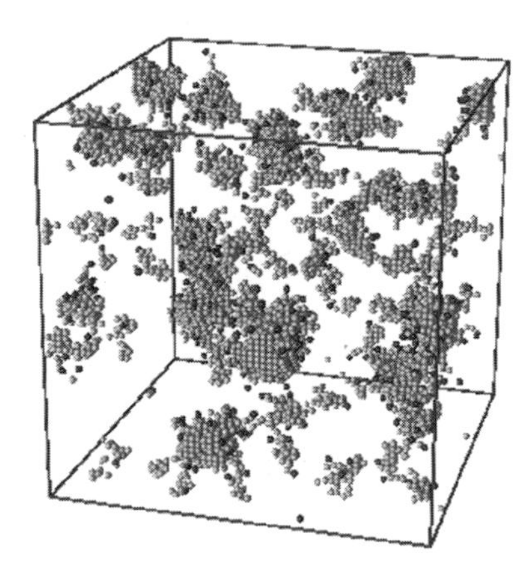

Figure 6.7 Snapshot of surfactant aggregates as obtained from a canonical Monte Carlo simulation in three dimensions. The dark spheres represent hydrophilic headgroups and bright spheres represent hydrophobic tailgroups in a surfactant molecule.

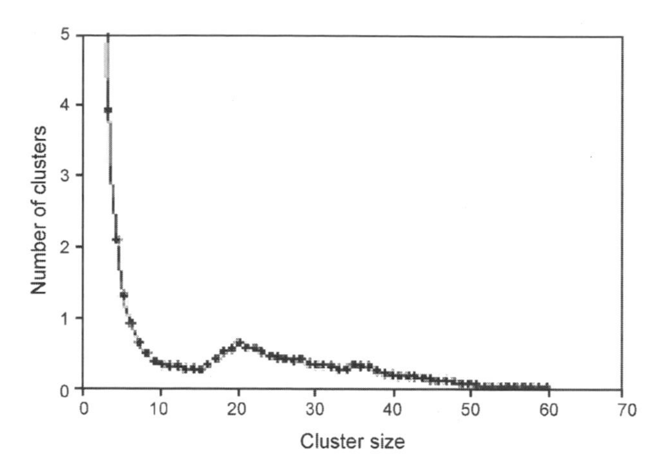

Figure 6.8 Cluster size distributions of the aggregates shown in Fig. 6.7.

6.3.5 Kinetic Monte Carlo

The MC methods described heretofore enable rapid relaxation of some system toward equilibrium configurations following physically unrealizable paths. For instance, the MC moves are selected without much relation to the hierarchy in their relaxation times. To model the kinetics of a process using MC methods, one must follow pathways that can be related to the actual path followed by the real system. The occurrence of an event is determined by rate constants or event frequencies. An event is defined as any single change in the configuration of the system. The implementation of a kinetic MC simulation therefore involves the selection of an event according to a uniform probability. Random numbers and the relative frequencies are used to accept or reject the event. The succession of events may be related to time. Several approaches have been employed to define time.[90]

6.3.6 Application of kinetic MC to self-assembly of protein subcellular nanostructures

In this section, we illustrate the kinetic MC method with an example borrowed from the realm of biological nanostructures, namely, the dynamical behavior of microtubules. Microtubules (MTs) are naturally formed proteinaceous nanotubes, 24 nm in diameter and up to hundreds of microns in length. MTs are biopolymers assembled from two related protein monomers; α and β tubulins.[91] In the presence of the small molecule guanosine 5′-triphosphate (GTP), these tubulin monomers form a heterodimer, which self-assembles into the microtubule structure. Due to the geometry of self-assembly and differences in addition rates, a MT is polarized containing (−) and (+) ends. The (−) end contains an exposed α tubulin and undergoes slower heterodimer addition rates than the (+) end, which consists of an exposed β tubulin. Therefore, net MT polymerization occurs from the (+) end of the growing polymer or nucleation complex. MTs generated from pure tubulins exist in a dynamic state with net addition of monomers to the (+) end and net removal of monomers from the (−) end.[92]

Dynamic instability is an intrinsic property of MTs. For $\alpha\beta$-tubulin concentration above a critical value C_c, tubulin dimers polymerize into MTs; while below C_c, MTs depolymerize.[91] Near C_c, MTs exhibit dynamic instability during which a single MT undergoes apparently random successive periods of assembly (slow growth) and disassembly (rapid depolymerization). Computer simulations of MT assembly/disassembly have recreated many experimentally observed aspects of MT behavior and have given strong support to a lateral cap model of MT dynamics. This model utilizes a coarse-grained representation of the protein tubulin heterodimers.

Early kinetic MC studies of models based on simplified single-helix[93] and multihelices[94] generated phase change between a slow-growing GTP-capped MT end and rapidly shortening uncapped MT end. Subsequent simulations by Bayley et al.[95,96] were based on a simplified helical lattice model with only longitudinal and single lateral interactions between $\alpha\beta$-tubulin subunits (the "lateral cap model"). Bayley's model differs from the model of Chen and Hill[94] in that it gives a molecular description to the switching of MT between assembling and disassembling states in terms of a fully coupled mechanism linking tubulin-GTP (Tu-GTP) addition and GTP hydrolysis (conversion of Tu bound GTP into the diphosphate GDP). Bailey's model focused on the "5-start" helical 13-protofilament MT lattice. MTs are known to readily form different lattices, some having a "seam" in which the lateral interactions between adjacent protofilaments are misaligned.[97] Martin et al.[98] have developed a more rigorous lattice model that accounts for MT lattice variations and seams. In this latter model, association and dissociation rate constants are obtained from estimates of the free energies of specific protein-protein interactions in terms of the basic MT lattice. The performance of kinetic MC simulations of MTs does not appear to be too sensitive to the detailed numerical values assigned to the intersubunit bond energies.[99] Martin's model rationalizes the dynamic properties in terms of a metastable MT lattice of T-GDP stabilized by the kinetic process of T-GTP addition. Furthermore, with this model, the effects of small tubulin-binding molecules are readily treated. The lateral cap model provides a basis for the examination of the effect of antimitotic drugs (e.g., colchicine, taxol, etc.) on MT dynamics. In particular, it was used to study the control of MT dynamics by substoichiometric concentration of drugs. The lateral cap model was further modified to simulate the effect of MT assembly/disassembly on transport of a motor protein-coated bead that moves along a protofilament.[100]

To illustrate the application of the kinetic MC method to the dynamical assembly/disassembly of MT we briefly review the five-start helix lattice model of Chen and Hill.[94] With this model a single MT consists of a 2D helical lattice composed of 13 grid sites (13 tubulin heterodimer protofilaments) perpendicular to the direction of growth (MT principal axis). The lattice is infinite along the principal direction of the MT. Helical periodic boundary conditions are applied to wrap the lattice into a tubular structure with a helicity of five lattice points. The steps involved in the kinetic MC simulation of MT dynamics at fixed tubulin-GTP concentration ([Tu-GTP]) are as follows:

- Step 1: Identify at the ends of each protofilament along the jagged helical surface of the tip of a MT the sites "i" for dissociation (occupied grid site at the top of a step) and association (empty grid site at the bottom of a step).
- Step 2: Assign a rate constant k_i for dissociation or association events at every site "i." These rate constants depend on the physical structure of the binding site, the nucleotide content of the unit in adjacent protofilaments (i.e., both relate to the binding free energy) and [Tu-GTP] in the case of association.
- Step 3: Calculate the time t_i for dissociation or association at every site "i" at which an event would occur statistically, using the relationship $t_i = -\ln(1 - R_i)/k_i$, where R_i is a uniformly distributed random number between 0 and 1.
- Step 4: Implement the event with the shortest time ($t_{\min}$) and modify the lattice. For addition events, implement a hydrolysis rule for conversion of Tu-GTP completely embedded into the MT lattice into Tu-GDP.
- Step 5: Increment the total time by $t_{\min}$.

Using the association and dissociation rate constants of Bayley et al.[96] and a kinetic MC program based on the lateral cap model, we reproduce in Fig. 6.9 the results on the effect of [Tu-GTP] on the dynamical instability of a single "5-start"

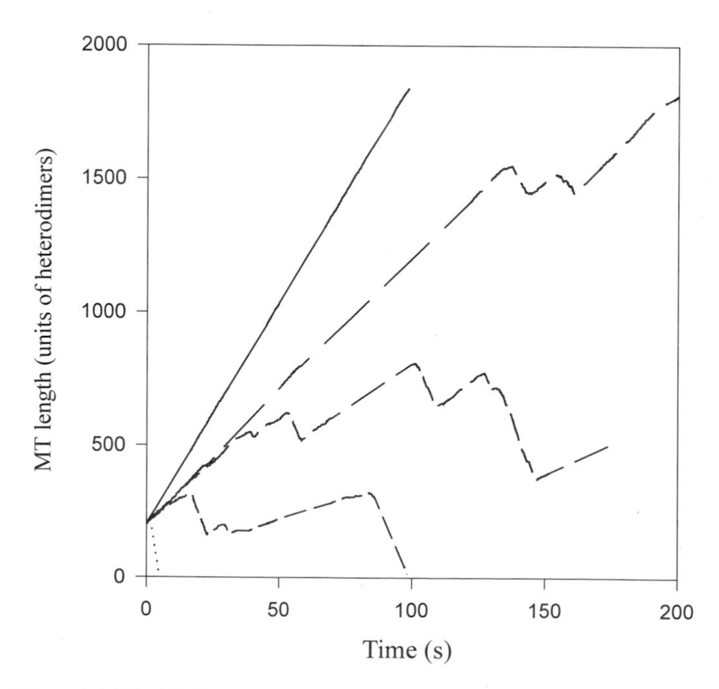

Figure 6.9 Effect of [Tu-GTP] on the dynamical behavior of a single "5-start" helix, 13 protofilaments microtubule. The plots from top to bottom correspond to concentrations, [Tu-GTP], amounting to 2.45, 165, 1.45, 1.25, and 0.45 $\times$ 10^{-5} M, respectively.

helix, 13-protofilaments MT. The initial MT length is 200 $\alpha\beta$ tubulin heterodimers. The critical Tu-GTP concentration C_c is approximately 1.45×10^{-5} molar (M).

6.4 Multiscale simulation schemes

Multiscaling has recently received much attention in the simulation of nanoscale systems. Indeed for most practical cases, nanoscale structures are not isolated in a vacuum but are attached to substrates or embedded in a matrix. Since it is computationally prohibitive to simulate large systems at the atomic level, multiscale schemes have been proposed to reduce the computational effort associated with the material/environment that surrounds the nanoscale system. While the nanoscale system is modeled and simulated at the atomic level, the surrounding environment, in contrast, is treated with a smaller number of degrees of freedom while retaining some of the important physics and/or chemistry.

Existing multiscale simulation methodologies can be characterized as serial or concurrent. Within serial methods, a set of calculations at a fundamental level (small length scale) is used to evaluate parameters as input for a more phenomenological model that describes a system at longer length scales. For example, the quasi-continuum (QC) method is a zero-temperature technique with a formulation based on standard continuum mechanics [e.g., the finite element (FE) method] with the additional feature that the constitutive equations are drawn from calculations at the atomic scale.[101–104] Another example of a serial methodology enabling microscopic fluctuations to propagate to microscopic scales has been illustrated for biological membranes.[105] This approach couples nonequilibrium MD to a method that solves the large-deformation problem in continuum mechanics.

In contrast, concurrent methods build around the idea of describing the physics of different regions of a material with different models and linking them via a set of boundary conditions. The archetype of concurrent methods divides the space into atomistic regions coupled with a continuum modeled[106,107] via FE. Coarse graining has been proposed as a means to couple seamlessly an MD region to a FE mesh.[108] Coarse-grained MD produces equations of motion for a mean-displacement field at the nodes of a coarse-grained mesh partitioning the atomistic system.

Other algorithms to couple atomistic and continuum regions have also been proposed.[109–112] Broughton et al.[109] presented an algorithm involving handshaking between FE and MD. This algorithm was able to dynamically track a crack propagating through silicon. The handshaking between the MD and FE regions was achieved by drawing an imaginary surface between them. Within the range of the MD interatomic potential from this surface, FE mesh points were located at equilibrium atomic sites. Any FE element that crossed the interface contributed half its weight to a conservative Hamiltonian. Any MD interaction that crossed the interface also contributed half its weight to this Hamiltonian. Kohlhoff et al.[110] introduced a similar transition region between the atomic and continuous regions. They also scaled down the FE size to the atomic scale in this transition region, but the interface was of finite size and not sharp. Abraham et al.[111] combined the foregoing two techniques by constructing an explicit Hamiltonian for the atoms and the FE nodes in the transition region by weighing their contributions with respect to their distance away from the middle of the interface. Ogata and coworkers[112] used a similar algorithm to study chemical reactions and their interplay with mechanical phenomena in materials, such as in the oxidation of Si (111) surface.

There are several issues associated with the coupling of a nanoscale system to a system with larger scales. For instance in linear elasticity, the fundamental properties such as stress, strain, and the elastic moduli are thermomechanical quantities; i.e., they satisfy the thermodynamic and the long-time limits. The calculation of some of these quantities from atomistic models does not present significant difficulties, as long as large enough systems and long enough times are used. This constitutes the basis for coarse graining that enables the extension of atomistic systems into the realm of continuous models with seamless coupling between length scales.[108] However, spatial coupling becomes a problem when dealing with atomistic nanoscale systems. A condition necessary to achieve reasonable coupling between an atomistic system and a continuum is that there are spatial and time scales over which the two systems overlap. This is not the case in many of the methodologies reviewed in the previous paragraph where the FEs coupled to an MD region are reduced to "atomic" dimensions. The spatial coupling between unphysically small FEs and atoms implies also that the long-time limit may not be satisfied. In addition, an elastic continuum does not obey the same physics over all possible wavelengths as that of a discrete atomic system. This physical mismatch is easily noted in the dispersion relations of both systems that overlap only in the long-wavelength limit. Therefore, one can expect an elastic impedance mismatch between a continuum and an atomic simulation when an attempt is made to couple them.[113] Depending on the phenomenon to be investigated, the behavior of the atomistic system may be altered detrimentally, should the physics of the nanoscale system be much different from the physics of the medium to which it is coupled.

In the remainder of this section, we illustrate recent methods that enable the coupling between an atomistic system and another system with a coarser scale. First, a nanograin polycrystalline MD system is coupled to a coarser lattice MC model, and overlap of spatial scales is stressed. In a second illustrative example, we point out the importance of overlap of time scales by coupling an elastic continuous model to an atomistic one.

6.4.1 Coupling of MD and MC simulations

By analogy with its use in signal and image processing, the wavelet transform has been used to analyze MD outputs.[114] Wavelet transforms can be seen as a mathematical microscope that provides ready information on the intricate structure of a "pattern." The wavelet coefficients provide local information on the nature of any function at various scales (ranging from the finest to the coarsest), and one can identify the "important" scale by examining the coefficients at every scale. Therefore it constitutes an ideal tool for multiscale modeling.

The compounded wavelet matrix (CWM) method[115] has been used to bridge two computational methodologies (atomistic MD simulation and coarse MC simulation) applied to a small region of a nanograin-sized polycrystalline material. The CWM method possesses several advantages. First, it does not assume *a priori* that a collection of small atomic-scale systems is equivalent to a microscale-based model of a large system. Second, the simulation time of the coarsest methodology is not controlled by the methodology with the slowest dynamics.

An illustrative example of the CWM method is provided via the problem of 2D grain growth in a nanograin polycrystalline material. This example is based on a MD simulation of a 2D Lennard-Jones (L-J) system[112] and a MC simulation of a Q-states Potts model[116] that can overlap over a range of spatial and time scales. These two models are bridged in the spatial domain. Atoms in the MD system interact via a simple 2D 6–12 L-J potential[30] with parameters $\varepsilon = 119.79$ K and $\sigma = 3.405$ Å. The MD simulation cell contains 90,000 particles in a cell with edge length ~0.106 μm. Interactions between atoms are extended up to third nearest neighbors. In addition to this large system, the grain growth process is also simulated in a smaller L-J system. The small system is one-quarter the size of the former one. This smaller atomic system consists of 22,500 particles in a cell with edge length ~0.0503 μm.

For both MD systems, polycrystalline microstructures with fine grains are initially obtained by quenching a liquid. The initial microstructures are then evolved with a constant temperature (momentum rescaling thermostat)-constant volume MD algorithm. The temperature is maintained at approximately 70% of the melting point. Periodic boundary conditions are used for about 400,000 MD integration time steps or nearly 1.7×10^{-9} s. During that period of time, the total energy of both systems drops by nearly 63%; thus, coarser microstructures are obtained. These microstructures are then characterized by calculating the excess atomic potential energy of each individual atom (relative to the potential energy of an atom in a perfect lattice at the same temperature). The excess atomic energy is then normalized by the total excess energy of the microstructure at $t = 0$ s. The spatial distribution of the normalized excess atomic energy is then mapped onto a 512×512 square matrix for the large system (Fig. 6.10) and 256×256 matrix for the smaller one to obtain what will be referred to in the rest of the section as energy maps. We have used energy maps from microstructures quenched at low temperatures to minimize the noise due to the contribution of thermal vibration.

In an MC simulation of grain growth with a Potts model, both spatial and "MC time" scales are coarser than those in MD. The Potts model maps the microstructure onto a discrete lattice coarser than the atomic scale, and the "spin" state $S = 1, \ldots, Q$ of each lattice site represents the orientation of the grain in which it is embedded.[116] A grain boundary exists between two adjacent lattice sites with different orientations. An interaction energy J_{int} is then assigned to a pair of neighboring sites with different orientations. We employ a Potts model with a square lattice containing 128×128 sites and $Q = 10$ with nearest-neighbor interactions. This model is designed to represent a piece of material with dimensions

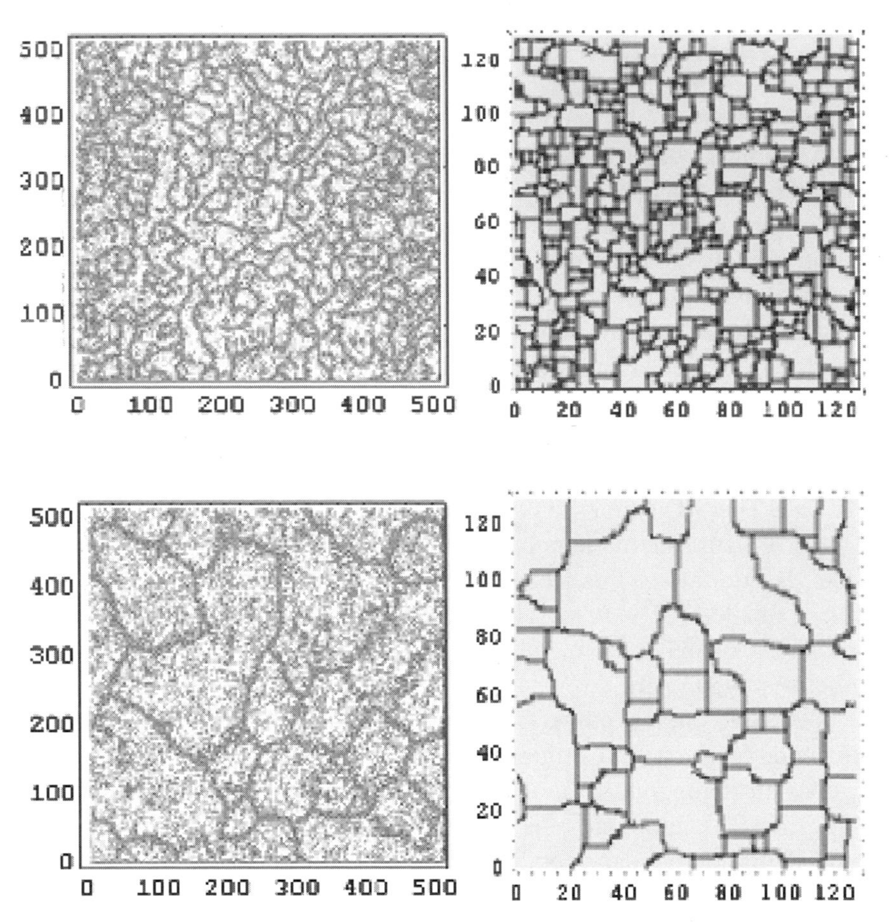

Figure 6.10 Gray-scale representation of the energy maps for the initial large MD system (upper left), annealed MD system (lower left), one initial Potts/MC system (upper right), and annealed MC system (lower right). Energy increases from white to black.

similar to those of the large MD system. Periodic boundary conditions are applied onto the Potts model. A canonical Monte Carlo algorithm is used to evolve this model. The thermal energy $k_B T = 0.2$ J. Initial microstructures are produced from totally random configurations after 4×10^6 MC moves.

A total of four MC initial configurations corresponding to microstructures optically similar to the initial configuration of the large L-J system are thus obtained. Subsequently, long MC simulations are performed to anneal the initial microstructures until the total energy averaged over the four systems decreases to nearly 63% of the average energy of the initial configurations. The 128×128 matrices containing the value of energy at every lattice site characterize the final MC microstructures (see Fig. 6.10). Note that the energy in the Potts model represents an excess energy relative to a perfectly ordered system (perfect crystal). Normalization of the energy at each lattice site by the total excess energy of the initial microstructure allows a direct comparison with the energy maps produced with the MD simulations.

In one dimension, a wavelet $\psi(x)$ transforms a fluctuating function[117] $f(x)$ as follows:

$$W_f(a,b) = \int_{-\infty}^{\infty} f(x)\,\psi_{a,b}(x)\,dx. \tag{6.18}$$

The two-parameter family of functions, $\psi_{a,b}(x) = (1/\sqrt{a})\psi(xb/a)$ is obtained from the mother wavelet function $\psi(x)$ through dilations by the scaling factor a and translations by the factor b. The factor $1/\sqrt{a}$ is included for normalization. The parameter a can take any positive real value, and the fluctuations of $f(x)$ at position b are measured at the scale a.

When discretized, wavelet analysis can be performed with fast algorithms. Given the wavelet coefficients $W_f(a,b)$ associated with a function f, it is possible to reconstruct f at a range of scales between s_1 and s_2 ($s_1 \leq s_2$) through the inversion formula

$$f_{s_1,s_2}(x) = \frac{1}{c_\psi}\int_{s_1}^{s_2}\int_{-\infty}^{\infty} W_f(a,b)\psi_{a,b}(x)\,db\,\frac{da}{a^2}. \tag{6.19}$$

The limits $s_1 \to 0$ and $s_2 \to \infty$ reconstruct the original function over all scales.

A 2D wavelet transform includes transforms in the x direction, the y direction, and in the diagonal x, y direction[117]. For example, given an energy map of 524×524 points such as that generated in the large MD simulation, the wavelet transform consists of three 256×256 matrices (one in each direction), three 128×128 matrices, and so on. Each decomposition level is at half the resolution from the previous one. The final level of decomposition represents the map at the coarsest resolution. Wavelet analysis of a MD energy map provides its wavelet transform coefficients from the atomic to its coarsest scale (corresponding to the physical dimensions of the system). Similarly, the wavelet coefficients of an MC energy map extend over scales ranging from the lattice spacing of the grid to the system size.

Let us consider the Q-states Potts model and the small L-J system which have different physical dimensions. The range of scales for both systems overlap provided that the Potts model is larger than the L-J system and the lattice spacing of the Potts model is not too coarse. The coarser scales of the L-J model may then correspond to the finer scales of the Potts model. A compound matrix of wavelet coefficients is then formed such that, at those scales common to the small L-J and Potts systems, the statistical properties of the coefficients are those of the small L-J system; at coarser scales, the statistical properties of the coefficients are those of the Potts model. This yields a compound wavelet matrix representing the phenomenon of grain growth over the interval of scales now being the union of the intervals treated individually by the two models.

Figure 6.11 shows the energy with respect to scale for the three systems considered here, small and large MD systems and the MC Potts model. The energy associated with a given scale is evaluated from the wavelet representation of the energy maps at that same scale. For the wavelet representation of the energy map at scale s, the wavelet coefficients at all scales except those at s are set to zero; with this set of coefficients, the inverse wavelet transform is performed [see Eq. (6.18)]. This inverse wavelet transform represents the spatial distribution of the contribution of scale s to the energy map. The total energy associated with scale s is then calculated as the sum of the energies in the representation at scale s. We have also averaged the energy versus scale plots of the four final Potts systems. For the small MD system (designated as "1/4 L-J" in the figure), the plot contains eight points (scales 2 through 9) corresponding to resolutions of 1×1 (2 on the horizontal axis), ..., 64×64 (8) and 128×128 (9). Since this system is physically ¼ of the MC system, these scales correspond to the large MD system with resolution of 2×2 (scale 2), ..., 128×128 (8) and 256×256 (9). All three plots show very similar behavior in the interval of scales from 1 to 7: a maximum between 5 and 6 representing the mean grain size, a monotonous decrease toward the continuum limit. The small L-J and large L-J agree quite well with each other.

The energy of the average Potts system is also slightly larger than the energy of the L-J system. Since this difference is particularly significant near scales 5 and 6, it can be attributed to differences in grain boundary energies. These differences are not due to a weakness of the wavelet analysis, but only of the incomplete

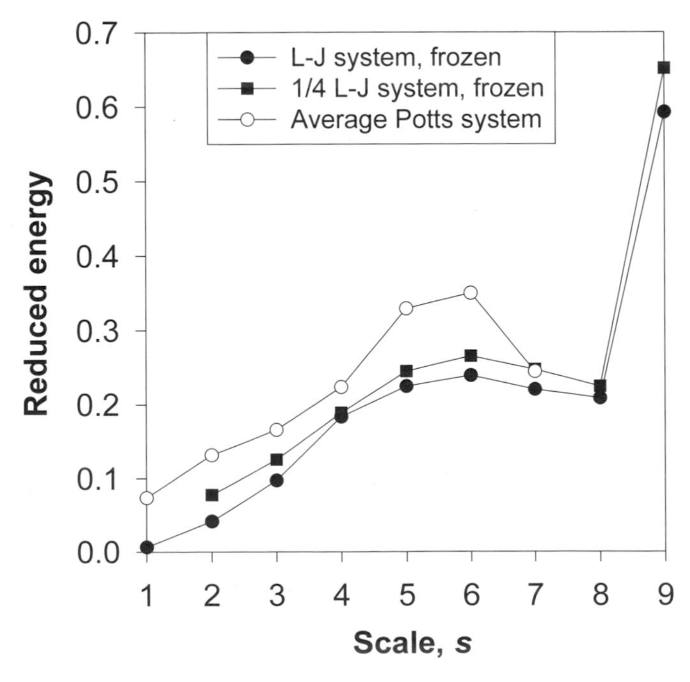

Figure 6.11 Scale dependency of the energy of the small and large MD systems and MC/Potts model.

quantitative correspondence between the physics of the Potts and L-J models. The average Potts system and the small L-J system overlap at scales $s \in [2, 7]$, The two systems show similar trends in the evolution of energy versus scale. However, the Potts model lacks information on small-scale features ($s = 8, 9$), which are now provided by the small L-J system. The small L-J system lacks information on large-scale features. Construction of a CWM from the small L-J system and the Potts model leads to a description of the microstructure (over a combined range of scales $s \in [1, 9]$) statistically equivalent to that of the wavelet transform of the large L-J system.

6.4.2 Coupling of an atomistic system with a continuum

In this section, we quantify the impedance mismatch between an elastic continuum and an atomistic region as the continuum spatial and temporal scales are forced toward atomic scales. We have coupled dynamically an elastic continuum modeled with the finite-difference time-domain (FDTD) method[118,119] and an atomistic system modeled with MD. The impedance mismatch between the MD and the FDTD systems is probed with an incoming elastic wave packet with broadband spectral characteristics centered on a predetermined central frequency. Reflection of part of the probe wave packet is a sign of impedance mismatch between the two systems. The FDTD method solves numerically the elastic wave equation in homogeneous or inhomogeneous media.[118,119] The elastic wave equations are integrated by means of discretization in both the spatial and the temporal domains. More specifically, real space is discretized into a grid on which all the variables and parameters are defined. The main variables are the acoustic displacement and the stress tensor at every site on the grid. The relevant parameters of the system are the mass densities and the stiffness/compliance coefficients for each constitutive element. The relevant parameters of the FDTD simulation are the grid spacing and the size of the time step. Appropriate boundary conditions such as periodic boundary conditions or absorbing boundary conditions are applied.

The FDTD scheme discretizes the wave equation:

$$\frac{\partial^2 u_i}{\partial t^2} = \frac{1}{\rho}\frac{\partial T_{ij}}{\partial x_j}, \quad i \in [1, 3], \; j \in [1, 3], \tag{6.20}$$

in both the spatial and time domains and explicitly calculates the evolution of the displacement u in the time domain. Here, T_{ij} are the components of the stress tensor, and ρ is the mass density. For the sake of simplicity, we limit the consideration to 1D propagation. The FDTD region is discretized into N 1D elements of length Δx. It is assumed that the FDTD region is infinitely stiff in the other two directions. The elastic wave equations are approximated using center differences in both time (time step Δt) and space. The displacement u_n of any element n at each time step is a function of the stress gradient across that element. Thus, in this technique, one can predict the displacement of every element after knowing the stress on that element. The stress on any element is assumed to be uniform. Absorbing boundary conditions[118] are implemented in order to prevent reflection from the end elements of the FDTD mesh.

First, one establishes the physical correspondence between the continuum and the atomistic system. The elastic continuous system to be probed is chosen to have the physical properties of a L-J model for solid argon. The 1D compliance was found from a series of MD simulations carried out under the following conditions: the model for the atomic system was a 3D FCC crystal with periodic boundary conditions containing 500 particles interacting through the 6–12 L-J potential with parameters chosen to simulate argon. The interatomic potential was truncated at a distance of 8.51 Å. The uniaxial long-time limit stress-strain relationship for that crystal was obtained with the temperature maintained at 46 K via a momentum-rescaling scheme. For these calculations, a strain was applied in one direction while maintaining the length of the other edges of the rigid simulation cell. The strain was applied in increments of 2×10^{-4} in the interval $[-0.1, 0.1]$ and the resulting stress was then calculated from a virial-like equation[21] by averaging over 5000 MD time steps. An MD time step (δt) equals 10.0394 fs. The curve was then fitted to a third-degree polynomial with the coefficient of the linear term representing the linear elastic coefficient of the medium.

In the second step, the coupling between the continuum and an atomistic system is handled by replacing one FDTD element by a dynamical 3D L-J MD cell (Fig. 6.12). The number of FDTD elements is 10,000. The MD cell is located at element 6000. The length of each FDTD element (Δx) is equal to the zero pressure box length of the MD cell (26.67 Å).

As shown in Eq. (6.20), the calculation of the displacement of the acoustic wave throughout the medium requires the knowledge of the stress fields for every FDTD element at every FDTD time step. Thus, when a FDTD element is replaced by an MD cell, the equivalent stress for the element is calculated by uniaxially straining (according to the FDTD displacement) the MD cell along the direction of the wave propagation. The condition of rigidity in the other two directions is satisfied by keeping the length of the edges of the MD cell constant in those directions. The average value of the MD stress is evaluated for every FDTD time step with the final configuration of the MD atoms obtained at the previous FDTD time step serving as the initial state for the current MD calculation.

The coupled continuum and atomistic hybrid system is probed with a 1D wave packet of the form $a_0 \cos(-kx) \exp[-(kx)^2/2]$, where k is the wave number, and a_0 is the maximum amplitude of the wave. The probing signal is initially cen-

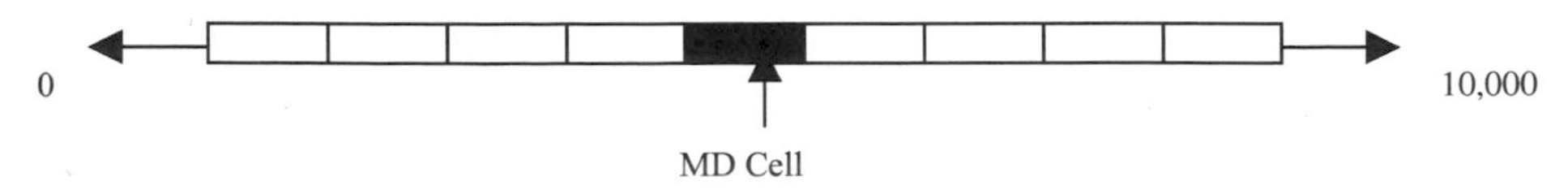

Figure 6.12 An illustrative representation of the system consisting of 10,000 elements; the open boxes represent FDTD elements and the darkened box corresponds to the MD cell.

tered about the 5000th element and is propagated along the positive x direction. The wave is propagated through the medium with an initial longitudinal velocity of the elastic wave through the medium, c_0. The signal's frequency spectrum is broadband and the central frequency of the wave packet ν equals c_0k. The central wavelength of the wave packet was chosen to be an integral multiple of Δx, to ensure stability of the FDTD algorithm. A preliminary study[120] of the wave propagation characteristics indicated that the FDTD time step (Δt_{crit}) had to be smaller than ($\Delta x/2c_0$) for a stable algorithm.

At every FDTD time step, the MD stress is calculated by averaging over $N_{\mathrm{MD}} = \Delta t/dt$ time steps, with $\Delta t \leq \Delta t_{\mathrm{crit}}$. A reduction in Δt automatically leads to a decrease in the number of MD time steps over which stress is averaged (for every FDTD time step). It is possible to push the limit of time coupling between the two simulation techniques toward one to one correspondence between the two time steps (i.e., $\Delta t = \delta t$). The FDTD/MD hybrid method, therefore, enables us to test a range of time-scaling conditions from coarse graining to time matching between a continuum and an atomic system.

The coupling between the continuum and atomistic systems is examined by analyzing the reflected signal at an element some distance away from the MD cell. This signal is compared and contrasted with the signal that is reflected in the case when the MD cell behaves as an FDTD element with a nonlinear elastic coefficient as determined previously from the long-time third-order stress/strain relationship. The latter case is referred to as the "pseudo MD-FDTD coupling (PC)" while the former is referred to as "real-time MD-FDTD coupling (RTC)." Discrete fast Fourier transforms (FFT) are used to obtain the frequency spectrum of all of the signals.

The impedance mismatch between the continuum system and the MD system is very small and most of the probing signal passes through the MD cell. However, there is a small amount of reflected signal. The Fourier spectrum of this reflected signal for a probing signal with frequency equal to 3.930 GHz is illustrated in Fig. 6.13. The PC reflected signal is essentially limited to low frequencies. The difference in reflected signal between the PC and RTC simulation clearly illustrates the fact that the RTC simulation does not satisfy the long-time limit. Indeed, the PC simulation corresponds to the long-time limit as the stress of the pseudo-MD cell is calculated using the predetermined elastic coefficients. The RTC calculation with stress averaged over only 23 MD steps is unable to achieve that limit as $N_{\mathrm{MD}} = 23$ does not even last the time of one atomic vibration.

The RTC and the PC signals have distinct frequency cutoffs, with the cutoff for the PC signals being much smaller than that of the RTC signals. This can be explained again on the basis that the PC signal represents the long-time limit of the coupling, where the high-frequency (short-wavelength) modes are averaged out; while the abrupt cutoff for the RTC signal represents an upper limit in the frequencies that can be supported by the FDTD system. The discretization of the continuum into small elements modifies its dispersion relation by introducing an upper limit on the frequencies (a Debye-like frequency) that can be resolved numerically. This upper limit on frequency for traveling waves depends on the extent

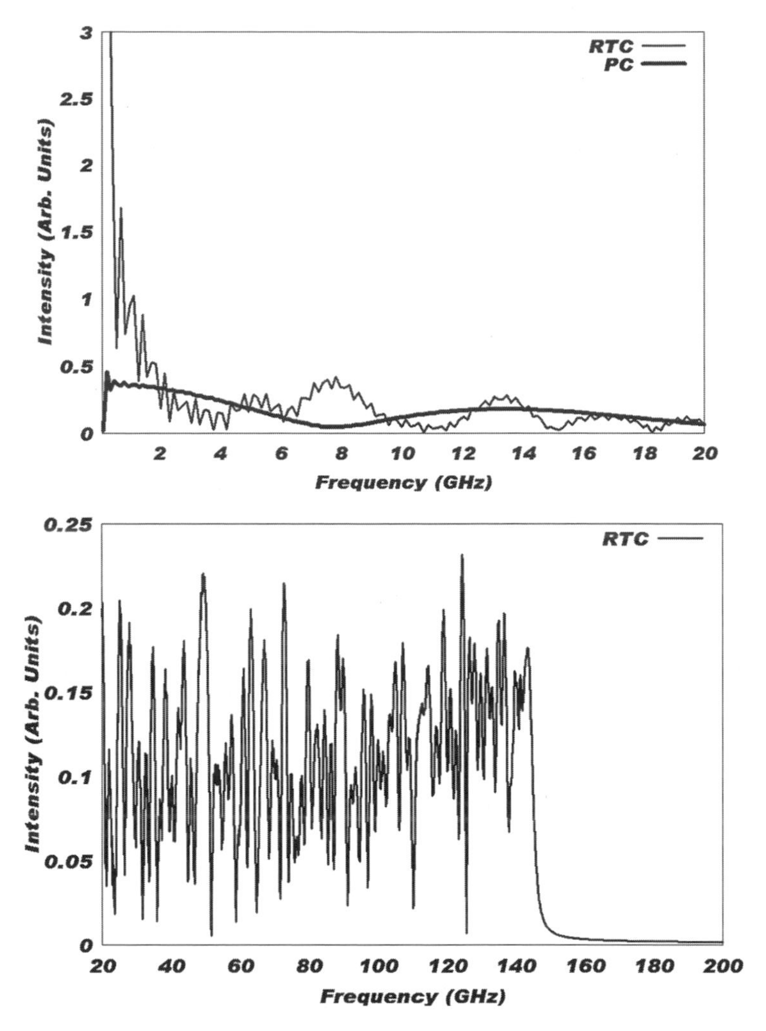

Figure 6.13 Frequency spectrum of the reflected signals at 3.930 GHz; $N_{\mathrm{MD}} = 23$. The upper figure represents the low-frequency range of the reflected signal, while the lower figure corresponds to the high-frequency range of the signal. Here RTC stands for a real-time coupling between the FDTD and the MD region and PC corresponds to a pseudo-coupling between the two regions (see text for details).

of discretization of the continuum, i.e., the size of the FDTD element. Part of the high-frequency signal of the RTC system is not due to reflection of the probing signal by the MD cell. Since the temperature of the MD system is maintained constant with a momentum-rescaling thermostat, the internal stress of the MD cell fluctuates, this even in absence of the probing signal. In absence of the probing wave, the fluctuating stress averages to zero in the long-time limit and does not affect the neighboring FDTD elements. However, due to the RTC conditions, the stress averaged over a small number of MD steps (N_{MD}) does not vanish. The thermalized MD system becomes a source of elastic energy. High-frequency elastic waves propagate along the FDTD system outward from the MD cell.

6.5 Concluding remarks

Atomistic computer simulation techniques offer opportunities for nanoscale science and engineering. Nanoscale structures are ideal for computational studies using MD or MC methods. Individual nanostructures can be modeled and simulated effectively because their behavior is limited to a small number of spatial and temporal scales. Simulations of nanoscale systems have, therefore, the potential of being predictive. However, as illustrated by the example of the thermal stability of a C_{60} molecule, the choice of a thermostat for coupling the individual nanostructure to a "macroscopic" heat bath is important as it influences the dynamics of the nanostructure. Furthermore, the predictive capability of an atomistic simulation will be limited by the degree of realism of the input interatomic potential.

Nanoscale composite systems constituted of several nanostructures (or composed of nanostructures embedded in a matrix or lying on a substrate) exhibit greater interactions between vastly different spatial and temporal scales. The simulation of a collection of nanoscale structures may include time scales ranging from the characteristic time of atomic vibration to the characteristic time of molecular conformational change to the characteristic time of molecular diffusion etc. MC simulations are not constrained by a hierarchy of characteristic time constants. MC methods offer an alternative to achieve fast exploration of phase space. The kinetic MC method reinstates the hierarchy of time scales and provides kinetic information on the simulated process.

Multiscale MD and MC methods are emerging as effective simulation approaches for composite nanostructures. The predictive capability of a multiscale MD or MC method depends on how well one achieves scale parity.[121] Many modeling and simulation methodologies are developed at one primary length or time scale. For instance, continuum mechanics limits the representation of vibrations to long wavelengths and low frequencies. Dynamical atomistic simulations naturally include high-frequency and short-wavelength vibrational modes. Successful coupling between simulation methods should not give the priority to any one scale. Overcoming this built-in bias should be of primary concern in the simulation of composite nanoscale system via multiscale methods. In Sec. 6.4.1, we showed that overlap of spatial scales can be used advantageously to bridge simulation methodologies modeling grain growth of a nanograin polycrystalline material over different intervals of scales to achieve a representation of the phenomenon over a range of scales union of the individual intervals. The coupling between a small atomistic system and an elastic continuum served as an example of a system for which scale parity in time may not be satisfied.

Future research directions in atomistic computer simulations of nanoscale structures will be driven by the needs to (1) expand the range of accessible spatial and temporal scales, (2) develop realistic transferable models, (3) improve compatibility between models, and (4) establish scale parity. These needs require a research effort in the development of improved simulation methodologies, of first-principle models that can be easily input into simulation software, and of numerical means of interfacing quantum/atomistic/continuum models without imposing model-related bias. More specifically, it is important to pursue the development of accelerated MD methods that enable the simulation of atomistic systems over very long times (seconds to hours). Accelerated MD simulation of surface diffusion has been addressed recently by Voter and coworkers[122–124] who, based on transition state theory and several methodologies (accelerated dynamics methods such as hyperdynamics, parallel replica dynamics, temperature-accelerated dynamics), were able to achieve the simulation of diffusion over extended time intervals. An intermediate-resolution protein folding model combined with constant-temperature discontinuous MD has enabled the simulation of protein folding and protein aggregation over relatively long time scales.[125,126] Finally, large-scale MD simulations that include chemical reactivity have been made possible by using quantum chemical transfer Hamiltonians of the semiempirical type. These methods provide a quantum chemical treatment of interatomic forces while requiring several orders of magnitude less time than *ab initio* calculations.[127,128]

References

1. G. Ciccotti, D. Frenkel, and I. R. McDonald, Eds., *Simulation of Liquids and Solids: Molecular Dynamics and Monte Carlo Methods in Statistical Mechanics*, North-Holland, Amsterdam (1987).
2. C. L. Rountree, R. K. Kalia, E. Lidorikis, A. Nakano, L. Van Brutzel, and P. Vashishta, "Atomistic aspects of crack propagation in brittle materials: multimillion atom molecular dynamics simulations," *Ann. Rev. Mater. Res.* **32**, 377–400 (2002).
3. F. Shimojo, T. J. Campbell, R. K. Kalia, A. Nakano, S. Ogata, P. Vashishta, and K. Tsuruta, "A scalable molecular-dynamics algorithm suite for materials simulations: design-space diagram on 1024 Cray T3E processors," *Future Gen. Comp. Sys.* **17**, 279–291 (2000).
4. J. D. Schall and D. W. Brenner, "Molecular dynamics simulations of carbon nanotubes rolling and sliding on graphite," *Mol. Simulat.* **25**, 73–79 (2000).
5. J. Han, A. Globus, R. Jaffe, and G. Deardorff, "Molecular dynamics simulations of carbon nanotube-based gears," *Nanotechnology* **8**, 95–102 (1997).
6. M. Huhtala, A. Kuronen, and K. Kashu, "Carbon nanotube structures: molecular dynamics simulation at realistic limit," *Comp. Phys. Commun.* **146**, 30–37 (2002).
7. D. Srivastava, M. Menon, and K. Cho, "Nanoplasticity of single-wall carbon nanotubes under uniaxial compression," *Phys. Rev. Lett.* **83**, 2973–2976 (1999).
8. Y. Yamaguchi and S. Maruyama, "A molecular dynamics simulation of the fullerene formation," *Chem. Phys. Lett.* **286**, 336–342 (1998).
9. S. G. Kim and D. Tomanek, "Melting the fullerenes: a molecular dynamics study," *Phys. Rev. Lett.* **72**, 2418–2421 (1994).

10. K. Fukui, B. G. Sumpter, M. D. Barnes, D. W. Noid, and J. U. Otaige, "Molecular dynamics simulation of the thermal properties of nanoscale polymer particles," *Macromol. Theory Simul.* **8**, 38–45 (1999).
11. H. A. Wu, X. G. Ni, Y. Wang, and X. X. Wang, "Molecular Dynamics simulation on bending behavior of metal nanorod," *Acta Phys. Sin.* **51**, 1412–1415 (2002).
12. P. Walsh, R. K. Kalia, A. Nakano, P. Vashishta, and S. Saini, "Amorphization and anisotropic fracture dynamics during a nanoindentation of silicon nitride: a multimillion-atom molecular dynamics study," *Appl. Phys. Lett.* **77**, 4332–4334 (2000).
13. X. T. Su, R. K. Kalia, A. Nakano, P. Vashishta, and A. Madhukar, "Critical lateral size for stress domain formation in InAs/GaAs square nanomesas: a multimillion-atom molecular dynamics study," *Appl. Phys. Lett.* **79**, 4577–4579 (2001).
14. P. Walsh, W. Li, R. K. Kalia, A. Nakano, P. Vashishta, and S. Saini, "Structural transformation, amorphization, and fracture in nanowires: a multimillion-atom molecular dynamics study," *Appl. Phys. Lett.* **78**, 3328–3330 (2001).
15. R. Rajagopalan, "Review: simulations of self-assembling system," *Curr. Opin. Colloid Interf. Sci.* **6**, 357–365 (2001).
16. R. Rajagopalan, L. A. Rodriguez-Guadarrama, et al., "Lattice Monte Carlo simulations of micellar and microemulsion systems," in *Handbook of Microemulsion Science and Technology*, P. Kumar and K. L. Mittal, Eds., 105–137, Marcel Dekker, New York (1999).
17. S. Bandyopadyay, J. C. Shelly, M. Tarek, P. B. Moore, and M. L. Klein, "Surfactant aggregation at a hydrophobic surface," *J. Phys. Chem. B* **102**, 6318–6322 (1998).
18. L. Zhang, K. Wesley, and S. Jjiang, "Molecular simulation study of alkyl monolayers on Si(111)," *Langmuir* **17**, 6275–6281 (2001).
19. C. M. Wijmans and P. Linse, "Monte Carlo simulations of the adsorption of amphiphilic oligomers at hydrophobic interfaces," *J. Chem. Phys.* **106**, 328–338 (1997).
20. C. M. Wijmans and P. Linse, "Surfactant self assembly at a hydrophilic surface: a Monte Carlo study," *J. Phys. Chem.* **100**, 12583–12591 (1996).
21. D. A. McQuarrie, *Statistical Mechanics*, Harper&Row, New York (1976).
22. B. J. Alder and T. E. Wainwright, "Studies in molecular dynamics. I. General method," *J. Chem. Phys.* **31**, 459–466 (1959).
23. N. Metropolis, A. W. Rosenbluth, M. N. Rosenbluth, A. Teller, and E. Teller, "Equation of state calculations by fast computing machines," *J. Chem. Phys.* **21**, 1087–1092 (1953).
24. A. P. Sutton and R. W. Balluffi, *Interfaces in Crystalline Materials*, Clarendon Press, Oxford (1995).
25. S. Sarman, D. J. Evans, and P. T. Cummings, "Recent developments in nonequilibrium molecular dynamics," *Phys. Rep.* **305**, 1–92 (1998).

26. P. Stoltze, "Microkinetic simulation of catalytic reactions," *Prog. Surf. Sci.* **65**, 65–150, (2000).
27. V. A. Schneidman, K. A. Jackson, and K. M. Beatty, "Nucleation and growth of a stable phase in an Ising-type system," *Phys. Rev. B* **59**, 3579–3589 (1999).
28. Y. C. Tao and C. S. Peskin, "Simulating the role of microtubules in depolymerization-driven transport: a Monte Carlo approach," *Biophys. J.* **75**, 1529–1540 (1998).
29. L. Verlet, "Computer 'experiments' on classical fluids. I. Thermodynamical properties of Lennard-Jones molecules," *Phys. Rev.* **159**, 98–103 (1967).
30. D. W. Heermann, *Computer Simulation Methods in Theoretical Physics*, Springer, Berlin (1986).
31. D. C. Rapaport, *The Art of Molecular Dynamics Simulation*, Cambridge University Press, Cambridge (1995).
32. K. Ohno, K. Esfarjani, and Y. Kawazoe, *Computational Materials Science from ab initio to Monte Carlo Methods*, Springer, Berlin (1999).
33. D. Frenkel and B. Smit, *Understanding Molecular Simulation from Algorithm to Applications*, Academic Press, San Diego, CA (2002).
34. H. C. Andersen, "Molecular dynamics simulations at constant pressure and/or temperature," *J. Chem. Phys.* **72**, 2384–2393 (1980).
35. W. G. Hoover, A. J. C. Ladd, and B. Moran, "High strain rate plastic flow studied via non-equilibrium molecular dynamics," *Phys. Rev. Lett.* **48**, 1818–1820 (1982).
36. W. G. Hoover, "Canonical dynamics: equilibrium phase-space distributions," *Phys. Rev. A* **31**, 1695–1697 (1985).
37. S. Nosé, "A molecular dynamics method for simulation in the canonical ensemble," *Mol. Phys.* **52**, 255–268 (1984).
38. G. J. Martyna, M. L. Klein, and M. E. Tuckerman, "Nosé-Hoover chains: the canonical ensemble via continuous dynamics," *J. Chem. Phys.* **97**, 2635–2643 (1992).
39. D. J. Evans and G. P. Morriss, "Non-Newtonian molecular dynamics," *Comp. Phys. Rep.* **1**, 297–343 (1984).
40. M. E. Tuckerman and G. J. Martyna, "Understanding modern molecular dynamics: techniques and applications," *J. Phys. Chem. B* **104**, 159–178 (2000).
41. A. MacKerell, Jr., D. Bashford, M. Bellott, R. L. Dunbrack, Jr., J. D. Evanseck, M. J. Field, S. Fischer, J. Gao, H. Guo, S. Ha, D. Joseph McCarthy, L. Kuchnir, K. Kuczena, F. T. K. Lau, C. Mattos, S. Michnick, T. Ngo, D. T. Nguyen, B. Prodhom, W. E. Reiher, III, B. Roux, M. Schlenkrich, J. C. Smith, R. Stote, J. Straub, M. Watanabe, J. W. Corkiewicz-Kuczera, D. Yin, and M. Karplus, "All-atom empirical potential for molecular modeling and dynamics studies of proteins," *J. Phys. Chem. B* **102**, 3586–3616 (1998).
42. M. Parrinello and A. Rahman, "Study of an F center in KCl," *J. Chem. Phys.* **80**, 860–867 (1984).

43. R. W. Hall and B. J. Berne, "Nonergodicity in path integral molecular dynamics," *J. Chem. Phys.* **81**, 3641–3643 (1984).
44. M. E. Tuckerman, B. J. Berne, G. J. Martyna, and M. L. Klein, "Efficient molecular dynamics and hybrid Monte Carlo algorthims for path integrals," *J. Chem. Phys.* **99**, 2796–2808 (1993).
45. S. Miura and S. Okazaki, "Path integral molecular dynamics for Bose-Einstein and Fermi-Dirac statistics," *J. Chem. Phys.* **112**, 10116–10124 (2000).
46. M. Parrinello and A. Rahman, "Polymorphic transition in single crystals: a new molecular dynamics method," *J. Appl. Phys.* **52**, 7182–7190 (1981).
47. S. R. Phillpot and J. M. Rickman, "Reconstrution of a high-angle twist grain boundary by grand canonical simulated quenching," *Mater. Res. Symp. Proc.* **238**, 183–188 (1992).
48. J. O. Hirschfelder, C. F. Curtiss, and R. B. Bird, *Molecular Theory of Gases and Liquids*, Wiley, New York (1964).
49. A. M. Stoneham, "Handbook of interatomic potentials I: ionic crystals," UK Atomic Energy Authority, AERE report R9598 (1981).
50. A. M. Stoneham and R. Taylor, "Handbook of interatomic potentials II: metals," UK Atomic Energy Authority, AERE report R10205 (1981).
51. V. Vitek and D. J. Srolovitz, Eds., *Atomistic Simulation of Materials: Beyond Pair Potentials*, Plenum Press, New York (1988).
52. F. H. Stillinger and T. A. Weber, "Computer simulation of local order in condensed phases of silicon," *Phys. Rev. B* **31**, 5262–5271 (1985).
53. A. Rahman and F. H. Stillinger, "Molecular dynamics study of liquid water," *J. Chem. Phys.* **55**, 3336–3359 (1971).
54. J. Tersoff, "Empirical interatomic potential for carbon, with applications to amorphous carbon," *Phys. Rev. Lett.* **61**, 2879–2882 (1988).
55. D. W. Brenner, "Empirical potential for hydrocarbons for use in simulating the chemical vapor deposition of diamond films," *Phys. Rev. B* **42**, 9458–9471 (1990).
56. D. W. Brenner, "Erratum: empirical potential for hydrocarbons for use in simulating the chemical vapor deposition of diamond films," *Phys. Rev. B* **46**, 1948 (1992).
57. S. J. Stuart, A. B. Tutein, and J. A. Harrison, "A reactive potential for hydrocarbons with intermolecular interactions," *J. Chem. Phys.* **112**, 6472–6486 (2000).
58. A. Alavi, L. J. Alvarez, S. R. Elliott, and I. R. McDonal, "Charge transfer molecular dynamics," *Phil. Mag. B* **65**, 489–500 (1992).
59. M. S. Daw and M. I. Baskes, "Semi empirical, quantum mechanical calculation of hydrogen embrittlement in metals," *Phys. Rev. Lett.* **50**, 1285–1288 (1983).
60. M. S. Daw and M. I. Baskes, "Embedded-atom method: derivation and application to impurities, surfaces, and other defects in metals," *Phys. Rev. B* **29**, 6443–6453 (1984).

61. M. I. Baskes, "Determination of modified embedded atom method parameters for nickel," *Mater. Chem. Phys.* **50**, 152–158 (1997).
62. M. M. G. Alemany, O. Dieguez, C. Rey, and L. J. Gallego, "Molecular dynamics study of the dynamic properties of FCC transition and simple metals in the liquid phase using the second moment approximation to the tight binding method," *Phys. Rev. B* **60**, 9208–9211 (1993).
63. L. D. Phuong, A. Pasturel, and D. N. Manh, "Effect of s-d hybridization on interatomic pair potentials of the 3d liquid transition metals," *J. Phys. Condens. Matter* **5**, 1901–1918 (1993).
64. R. Car and M. Parrinello, "Unified approach for molecular dynamics and density-functional theory," *Phys. Rev. Lett.* **55**, 2471–2474 (1985).
65. Richard J. Sadus, *Molecular Simulations of Fluids: Theory, Algorithms, and Object-Orientation*, Elsevier, New York (1999).
66. K. Binder and D. W. Heermann, *Monte Carlo Simulation in Statistical Physics*, Springer, Berlin (1988).
67. E. Kreyszig, *Advanced Engineering Mathematics*, Wiley, New York (2002).
68. J. L. Soto and A. L. Myers, "Monte Carlo studies of adsorption in molecular sieves," *Molec. Phys.* **42**, 971–983 (1981).
69. R. F. Cracknell, D. Nicholson, and N. Quirke, "A grand canonical Monte Carlo study of Lennard-Jones mixtures in slit pores; 2: mixtures of two center ethane with methane," *Molec. Simulat.* **13**, 161–175 (1994).
70. P. R. van Tassel, H. T. Davis, and A. V. McCormick, "Open system Monte Carlo simulations of Xe in NaA," *J. Chem. Phys.* **98**, 8919–8928 (1993).
71. J. I. Siepmann and D. Frenkel, "Configurational bias Monte Carlo: a new sampling scheme for flexible chains," *Molec. Phys.* **75**, 59–70 (1992).
72. M. N. Rosenbluth and A. W. Rosenbluth, "Monte Carlo simulation of the average extension of molecular chains," *J. Chem. Phys.* **23**, 356–359 (1955).
73. P. H. Verdier and W. H. Stockmayer, "Monte Carlo calculations on the dynamics of polymers in dilute solutions," *J. Chem. Phys.* **36**, 227–235 (1962).
74. P. De Gennes, "Reptation of a polymer chain in the presence of fixed obstacles," *J. Chem. Phys.* **55**, 572–579 (1971).
75. M. A. Floriano, E. Caponetti, and A. Z. Panagiotopoulos, "Micellization in model surfactant systems," *Langmuir* **15**, 3143–3151 (1999).
76. P. C. Hiemenz and R. Rajagopalan, *Principles of Colloid and Surface Chemistry*, Marcel Dekker, New York (1997).
77. J. N. Israelachvili, *Intermolecular and Surface Forces*, Academic Press, New York (1991).
78. A. M. Almanza-Workman, S. Raghavan, P. Deymier, D. J. Monk, and R. Roop, "Water dispersible silane for wettability modification of polysilicon," *J. Electrochem. Soc.* **149**, H6–H11 (2002).
79. A. M. Almanza-Workman, S. Raghavan, P. Deymier, D. J. Monk, and R. Roop, "Wettability modification of polysilicon for stiction reduction in silicon based micro-electromechanical structures," in *Proceedings of the Fifth International Symposium on Ultra Clean Processing of Silicon Surfaces* (UCPSS 2000), Diffus. Defect Data, Pt. B, Oostende, Belgium (2000).

80. R. G. Larson, "Monte Carlo lattice simulation of amphiphilic systems in two and three dimensions," *J. Chem. Phys.* **89**, 1642–1650 (1988).
81. R. G. Larson, "Monte Carlo simulation of model amphiphile-oil-water systems," *J. Chem. Phys.* **83**, 2411–2420 (1985).
82. R. G. Larson, "Self-assembly of surfactant liquid crystalline phases by Monte Carlo simulation," *J. Chem. Phys.* **91**, 2479–2488 (1989).
83. R. G. Larson, "Monte Carlo simulation of microstructural transitions in surfactant systems," *J. Chem. Phys.* **96**, 7904–7918 (1992).
84. R. G. Larson, "Simulation of lamellar phase transitions in Block copolymers," *Macromolecules* **27**, 4198–4203 (1994).
85. V. Kapila, J. M. Harris, P. A. Deymier, and S. Raghavan, "Effect of long-range and steric hydrophilic interactions on micellization of surfactant systems: a Monte-Carlo study in 2D," *Langmuir* **18**, 3728–3736 (2002).
86. D. Brindle and C. M. Care, "Phase-diagram for the lattice model of amphiphile and solvent mixtures by Monte-Carlo-simulation," *J. Chem. Soc. Farday Trans.* **88**, 2163–2166 (1992).
87. C. M. Care, "Cluster size distribution in a Monte Carlo simulation of the micellar phase of an amphiphile and solvent mixture," *J. Chem. Soc. Faraday Trans. I* **83**, 2905–2912 (1987).
88. J. C. Desplat and C. M. Care, "A Monte Carlo simulation of the micellar phase of an amphiphile and solvent mixture," *Molec. Phys.* **87**, 441–453 (1996).
89. S. K. Talsania, Y. Wang, R. Rajagoplan, and K. K. Mohanty, "Monte Carlo simulations of micellar encapsulation," *J. Colloid Interface Sci.* **190**, 92–103 (1997).
90. K. Binder, *Monte Carlo Methods in Statistical Physics*, 2nd ed., Springer, New York (1986).
91. H. Lodish, A. Berk, S. L. Zipursky, P. Matsudaira, D. Baltimore, and J. E. Darnell, *Molecular Cell Biology*, 4th ed., Freeman, New York (2000).
92. S. C. Schuyler and D. Pellman, "Microtubule 'plus-end-tracking proteins': the end is just the beginning," *Cell* **105**, 421–424 (2001).
93. T. L. Hill and Y. Chen, "Phase changes at the end of a microtubule with a GTP cap," *Proc. Natl. Acad. Sci. USA* **81**, 5772–5776 (1984).
94. Y. Chen and T. L. Hill, "Monte Carlo study of the GTP cap in a five-start helix model of microtubule," *Proc. Natl. Acad. Sci. USA* **82**, 1131–1135 (1985).
95. P. Bayley, M. Schilstra, and S. Martin, "A lateral cap model of microtubule dynamic instability," *FEBS Letts.* **259**, 181–184 (1989).
96. P. M. Bayley, M. J. Schilstra, and S. R. Martin, "Microtubule dynamic instability: numerical simulation of microtubule transition properties using a lateral cap model," *J. Cell. Sci.* **95**, 33–48 (1990).
97. E. M. Mandelkow, R. Schultheiss, R. Rapp, M. Muller, and E. Mandelkow, "On the surface lattice of microtubules: helix starts, protofilament number, seam, and handedness," *J. Cell. Biol.* **102**, 1067–1073 (1986).

98. S. R. Martin, M. J. Schilstra, and P. M. Bayley, "Dynamic instability of microtubules: Monte Carlo simulation and application to different types of microtubule lattice," *Biophys. J.* **65**, 578–596 (1993).
99. R. A. B. Keater, "Dynamic microtubule simulation, seams and all," *Biophys. J.* **65**, 566–567 (1993).
100. Y. Tao and C. S. Peskin, "Simulating the role of microtubules in depolymerization-driven transport: a Monte Carlo approach," *Biophys. J.* **75**, 1529–1540 (1998).
101. E. B. Tadmor, M. Ortiz, and R. Phillips, "Quasicontinuum analysis of defects in solids," *Phil. Mag. A* **73**, 1529–1563 (1996).
102. E. B. Tadmor, R. Phillips, and M. Ortiz, "Mixed atomistic and continuum models of deformation in solids," *Langmuir* **12**, 4529–4534 (1996).
103. V. B. Shenoy, R. Miller, E. B. Tadmor, R. Phillips, and M. Ortiz, "Quasicontinuum models of interfacial structure and deformation," *Phys. Rev. Lett.* **80**, 742–745 (1998).
104. G. S. Smith, E. B. Tadmor, and E. Kaxinas, "Multiscale simulation of loading and electrical resistance in silicon nanoindentation," *Phys. Rev. Lett.* **84**, 1260–1263 (2000).
105. G. S. D. Ayton, S. Bardenhagen, P. McMurtry, D. Sulsky, and G. A. Voth, "Interfacing molecular dynamics with continuum dynamics in computer simulation: toward an application to biological membranes," *IBM J. Res. Dev.* **45**, 417–426 (2001).
106. M. Mullins and M. A. Dokainish, "Simulations of the (001) plane crack in alpha-iron employing a new boundary scheme," *Phil. Mag. A* **46**, 771–787 (1982).
107. H. Kitagawa, A. Nakatami, and Y. Sibutani, "Molecular dynamics study of crack process associated with dislocation nucleated at the tip," *Mater. Sci. Engr. A* **176**, 263–269 (1994).
108. R. E. Rudd and J. Q. Broughton, "Coarse-grained molecular dynamics and the atomic limit of finite elements," *Phys. Rev. B* **58**, R5893–R5896 (1998).
109. J. Q. Broughton, F. F. Abraham, N. Bernstein, and E. Kaxiras, "Concurrent coupling of length scales: methodology and application," *Phys. Rev. B* **60**, 2391–2403 (1999).
110. S. Kohlhoff, P. Gumbsch, and H. F. Fischmeister, "Crack propagation in BCC crystals studied with a combined finite-element and atomistic model," *Phil. Mag. A* **64**, 851–878 (1991).
111. F. F. Abraham, J. Q. Broughton, N. Bernstein, and E. Kaxiras, "Spanning the length scales in dynamic simulation," *Comp. Phys.* **12**, 538–546 (1998).
112. S. Ogata, E. Lidorikis, F. Shimojo, A. Nakano, P. Vashista, and R. K. Kalia, "Hybrid finite-element/molecular dynamics/electronic-density functional approach to materials simulations on parallel computers," *Comp. Phys. Commun.* **138**, 143–154 (2001).
113. P. A. Deymier and J. O. Vasseur, "Concurrent multiscale model of an atomic crystal coupled with elastic continua," *Phys. Rev. B* **66**, 134106(1-5) (2002).

114. A. Askar, A. E. Cetin, and H. Rabitz, "Wavelet transform for analysis of molecular dynamics," *J. Phys. Chem.* **100**, 19165–19173 (1996).
115. G. Frantziskonis and P. A. Deymier, "Wavelet methods for analyzing and bridging simulations at complementary scales—the compound wavelet matrix and application to microstructure evolution," *Model. Sim. Mater. Sci. Eng.* **8**, 649–664 (2000).
116. P. S. Sahni, G. S. Grest, M. P. Anderson, and D. J. Srolovitz, "Kinetics of the Q-state Potts model in 2 dimensions," *Phys. Rev. Lett.* **50**, 263–266 (1983).
117. I. Daubechies, *Ten Lectures on Wavelets*, SIAM, Philadelphia (1992).
118. M. M. Sigalas and N. Garcia, "Theoretical study of three dimensional elastic band gaps with the finite difference time domain method," *J. Appl. Phys.* **87**, 3122–3125 (2000).
119. D. Garcia-Pablos, M. M. Sigalas, F. R. M. de Espinosa, M. Torres, M. Kafesaki, and N. Garcia, "Theory and experiments on elastic band gaps," *Phys. Rev. Lett.* **84**, 4349–4352 (2000).
120. K. Muralidharan, P. A. Deymier, and J. H. Simmons, "Multiscale modeling of wave propagation: FDTD/MD hybrid method," *Model. Simulat. in MSE* **11**, 487–501 (2003).
121. S. M. Trickey, and P. A. Deymier, "Challenges and state of the art in simulation of chemo-mechanical processes," in *Proceedings of the 4th International Symposium on Chemical Mechanical Polishing* (CMP), 198th meeting of the Electrochemical Soc. Phoenix, AZ Oct. 23, 2000 (2001).
122. F. Montalenti and A. F. Voter, "Exploiting past visits or minimum-barrier knowledge to gain further boost in the temperature-accelerated dynamics method," *J. Chem. Phys.* **116**, 4819–4828 (2002).
123. A. F. Voter, "Hyperdynamics: accelerated molecular dynamics of infrequent events," *Phys. Rev. Lett.* **78**, 3908–3911 (1997).
124. M. R. Sorensen and A. F. Voter, "Temperature-accelerated dynamics for simulation of infrequent events," *J. Chem. Phys.* **112**, 9599–9606 (2000).
125. A. Voegler Smith and C. K. Hall, "α-Helix formation: discontinuous molecular dynamics on an intermediate-resolution protein model," *Proteins* **44**, 344–360 (2001).
126. A. Voegler Smith and C. K. Hall, "Assembly of a tetrameric α-helical bundle: computer simulations on an intermediate-resolution protein model," *Proteins* **44**, 376–391 (2001).
127. Y. W. Hsiao, K. Runge, M. G. Kory, and R. J. Bartlett, "Direct molecular dynamics using quantum chemical Hamiltonian: C-60 impact on a passive surface," *J. Phys. Chem. A* **105**, 7004–7010 (2001).
128. C. E. Taylor, M. G. Cory, R. J. Bartlett, and W. Theil, "The transfer Hamiltonian: a tool for large scale simulations with quantum mechanical forces," *Comput. Mater. Sci.* **27**, 204–211 (2003).

List of symbols

E	internal energy
V	volume
T	temperature
N	number of particles
N_i	number of particles of type i
P	pressure
μ	chemical potential
H	Hamiltonian
$\mathbf{p}_i$	momentum of particles i
$\mathbf{r}_i$	position of particle i
$\dot{\mathbf{r}}_i$	velocity of particle i
$\ddot{\mathbf{r}}_i$	acceleration of particle i
$\{\mathbf{p}\}$	momenta of all of the particles in a system
$\{\mathbf{r}\}$	positions of all of the particles in a system
m	mass
V	potential energy function
k_B	Boltzmann's constant
$\mathbf{F}_i$	force on particle i
α	damping factor for isothermal constraint method
η	Nosé-Hoover thermostat degree of freedom
p_η	momentum associated with η
Q	mass associated with the Nosé-Hoover thermostat degree of freedom
η_j	degree of freedom associated with jth thermostat in Nosé-Hoover chain of thermostats
p_{η_j}	momentum associated with η_j
Q_j	mass of the jth thermostat in a Nosé-Hoover chain of thermostats
d	dimensionality of a system
H	enthalpy
V_{ij}	bond energy of Tersoff's potential
$f_R(r_{ij})$	repulsive pair potential of Tersoff's potential
$f_A(r_{ij})$	attractive pair potential of Tersoff's potential
$f_c(r_{ij})$	cutoff function of Tersoff's potential
b_{ij}	many-body environment-dependent bond order term in Tersoff's potential
$\langle A \rangle$	ensemble average of A
$p(\{\mathbf{r}\})$	probability of occurrence of a configuration $\{\mathbf{r}\}$
p	acceptance probability
$Z_{\mu VT}$	grand canonical partition function
Λ	de Broglie thermal wavelength
Ω	phase space
ε_{IJ}	nearest-neighbor pair energies between chemical species I and J

k_i	rate constant for dissociation or association events at a site i
t_i	time for dissociation or association at a site i
R, R_i	uniformly distributed random number between 0 and 1
$t_{\min}$	shortest time among the t_i's
ε, σ	Lennard-Jones potential parameters
$S \in [1, Q]$	"spin" or orientation states of Q-states Potts model
J_{int}	interaction energy of Potts model
$\psi(x)$	1D mother wavelet function
$f(x)$	1D fluctuating function
$\psi_{a,b}(x)$	two-parameter family of wavelet functions
a	scaling factor
b	translating factor
$W_f(a, b)$	wavelet coefficient
s	scale
f_{s_1,s_2}	function f reconstructed with an inverse wavelet transformation limited to the interval of scales $[s_1, s_2]$
$u_i\ i \in [1, 3]$	components of the elastic displacement
$T_{ij}\ i, j \in [1, 3]$	components of the stress tensor
ρ	mass density
Δt	finite-difference time-domain time step
δt	molecular dynamics time step
k	wave number of 1D wave packet
a_0	amplitude of 1D wave packet
c_0	longitudinal velocity of an elastic wave
ν	central frequency of wave packet

Pierre A. Deymier is a professor of materials science and engineering at the University of Arizona, and received his "Diplôme d'Ingénieur" in materials science from the Université des Sciences et Techniques du Languedoc, Montpellier, France, and his PhD degree in ceramics from the Massachusetts Institute of Technology. He is currently associate head of the Department of Materials Science and Engineering and directs the recently initiated nanobiomolecular engineering, science, and technology (n-BEST) program. His primary research interests are in materials modeling and simulation. Deymier has authored or coauthored over 60 journal articles in the fields of multiscale modeling, simulation of wave propagation in inhomogeneous media, megasonic cleaning, quantum MD simulations of strongly correlated electrons, equilibrium and nonequilibrium MD simulation of molten metals, interfaces and grain boundaries, MC simulations of complex fluids and self-assembly processes.

Vivek Kapila was born in Sahnewal (Ludhiana), India. He received his BE degree in metallurgical engineering from the University of Roorkee (now IIT-R), India in 1996 and his MS degree in materials science and engineering from South Dakota School of Mines and Technology, Rapid City, USA in 1997. His master's research was focused on the development of a fiber optic Fourier-transform-IR evanescent wave strain sensor. He joined the PhD program in materials science and engineering at the University of Arizona, Tucson, USA in 1999. His doctoral research involves investigation of self-assembly of surfactants and silanes in aqueous solutions and on the surfaces by MC simulations. He is also studying the frictional behavior of the films of silanes deposited on hydrophilic surfaces via MD simulations. His research interests include molecular modeling of self-assembly of chain molecules, surface chemistry in microelectronics and microelectromechanical systems (MEMS), fiber optic sensors, and composite materials.

Krishna Muralidharan obtained his MSc degree in physics at the Indian Institute of Technology Madras, India. Later, he moved to University of Florida, Gainesville, where he graduated with an MS degree in materials science and engineering under the guidance of Prof. Joseph Simmons. His work involved the simualtion studies of bulk self diffusion in Beta-SiC. Currently, he is pursuing his PhD degree in materials science and engineering at the University of Arizona, Tucson and is based at the Los Alamos National Labs, Los Alamos, New Mexico. His present work involves the thermodynamic modeling of Actinide elements (with Dr. M. I. Baskes and Dr. M. Stan) and the multiscale modeling of brittle fracture (under the guidance of Prof. P. A. Deymier and Prof. J. H. Simmons).

Chapter 7

Nanomechanics

Vijay B. Shenoy

7.1. Overview 256
   7.1.1. Introduction 256
   7.1.2. Aim and scope 256
   7.1.3. Notation 261
7.2. Continuum concepts 261
   7.2.1. Forces, equilibrium, and stress tensor 262
   7.2.2. Kinematics: deformation and strain tensor 265
   7.2.3. Principle of virtual work 268
   7.2.4. Constitutive relations 269
   7.2.5. Boundary value problems and finite element method 270
7.3. Atomistic models 274
   7.3.1. Total energy description 274
   7.3.2. Atomistic simulation methods 287
7.4. Mixed models for nanomechanics 295
   7.4.1. The quasi-continuum method 295
   7.4.2. Augmented continuum theories 302
7.5. Concluding remarks 311
Acknowledgments 311
References 311
List of symbols 316

7.1 Overview

7.1.1 Introduction

The ability to manipulate matter at the atomic scale bears promise to produce devices of unprecedented speed and efficiency. The emerging area called nanoscience and nanotechnology has seen phenomenal growth in the past decade and is likely to be the frontal area of research for the next two decades. The outcome of this research is likely to revolutionize technology in ways that will enable humankind to manipulate even individual atoms so as to produce desired effects. The vision of nanotechnology is not new; it is now well over 40 years since Richard Feynman[1] made his foresightful speech at the winter meeting of the American Physical Society at Caltech. Eric Drexler is the one many would call the "father of nanotechnology." His vision was first outlined in his book, *Engines of Creation, The Coming Era of Nanotechnology*, the first few paragraphs of which are deeply insightful and worth quoting verbatim:

> Coal and diamonds, sand and computer chips, cancer and healthy tissue: throughout history, variations in the arrangement of atoms have distinguished the cheap from the cherished, the diseased from the healthy. Arranged one way, atoms make up soil, air and water; arranged another, they make up ripe strawberries. Arranged one way, they make up homes and fresh air; arranged another, they make up ash and smoke.
>
> Our ability to arrange atoms lies at the foundation of technology. We have come far in our atom arranging, from chipping flint for arrowheads to machining aluminum for spaceships. We take pride in our technology, with our lifesaving drugs and desktop computers. Yet our spacecraft are still crude, our computers are still stupid and the molecules in our tissues still slide into disorder, first destroying health, then life itself. For all our advances in arranging atoms, we still use primitive methods. With our present technology, we are still forced to handle atoms in unruly herds.
>
> But the laws of nature leave plenty of room for progress, and the pressures of the world competition are even now pushing us forward. For better or for worse, the greatest technological breakthrough in history is still to come.[2]

Many of the foresights of the these visionaries are a reality today. Capability to manipulate individual atoms (Fig. 7.1) exists, and micromachines or microelectromechanical systems (MEMS) are made routinely (Fig. 7.2). Indeed, researchers[4] have already moved to submicron dimensions and produced nanoelecromechanical systems (NEMS), which have the extremely small response times of the order of 10^{-9} s (Fig. 7.3). Given these spectacular advances in the experimental front, the key to conversion of these scientific achievements into useful devices and products hinges critically on our *predictive capability* of phenomena at the nanoscales that are essential for the design of devices. This chapter focuses on theoretical aspects of "nanomechanics," a subject that allows for the prediction of mechanical properties at the nanoscale.

7.1.2 Aim and scope

The main goal of this chapter is to provide an overview of theoretical approaches to understanding mechanics at the nanoscale. Focus lies on methods that enable

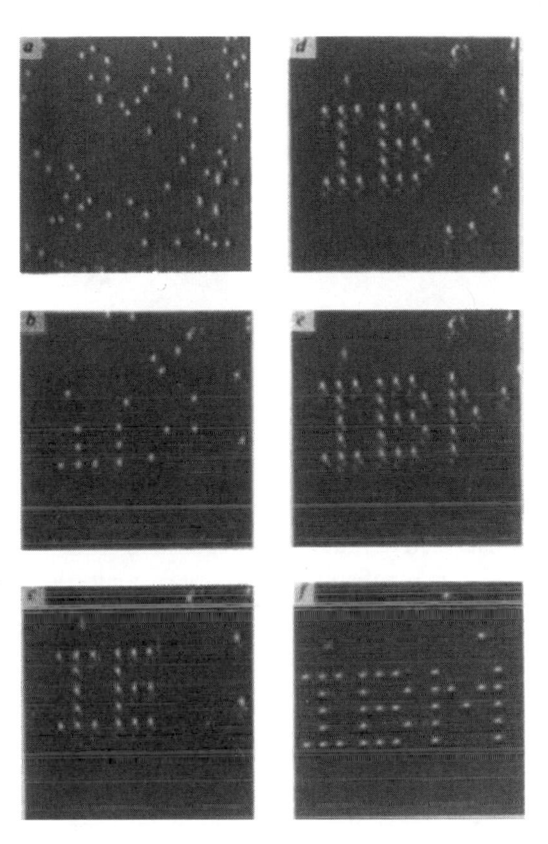

Figure 7.1 Xenon atoms were arranged in an array on a Ni (110) surface to create an "atomic name plate." The height of each letter is about 50 Å. The atomic manipulations were performed using an atomic force microscope (AFM), and the image was captured using a scanning tunneling microscope (STM). (Reprinted with permission from Ref. 3, © 1990 The Nature Publishing Group.)

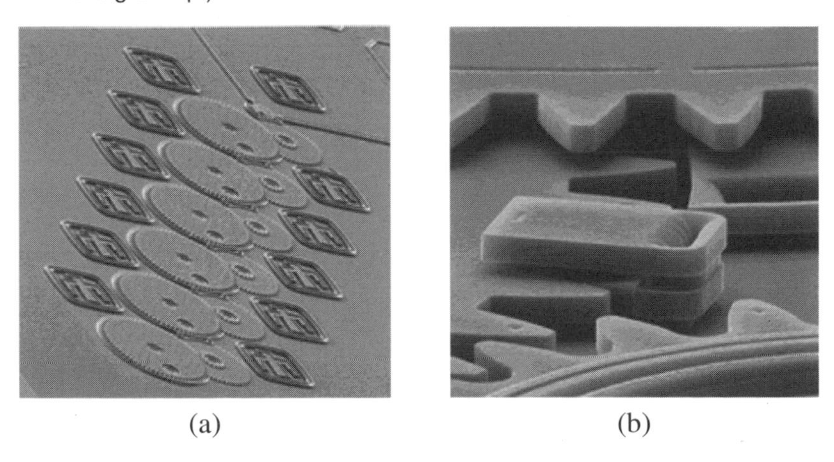

(a) (b)

Figure 7.2 (a) A 3 million:1 transmission made at Sandia labs, featuring six intermeshing gearing reduction units, each with gears in ratio of 1:3 and 1:4. Each single transmission assembly (six are shown) is capable of being duplicated and meshed with other assemblies. The gear wheels are each about the diameter of a human hair. (b) A close-up of the gear system. (From http://www.sandia.gov/media/microtrans.html.)

Figure 7.3 NEMS made of silicon carbide with response times on the order of 10^{-8} s. (After Roukes.[4])

predictive capability. The term "nanomechanics" itself is interpreted in a broad sense. The methods described herein are applicable to study two general classes of problems: (1) the nanoscale mechanical behavior of materials and (2) the mechanical behavior of nanostructures. A problem that falls into the first class, the area of computer-aided materials design, is the prediction of the fracture toughness of a new alloy from a knowledge of its atomic constituents alone. Prediction of the flexural rigidity of a nanorod of SiC (see Fig. 7.3) is an example of problems of the second type. Theoretical approaches to both classes of problems have much in common, and the methods described in this chapter shall find use in both classes. The grand challenge in the development of theoretical approaches to modeling mechanics at the nanoscales is the treatment of the *multiple length and time scales* that are present in phenomena at the meso- and nanoscales.

The issue of multiple scales in computer-aided materials design is now illustrated. Suppose that the aim is to calculate macroscopic physical properties (elastic modulus tensor, thermal expansion coefficient, specific heat, thermal conductivity, yield stress, etc.) of a given material when an atomistic description of the material of interest is known. The macroscopic properties of materials are governed by phenomena that have multitudes of length and time scales. This point can be clarified by considering the example of the prediction of stress-strain curve of a metallic single crystal. Roughly three of the macroscopic properties are related to the stress-strain curve of a single crystal: elastic modulus, yield stress, and hardening modulus. The elastic modulus is determined by the physics at the smallest length scales, i.e., directly by the bonding between the atoms; the length scale of interest is a few unit cells that make up the crystal (about 10 to 100 Å). Yield stress (of a single crystal) is governed by the stress required to make individual dislocations move—the Peierls stress; the appropriate length scales are about 100 to 1000 Å, roughly the size of the dislocation cores and distances between dislocations. The hardening modulus is governed by density of dislocations, dislocation-dislocation interactions etc.; i.e., the appropriate length scales are 1000 to 10,000 Å. Clearly, there are more than four decades of length scales involved in the problem of determination of the stress strain curve from an atomistic description of matter.

Modeling of nanostructures also presents the problem of multiple scales. A key issue is that properties of nanostructures are *size dependent*. An illustration of this point is provided by a set of beautiful experiments (see Fig. 7.4) on carbon nanotubes performed by Gao et al.[5] The experiments involve single carbon nanotubes excited by a sinusoidally time varying field. The tubes, held in a cantilever configuration, vibrate in response to the applied field, the amplitude of which is obtained as a function of the excitation frequency. This response curve is used to

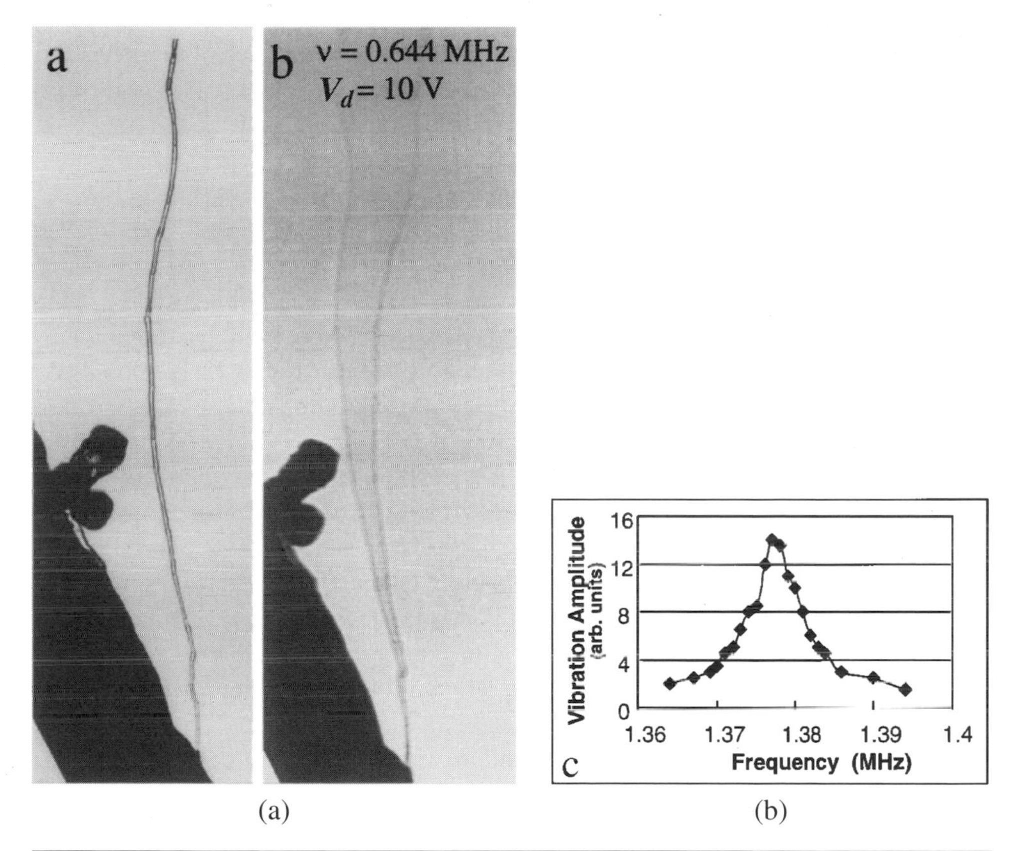

Nanotube	Outer diameter D (nm) (±1)	Inner diameter D_1 (nm) (±1)	Length L (μm) (±0.05)	Frequency ν (MHz)	E_b (GPa)
1	33	18.8	5.5	0.658	32 ± 3.6
2	39	19.4	5.7	0.644	26.5 ± 3.1
3	39	13.8	5	0.791	26.3 ± 3.1
4	45.8	16.7	5.3	0.908	31.5 ± 3.5
5	50	27.1	4.6	1.420	32.1 ± 3.5
6	64	27.8	5.7	0.968	23 ± 2.7

(c)

Figure 7.4 Nanomechanics of single carbon nanotubes. The tubes are excited by a sinusoidal time-varying field (a) and the amplitude of the response is measured as a function of excitation frequency (b). The elasticity of the tube is calculated from the response as shown in table (c). (Reprinted with permission from Ref. 5, © 2000 The American Physical Society.)

determine the elastic modulus of the carbon nanotube material using equations derived from standard continuum mechanics that relate the resonant frequency of a cantilever beam to its elastic modulus and other geometric parameters. The results tabulated in Fig. 7.4 show that the elastic modulus depends rather strongly on the cross-sectional dimensions of the tube. Observation of the size dependence of "material properties" is not limited to mechanical properties. Indeed, these observations of size-dependent material properties have resulted in popular phrases such as "at nanoscales, material = device," indicating that as size scales approach the atomic scale the conventional thinking of the structure and the material being distinct entities has to be abandoned. With diminishing size scales of the structure, the discrete nature of materials, i.e., the atomistic nature of matter, becomes increasingly important. Nanomechanics, therefore, requires a shift in the paradigm from conventional theoretical approaches to mechanics.

Conventional approaches to understanding mechanics of materials and structures exploit the *continuum* concept. The continuum mechanics approach is to treat the material or structure of interest as a continuum where quantities of interest such as the stress and strain tensors etc., are treated as *fields*. The fields satisfy certain basic physical relations such as equilibrium conditions and geometric compatibility relations. Material behavior is incorporated by means of constitutive relations between stress and strain. This theoretical framework results in a field theory, and specific problems are reduced to the solutions of sets of coupled partial differential equations or a *boundary-value problem*. The techniques for the solution of such differential equations are well established, and have become commonplace in engineering design. This traditional continuum approaches have severe limitations as the size scale of the the structure becomes close to the atomic dimensions. A key drawback of standard continuum mechanics is the absence of an intrinsic length scale in the theory, which is characteristic of matter (this length scale is roughly equal to the spacings between the atoms), which governs much of the phenomena at the nanoscale.

Atomistic models, alternative to continuum approaches in modeling mechanics at nanoscales, explicitly acknowledge the discrete nature of matter. The degrees of freedom in these class of models are the coordinates of the atoms that make up the solid. The dynamics of the collection of atoms is determined from the interactions between the atoms. A key step in the construction of an atomistic model is the description of the interactions between the atoms. There are various approaches to achieve this goal, and typically the approach chosen will be a compromise between accuracy and available computational resources. Once an atomistic description of the material or structure is available, standard tools of molecular dynamics and statistical mechanics can be brought to bear on the problem. The advantage of the atomistic description is that the intrinsic nonlocality (presence of an intrinsic length scale) and nonlinearly are automatically built into the model. It might therefore seem that atomistic models are the natural choice for the study of mechanics at the nanoscales. However, the advantages of the atomistic methods come at sometimes exorbitant computational price tags. Also, much of our understanding about mechanics has been shaped by continuum concepts, and learning from a large multibillion atom simulation presents other challenges.

Promising theoretical approaches to the problems of nanomechanics exploit advantages of both the continuum concept and the atomistic methods—the so-called "mixed methods" or "hybrid methods." Mixed methods involve augmenting standard continuum approaches to include an intrinsic scale (nonlocal continuum theories), and/or incorporating the presence of free surfaces that become important at smaller scales. Numerical approaches based on these ideas, such as the quasi-continuum method,[6] bear much potential as efficient models to address problems of nanomechanics.

The aim of this chapter is to present short descriptions of continuum methods, atomistic methods, and the more recent mixed methods. Attention is focused on the main ideas underlying these methods with details being available in the cited references. Section 7.2 contains a brief review of continuum mechanics including a discussion of equilibrium, kinematics, and constitutive relations. A short discussion of the finite element method, the standard numerical approach to solving boundary-value problems of continuum mechanics, is also presented to make the chapter self-contained. Section 7.3 is a summary of atomistic models. Topics treated are total energy descriptions, lattice statics, and molecular dynamics. Mixed models are discussed in Sec. 7.4, where the quasi-continuum method and augmented continuum models are presented.

7.1.3 Notation

The language of vectors and tensors[7,8] is used freely throughout the chapter. Invariant forms of vectors and tensors are denoted in bold font using either Latin or Greek symbols (for example, $\boldsymbol{x}$, $\boldsymbol{F}$, and $\boldsymbol{\sigma}$). Components of a vector $\boldsymbol{x}$ with respect to a Cartesian basis $\{\boldsymbol{e}_i\}$ are denoted as x_i. A second-order tensor $\boldsymbol{F}$ is expanded in the basis $\{\boldsymbol{e}_i \otimes \boldsymbol{e}_j\}$ with components F_{ij}. The gradient is denoted by $\boldsymbol{\nabla}$, and the divergence of a field is denoted by $\boldsymbol{\nabla}\cdot$. Summation convention is used wherever necessary. For example, the divergence of a second-order tensor $\boldsymbol{\sigma}$ denoted as $\boldsymbol{\nabla}\cdot\boldsymbol{\sigma}$ is expressed in component form using the summation convention as follows:

$$\boldsymbol{\nabla}\cdot\boldsymbol{\sigma} \equiv \sigma_{ij,j} \equiv \sum_{j=1}^{3} \sigma_{ij,j}. \tag{7.1}$$

7.2 Continuum concepts

All matter is made of atoms. In constructing theories to understand the mechanics of a solid body, it would be impossible to consider all of the atoms that make up the solid if the size of the body is much larger than the spacing between the atoms. The alternative is to "smear out" the atoms and consider the solid as a continuum. By continuum, it is understood that the body under consideration has the same topological structure as a subset of the 3D space that it exists in. Physical quantities of interest such as stress and strain are defined at every point in the body (i.e., they are represented as fields), and these satisfy conditions of equilibrium and compatibility. Specifics of material behavior are embodied into the constitutive equations that relate the stress and strain at every point. The combination of these three concepts (namely, equilibrium, compatibility, and constitutive relations) results in a boundary-value problem (stated as a set of partial differential equations) for the fields of interest. There are comprehensive accounts of continuum concepts applied to solids to be found in several excellent books.[7–9]

A solid body of interest (see Fig. 7.5) is considered to be a collection of material points that occupy a region $\mathcal{V}$ of space enclosed by a surface $\mathcal{S}$ with outward normal $\boldsymbol{n}$. The positions of material points in the body with respect to some chosen origin is described by a vector $\boldsymbol{x}$. Although the concept of the material point in continuum mechanics does not appeal to any intrinsic scale in the body, it is understood physically that each material point in reality represents a large collection of atoms over which all properties are averaged, i.e., each material point represents a representative volume element. Physical quantities of interest are represented as *fields*, i.e., for every value of $\boldsymbol{x}$ and therefore for each material point, the quantity will be defined (and assumed to be sufficiently smooth as $\boldsymbol{x}$ varies).

7.2.1 Forces, equilibrium, and stress tensor

A fundamental concept in the mechanics of point particles is the force. The concept is generalized in continuum mechanics to distributed forces. The *body force field* $\boldsymbol{b}(\boldsymbol{x})$ is a vector field defined in $\mathcal{V}$ such that a material point at $\boldsymbol{x}$ experiences a force $\boldsymbol{b}$ per unit volume. Similarly, forces can be distributed along the surface and

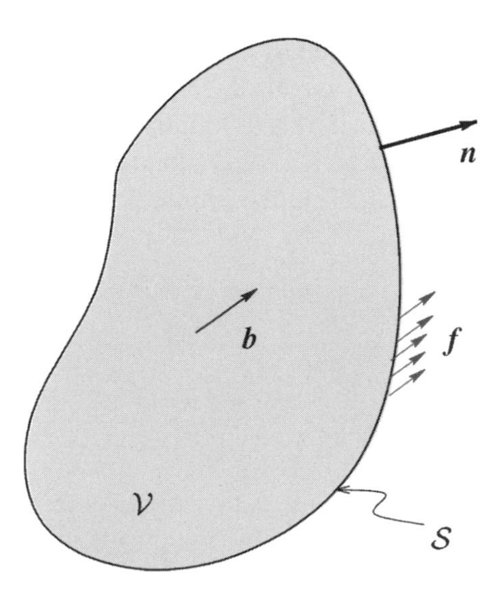

Figure 7.5 A solid body is considered as a continuum object occupying a region $\mathcal{V}$ enclosed by surface $\mathcal{S}$ with outward normal $\boldsymbol{n}$. Body forces $\boldsymbol{b}$ act in $\mathcal{V}$, and surface tractions on $\mathcal{S}$.

are called surface forces $\boldsymbol{f}(\boldsymbol{x})$, where $\boldsymbol{f}$ is the force per unit area of a point $\boldsymbol{x}$ on the surface $\mathcal{S}$ of the body. Examples of body forces include the force of gravity. Surface forces could arise due to a fluid contacting the surface of the body, contact with other solids, etc.

The condition for static equilibrium of a body under the influence of body forces $\boldsymbol{b}(\boldsymbol{x})$ and surface forces $\boldsymbol{f}(\boldsymbol{x})$ is given by the Euler–Cauchy law (which is the generalization of Newton's law for point particles to a continuous body). The Euler–Cauchy law states that the distributed forces $\boldsymbol{b}(\boldsymbol{x})$ and $\boldsymbol{f}(\boldsymbol{x})$ will allow the body to be in static equilibrium only if the net force and net moment acting on the body vanish. Mathematically these conditions can be expressed as

$$\int_{\mathcal{V}} \boldsymbol{b}\, dV + \int_{\mathcal{S}} \boldsymbol{f}\, dS = \boldsymbol{0}, \tag{7.2}$$

and

$$\int_{\mathcal{V}} \boldsymbol{x} \times \boldsymbol{b}\, dV + \int_{\mathcal{S}} \boldsymbol{x} \times \boldsymbol{f}\, dS = \boldsymbol{0}, \tag{7.3}$$

where $\times$ denotes the cross product. Equation (7.2) represents the condition for the translational equilibrium, while Eq. (7.3) specifies conditions on the distributed forces that enable the body to be in rotational equilibrium.

A fundamental concept in continuum mechanics is that of *traction*. To illustrate this concept, a point P in the interior of the body is considered. The body is imagined to be cut into two parts $\mathcal{A}$ and $\mathcal{B}$ by a surface $\mathcal{S}_{\mathcal{AB}}$ that passes through the point P (see Fig. 7.6). The normal $\boldsymbol{n}$ is the outward normal to the part $\mathcal{A}$ of the body. On investigation of the equilibrium of the part $\mathcal{A}$ of the body, it is evident that some forces must be exerted on the this part of the body by the other part (part $\mathcal{B}$) through the surface $\mathcal{S}_{\mathcal{AB}}$. Similarly, the part $\mathcal{A}$ exerts an equal but opposite force on part $\mathcal{B}$. The nature of the forces that are transmitted across the surface $\mathcal{S}_{\mathcal{AB}}$ can be understood by considering a small patch of the surface $\mathcal{S}_{\mathcal{AB}}$ passing through the point P (see right side of Fig. 7.6), of area ΔS with normal $\boldsymbol{n}$. The part $\mathcal{B}$ of the body exerts a force $\Delta \boldsymbol{T}$ on this area element. The traction $\boldsymbol{t}$ is defined as

$$\boldsymbol{t} = \lim_{\Delta S \to 0} \frac{\Delta \boldsymbol{T}}{\Delta S}. \tag{7.4}$$

The traction vector $\boldsymbol{t}$, in general, depends on the surface that passes through the point P, and the relationship between the $\boldsymbol{t}$ and the surface can be complex. Cauchy postulated (later proved by Noll[10]) that the traction vector on a surface that passes through a point depends *only on the normal to the surface*. Expressed mathematically, Cauchy's principle states the traction depends on the normal to the surface through a function $\boldsymbol{t}(\boldsymbol{n})$, which satisfies the condition

$$\boldsymbol{t}(-\boldsymbol{n}) = -\boldsymbol{t}(\boldsymbol{n}). \tag{7.5}$$

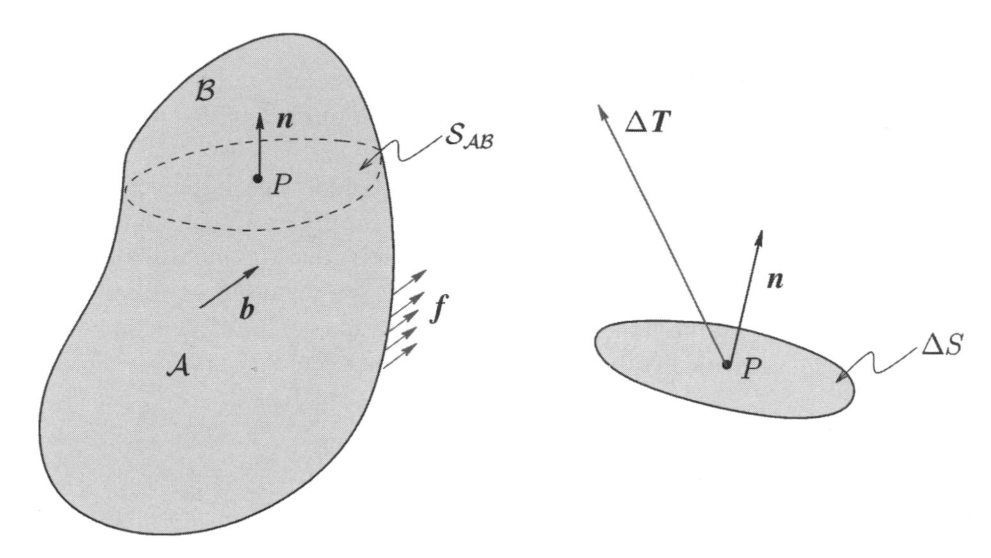

Figure 7.6 An imaginary surface $\mathcal{S}_{AB}$ passing through the point P in the body. The surface divides the body into two parts $\mathcal{A}$ and $\mathcal{B}$. The figure on the right shows a magnified view of a small patch of the surface that passes through the point P.

Cauchy further proved that if the traction depends *only* on the normal $\boldsymbol{n}$ and satisfies Eq. (7.5), then the relationship between the traction and the normal is linear, i.e.,

$$\boldsymbol{t}(\boldsymbol{n}) = \boldsymbol{\sigma}\boldsymbol{n}, \tag{7.6}$$

where $\boldsymbol{\sigma}$ is the stress tensor and $\boldsymbol{\sigma}\boldsymbol{n}$ represents the action of the tensor $\boldsymbol{\sigma}$ on the vector $\boldsymbol{n}$. The tensor $\boldsymbol{\sigma}$ is expressed in terms of a Cartesian basis as

$$\boldsymbol{\sigma} = \sigma_{ij}\boldsymbol{e}_i \otimes \boldsymbol{e}_j, \tag{7.7}$$

where σ_{ij} are the tensor components (summation convention over repeated indices is assumed). The symbol $\otimes$ stands for the tensor product as explained in the books already mentioned.[7,8] Thus, on application of distributed forces to a body, a stress tensor field $\boldsymbol{\sigma}(\boldsymbol{x})$ develops in the body.

It is evident from Eq. (7.6) that the stress tensor at the surface is related to the applied surface forces $\boldsymbol{f}$ via

$$\boldsymbol{\sigma}\boldsymbol{n} = \boldsymbol{f} \quad \text{on } \mathcal{S}, \tag{7.8}$$

where $\boldsymbol{n}$ is the outward normal to the body. On using Eq. (7.8) in Eq. (7.2) and insisting every part of the body be in equilibrium, the relationship between the distributed body forces and the stress tensor field is obtained as

$$\nabla \cdot \boldsymbol{\sigma} + \boldsymbol{b} = \boldsymbol{0} \quad \text{in } \mathcal{V}. \tag{7.9}$$

Also, the condition of rotational equilibrium provides a further condition on the stress tensor (in the absence of body moments)

$$\boldsymbol{\sigma} = \boldsymbol{\sigma}^T, \tag{7.10}$$

where $\boldsymbol{\sigma}^T$ is the transpose of the tensor given by

$$\boldsymbol{\sigma}^T = \sigma_{ji}\boldsymbol{e}_i \otimes \boldsymbol{e}_j. \tag{7.11}$$

Equations (7.8), (7.9), and (7.10) enforce the condition that the stress tensor field that develops in the body is in equilibrium with the applied distributed forces. These three relations, in fact, define a boundary value problem for the stress components σ_{ij}. The possibility of a unique solution in various spatial dimensions of this boundary value problem can be understood from the information collected in Table 7.1. It is evident that in dimensions higher than one, there is no possibility of a unique solution for the boundary value problem for the stress components obtained from the equilibrium conditions alone. Expressed in other words, *there are no statically determinate problems* in two and more spatial dimensions.

The stress tensor $\boldsymbol{\sigma}$ developed in this section is called the Cauchy stress tensor or the true stress tensor. Other stress measures are required with dealing with nonlinear problems such as the first and second Piola–Kirchoff stress tensors. Details regarding these may be found in the books by Chadwick[7] or Ogden.[8]

7.2.2 Kinematics: deformation and strain tensor

Development of continuum mechanics proceeds with the study of kinematics and deformation. Conceptually, kinematic quantities are distinct from dynamic quantities such as the stress tensor, i.e., kinematics is developed without any reference to the cause of deformation or strain.

Consider a continuous body occupying a region $\mathcal{V}$ enclosed by a surface $\mathcal{S}$ (see Fig. 7.7). Points P in the body are described by a position vector $\boldsymbol{x}$ with respect to an origin O, as shown in Fig. 7.7—this configuration of the body is called the *reference* or *undeformed configuration*. Due to some causes, the body deforms and occupies a new region $\mathcal{V}'$ enclosed by the surface $\mathcal{S}'$—a configuration called the *deformed configuration*. Every point in the undeformed configuration moves to a new point in the deformed configuration described by the position vector $\boldsymbol{y}$. In particular, the point P' is the point to which a material particle at point P in the un-

Table 7.1 Possibility of a unique solution to the boundary value problem for stresses in 1D, 2D, and 3D space.

Spatial Dimension	No. of Independent Stress Components	Number of Equations	Unique Solution Possible?
1	1	1	Yes
2	3	2	No
3	6	3	No

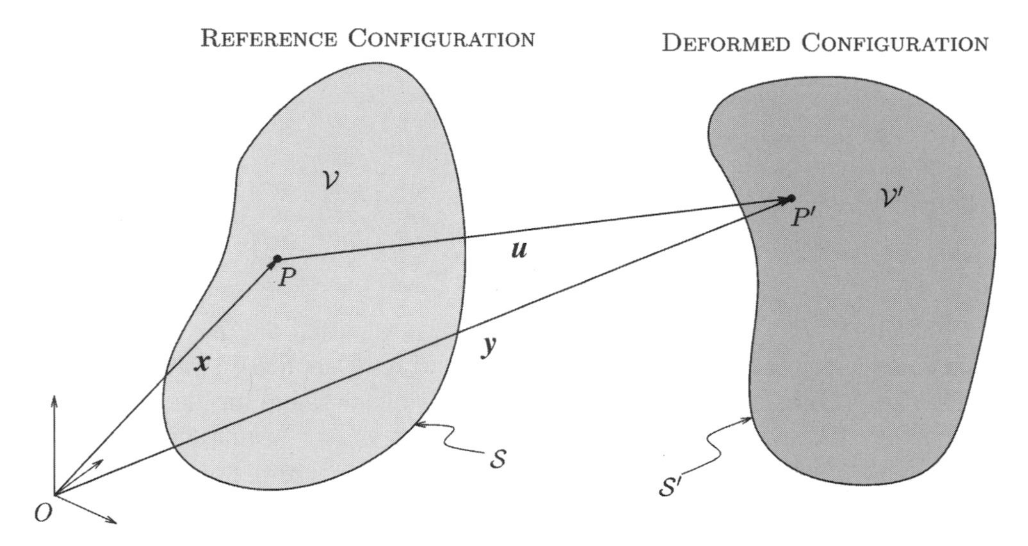

Figure 7.7 Deformation of a continuous body.

deformed configuration arrives after deformation. The mathematical description of deformation uses the idea of functions. The deformation function, more commonly known as the *deformation map*, is defined as

$$\boldsymbol{y} = \boldsymbol{y}(\boldsymbol{x}). \tag{7.12}$$

It prescribes a rule to obtain the image of any material point $\boldsymbol{x}$ in the reference configuration. It is assumed that the deformation map is well defined in that it preserves the topology of the body. Associated with the deformation map is the displacement field

$$\boldsymbol{u}(\boldsymbol{x}) = \boldsymbol{y}(\boldsymbol{x}) - \boldsymbol{x}, \tag{7.13}$$

which describes the displacement suffered by the material point at $\boldsymbol{x}$ in the reference configuration.

The deformation map of Eq. (7.12) is, in general, a nonlinear vector-valued function. A more "local" description is mathematically better managed. To this end, attention is focused on a point Q with position vector $\boldsymbol{x} + d\boldsymbol{x}$ in the neighborhood of the point P (whose position vector is $\boldsymbol{x}$) in the undeformed configuration (the vector that connects P to Q is the vector $d\boldsymbol{x}$, see Fig. 7.8). The point P maps to point P' under the deformation map [Eq. (7.7)], and the point Q maps to the point Q' in the neighborhood of P'. Since the position vector of P' is $\boldsymbol{y}$, the position vector of Q' is $\boldsymbol{y} + d\boldsymbol{y}$. It is evident that

$$\begin{aligned} d\boldsymbol{y} = \boldsymbol{y}(\boldsymbol{x} + d\boldsymbol{x}) - \boldsymbol{y}(\boldsymbol{x}) &\approx \underbrace{\frac{\partial \boldsymbol{y}}{\partial \boldsymbol{x}}}_{\boldsymbol{F}} d\boldsymbol{x} \\ \Longrightarrow \quad d\boldsymbol{y} &= \boldsymbol{F} d\boldsymbol{x}, \end{aligned} \tag{7.14}$$

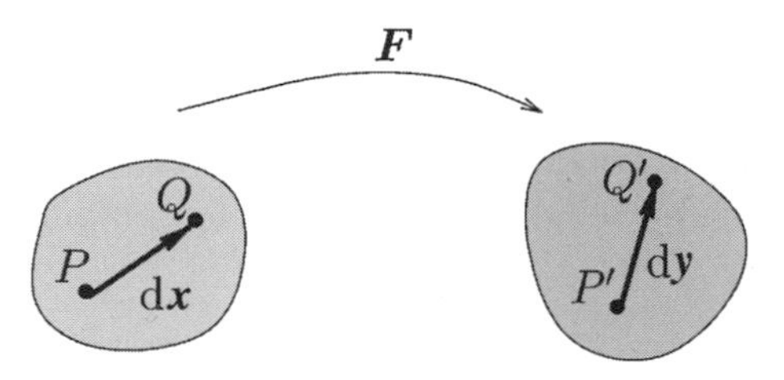

Figure 7.8 Deformation of a local neighborhood of a point P.

where $\boldsymbol{F}$ is the *gradient of deformation tensor* or the *deformation gradient tensor*. The gradient of deformation tensor at a point in the body describes the image of a "small" line element $d\boldsymbol{x}$ originating from that point. This tensor is related to the deformation map via

$$\boldsymbol{F} = \nabla \boldsymbol{y} = \nabla(\boldsymbol{x} + \boldsymbol{u}) = \boldsymbol{I} + \nabla \boldsymbol{u}. \tag{7.15}$$

Deformation maps that are generally considered in continuum mechanics satisfy the condition

$$\det \boldsymbol{F} > 0 \tag{7.16}$$

everywhere in the body, where $\det \boldsymbol{F}$ is determinant of the tensor $\boldsymbol{F}$.

The concept of strain is now introduced. The Green–Lagrange strain tensor $\boldsymbol{E}$ at a point defines the change in the squares of lengths of every material fiber that originates at the point by

$$\begin{aligned} |d\boldsymbol{y}|^2 - |d\boldsymbol{x}|^2 &= d\boldsymbol{x} \cdot \boldsymbol{F}^T \boldsymbol{F}\, d\boldsymbol{x} - \boldsymbol{x} \cdot \boldsymbol{I}\boldsymbol{x} \\ &= 2\, d\boldsymbol{x} \cdot \underbrace{\frac{1}{2}(\boldsymbol{F}^T \boldsymbol{F} - \boldsymbol{I})}_{\boldsymbol{E}}\, d\boldsymbol{x} \\ \Longrightarrow \quad |d\boldsymbol{y}|^2 - |d\boldsymbol{x}|^2 &= 2\, d\boldsymbol{x} \cdot \boldsymbol{E}\, d\boldsymbol{x}, \end{aligned} \tag{7.17}$$

where $\cdot$ stands for the dot product. The Green–Lagrange strain tensor can be written in terms of displacement fields as

$$\boldsymbol{E} = \frac{1}{2}\left(\nabla \boldsymbol{u} + \nabla \boldsymbol{u}^T + \nabla \boldsymbol{u}^T \nabla \boldsymbol{u}\right). \tag{7.18}$$

It is clear that the Green–Lagrange strain tensor has a nonlinear dependence on the gradient of displacement. Just as for the stress tensor, there are several other measures of strain, depending on the configuration in which they are defined. These are considered in great detail by Ogden.[8]

When the deformation is "not too severe," a condition that is mathematically characterized by $|\nabla \boldsymbol{u}| \ll 1$, the Green–Lagrange strain tensor can be linearized as

$$\boldsymbol{E} \approx \boldsymbol{\epsilon} = \frac{1}{2}(\nabla \boldsymbol{u} + \nabla \boldsymbol{u}^T) = \mathrm{sym}(\nabla \boldsymbol{u}), \tag{7.19}$$

where $\boldsymbol{\epsilon}$ is the "small-strain" tensor (called simply "strain tensor" in this chapter), which is equal to the symmetric part of the gradient of displacements as denoted by "sym." The small strain tensor is the tensorial generalization of the elementary concept of strain as "change in length by original length"; indeed

$$\frac{|d\boldsymbol{y}| - |d\boldsymbol{x}|}{|d\boldsymbol{x}|} = \frac{d\boldsymbol{x} \cdot \boldsymbol{\epsilon}\, d\boldsymbol{x}}{|d\boldsymbol{x}|^2}. \tag{7.20}$$

The antisymmetric part of the gradient of displacements is called the small rotation tensor

$$\boldsymbol{\omega} = \frac{1}{2}(\nabla \boldsymbol{u} - \nabla \boldsymbol{u}^T) = \text{asym}(\nabla \boldsymbol{u}). \tag{7.21}$$

There are six independent components for the strain tensor. These components cannot be specified independently [since the strains are related to displacements via Eq. (7.19)], and satisfy the compatibility equation (expressed in indicial notation)

$$\epsilon_{ij,kl} + \epsilon_{kl,ij} - \epsilon_{ik,jl} - \epsilon_{jl,ik} = 0. \tag{7.22}$$

7.2.3 Principle of virtual work

The principle of virtual work is a means to state the ideas of equilibrium and geometric compatibility under a single principle. The principle hinges on two key ideas. First, the set of three fields $\{\boldsymbol{b}(\boldsymbol{x}), \boldsymbol{f}(\boldsymbol{x}), \boldsymbol{\sigma}(\boldsymbol{x})\}$ is said to be *statically admissible state of stress* (SASS), if the conditions of Eqs. (7.8), (7.9), and (7.10) are satisfied. Second, the set of three fields $\{\boldsymbol{u}^0(\boldsymbol{x}), \boldsymbol{u}(\boldsymbol{x}), \boldsymbol{\epsilon}(\boldsymbol{x})\}$, where $\boldsymbol{u}^0(\boldsymbol{x})$ is a vector field of displacements specified on the surface of the body, is said to be a *kinematically admissible state of strain* (KASS) if $\boldsymbol{u}(\boldsymbol{x}) = \boldsymbol{u}^0(\boldsymbol{x})$ on $\mathcal{S}$ and $\boldsymbol{u}(\boldsymbol{x})$ and $\boldsymbol{\epsilon}(\boldsymbol{x})$ satisfy Eq. (7.19).

The principle of virtual work states that any given SASS and KASS will satisfy the virtual work equation

$$\int_V \boldsymbol{\sigma} : \boldsymbol{\epsilon}\, dV = \int_V \boldsymbol{b} \cdot \boldsymbol{u}\, dV + \int_S \boldsymbol{f} \cdot \boldsymbol{u}\, dS. \tag{7.23}$$

The significance of this principle is best understood from the following two theorems.

Theorem of Equilibrium: A state of stress defined by $\{\boldsymbol{b}(\boldsymbol{x}), \boldsymbol{f}(\boldsymbol{x}), \boldsymbol{\sigma}(\boldsymbol{x})\}$ is an equilibrium state (i.e., an SASS), if it satisfies the virtual work Eq. (7.23) for *every* KASS defined on the body.

Theorem of Compatibility: A state of deformation defined by $\{\boldsymbol{u}^0(\boldsymbol{x}), \boldsymbol{u}(\boldsymbol{x}), \boldsymbol{\epsilon}(\boldsymbol{x})\}$ is a geometrically compatible state (i.e., a KASS), if it satisfies the virtual work Eq. (7.23) for *every* SASS defined on the body.

The principle of virtual work in the form of the theorem of equilibrium is used widely to construct numerical methods, an example being the finite element method briefly discussed in Sec. 7.2.5. Although the discussion here has been based on a linear formulation, the principle of virtual work is applicable to fully nonlinear situations as well.

7.2.4 Constitutive relations

The theory thus far has introduced (in three spatial dimensions) six unknown components of the stress tensor, six components of the strain tensor, and three displacement components—a total of 15 quantities. In all, the equations available to solve these are nine, three equilibrium equations [Eq. (7.9)] and six strain-displacement relations [Eq. (7.19)]. Additional relations are necessary for the solution of the unknowns. These are material-specific relations called *constitutive relations* to relate the stress tensor to the strain tensor. The study of constitutive relations is a vast one; and the account presented here is brief.

The constitutive relations are based on three general principles.[7] First, the principle of determinism states that the stress at any point in the body is determined uniquely by the entire history of deformation. Second, the principle of local action states that the stress at any point is determined only by the strain history at that point, but not by that of any neighboring point. Finally, according to the principle of objectivity, the constitutive relation must provide for the same material response for equivalent observers. While the constitutive equations of macroscopic continuum mechanics are developed based on these principles, the principle of local action has only a limited validity at the nanoscales. Nonlocal constitutive relations will be discussed in Sec. 7.4.2.

Present attention is restricted to elastic constitutive relations that satisfy the principle of local action. A material is said to be hyperelastic if there exists a scalar function $W(\epsilon)$ such that the stress is given by

$$\sigma_{ij} = \frac{\partial W}{\partial \epsilon_{ij}}. \tag{7.24}$$

A material is said to be linearly hyperelastic if there is a fourth-order tensor C_{ijkl} called the elastic modulus tensor such that

$$W(\epsilon) = \frac{1}{2} C_{ijkl} \epsilon_{ij} \epsilon_{kl}, \tag{7.25}$$

and

$$\sigma_{ij} = C_{ijkl} \epsilon_{kl}. \tag{7.26}$$

The elastic modulus tensor has the symmetries

$$C_{jikl} = C_{ijkl}, \tag{7.27}$$

$$C_{ijlk} = C_{ijkl}, \tag{7.28}$$

$$C_{klij} = C_{ijkl}, \tag{7.29}$$

which reduce the number of independent components to 21.

Further reduction in the number of elastic constants is brought about by specific symmetries of the solid. The number of independent elastic constants for solids made of different crystal classes[9] is given in Table 7.2.

7.2.5 Boundary value problems and finite element method

The key ideas that enable the definition of a boundary value problem (namely, equilibrium, compatibility, and constitutive relations) have been described in the previous section. The aim is the determination of the stress, strain, and displacement fields that develop in the body in response to applied forces and displacement constraints.

The body occupies a region $\mathcal{V}$ (as shown in Fig. 7.5) enclosed by the surface $\mathcal{S}$. The surface $\mathcal{S}$ is split into two disjoint parts $\mathcal{S}_\sigma$ on which distributed surface forces are applied, and $\mathcal{S}_u$ on which displacements $\boldsymbol{u}^0$ are prescribed. The complete statement of the boundary value problem (for linear elastic materials) is

$$\sigma_{ij,j} + b_i = 0 \quad \text{(Equilibrium)}, \tag{7.30}$$

$$\epsilon_{ij} - \frac{1}{2}(u_{i,j} + u_{j,i}) = 0 \quad \text{(Strain displacement)}, \tag{7.31}$$

$$\sigma_{ij} = C_{ijkl}\epsilon_{kl} \quad \text{(Constitutive relations)}, \tag{7.32}$$

with boundary conditions on $\mathcal{S}$ being

$$\sigma_{ij}n_j = f_i \quad \text{on } \mathcal{S}_\sigma \tag{7.33}$$

$$u_i = u_i^0 \quad \text{on } \mathcal{S}_u. \tag{7.34}$$

Even for the case of the linear elastic boundary-value problem presented here, analytical solutions are impossible in all but special cases. Powerful numerical methods have been developed to solve the class of elliptic boundary-value problems that arise in linear elasticity, using digital computers. The finite element method has especially found favor in solid mechanics owing to its advantages both as an approximation scheme for the fields and for representation of the geometry of the solid. There are several excellent accounts of the finite element methods (for example, Hughes[11] or Zienkiewicz[12]), and a brief outline is presented here.

Table 7.2 Number of independent elastic constants for different crystal classes.

Crystal	No. of Independent Elastic Constants
Triclinic	21
Monoclinic	13
Orthorhombic	9
Tetragonal	7
Rhombohedral	7
Hexagonal	5
Cubic	3

The first step in the finite element method is the discretization of the domain of definition of the boundary value problem. This is achieved by selecting a set of points called *nodes*. The nodes are used to generate elements such that the resulting ensemble of nodes and elements closely approximates geometry of the body under consideration. One possible route is to select nodes and use a Delaunay tetrahedrization (in three dimensions) or triangulation (in two dimensions)[13] to generate the elements. A collection of nodes and elements is called a finite element mesh (see Fig. 7.9).

Associated with the node a is a *shape function* $N_a(\boldsymbol{x})$, which satisfies the condition

$$N_a(\boldsymbol{x}_b) = \delta_{ab}, \tag{7.35}$$

where $\boldsymbol{x}_b$ is the position of node b and δ_{ab} is the Kronecker delta symbol ($\delta_{ab} = 1$ if $a = b$ and vanishes if $a \neq b$). In particular, $N_a(\boldsymbol{x})$ are chosen such that they vanish in all elements for which a is not a node. There are several possible choices for N_a, a common choice being a piecewise linear dependence on $\boldsymbol{x}$ (see Fig. 7.9 for an illustration in 1D space). The essential idea of the finite element method is to use the shape functions to construct an approximate form of the displacement field

$$\boldsymbol{u}^h(\boldsymbol{x}) = \sum_a N_a(\boldsymbol{x})\boldsymbol{u}_a, \tag{7.36}$$

where $\boldsymbol{u}^h$ represents an approximate solution and $\boldsymbol{u}_a$ are the values of nodal displacements to be determined. Thus, the finite element method reduces the determination of the displacement field to the determination of a finite number of unknown quantities. Once the nodal displacements are determined, the approximate value of the displacement can be obtained at any point in the body using Eq. (7.36). It is advantageous to represent Eq. (7.36) in matrix form as

$$\boldsymbol{u}^h(\boldsymbol{x}) = \{\boldsymbol{N}\}^T\{\boldsymbol{U}\}, \tag{7.37}$$

where $\{\boldsymbol{N}\}^T$ is the array of element shape functions (the elements of which are spatially dependent functions) and $\{\boldsymbol{U}\}$ is the array of nodal displacements $\boldsymbol{u}_a$. The

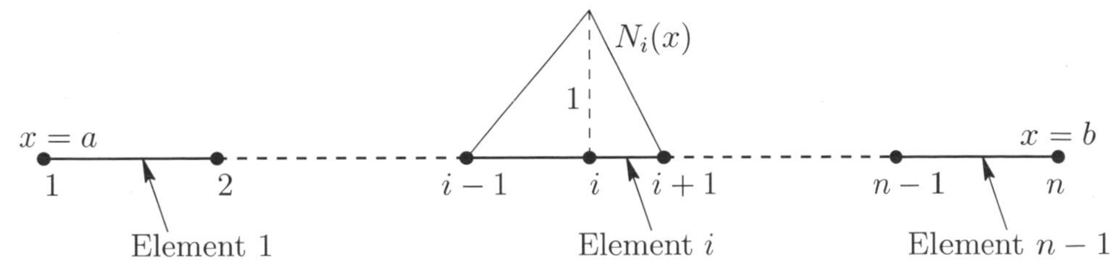

Figure 7.9 A (1D) illustration of a finite element mesh for the interval $[a, b]$ with n nodes and $n-1$ elements; N_i is the piecewise linear shape function centered around the node i.

strains due to the approximate displacement field can be expressed as

$$\boldsymbol{\epsilon}^h(\boldsymbol{x}) = [\boldsymbol{B}]\{\boldsymbol{U}\}, \tag{7.38}$$

where $[\boldsymbol{B}]$ is the strain-displacement matrix whose elements are spatial gradients of the shape functions. Equation (7.38) ensures that the approximate strain fields are related to the approximate displacement fields via Eq. (7.19). The stresses corresponding to these strain fields can be obtained using the constitutive relations

$$\boldsymbol{\sigma}^h(\boldsymbol{x}) = \boldsymbol{C}[\boldsymbol{B}]\{\boldsymbol{U}\}, \tag{7.39}$$

where $\boldsymbol{C}$ is the constitutive matrix.

The strategy to determine the unknown nodal displacements $\{\boldsymbol{U}\}$ is to enforce the condition that the approximate stress field [Eq. (7.39)] is an equilibrium field. This is achieved by exploiting the theorem of equilibrium version of the principle of virtual work (see Sec. 7.2.3). All possible KASS on the body are defined using the finite element approximation as

$$\boldsymbol{v}^h(\boldsymbol{x}) = \{\boldsymbol{N}\}^T\{\boldsymbol{V}\}, \tag{7.40}$$

where $\{\boldsymbol{V}\}$ is an array of arbitrary nodal displacements. The strain field $\boldsymbol{e}^h(\boldsymbol{x})$ defined by

$$\boldsymbol{e}^h(\boldsymbol{x}) = [\boldsymbol{B}]\{\boldsymbol{V}\}, \tag{7.41}$$

and the displacement field $\boldsymbol{v}^h(\boldsymbol{x})$ form a KASS. Thus, if the virtual work Eq. (7.23)

$$\int_{\mathcal{V}} \boldsymbol{e}^h(\boldsymbol{x}) : \boldsymbol{\sigma}^h(\boldsymbol{x})\, dV = \int_{\mathcal{V}} \boldsymbol{v}^h \cdot \boldsymbol{b}\, dV + \int_{\mathcal{S}} \boldsymbol{v}^h \cdot \boldsymbol{f}\, dS, \tag{7.42}$$

for every choice of $\{\boldsymbol{V}\}$ (i.e., for every KASS) where $\boldsymbol{e} : \boldsymbol{\sigma} = e_{ij}\sigma_{ij}$, then the unknown nodal displacements $\{\boldsymbol{U}\}$ satisfy the condition

$$[\boldsymbol{K}]\{\boldsymbol{U}\} = \{\boldsymbol{P}\}, \tag{7.43}$$

where

$$[\boldsymbol{K}] = \int_{\mathcal{V}} [\boldsymbol{B}]^T \boldsymbol{C}[\boldsymbol{B}]\, dV \tag{7.44}$$

is the called the *stiffness matrix*, and

$$\{\boldsymbol{P}\} = \int_{\mathcal{V}} \{\boldsymbol{N}\}\boldsymbol{b}\, dV + \int_{\mathcal{S}} \{\boldsymbol{N}\}\boldsymbol{f}\, dS \tag{7.45}$$

is the *force vector*. Although the expressions for the stiffness matrix and the force vector involve integrals over the entire body, these integrals can be efficiently evaluated element-wise, and the resulting stiffness matrix is generally sparse. This allows the use of fast solvers to obtain numerical solutions of Eq. (7.43).

The presentation here of the finite element method is restricted to linear elastic problems. The method, however, is capable of treating strongly nonlinear problems in solids and fluids.[12] An example of the application of the finite element method to a nonlinear problem of crack spalling in a misfitting epitaxial film-substrate system is shown in Fig. 7.10.

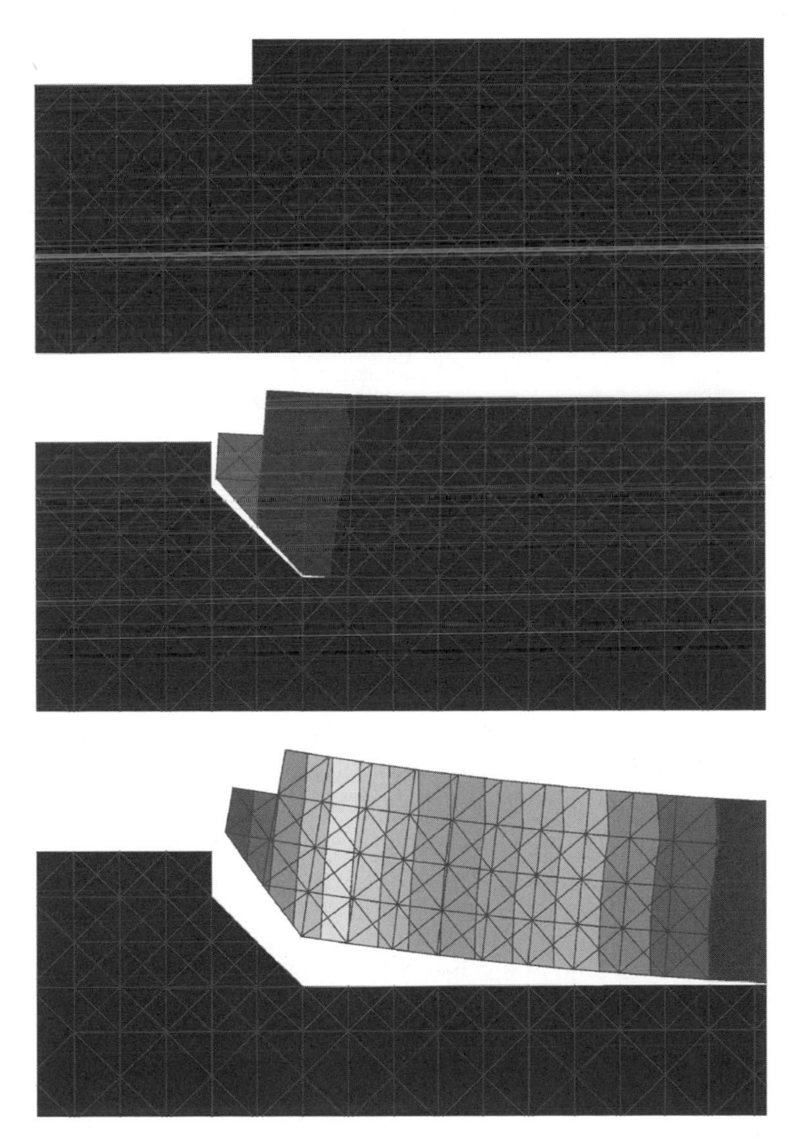

Figure 7.10 Finite element analysis of crack spalling in a misfitting epitaxial film-substrate system.

7.3 Atomistic models

Atomistic models explicitly acknowledge the discrete nature of matter, and the degrees of freedom of these models are the positions of the atoms that make up the material or structure of interest. The most important ingredient of atomistic models is the total energy function, i.e., a description to obtain the total energy of the system when the positions of the atoms are known. Once this function is known, standard methods of statistical mechanics and/or molecular dynamics are used to obtain the macroscopic properties of the material. These methods today have reached a level of sophistication that no input other than some fundamental constants (electron mass, Planck's constant, etc.) are required to predict properties, i.e., atomistic models promise to be truly predictive tools. This section is divided into two parts. In the first part, methods for calculating the total energy of the system are described, and simulation techniques (which require a total energy description) are presented in the second.

7.3.1 Total energy description

The goal of this section is to present methods that enable the calculation of the potential energy E_{tot} of a collection of atoms given their positions. A collection of N atoms is considered such that the position of the ith atom is given by $\boldsymbol{x}_i$, and the function $E_{\mathrm{tot}}(\boldsymbol{x}_1, \ldots, \boldsymbol{x}_N)$ is desired. This potential energy, in reality, is dependent on the bonding between the atoms (i.e., it depends on the electronic states in the atomic system) and falls in the realm of quantum mechanics. While such methods based on quantum mechanics have provided some of the most accurate predictive models, these are computationally expensive. There are other approaches that do not treat the electronic states explicitly (during the simulation), but the function $E_{\mathrm{tot}}(\boldsymbol{x}_1, \ldots, \boldsymbol{x}_N)$ is derived by approximate methods from the underlying quantum mechanics, or by recourse to empirical methods. Both types of methods, i.e., the ones that take recourse to explicit treatment of electrons or otherwise, are discussed. Examples of the former include density functional theory and tight-binding methods, while the latter include pair potentials, the embedded atom method, etc.

7.3.1.1 Quantum mechanical methods with explicit treatment of electrons

The basic relation of quantum mechanics** for a particle of mass m moving in a potential $V(\boldsymbol{r})$ is given by the Schrödinger equation

$$\mathcal{H}\psi = \mathbf{i}\hbar\frac{\partial\psi}{\partial t}, \tag{7.46}$$

**Analogous to Newton's law in classical mechanics; excellent treatment of quantum mechanics can be found in books by Schiff[14] or Sakurai.[15]

where $\boldsymbol{r}$ is the position vector, $\psi(\boldsymbol{r}, t)$ is the wave function of the particle,

$$\mathcal{H} = -\frac{\hbar^2}{2m_e}\nabla^2 + V(\boldsymbol{r}) \tag{7.47}$$

is the Hamiltonian operator, and $\hbar$ is Planck's constant. The wave function $\psi(\boldsymbol{r}, t)$ describes the state of the particle; and, in particular, there are states called *stationary states*, which satisfy the relationship

$$\mathcal{H}\psi = E\psi, \tag{7.48}$$

where E is the energy eigenvalue.

The body of interest is taken to consist of N ions and M electrons. The positions of the ions are $\{\boldsymbol{x}_i\}$ and those electrons are $\{\boldsymbol{r}_j\}$ The electronic states in this system (under the so-called Born–Oppenheimer approximation, see Szabo and Ostlund[16]) are described by the Hamiltonian operator

$$\mathcal{H} = \sum_{j=1}^{M} -\frac{\hbar^2}{2m_e}\nabla_j^2 + \frac{1}{2}\frac{e^2}{4\pi\epsilon_0}\sum_{jk}\frac{1}{|\boldsymbol{r}_j - \boldsymbol{r}_k|} - \sum_{j=1}^{M}\sum_{i=1}^{N}\frac{Z_i e^2}{4\pi\epsilon_0}\frac{1}{|\boldsymbol{r}_j - \boldsymbol{x}_i|}, \tag{7.49}$$

where ϵ_0 is the permittivity of free space, and Z_i is the atomic number of the ith nucleus. The stationary state of this system is given by

$$\mathcal{H}\Psi = E\Psi, \tag{7.50}$$

where $\Psi(\boldsymbol{r}_1, \ldots, \boldsymbol{r}_M)$ is the many-electron wave function, and E is the energy eigenvalue. The first term in the Hamiltonian of Eq. (7.49) is the kinetic energy of the electrons, the second term denotes the Coulombic interaction among the electrons, and the third term involves the interaction between the electrons and the ions. The state of the system Ψ_g with the lowest energy eigenvalue E_g that satisfies Eq. (7.50) is called the ground state. Clearly, the ground state energy depends parametrically on the positions $\{\boldsymbol{x}_i\}$ of the ions. The total potential energy of the system can be calculated when the ground state energy is known:

$$E_{\text{tot}}(\boldsymbol{x}_1, \ldots, \boldsymbol{x}_N) = E_g(\boldsymbol{x}_1, \ldots, \boldsymbol{x}_N) + \frac{1}{2}\sum_{i \neq j} V_{I-I}(|\boldsymbol{x}_i - \boldsymbol{x}_j|), \tag{7.51}$$

where V_{I-I} is the direct Coulombic interaction between the ions. Given an atomistic system, the determination of total energy as a function of the positions of the atoms requires the solution of Eq. (7.50). Analytical solution of Eq. (7.50) is seldom possible for more than one electron.

The mathematical difficulties encountered in the solution of Eq. (7.50) have prompted the development of several approximation techniques. The earliest of the approximate methods is due to Hartree and was later modified by Fock; it is now commonly known as the Hartree–Fock method.[17,18] Other approximate methods include the tight-binding method discussed in Sec. 7.3.1.3. Density functional theory, one of the most accurate methods from the point of view of modeling, is discussed in the next section. This method falls in the class of *ab initio* methods, in that only fundamental physical parameters such as electron mass and Planck's constant are the necessary inputs.

7.3.1.2 *Ab initio* density functional theory

A breakthrough in the calculation of the ground state energy of the many-electron system was achieved by Hohenberg and Kohn,[19] who proved a theorem that is the basis of density functional theory. The Hohenberg–Kohn theorem states that the ground state energy E_g of the many-electron system is a functional of the electron density $n(\boldsymbol{r})$. The electron density is related to the many-body wave function via

$$n(\boldsymbol{r}) = M \int |\Psi(\boldsymbol{r}, \boldsymbol{r}_2, \ldots, \boldsymbol{r}_M)|^2 d\boldsymbol{r}_2 \ldots d\boldsymbol{r}_M. \tag{7.52}$$

The electron density $n(\boldsymbol{r})$ plays a fundamental role in density functional theory in that, once the ground state density is specified, the ground state energy, the ground state wave function and even the Hamiltonian (up to a constant) are uniquely determined.

The Hohenberg–Kohn functional that determines the ground state energy is typically written as

$$E[n(\boldsymbol{r})] = T[n(\boldsymbol{r})] + U[n(\boldsymbol{r})] + V[n(\boldsymbol{r})], \tag{7.53}$$

where $T[n(\boldsymbol{r})]$ and $U[n(\boldsymbol{r})]$ are universal functions that represent the kinetic and Coulombic energies of the electrons, and $V[n(\boldsymbol{r})]$ is the functional defined as

$$V[n(\boldsymbol{r})] = \int n(\boldsymbol{r}) v(\boldsymbol{r})\, dV, \tag{7.54}$$

where $v(\boldsymbol{r})$ is the potential due to the ions. The function $n(\boldsymbol{r})$ that minimizes Eq. (7.53) is the ground state electron density, and the minimum value of the functional is the ground state energy. The potential $v(\boldsymbol{r})$ defined in Eq. (7.54) depends parametrically on the position of the ions, and the dependence of the ground state energy on the positions of the ions can be obtained by the minimization of Eq. (7.53). A difficulty arises at this stage in that the universal functionals $T[n(\boldsymbol{r})]$ and $U[n(\boldsymbol{r})]$ are not explicitly known.

Kohn and Sham[20] overcame this difficulty by introducing a single particle state defined by wave functions $\phi_i(\boldsymbol{r})$ as if the the collection of electrons are noninteracting. The electronic density is related to the single particle states via

$$n(\boldsymbol{r}) = \sum_{i=1}^{M'} |\phi_i|^2, \tag{7.55}$$

where M' is a number related to M determined by the fact that each single-particle state can accommodate two electrons of opposite spin. In terms of the single-particle states, the kinetic energy functional is approximated as

$$T_s[n] = -\frac{\hbar^2}{2m_e}\sum_{i=1}^{M'}\int \phi_i^*(\boldsymbol{r})\nabla^2\phi_i(\boldsymbol{r}). \tag{7.56}$$

This form of the kinetic energy functional neglects the correlations between electrons. The Coulombic interaction between the electrons is approximated using the Hartree approximation U_H [with Eq. (7.55) for the density] as

$$U_H[n] = \frac{1}{2}\frac{e^2}{4\pi\epsilon_0}\int\int \frac{n(\boldsymbol{r})n(\boldsymbol{r}')}{|\boldsymbol{r}-\boldsymbol{r}'|}\,dV\,dV'. \tag{7.57}$$

The Hartree approximation for the Coulombic interaction energy between the electrons neglects the exchange effects, i.e., the change in the electrostatic energies that arise due to the fact that electrons of the same spin are separated in space. The effects of electron correlations and exchange are clubbed together in one functional called the exchange-correlation functional $E_{\text{xc}}[n]$. The main approximation that allows density functional theory to be a viable method is the local-density approximation (LDA)

$$E_{\text{xc}}[n] = \int e_{\text{xc}}(\boldsymbol{r})n(\boldsymbol{r})\,dV, \tag{7.58}$$

where $e_{\text{xc}}(\boldsymbol{r})$ is the value of the the exchange-correlation energy in an electron gas with homogeneous density $n(\boldsymbol{r})$. There are several possible choices[21] for $e_{\text{xc}}(\boldsymbol{r})$. With the introduction of the single particle orbitals and the local density approximation, the energy functional of Eq. (7.53) reduces to

$$E[n] = T_s[n] + U_H[n] + E_{\text{xc}}[n] + V[n]. \tag{7.59}$$

Minimization of the functional with respect to the density n [keeping Eq. (7.55) in view] results in

$$-\frac{\hbar^2}{2m_e}\nabla^2\phi_i + v_{\text{eff}}(\boldsymbol{r})\phi_i = \epsilon_i\phi_i, \tag{7.60}$$

where v_{eff} is a single-particle effective potential that contains the Hartree potential, exchange-correlation potential and the potential from the ions [defined in Eq. (7.54)], i.e.,

$$v_{\text{eff}} = v_H + v_{\text{xc}} + v. \tag{7.61}$$

The terms $v_H = \delta U_H/\delta n$ and $v_{xc} = \delta E_{xc}/\delta n$ depend on the density. Equation (7.60) for the single particle states of Eq. (7.60) is therefore nonlinear and solved in a self-consistent fashion.[21] Once the single-particle levels are obtained, the ground state energy, which is the total energy that depends on the positions of the ions via $v(\boldsymbol{r})$ defined in Eq. (7.54), can be obtained as

$$E_g(\boldsymbol{x}_1, \ldots, \boldsymbol{x}_N) = \sum_{i=1}^{M'} \epsilon_i - \frac{1}{2}\frac{e^2}{4\pi\epsilon_0} \int\int \frac{n(\boldsymbol{r})n(\boldsymbol{r}')}{|\boldsymbol{r} - \boldsymbol{r}'|} dV\, dV' + \int n(\boldsymbol{r})[e_{xc}(\boldsymbol{r}) - v_{xc}(\boldsymbol{r})]\, dV. \tag{7.62}$$

The practical implementation of density functional theory involves many techniques that have been developed with years of experience.[21] One of the standard methods is to choose a plane wave basis to construct the single particle wave functions. As the consideration of all electrons involved in the solid can be prohibitive, an alternative approach is to replace the bare ionic potential by a pseudo-potential that accounts for the core electrons. There are also several corrections and improvements to local density approximations, a short review of which may be found in a paper by Capelle.[22]

Density functional theory has found wide use in computer-aided materials design and nanomechanics. For example, it has been used to understand the mechanics of carbon nanotubes; Fig. 7.11 shows the results of density functional theory calculations[23] that predict symmetry-driven phase transitions in bundles of carbon nanotubes. Another example[24] of its use is the determination of the ultimate tensile strength of MoSe nanowires (Fig. 7.12). An added advantage of using density functional theory is that the electronic properties can also be simultaneously determined, as shown in Fig. 7.13 for MoSe nanowires.

The main drawback of the density functional theory is the large computational resources necessary to carry out meaningful simulations. With the advent of faster computers, ever larger problems will be addressed with density functional theory. In the view of the author, density functional theory will be one of the main tools for theoretical nanomechanics in the years to come.

7.3.1.3 Tight-binding method

Tight-binding approaches to calculating the total energies of a many-electron system are another example of methods that provide for an explicit treatment of the electronic states. The tight-binding method was pioneered by Slater and Koster.[25] The basic idea of the tight-binding method is the same as that of the Rayleigh–Ritz method that is used to determine vibrational frequencies of structures.[12]

The model consists of ions located at $(\boldsymbol{x}_i, \ldots, \boldsymbol{x}_N)$ and M electrons (not all electrons need to be considered, many calculations only treat the valance electrons), and the first step in the model consists of choosing a basis set for describing the single particle electronic states. Usually, the basis set is made up of

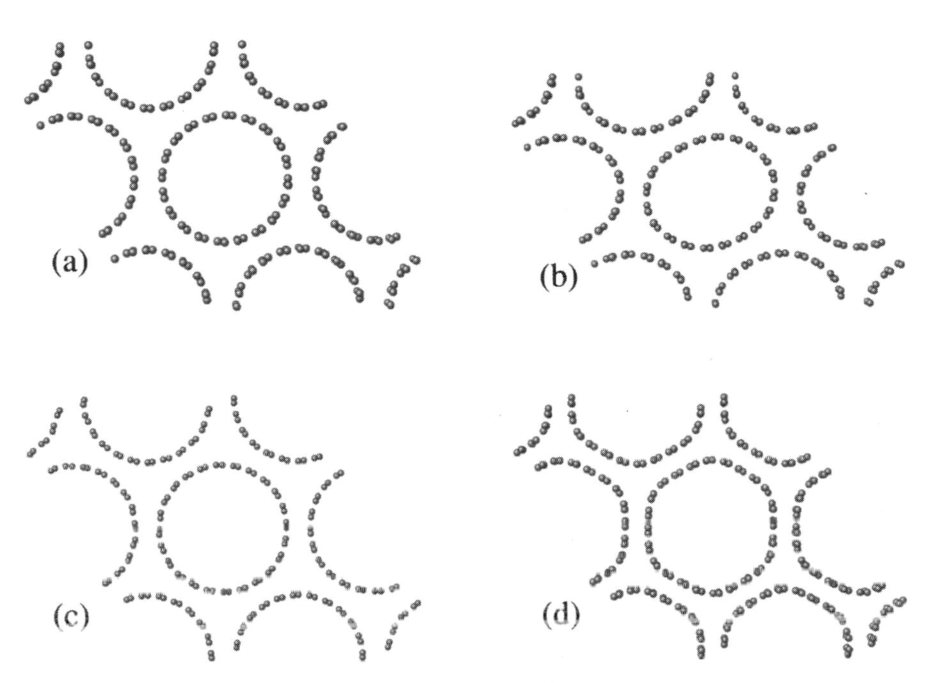

Figure 7.11 Geometric phase transitions in bundles of carbon nanotubes under hydrostatic pressure simulated using density functional theory: (a) (10,10) single-walled carbon nanotube bundle under zero hydrostatic stress, (b) monoclinic structure with elliptic cross section under 2 GPa external pressure, (c) (12,12) single-walled tube under zero stress, and (d) nanotubes showing polygonalization under a pressure of 6 GPa. (Reprinted with permission from Ref. 23, © 2002 The American Physical Society.)

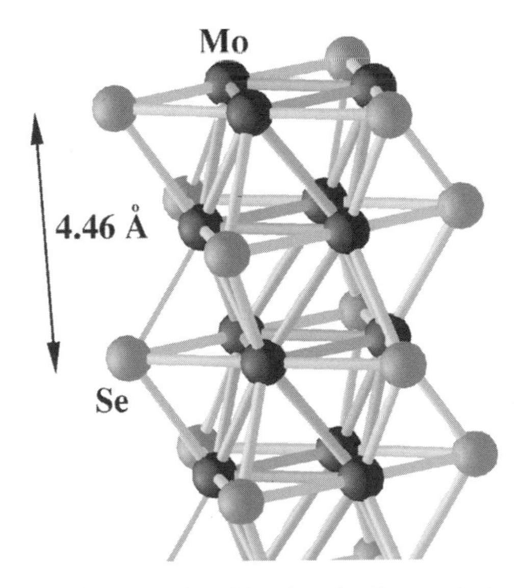

Figure 7.12 Model of a MoSe nanowire. (Reprinted with permission from Ref. 24, © 2002 The American Physical Society.)

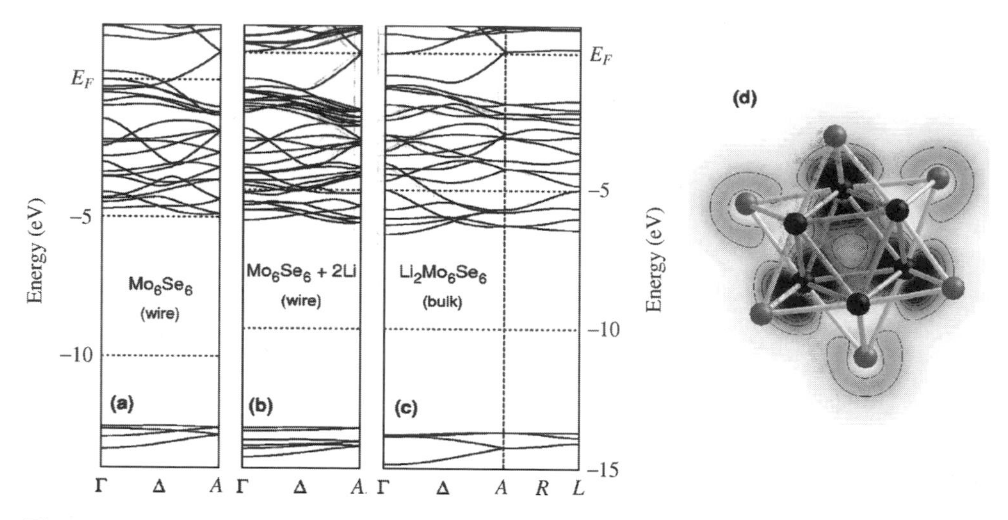

Figure 7.13 Electronic structure of MoSe nanowire: (a) band structure of MoSe nanowire; (b) and (c) band structure with Li adsorbed; (d) charge density plot. (Reprinted with permission from Ref. 24, © 2002 The American Physical Society.)

atomic orbitals. For example, to study the diamond structure of carbon, a possible choice of the basis set would be $\{2s, 2p_x, 2p_y, 2p_z\}$ centered at each carbon atom in the model. In general, a basis set is assumed $\{\phi_\alpha(\boldsymbol{r} - \boldsymbol{x}_i)\}$, $(\alpha = 1, \ldots, B$, $i = 1, \ldots, N)$, where α is the index of the atomic orbital, i.e., the basis set consists of atomic orbitals centered at each of the atoms. Single particle stationary states $\psi(\boldsymbol{r})$ are expressed as a linear combination of atomic orbitals (LCAO) as

$$\psi(\boldsymbol{r}) = \sum_{i=1}^{N} \sum_{\alpha=1}^{B} c_{i\alpha} \phi_\alpha(\boldsymbol{r} - \boldsymbol{x}_i), \tag{7.63}$$

where $c_{i\alpha}$ are constants yet to be determined.

If $\mathcal{H}$ is the Hamiltonian operator of the system, then the single-particle states are stationary states of functional

$$\int \psi^* \mathcal{H} \psi \, dV, \tag{7.64}$$

subject to the normalization condition

$$\int \psi^* \psi \, dV = 1. \tag{7.65}$$

The unknown constants $c_{i\alpha}$ are determined by extremization of the functional

$$F[\{c_{i\alpha}\}] = \int \psi^* \mathcal{H} \psi \, dV - E \int \psi^* \psi \, dV, \tag{7.66}$$

with respect to the coefficients $\{c_{i\alpha}\}$. The procedure results in the eigenvalue problem

$$\sum_{j=1}^{N}\sum_{\beta=1}^{B} H_{i\alpha j\beta} c_{j\beta} = E \sum_{j=1}^{N}\sum_{\beta=1}^{B} J_{i\alpha j\beta} c_{j\beta}, \tag{7.67}$$

where

$$H_{i\alpha j\beta} = \int \phi_{\alpha}^{*}(\boldsymbol{r} - \boldsymbol{x}_i)\mathcal{H}\phi_{\beta}(\boldsymbol{r} - \boldsymbol{x}_j)\, dV \tag{7.68}$$

are elements of the Hamiltonian matrix, and

$$J_{i\alpha j\beta} = \int \phi_{\alpha}^{*}(\boldsymbol{r} - \boldsymbol{x}_i)\phi_{\beta}(\boldsymbol{r} - \boldsymbol{x}_j)\, dV \tag{7.69}$$

are the elements of the so-called overlap matrix. In orthogonal tight-binding methods, the overlap matrix is the identity matrix, or $J_{i\alpha j\beta} = \delta_{ij}\delta_{\alpha\beta}$.

Once the energy eigenstates are determined, the total electronic energy can be obtained by filling up the states (accounting for spin and Pauli's exclusion principle). Thus,

$$E_{\text{tot}} = \sum_{k=1}^{M'} E_k + \frac{1}{2}\sum_{i,j} V_{I-I}(|\boldsymbol{x}_i - \boldsymbol{x}_j|), \tag{7.70}$$

where M' is a number dependent on M (determined to account for spin and Pauli's principle), E_k are energy eigenvalues obtained from the solution of Eq. (7.67), and V_{I-I} is the term that is included to account for the "direct" ion-ion interaction.

The tight-binding method outlined here is what is commonly called "empirical tight binding." The main ingredients of the tight binding model is the Hamiltonian matrix elements $H_{i\alpha j\beta}$, which depend on the positions of the ions. Usually, these are constructed using the "two-center approximation," which assumes that $H_{i\alpha j\beta}$ depends only on the positions and orientations of the atoms i and j. Tight-binding methods also have a further "hidden" approximation, in that the Hamiltonian matrix elements do not depend on the electronic states, i.e., self-consistency is neglected. More sophisticated tight-binding methods that account for self-consistency have been developed; for a review of the state of the art in tight-binding methods, the reader is referred to Goringe et al.[27]

Tight-binding methods have found wide application in the material modeling.[27] Examples of the use of tight binding methods in the area of nanomechanics include the determination of the structure of fullerenes,[26] as shown in Fig. 7.14, and the formation of carbon nanotubes[28] from graphite patches (see Fig. 7.15).

The advantage of tight-binding method is that it enables simulation of larger systems as compared with density-functional methods. While density functional

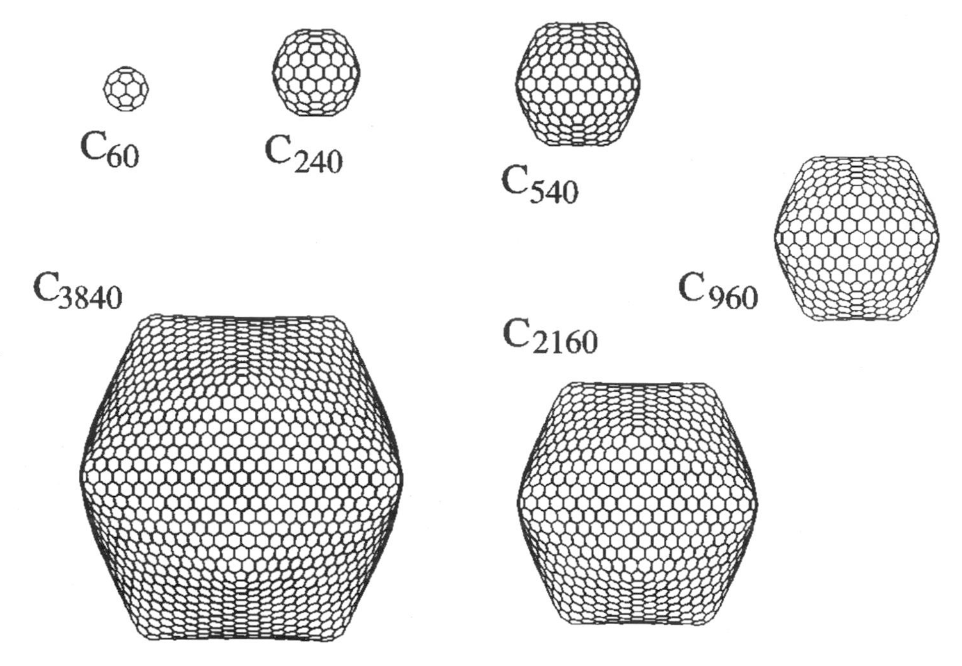

Figure 7.14 Structure of large fullerene balls obtained using the tight binding method. (Reprinted with permission from Ref. 26, © 1996 The American Physical Society.)

methods are definitely more accurate, tight-binding methods offer an attractive route to study problems of nanomechanics in the near term.

7.3.1.4 Total energy formulations without explicit treatment of electrons

While the methods that explicitly involve electrons are accurate, they are computationally intensive. There are several approaches that do not require a direct evaluation of the electronic states. In such approaches, the total energy of a system is expressed directly as a function of the positions of the atoms. These functions can be derived either from basic quantum mechanics by "integrating out" the electronic degrees of freedom or by resorting to empirical methods. Much effort has been invested in the calculation of effective total energy descriptions—generally called "interatomic potentials"—that do not require the explicit evaluation of electronic states. Only representative potential types that have possible application in nanomechanics are covered here; an excellent summary of such potentials has been provided by Voter.[29] From the point of view of nanomechanics, these methods are likely to be very coarse, i.e., they should be used to study essential physics and/or trends but not to obtain quantitative results.

7.3.1.5 Pair potentials

In the pair-potential description, the total energy of the system is considered to consist of purely pairwise interactions between the atoms. The total energy is written

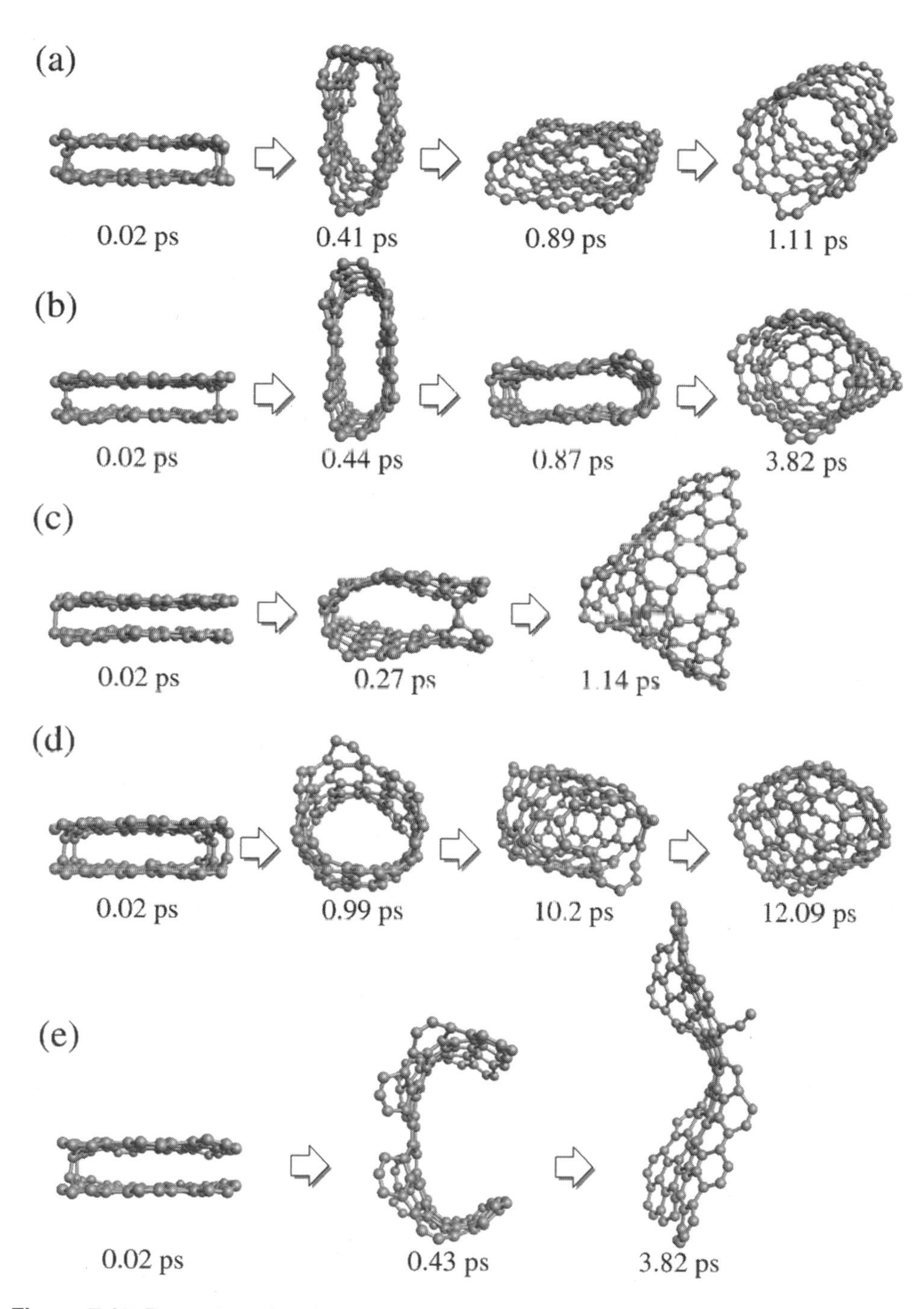

Figure 7.15 Formation of carbon nanotubes from graphite patches. These figures show the temporal evolution under various temperature conditions and for different structures of graphite patches. (Reprinted with permission from Ref. 28, © 2002 The American Physical Society.)

as (assuming all atoms are of the same type)

$$E_{\text{tot}}(\boldsymbol{x}_1, \ldots, \boldsymbol{x}_N) = \frac{1}{2} \sum_{ij, i \neq j} V(r_{ij}), \qquad r_{ij} = |\boldsymbol{x}_i - \boldsymbol{x}_j|. \tag{7.71}$$

Examples of the pairwise interaction include the Lennard-Jones potential

$$V(r) = \frac{A}{r^{12}} - \frac{B}{r^6}, \tag{7.72}$$

where the parameters A and B are determined either by quantum mechanical calculations or by a fit of some experimentally measured properties such as the cohesive energy and the lattice parameter. The Lennard-Jones potential is especially useful in describing inert gas solids. Potentials with other analytical forms and potentials without direct analytical form are used widely to simulate metals. Typically, these potentials have a finite cutoff distance r_{cut}, i.e., $V(r) = 0, r \geq r_{\text{cut}}$. If, however, a part of the potential is due to Coulombic interactions (which is true for ionically bonded materials such as ceramics), then the potentials are infinite ranged and specialized techniques such as Ewald summation[30] is necessary to compute the sum in Eq. (7.71).

Pair potentials used to describe metals do not have simple analytical formulas such as Eq. (7.72). They are derived by considering second-order corrections to electronic states due to the presence of the ions.[31] Early studies using such pair potentials were key to providing important clues to the understanding of plastic behavior of metals with body-centered cubic (BCC) structure. It was known experimentally that the yield stress of single crystal of BCC metals is roughly a hundredth of the shear modulus, while that for the case of face-centered cubic (FCC) crystals is the order of a thousandth of the shear modulus, i.e., the yield stress as a fraction of the shear modulus in BCC metals is an order of magnitude larger than that in FCC metals. Pioneering computer simulations[32,33] using pair potentials in a Na crystal revealed that the core structure (Fig. 7.16) of a screw dislocation in BCC crystals is nonplanar, i.e., a larger stress is required to move such a dislocation, resulting in a larger yield stress. These simulations are not only important from the point of view of understanding the plastic behavior of body-centered cubic materials, but also stand out as an important example the use of computer simulations in understanding material behavior even with simple descriptions such as pair potentials.

Although pair potentials have been extensively used to study materials, they provide only a very crude description of the total energy. Pair potentials are notorious in underestimating the stacking fault energies,‡ and hence can give erroneous results for defect cores. Even at the level of elasticity, a purely pairwise description

‡Indeed, a near-neighbor pair potential cannot distinguish between a FCC and a hexagonal close-packed (FCC) structure!

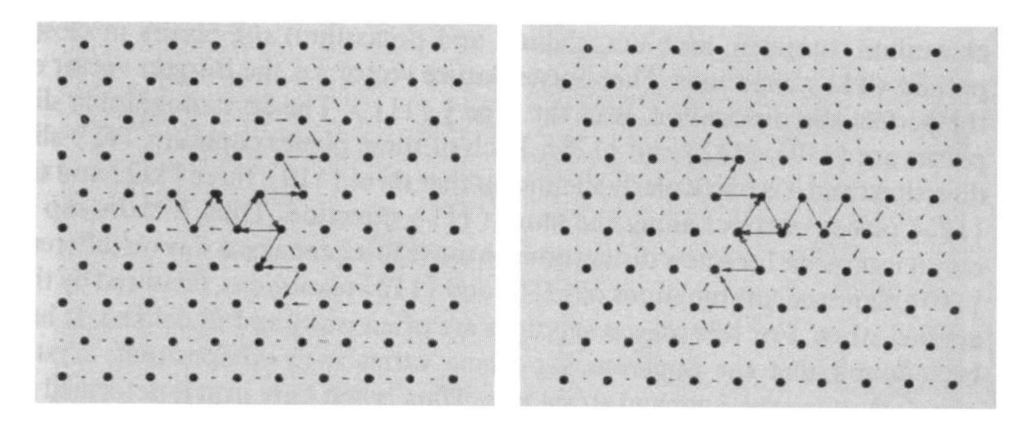

Figure 7.16 Nonplanar core structure of a screw dislocation in a body-centered cubic crystal simulated using pair potentials. (Reprinted with permission from Ref. 32, © 1976 The Royal Society.)

of energy gives rise to the so-called "Cauchy relation" between the elastic constants. In the case of cubic crystals, the Cauchy relation implies that $C_{1212} = C_{1122}$ [see Eq. (7.26)].

7.3.1.6 Embedded atom method

The embedded atom method (EAM) was developed by Daw and Baskes[34] (see also, Finnis and Sinclair[35]) with the aim of alleviating some of the crippling difficulties of pair potentials. The main goal of this method is the inclusion of "many-body effects" in an approximate fashion.

The construction of the embedded-atom potential is based on the following *ansatz*. To a collection of $N-1$ atoms, an additional atom is thought to be placed at the point $\boldsymbol{r}_N$. This new atom interacts with the other atoms (called the "host") via the pair potential. In addition, there is an additional energy called the embedding energy that arises due to the interaction of the new atom with the electron density at the site $\boldsymbol{r}_N$ due to the host. The total energy of the system is equal to the sum of the pairwise interaction and the embedding energies of all atoms. The total energy given by the embedded atom method for a collection of atoms (of the same type) is

$$E_{\text{tot}}(\boldsymbol{x}_1, \ldots, \boldsymbol{x}_N) = \frac{1}{2} \sum_{ij, i \neq j} V(r_{ij}) + \sum_i F(\rho_i), \qquad r_{ij} = |\boldsymbol{x}_i - \boldsymbol{x}_j|, \qquad (7.73)$$

where V is a pairwise interaction, F is the embedding function, and ρ_i is the electron density at the site $\boldsymbol{x}_i$. The electron density at the site $\boldsymbol{x}_i$ is given by

$$\rho_i = \sum_{j \neq i} \rho(r_{ij}), \qquad (7.74)$$

where $\rho(r)$ is the electron density at a distance r from the nucleus of an atom. The total electron density at site $\boldsymbol{x}_i$ is the sum of electron densities of all of the atoms except the ith atom. While the pair-potential description (for atoms of a single type) requires specification of but one function $V(r)$, the embedded atom method for the same case requires specification of three functions: $V(r)$, $F(\rho)$, and $\rho(r)$. There are several approaches to the determination of these three functions. Of particular importance is the embedding function $F(\rho)$. The approach by Finnis and Sinclair,[35] motivated by the tight-binding method, is to consider $F(\rho) \sim \sqrt{\rho}$, and fit the other two functions to experimental properties such as the lattice parameter, cohesive energies, and phonon frequencies. Foiles et al.[36] used an alternative approach where, in addition to the fit of the pair potential and electron density to several experimental parameters, the embedding function is determined so as to reproduce the universal binding energy relation (UBER) of Rose et al.[37] Density functional theory has also been used to derive embedded atom potentials. For example, Ercolessi and Adams[38] have used forces from density functional calculations to develop embedded atom potentials for aluminum.

Embedded atom methods have found wide use in materials simulations. The problems of pair-potential formulations (such as low stacking fault energy and the Cauchy relations between elastic constants) are absent. In the embedded atom method, the violation of the Cauchy relation occurs as $C_{1212} - C_{1122} \sim F''(\rho_0)$, where ρ_0 is the electron density at an atomic site in the crystal. Thus, as long as F is not a linear function, Cauchy relations are violated. Owing to their ease of use and reasonable accuracy, embedded atom potentials have been used to study defect cores,[33] fracture,[39] etc. A review by Voter[39] also contains a lucid account of the embedded atom method.

7.3.1.7 Three-body potentials

Nanostructures made of carbon (such as carbon nanotubes) are poorly described by pair potentials or the embedded atom method. This is due to the directional nature of bonding in these systems. For example, a graphite sheet contains sp^2 hybridized carbon atoms, where bonds make an angle of 120 deg. The change in angles of these bonds affects the total energy of the system. To describe such nanostructures using simple potentials (which do not require explicit treatment of the electrons), three-body interactions have to be introduced in addition to the pairwise term. Conceptually, the total energy in a three-body formulation is given by

$$E_{\text{tot}}(\boldsymbol{x}_1, \ldots, \boldsymbol{x}_N) = \frac{1}{2} \sum_{i \neq j} V(|\boldsymbol{x}_i - \boldsymbol{x}_j|) + \frac{1}{6} \sum_{i,j,k} V_3(\boldsymbol{x}_i, \boldsymbol{x}_j, \boldsymbol{x}_k), \tag{7.75}$$

where V_3 is the three-body potential. Generalizations beyond the three-body terms are possible (see Carlsson[40]).

A commonly used three-body potential is the potential developed by Stillinger and Weber[41] for silicon. The potential contains a pairwise interaction term and a second term that accounts for the additional energies due to the changes in the angle between the bonds. The potential takes the form

$$E_{\text{tot}}(\boldsymbol{x}_1,\ldots,\boldsymbol{x}_N)=\frac{1}{2}\sum_{i\neq j}V(r_{ij})+\sum_{ijk}h(r_{ij})h(r_{ij})\left(\cos\theta_{ijk}+\frac{1}{3}\right)^2, \tag{7.76}$$

where θ_{ijk} is the angle between the $i-j$ bond and the $i-k$ bond. The function $h(r)$, which is a pairwise function, decays with a finite cutoff. The potential strongly favors tetrahedral bonding found in silicon; indeed the potential is not useful in studying other phases of silicon that are not tetrahedral. This difficulty arises due to the fact that the potential does not account for the local environments.

This difficulty was overcome by Tersoff[42] and Brenner.[43] The so-called Tersoff–Brenner potentials may be thought of as the generalization of the embedded atom potentials or glue potentials in that the local environment of the atom is accounted for. The functional form of the potential is given by two parts, an attractive part and a repulsive part, as

$$E_{\text{tot}}(\boldsymbol{x}_1,\ldots,\boldsymbol{x}_N)=\frac{1}{2}\sum_{i\neq i}V_R(r_{ij})+\frac{1}{2}\sum_{i\neq j}B_{ij}V_A(r_{ij}), \tag{7.77}$$

where $B_{ij}=B(C_{ij})$ is the bond order with

$$C_{ij}=\sum_k h_c(r_{ik})f(\theta_{ijk})h(r_{ij}-r_{ik}), \tag{7.78}$$

where θ_{ijk} is the angle between the i—j and i—k bond, and $V_R(r)$, $h_c(r)$, $f(\theta)$, and $h(r)$ are functions to be determined. This approach works for a larger class of environments. However, difficulty arises in determining all the functions involved. The Tersoff–Brenner potential has been used to study[44] mechanisms of plasticity in single-walled carbon nanotubes (Fig. 7.17).

7.3.2 Atomistic simulation methods

The previous section reviewed the methods for describing the total energies of a collection of atoms. These methods ultimately provide a function $E_{\text{tot}}(\boldsymbol{x}_1,\ldots,\boldsymbol{x}_N)$ for the total energy that depends on the positions of the atoms, either with explicit treatment of the electronic degrees of freedom or without. Quantities of interest such as elastic modulus, fracture toughness, thermal conductivity, and defect core parameters can be obtained using these techniques for describing total energy.

The atomistic simulation methods are broadly classified into three classes for the sake of discussion. First is a class of methods that have come to be called lattice statics. Methods of the second class have their basis in statistical mechanics and are called Monte Carlo methods. Finally, molecular dynamics methods form the third class. Lattice statics methods are useful to study properties that are not temperature dependent or have a weak temperature dependence. Monte Carlo methods

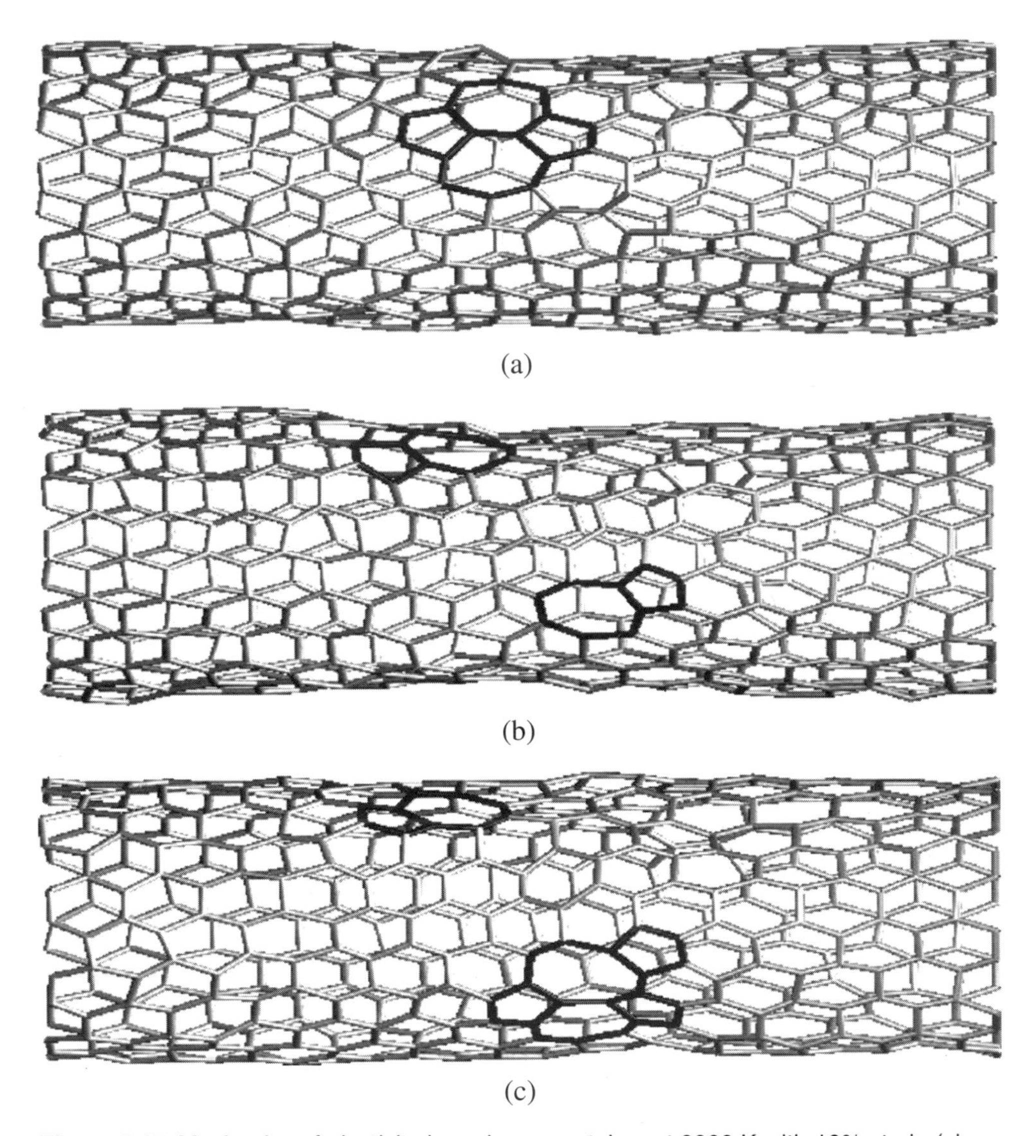

Figure 7.17 Mechanics of plasticity in carbon nanotubes at 2000 K with 10% strain (simulated with Tersoff–Brenner potentials): (a) formation of a pentagon-heptagon defect, (b) splitting and diffusion of a defect, and (c) formation of a more complex defect. (Reprinted with permission from Ref. 44, © 1998 The American Physical Society.)

are extensively used to obtain finite temperature equilibrium properties. Molecular dynamics methods, based on tracing the temporal evolution of the system, are a powerful class of methods that find applications in both finite-temperature equilibrium and nonequilibrium problems.

7.3.2.1 Lattice statics

When physical quantities of interest are weakly temperature dependent, the lattice statics method provides the most useful simulation tool. The basic principle of this method is the principle of minimum potential energy, i.e., an isolated system kept under a set of constraints chooses its degrees of freedom such that the total potential energy of the system is least. Thus, the positions of atoms are determined such that

$$\frac{\partial E_{\text{tot}}}{\partial \boldsymbol{x}_i} = \mathbf{0}, \quad i = 1, \ldots, N. \tag{7.79}$$

The minimization of the total energy of the system is achieved by the conjugate gradient method and other such methods.[21] These are iterative methods that require the knowledge of the derivatives $\partial E_{\text{tot}}/\partial \boldsymbol{x}_i$ (negative of forces) of the total energy, in addition to the total energy E_{tot} for a given configuration of atoms; and these derivatives are used to update the configuration $\{\boldsymbol{x}_i\}$ until the condition of Eq. (7.79) is satisfied. For *ab initio* density functional theory and tight binding methods, the derivatives are evaluated with the aid of the Hellman-Feynman theorem.[21] Analytical expressions for the derivatives are available for non-electronic-structure methods. For example, the derivative of the energy is given as

$$\frac{\partial E_{\text{tot}}}{\partial \boldsymbol{x}_i} = \sum_{j \neq i} \{V'(r_{ji}) + [F'(\rho_i) + F'(\rho_j)]\rho'(r_{ji})\} \frac{\boldsymbol{x}_j - \boldsymbol{x}_i}{r_{ij}}, \tag{7.80}$$

for the embedded atom method of Eq. (7.73).

Lattice statics methods have been used extensively to study defect cores. Figure 7.18 shows the core structure of an edge dislocation in aluminum obtained using the embedded atom potentials of Ercolessi and Adams.[38]

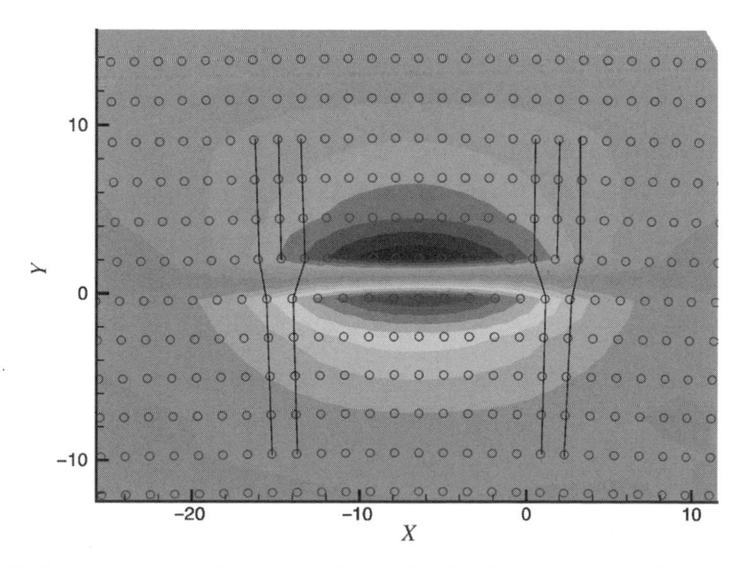

Figure 7.18 Core structure of an edge dislocation in aluminum, described by embedded atom potentials, obtained by lattice statics. Contours represent values of out-of-plane displacements.

7.3.2.2 Statistical mechanics methods—Monte Carlo method

The determination of temperature-dependent properties of materials and nanostructures is important for predictive capability. Once the total energy function E_{tot} that depends on the positions of the atoms is known, the temperature-dependent properties can be calculated by the well-known techniques of statistical mechanics. A brief discussion of statistical mechanics is included for the sake of completeness. Detailed expositions can be found in the books by Huang[45] or Pathria.[46]

Equilibrium statistical mechanics deals with systems that have a large number of microscopic degrees of freedom. The *equilibrium* state of such a system is considered, and predictions are made about the quantities that can be measured experimentally. In particular, different *ensembles* are used to describe the macroscopic conditions experienced by the system. The ensemble of typical interest is called the *canonical ensemble*, where the system of interest is kept at a *fixed volume* (or, more generally, in a fixed kinematic state) and in contact with a thermal reservoir at a prespecified temperature. Thus, the system can exchange energy with the reservoir, and therefore the total energy of the system fluctuates. The tools of statistical mechanics predict quantities such as the expected value of the energy, the specific heat, etc.

For the classical canonical ensemble, the probability of a *microscopic* state described by a set of parameters $\mathcal{C}$ is

$$P(\mathcal{C}) = \frac{1}{Z} e^{-\beta H(\mathcal{C})}, \tag{7.81}$$

where $P(\mathcal{C})$ refers to the probability of realizing the configuration $\mathcal{C}$, $H(\mathcal{C})$ is the Hamiltonian or the total energy of the system when it is in configuration $\mathcal{C}$, and β is equal to the reciprocal of the product of the Boltzmann constant k and the absolute temperature T, i.e., $\beta = 1/k_B T$. The quantity Z is called the *partition function* and is equal to

$$Z = \int e^{-\beta H(\mathcal{C})}\, d\mathcal{C}. \tag{7.82}$$

The partition function contains information about all of the observable macroscopic quantities. For example, the expected value of the energy U of the system is given by

$$U = \int H(\mathcal{C}) \frac{e^{-\beta H(\mathcal{C})}}{Z}\, d\mathcal{C} = -\frac{1}{\beta} \frac{\partial \ln Z}{\partial \beta}. \tag{7.83}$$

A very interesting and useful expression relating the partition function to a thermodynamic potential (see Huang[45] for details) is

$$Z = e^{-\beta A}, \tag{7.84}$$

where A is the *Helmholtz free energy*. Thus, statistical mechanics provides a link between the microscopic configurations that a system can realize and a thermodynamic potential like the Helmholtz free energy.

For a solid made of N atoms, a microscopic configuration is described by the positions and momenta of the atoms $(\boldsymbol{x}_1, \ldots, \boldsymbol{x}_N, \boldsymbol{p}_1, \ldots, \boldsymbol{p}_N)$. The total energy (or Hamiltonian) of such a configuration is given by

$$H(\boldsymbol{x}_1, \ldots, \boldsymbol{x}_N, \boldsymbol{p}_1, \ldots, \boldsymbol{p}_N) = E_{\text{tot}}(\boldsymbol{x}_1, \ldots, \boldsymbol{x}_N) + \sum_i \frac{p_i^2}{2m}, \tag{7.85}$$

where it is assumed that all of the atoms in the solid have mass m, and E_{tot} denotes the potential energy of the atoms. The second term on the right side of Eq. (7.85) is the kinetic energy of the atoms. The partition function for this system may now be evaluated using Eq. (7.85) in Eq. (7.82) to get

$$Z = \frac{1}{N!h^{3N}} \int e^{-\beta H(\boldsymbol{x}_1, \ldots, \boldsymbol{x}_N, \boldsymbol{p}_1, \ldots, \boldsymbol{p}_N)} d\boldsymbol{x}_1 \ldots d\boldsymbol{x}_N \, d\boldsymbol{p}_1 \ldots d\boldsymbol{p}_N. \tag{7.86}$$

The factor $1/(N!h^{3N})$ first arose to resolve classical paradoxes arising from the indistinguishability of atoms and to make the theory a correct high-temperature limit of quantum statistical mechanics, $d\Omega$ is a volume element in the phase space of the atomic system. The partition function in Eq. (7.86) gives

$$Z = \frac{1}{N!} \left(\frac{\sqrt{2\pi m k_B T}}{h} \right)^{3N} Q, \tag{7.87}$$

where

$$Q = \int e^{-\beta E_{\text{tot}}(\boldsymbol{x}_1, \ldots, \boldsymbol{x}_N)} d\boldsymbol{x}_1 \ldots d\boldsymbol{x}_N \tag{7.88}$$

is called the *configurational integral*. Except for the factor Q in Eq. (7.87), all terms are what would appear in the partition function for an *ideal gas*; and therefore it is Q that has all of the contribution due to the interatomic interactions that make up the solid.

To illustrate the Monte Carlo method, the problem of determining a property g is considered (for example, g could be the the set of expected values of positions of atoms near a grain boundary in a problem determining the grain boundary structure at a finite temperature). It is evident that the expected value of any quantity g that depends *only* on the positions of the atoms is given as

$$\langle g \rangle = \frac{1}{Q} \int g(\boldsymbol{x}_1, \ldots, \boldsymbol{x}_N) e^{-\beta E_{\text{tot}}(\boldsymbol{x}_1, \ldots, \boldsymbol{x}_N)} d\boldsymbol{x}_1 \ldots d\boldsymbol{x}_N. \tag{7.89}$$

The main idea in the Monte Carlo scheme is that of *importance sampling*, where a configuration is accepted or rejected based on its probability of realization. This enables an extremely efficient computation of averages, such as that in Eq. (7.89). Specifically, the following Metropolis algorithm[47] is adopted for a *single* Monte Carlo step:

1. Select an atom at random.
2. Compute a random vector $\boldsymbol{\delta}$ whose magnitude does not exceed a prespecified value.
3. Compute the change in the *potential energy* $\delta E_{\rm tot}$ of the system when the randomly chosen atom in step 1 is displaced by the random vector $\boldsymbol{\delta}$.
4. If $\delta E_{\rm tot} < 0$ then
   Accept this configuration;
   else
   (a) Generate a random number σ uniformly distributed on $[0, 1]$.
   (b) If $e^{\beta\delta E_{\rm tot}} > \sigma$, then accept the new configuration, else keep the old configuration.

A predetermined number $N_{\rm MC}$ of Monte Carlo steps are taken, and averages are computed. The main point is that the average in Eq. (7.89) can be evaluated as a simple average over the Monte Carlo steps, i.e.,

$$\langle g\rangle = \frac{1}{Q}\int g(\boldsymbol{x}_1,\ldots,\boldsymbol{x}_N)e^{-\beta E_{\rm tot}(\boldsymbol{x}_1,\ldots,\boldsymbol{x}_N)}\,d\boldsymbol{x}_1\ldots d\boldsymbol{x}_N \approx \frac{1}{N_{\rm MC}}\sum_{i=1}^{N_{\rm MC}} g_i, \quad (7.90)$$

where g_i is the quantity g evaluated using the configuration at the ith Monte Carlo step. Equation (7.90) is a very powerful tool in evaluating averages, and this is what makes the Monte Carlo method useful. The Monte Carlo method, therefore, provides by far the most accurate method to evaluate averages based on statistical mechanics.

The Monte Carlo method is particularly attractive in that it does not require the evaluation of the derivatives of energies, which can be computationally intensive. The main problem in the Monte Carlo method is the obtaining of sufficient statistical accuracy. For example, the thermodynamics of a defect is governed by the additional energy that it possesses over the perfect crystal; this would involve taking a statistical average of the difference of two large numbers, and this excess energy can be smaller than the fluctuations in the energy. It will require very long simulations to average out the effect of fluctuations. This problem is even more serious when computing quantities such as the elastic modulus, which depends directly on the fluctuations. One other problem involved in using the Monte Carlo method is that it is difficult to use in a situation that requires the use of nonperiodic boundary conditions, thus limiting the problems that are accessible via this method. For further details on the Monte Carlo method, see Allen and Tildesley.[48] A recent example of the use of the Monte Carlo method is the calculation[49] of the temperature dependence of elastic moduli of carbon nanotubes, as shown in Fig. 7.19.

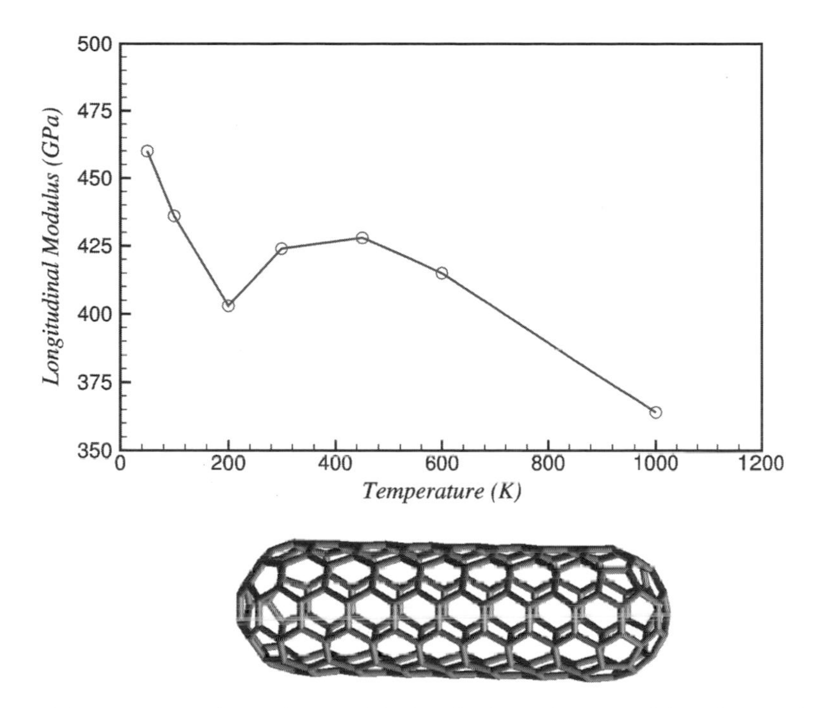

Figure 7.19 Temperature dependence of the longitudinal elastic modulus of the carbon nanotube shown obtained using Monte Carlo methods from strain fluctuations. (After Grigoras et al.[49])

7.3.2.3 Molecular dynamics

The aim of molecular dynamics methods is to trace the trajectory of the collection of atoms in its phase space. The desired properties of the material are expressed as functions of the positions and velocities of the atoms, and evaluated using the trajectories calculated. The trajectories are calculated by integrating the equations of motion obtained from the Hamiltonian of Eq. (7.85) as

$$\begin{aligned} \frac{\partial \boldsymbol{x}_i}{\partial t} &= \frac{\partial H}{\partial \boldsymbol{p}_i} = \frac{\boldsymbol{p}_i}{m}, \\ \frac{\partial \boldsymbol{p}_i}{\partial t} &= -\frac{\partial H}{\partial \boldsymbol{p}_i} = -\frac{\partial E_{\text{tot}}}{\partial \boldsymbol{x}_i}, \end{aligned} \tag{7.91}$$

where t denotes time. The integration of these equations involve the evaluation of the forces $-\partial E_{\text{tot}}/\partial \boldsymbol{x}_i$ as outlined in Sec. 7.3.2.1.

The most common method for the integration of the equations of motion [Eq. (7.91)] is the Verlet algorithm.[48,50,51] The basic time-stepping scheme in the Verlet algorithm enables the computation of the positions of the atoms at time $t + \Delta t$ from previous positions as

$$\boldsymbol{x}_i(t + \Delta t) = 2\boldsymbol{x}_i(t) - \boldsymbol{x}_i(t - \Delta t) + (\Delta t)^2 \boldsymbol{a}_i(t), \tag{7.92}$$

where $\boldsymbol{a}_i(t)$ is the acceleration of the atom i given by

$$\boldsymbol{a}_i = -\frac{1}{m}\frac{\partial E_{\text{tot}}}{\partial \boldsymbol{x}_i}. \tag{7.93}$$

The disadvantage of the Verlet algorithm is that although the positions are calculated accurately $O(\Delta t^4)$, the velocities, which do not explicitly enter the algorithm, are accurate only to $O(\Delta t^2)$. An improved version of the Verlet algorithm called the "velocity Verlet algorithm" is

$$\begin{aligned}
\boldsymbol{x}_i(t+\Delta t) &= \boldsymbol{x}_i(t) + \boldsymbol{v}_i(t)\Delta t + \frac{\Delta t^2}{2}\boldsymbol{a}_i(\Delta t),\\
\boldsymbol{v}_i\left(t+\frac{\Delta t}{2}\right) &= \boldsymbol{v}_i(t) + \frac{\Delta t}{2}\boldsymbol{a}_i(t),\\
\boldsymbol{a}(t+\Delta t) &= -\frac{1}{m}\frac{\partial E_{\text{tot}}}{\partial \boldsymbol{x}_i}[\{\boldsymbol{r}_j(t+\Delta t)\}],\\
\boldsymbol{v}_i(t+\Delta t) &= \boldsymbol{v}_i\left(t+\frac{\Delta t}{2}\right) + \frac{\Delta t}{2}\boldsymbol{a}_i(t+\Delta t).
\end{aligned} \tag{7.94}$$

Performing molecular dynamics simulations involves specifying the initial ($t = 0$) positions and velocities of the atoms.

An important point to be noted is that integration of Eq. (7.91) using either Eq. (7.92) or Eq. (7.94) conserves the total energy of the system (up to, of course, numerical errors). In the language of statistical mechanics, such an isolated system at constant energy is called the *microcanonical ensemble*. To simulate other ensembles that correspond to systems of interest, such as the canonical ensemble discussed in Sec. 7.3.2.2 or a constant pressure ensemble, other techniques are required. A detailed description of such methods may be found in Frenkel and Smit.[52] Attention is focused here on the constant-temperature method due to Nosé[53] and Hoover.[54]

The so-called Nosé–Hoover thermostat is a means to keep the temperature of the system as close to the desired temperature T_0 as possible. The main idea of the Nosé–Hoover thermostat is to consider the heat bath as an additional degree of freedom. In fact, Nose[55] proved that the microcanonical ensemble for this extended system (consisting of the atoms and the heat bath) implies a canonical ensemble for the collection of atoms. In an equilibrium system, the temperature T of the collection of N atoms at any time is related to their momenta via the total kinetic energy

$$T = \frac{2}{3}\frac{1}{Nk_B}\sum_{i=1}^{N}\frac{\boldsymbol{p}_i^2}{2m}. \tag{7.95}$$

The thermostatted equations modify the second equation of Eq. (7.91) as

$$\frac{\partial \boldsymbol{p}_i}{\partial t} = -\frac{\partial E_{\text{tot}}}{\partial \boldsymbol{x}_i} - \zeta \boldsymbol{p}_i, \tag{7.96}$$

where ζ is called the Nosé–Hoover drag coefficient. The drag coefficient has a temporal evolution given by

$$\frac{\partial \zeta}{\partial t} = \frac{1}{\tau^2}\left(\frac{T}{T_0} - 1\right), \tag{7.97}$$

where τ is a time constant associated with the heat bath. The integration of Eqs. (7.96) and (7.97) produces an ensemble closely approximating a canonical ensemble.

Molecular dynamics methods are widely used to study a large range of problems. One of the largest simulations ever performed is a billion atom stimulation by Abraham[56] and coworkers who studied brittle and ductile failure in nanocrystals (Fig. 7.20).

7.4 Mixed models for nanomechanics

The continuum approach discussed in Sec. 7.2 has the advantages of both conceptual simplicity (easier to interpret results) and computational efficiency. However, continuum theories are of limited use when applied to the atomic scale. Atomistic models of Sec. 7.3 are accurate and have all of the essential physics necessary for a complete description of phenomena at the atomic scale. Atomistic models are computationally intensive and require elaborate postprocessing to obtain the desired physical output. It is therefore advantageous to construct methods that have efficiency of the continuum approach and the accuracy of the atomistic models.

There are two main approaches to construct mixed models. The first is to start from a purely atomistic model and to apply continuum approximations. The second is to modify existing continuum theories so as to have in them essential physics to capture atomic scale phenomena. This section of the chapter briefly reviews both approaches. An example of the first approach is the quasi-continuum method, which uses continuum concepts in an atomistic model to achieve an effective reduction in the number of degrees of freedom to be considered. The second part of this section treats augmented continuum theories, which include nonlocal continuum theories and continuum theories that include effects of free surfaces.

7.4.1 The quasi-continuum method

The quasi-continuum method was developed by Tadmor et al.[6] as a nanoscale simulation method for materials physics. The method is thought of as an approximation scheme for the atomistic method—quasi-continuum is to atomistics as the finite element method is to continuum field theories. The guiding philosophy of the

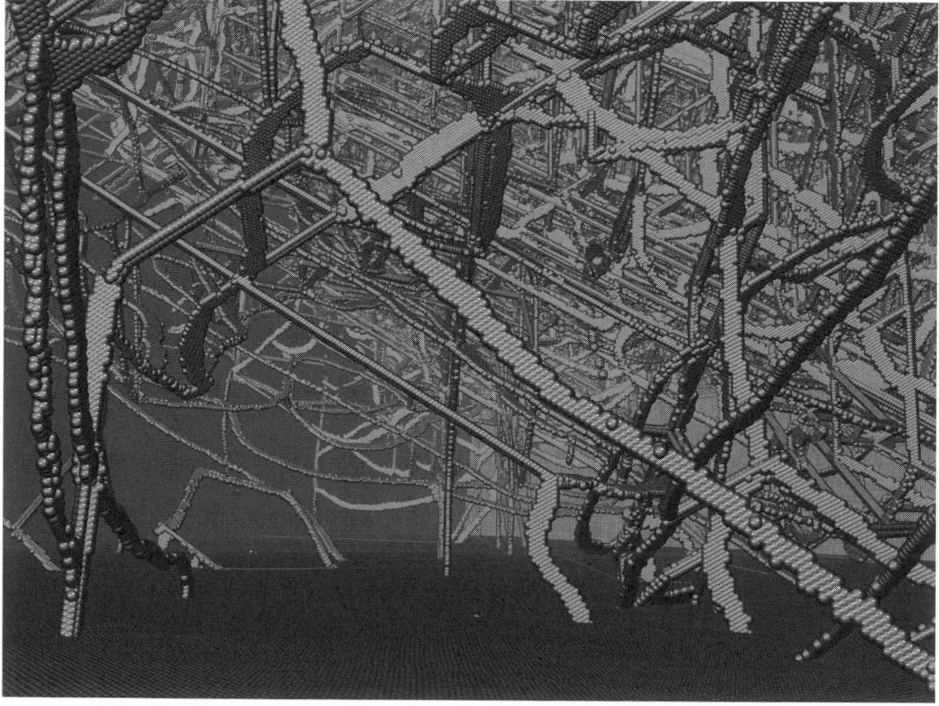

Figure 7.20 Ductile failure and dislocation patterns near a crack tip in a billion-atom molecular dynamics simulation. (From http://www.llnl.gov/largevis/atoms/ductile-failure/. See also Abraham.[56] Courtesy of University of California, Lawrence Livermore National Laboratory, and the Department of Energy under whose auspices the work was performed.)

finite element method (see Sec. 7.2.5) is to create a discrete model from a continuum field theory, while the quasi-continuum method aims to construct a discrete model with far fewer degrees of freedom than the original atomistic model. The common aim in this class of methods is to achieve the required reduction in the number of degrees of freedom to solve the problem at hand. The method described here[57,58] applies only when thermal effects can be neglected, i.e., the method is an approximation for lattice statics described in Sec. 7.3.2.1

The body under consideration is made of a large number of atoms N (see Fig. 7.21) to be built up of a variety of different grains with Bravais lattice vectors schematically indicated. A crystalline reference state is assumed to exist, which obviates the necessity to store the positions of all of the atoms in the solid. A given atom in the reference configuration is specified by a triplet of integers $\boldsymbol{l} = (l_1, l_2, l_3)$ and the grain to which it belongs. The position of the atom in the reference configuration is then given as

$$\boldsymbol{X}(\boldsymbol{l}) = \sum_{a=1}^{3} l_a \boldsymbol{B}_a^{\mu} + \boldsymbol{R}^{\mu}, \tag{7.98}$$

where $\boldsymbol{B}_a^{\mu}$ is the ath Bravais lattice vector associated with grain G_{μ} and $\boldsymbol{R}^{\mu}$ is the position of a reference atom in grain G_{μ}, which serves as the origin for the atoms in grain G_{μ}.

Once the deformed positions $\{\boldsymbol{x}_i\}$ of atoms are specified, the total energy is given by the function (see Sec. 7.3)

$$E_{\text{tot}} = E_{\text{exact}}(\boldsymbol{x}_1, \boldsymbol{x}_2, \boldsymbol{x}_3, \ldots, \boldsymbol{x}_N) = E_{\text{exact}}(\{\boldsymbol{x}_i\}). \tag{7.99}$$

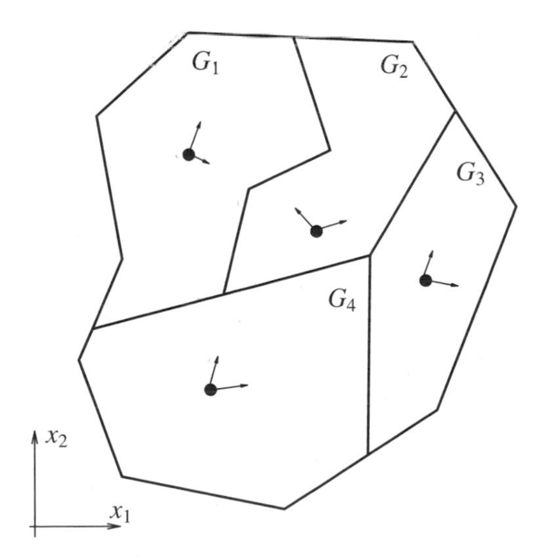

Figure 7.21 A schematic of a crystalline solid made up of grains G_{μ} with a reference atom in each grain and an associated set of Bravais lattice vectors. (After Shenoy et al.[58])

If the number of atoms is large, then the problem can become computationally intractable.

The first step in the quasi-continuum methodology is the selection of a subset of atoms called "representative atoms" whose positions are treated as the degrees of freedom of the system. The second step involves the construction of a finite element mesh with the representative atoms as the nodes. The (approximated) position of any other atom can be obtained from the positions of the representative atoms via the finite element interpolation (see Sec. 7.2.5) as

$$\boldsymbol{x}_i^h = \sum_{\alpha} N_\alpha(\boldsymbol{X}_i)\boldsymbol{x}_\alpha, \tag{7.100}$$

where $N_\alpha(\boldsymbol{X}_i)$ is the finite element shape function centered around the representative atom α [which is also a Finite element method (FEM) node] evaluated at the undeformed position $\boldsymbol{X}_i$ of the ith atom. The kinematics of the collection of atoms is completely described in that, on knowing the positions of the representative atoms, the positions of any other atom in the model can be obtained using Eq. (7.100).

After necessary kinematic approximation via the selection of the representative atoms and the construction of the finite element method, the next step in the process is the construction of an approximate method to evaluate the total energy of the atomic system that depends on the positions of the representative atoms alone. Further progress at the present state of development of the method hinges on a crucial assumption. It is assumed that the total energy of the system can be additively decomposed into energies of individual atoms as follows:

$$E_{\text{tot}} = \sum_{i=1}^{N} E_i. \tag{7.101}$$

Such a decomposition is allowed in the embedded atom method and the pair-potential formulations discussed in Secs. 7.3.1.5 and 7.3.1.6, respectively, but not in the case of more sophisticated formulations such as density functional theory (Sec. 7.3.1.2). Although this decomposition restricts the class of energy functionals that allow for the approximations discussed herein, the method developed is nevertheless useful in treating very large systems using the simpler atomistic formulations such as the EAM and pair potentials that would otherwise require the use of supercomputers. If Eq. (7.101) is used in the computation of the energy runs over all the atoms in the body, there is no gain in computational time. To achieve a true reduction in the number of degrees of freedom, the following approximation is made:

$$E_{\text{tot}} \approx \sum_{\alpha=1}^{R} n_\alpha E_\alpha. \tag{7.102}$$

The main idea embodied in Eq. (7.102) surrounds the selection of a set of representative atoms, each of which, in addition to providing a complete kinematic description of the body, are intended to characterize the energetics of a spatial neighborhood within the body as indicated by the weight n_α. In other words, n_α can be thought of as the number of atoms represented by the representative atom α. The statement of the approximate energy Eq. (7.102) is complete only with the specification summation weights n_α. The problem of the determination of n_α is treated in a manner similar to determination of quadrature weights in the approximate computation of definite integrals.[59] In this context, the goal is to approximate a finite sum ("definite integral" on the lattice) by an appropriately chosen quadrature rule where the quadrature points are the sites of the representative atoms. The quadrature rule of Eq. (7.102) is designed such that, in the limit in which the finite element mesh is refined all the way down to the atomic scale (a limit that is denoted as fully refined), each and every atomistic degree of freedom is accounted for, and the quadrature weights are unity (each representative atom represents only itself). On the other hand, in the far field regions where the fields are slowly varying in space, the quadrature weights reflect the volume of space (which is now proportional to the number of atoms) that is associated with the representative atom, and *this is where the continuum assumption is made.* The details of this procedure may be found in Shenoy et al.[58]

A further energetic approximation in the computation of Eq. (7.102) is made to simplify the energy calculations. This approximation also makes possible the formulation boundary conditions that mimic those expected in an elastic continuum. Figure 7.22, which depicts the immediate neighborhood of a dislocation core, motivates the essential idea of the approximation. The figure shows the atomic structure near the core of a Lomer dislocation characterized by the pentagonal group of atoms. If the environments of two of the atoms in this figure, one (labeled A) in the immediate core region, and the other (labeled B) in the far field of the defect, are considered, it is evident that the environment of atom A is nonuniform and that each of the atoms in that neighborhood experiences a distinctly different environment. On the other hand, atom B has an environment that is closely approximated as emerging from a uniform deformation, and each of the atoms in its vicinity experiences a nearly identical geometry.

These geometric insights provide for the computation of the energy E_α from an atomistic perspective in two different ways, depending upon the nature of the atomic environment of the representative atom α. Far from the regions of strong nonhomogeneity such as defect cores, the fact that the atomic environments are nearly uniform is exploited by making a *local* calculation of the energy in which it is assumed that the state of deformation is homogeneous and is well-characterized by the local deformation gradient $\boldsymbol{F}$ [see Eq. (7.14)]. To compute the total energy of such atoms, the Bravais lattice vectors of the deformed configuration $\boldsymbol{b}_a$ are obtained from those in the reference configuration $\boldsymbol{B}_a$ via $\boldsymbol{b}_a = \boldsymbol{F}\boldsymbol{B}_a$. The gradient of deformation is obtained from the finite-element interpolation of the positions of the atoms. Once the Bravais lattice vectors are specified, this reduces the computation of the energy to standard lattice statics.

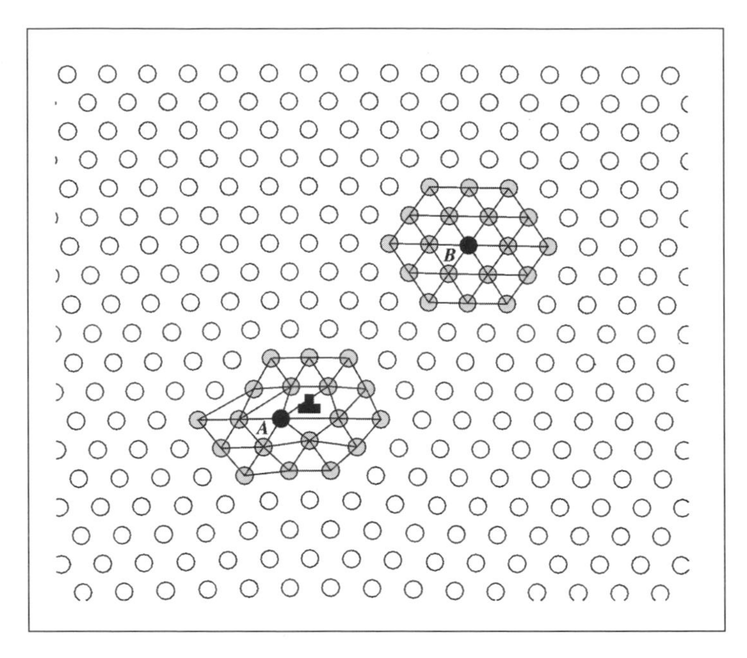

Figure 7.22 Atomic structure near the core of a Lomer dislocation in aluminum. The atom A in the core region experiences an inhomogeneous environment while the environment of atom B is nearly homogeneous. (After Shenoy et al.[58])

In regions that suffer a state of strongly nonuniform deformation, i.e., the deformations change on a scale smaller than the intrinsic atomistic scales, such as the core region around atom A in Fig. 7.22, the energy is computed by constructing a crystallite that reproduces the deformed neighborhood from the interpolated displacement fields. The atomic positions of each and every atom are given exclusively by $\boldsymbol{x} = \boldsymbol{X} + \boldsymbol{u}(\boldsymbol{X})$, where the displacement field $\boldsymbol{u}$ is determined from finite-element interpolation. This ensures that a fully nonlocal atomistic calculation is performed in regions of rapidly varying $\boldsymbol{F}$. An automatic criterion for determining whether to use the local or nonlocal rule to compute a representative atom's energy based on the variation of deformation gradient is available.[58] The distinction between local and nonlocal environments has the unfortunate side effect of introducing small spurious forces, referred to as "ghost" forces at the interfaces between the local and nonlocal regions. A correction for this problem is discussed by Shenoy et al.[58]

With the prescription to describe the kinematics with reduced degrees of freedom, and a method to calculate the total energy that depends only on the reduced degrees of freedom, the quasi-continuum method can be applied to obtain approximate solutions to lattice statics problems by use of standard energy minimization techniques such as conjugate gradients and Newton-Raphson techniques. There are several of technical issues that surround the use of either conjugate gradient or Newton-Raphson techniques, which are discussed in detail in Shenoy et al.[58]

An essential prerequisite in constructing the quasi-continuum formulation is an adaptive capability that enables the targeting of particular regions for refinement in response to the emergence of rapidly varying displacement fields. As an example, during the simulation of nanoindentation, the indentation process leads to the nucleation and subsequent propagation of dislocations into the bulk of the crystal. To capture the presence of slip that is tied to these dislocations, it is necessary that the slip plane be refined down to the atomic scale (see Fig. 7.23). The adaption

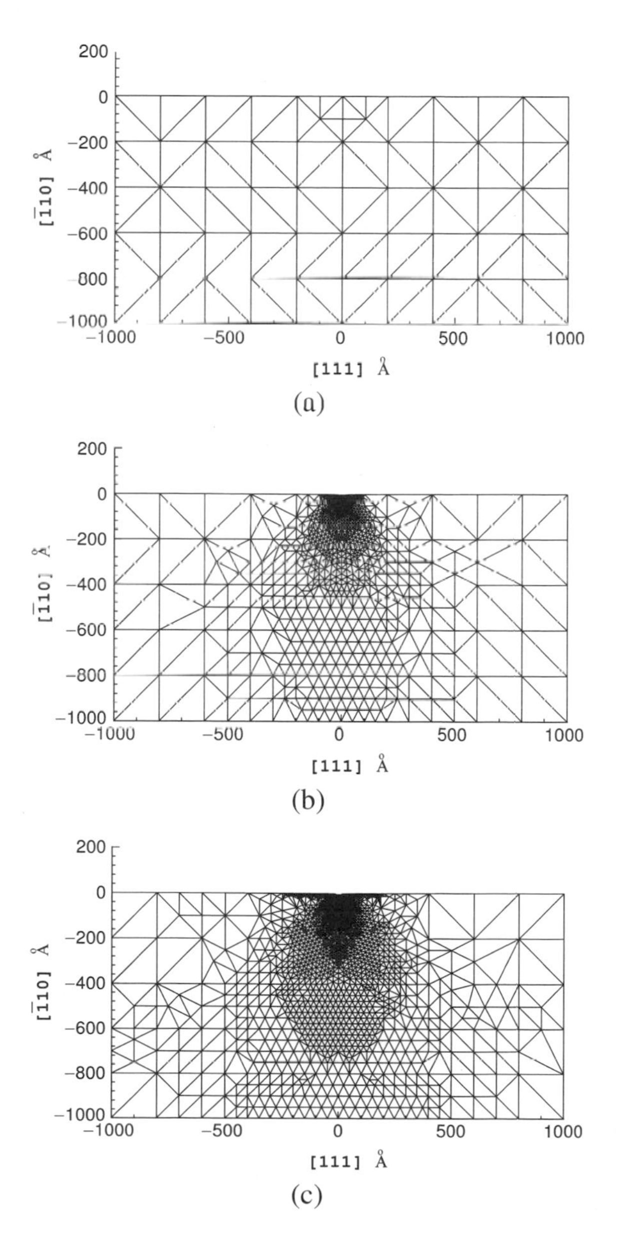

Figure 7.23 Automatic adaption process in action for the problem of nanoindentation. (After Shenoy et al.[58])

scheme enables the natural emergence of such mesh refinement as an outcome of the deformation history.

The essential points of the approximation scheme presented in this section are

1. A subset of the total number of atoms that make up the body is selected (representative atoms) and the atoms' positions are treated as the only unknowns. The position of any other atom in the body is then obtained from a finite element mesh, the nodes of which correspond to the representative atoms.
2. The energy of the system is also computed with the knowledge of energies of *only* the representative atoms. This is accomplished by the rule of Eq. (7.102).
3. A further approximation in the computation of the energies of the representative atoms is made where the deformations are approximately homogeneous on the scale of the lattice.
4. An adaptive scheme is included to capture evolving deformation.

The quasi-continuum method has been used to study defect nucleation, defect migration, fracture, and dislocation interaction. Figure 7.24 shows a quasi-continuum simulation of dislocation grain boundary interaction. The model consists of two grains with the top grain bounded by a free surface. A rigid indentor AB is used to generate dislocations at A. These dislocations move toward the grain boundary and "react" with it. These reactions and further details of the simulation may be found elsewhere.[58] The significant computational savings obtained by the use of the quasi-continuum method for this problem is worth noting. The number of degrees of freedom used in the quasi-continuum method was about 10^4 while a complete atomistic model of this problem would have required more than 10^7 degrees of freedom. The quasi-continuum simulation required about 140 h on a DEC-Alpha workstation, while a purely atomistic model would have required a parallel supercomputer. These simulations are based on a generalized 2D formulations (three components of displacements are considered to depend only on two coordinates). A fully 3D version of the quasi-continuum method has been used to study[60] dislocation junctions (see Fig. 7.25).

The methodology presented here is useful for only zero-temperature simulations. In the present form, the quasi-continuum method can be thought of as an approach to bridge multiple length scales. Generalizations of this method to include dynamics must bridge multiple time scales in addition to multiple length scales. There are several preliminary attempts to solve this problem,[61,62] where a subcyling algorithm is used to coarse grain over time in addition to length scales. Results of dynamic nanoindentation studied with this technique are shown in Fig. 7.26.

7.4.2 Augmented continuum theories

A very important phenomenon in nanostructures is the occurrence of the so-called "size dependence" of properties. In other words, if a property of the material that

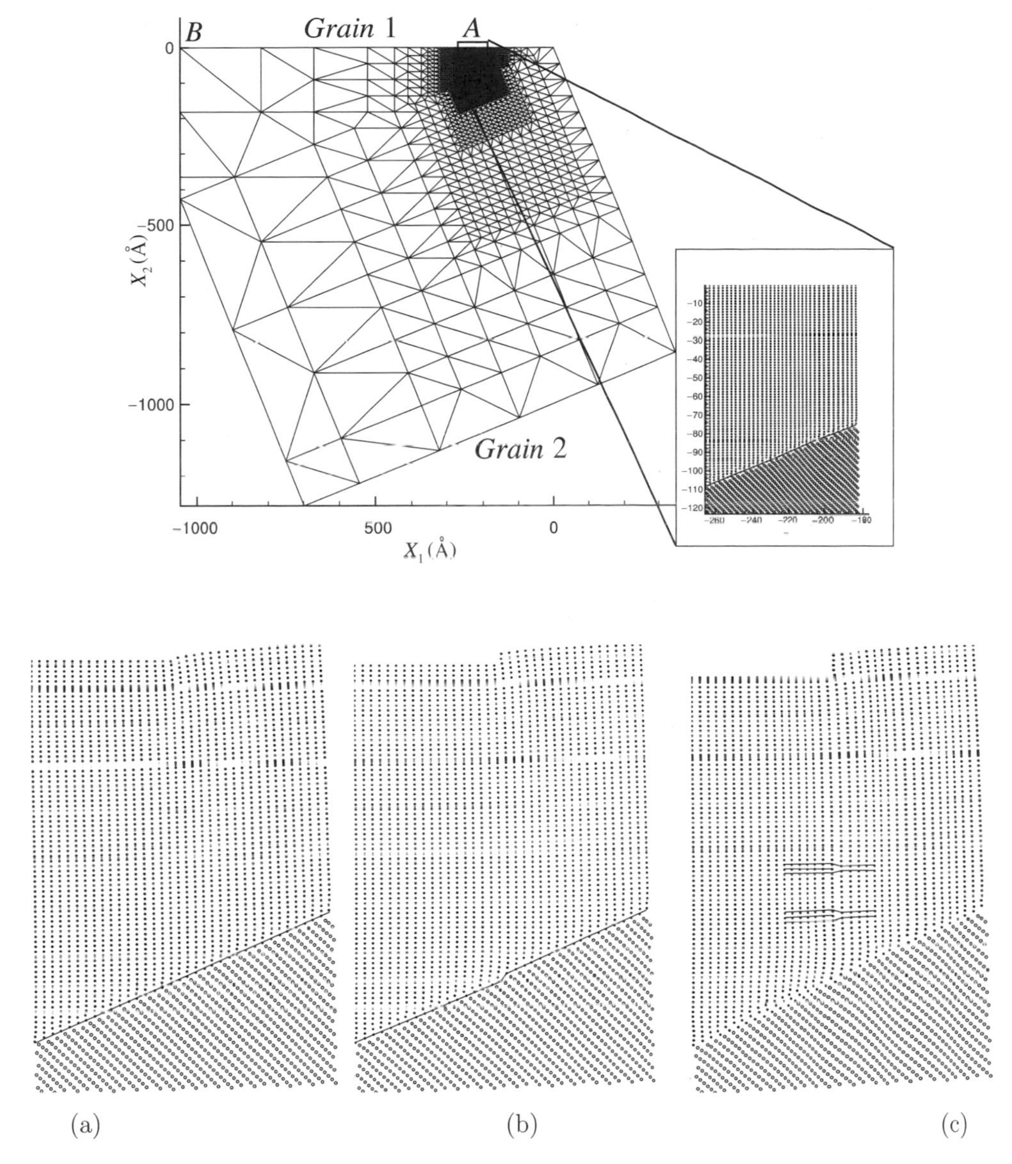

Figure 7.24 Top: Mesh designed to model the interaction of dislocations and a grain boundary. Dislocations are generated at point A by rigidly indenting on face AB of the crystal. Bottom: Snapshots of atomic configurations depicting the interaction of dislocations with a grain boundary: (a) atomic configuration immediately before the nucleation of the partial dislocations, (b) atomic configuration immediately after the nucleation of the first set of partial dislocations that have been absorbed into the boundary, and (c) the second pair of nucleated partial dislocations form a pile-up. (After Shenoy et al.[58])

makes up the nanostructure is calculated from a measured response using relations of standard continuum mechanics, the properties turn out to depend on the size of the nanostructure (see Fig. 7.4). A further example of this can be seen from the calculation of the tensile modulus of a plate D from lattice statics simulations

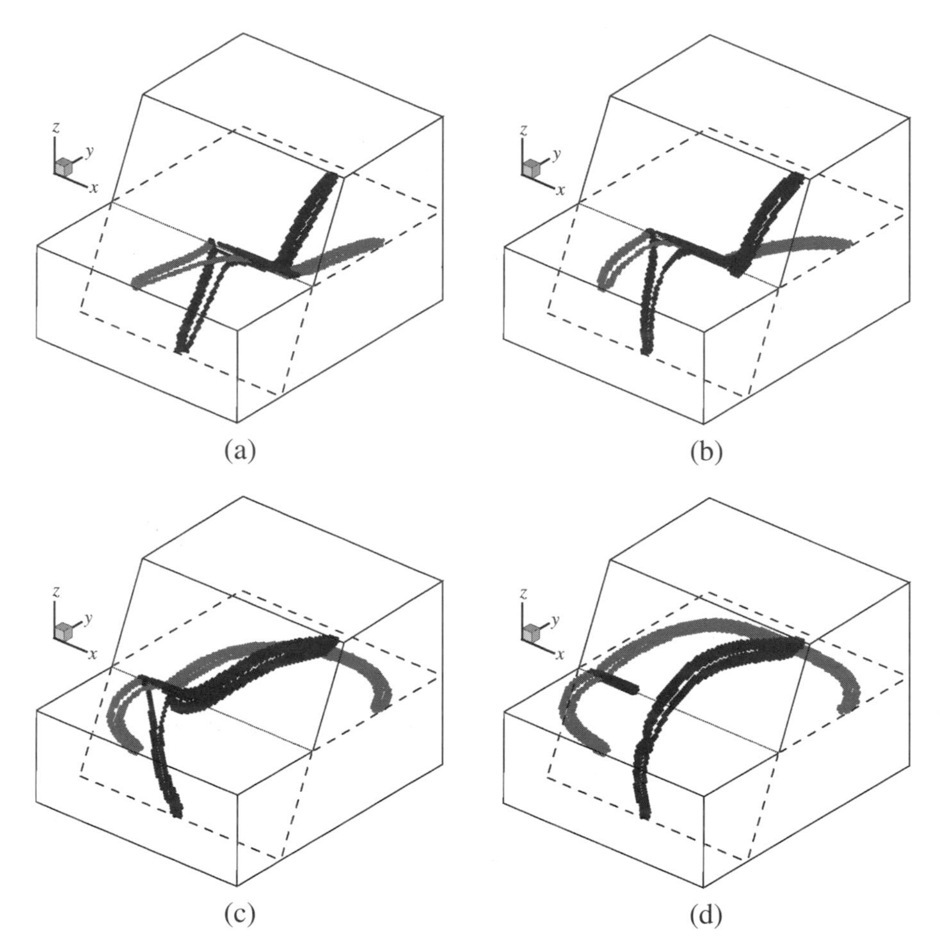

Figure 7.25 Evolution of a dislocation junction under stress simulated by the 3D quasi-continuum method. (Reprinted with permission from Ref. 60, © 1999 The American Physical Society.)

(outlined in Sec. 7.3.2.1) compared with the prediction of the continuum theory D_c, shown in Fig. 7.27. It is evident that the modulus predicted by continuum theory differs from that of the full atomistic simulations *in a very regular fashion*; in fact, the nondimensional difference $(D - D_c)/D_c$ scales as $1/d$, where d is the thickness of the plate. Since $(D - D_c)/D_c$ is a nondimensional quantity, this observation also implies the presence of an intrinsic length scale d_0.

The size dependence of properties can arise from two main causes. The first cause comprises nonlocal effects in the bulk, and the second is due to the presence of free surfaces that become increasingly important as the size of the structure reduces. These two effects are now discussed in the context of elasticity.

Nonlocal continuum theories were pioneered by Eringen[63] who relaxed the principle of local action (see Sec. 7.2.4) that is, sometimes tacitly, assumed in standard constitutive equations of continuum mechanics. Thus, the stress at a point in the body depends not on the strain at that point alone, but possibly on the strains at

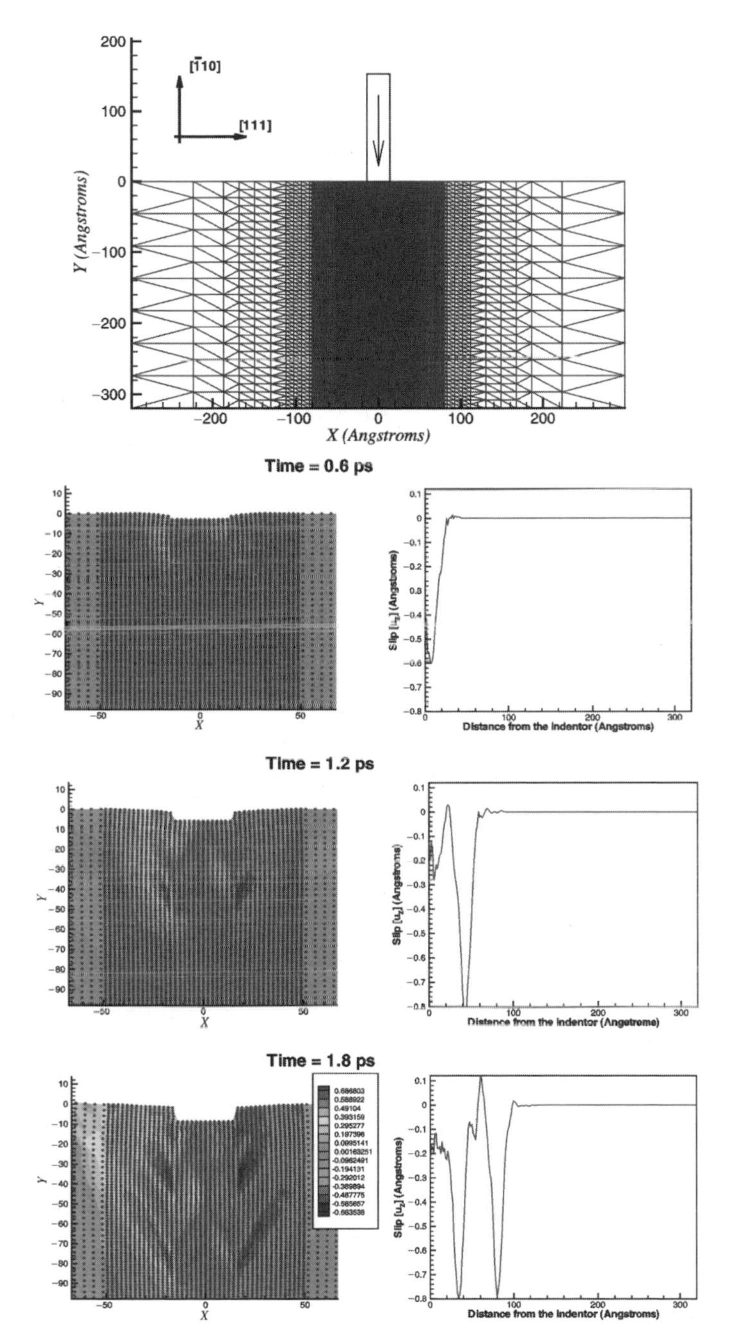

Figure 7.26 Top: Finite-element mesh for dynamic nanoindentation. Bottom: Results of dynamic nanoindentation with subcycling. Contours show the presence of a supersonic dislocation. (After Shenoy.[61])

all points in the body. Eringen's generalization to Eq. (7.26) can be expressed as

$$\sigma_{ij}(\boldsymbol{x}) = \int_{\mathcal{V}} C_{ijkl}(\boldsymbol{x}, \boldsymbol{x}')\epsilon_{kl}(\boldsymbol{x}')\,dV', \tag{7.103}$$

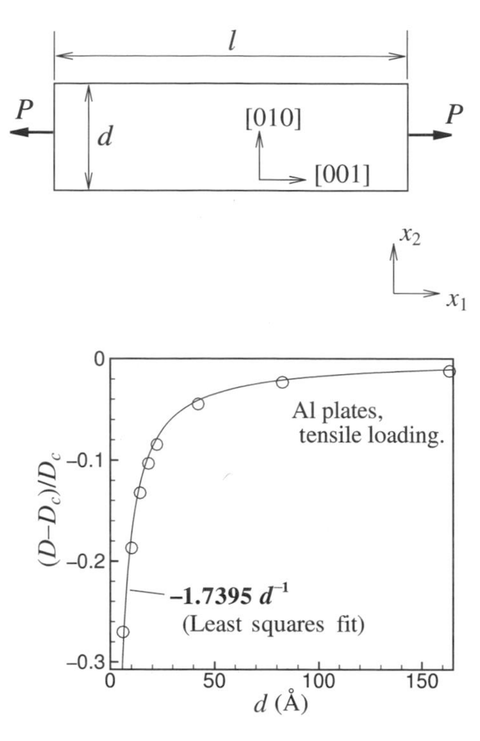

Figure 7.27 Top: Schematic geometry of the plane strain plate. Bottom: Nondimensional difference between plate modulus computed atomistically and that predicted by standard continuum theory. The coefficient of $1/h$ is obtained by a least-squares fitting procedure to be -1.7395 Å. (After Miller and Shenoy.[64])

where $C_{ijkl}(\boldsymbol{x}, \boldsymbol{x}')$ is the nonlocal elastic modulus tensor. Note that nonlocality is inherent in atomistic models because the energy of any given atom depends on the positions of other atoms in its neighborhood. In practice, the tensor $C_{ijkl}(\boldsymbol{x}, \boldsymbol{x}')$ vanishes if $|\boldsymbol{x} - \boldsymbol{x}'| > r_c$, and the length r_c provides an intrinsic length scale. Nonlocal effects become increasingly important if the deformations of the structure vary strongly over the length scale r_c. Using this formation, Eringen[63] showed, among other things, that the stresses near a cracked tip do not go to infinity in magnitude, as predicted by standard linear elasticity, even while being in agreement with linear elasticity at large distances from the cracked tip. These nonlocal approaches are likely to prove useful in the development of simple yet useful theories of nanomechanics.

Attention is now turned to the effect of free surfaces that are of increasing importance with the reduction in the size of the structure. A general continuum theory based on this observation has been developed.[64,65] Several authors have previously utilized continuum theories, (although not in the context of nanomechanics, of solids with surface effects[66–68]) to study a variety of problems ranging from diffusive cavity growth in stressed solids to the stability of stressed epitaxial films.

The body $\mathcal{V}$, described by coordinates x_i, considered in the augmented continuum theory is bounded by a surface $\mathcal{S}$. It is assumed that the surface $\mathcal{S}$ is piecewise flat (this assumption eliminates the need to consider contravariant and covariant components of surface tensors) and is described by coordinates x_α for each flat face. The bulk stress tensor in the body $\mathcal{V}$ is denoted by σ_{ij} and the surface stress tensor by $\tau_{\alpha\beta}$. Mechanical equilibrium of a bulk material element implies that the bulk stress tensor satisfies [Eq. (7.9) with no body forces]

$$\sigma_{ij,j} = 0.$$

Equilibrium of a surface element necessitates that

$$\tau_{\alpha\beta,\beta} + f_\alpha = 0, \tag{7.104}$$

and

$$\tau_{\alpha\beta}\kappa_{\alpha\beta} = \sigma_{ij} n_i n_j, \tag{7.105}$$

where n_i is the outward normal to the surface, f_α is the negative of the tangential component of the traction $t_i = \sigma_{ij} n_j$ along the α direction of surface $\mathcal{S}$, and $\kappa_{\alpha\beta}$ is the surface curvature tensor. The assumption of the piecewise flat surfaces implies that the surface curvature vanishes everywhere along the surface except at corners and edges, which must be treated separately. Note that the assumption of a piecewise flat surface is merely for the sake of mathematical simplicity; the present theoretical framework is valid for curved surfaces as well.

The kinematics of the body are described by the displacement field u_i defined at every point in the body. The strain tensor ϵ_{ij} in the body is obtained using a small strain formulation as given in Eq. (7.19). The surface strain tensor $\epsilon_{\alpha\beta}$ is derived from the bulk strain tensor ϵ_{ij} such that every material fiber on the surface has the same deformation whether it is treated as a part of the surface or as a part of the bulk, i.e., the surface strain tensor is compatible with the bulk strain tensor.

The final ingredient of the augmented continuum theory is the constitutive relations that relate the stresses to strains. The bulk is considered to be an anisotropic linear hyperelastic solid [see Eq. (7.25)] with a free energy density W defined and the stresses derived as in Eq. (7.26). Constitutive relations for the surface stress tensor are more involved. The surface stress tensor is related to the surface energy γ as

$$\tau_{\alpha\beta} = \gamma\delta_{\alpha\beta} + \frac{\partial\gamma}{\partial\epsilon_{\alpha\beta}}, \tag{7.106}$$

a relation that is generally attributed to Gibbs,[68] also called the Shuttleworth relation.[69] The surface stress tensor can be expressed as a linear function of the strain tensor as

$$\tau_{\alpha\beta} = \tau^0_{\alpha\beta} + S_{\alpha\beta\gamma\delta}\epsilon_{\gamma\delta}, \tag{7.107}$$

where $\tau^0_{\alpha\beta}$ is the surface stress tensor when the bulk is unstrained [obtained from Eq. (7.106) with $\epsilon_{\alpha\beta} = 0$] and $S_{\alpha\beta\gamma\delta}$ is the *surface elastic modulus tensor*. This is an important quantity in that the size dependence of elastic properties will be shown to be determined by the ratio of a surface elastic constant and a bulk elastic constant. The constitutive constants C_{ijkl} and $S_{\alpha\beta\gamma\delta}$ are external to the augmented continuum theory; these are determined *from atomistic models* of materials considered (Sec. 7.3). Thus, continuum mechanics is augmented to have surface effects, and surface properties are determined from atomistic simulations.

Values of properties of selected surfaces in two elemental materials are given in Table 7.3. Note that in some cases the values of surface elastic constants are negative. While this might be counterintuitive, it must be remembered that the surface elastic energy need not be positive definite; only the combined energy of the surface and the bulk must be positive definite.

The theory is illustrated with an example of tensile moduli of prismatic single crystal bars. Bars of aluminum and silicon are considered with cross-sectional geometries, as shown in Fig. 7.28. For the square silicon bar in the aforementioned orientation, standard continuum theory gives the relationship between the bar force P and the strain as

$$P = D_c\varepsilon = EA\varepsilon = Ed^2\varepsilon, \tag{7.108}$$

Table 7.3 Surface elastic constants and surface stresses for Al and Si. Units are electron volts per square angstrom. (From Miller and Shenoy.[64])

Surface	S_{1111} $(= S_{2222})$	S_{1122}	τ_1^0 $(= \tau_2^0)$
Al [100]	−0.495	0.254	0.036
Al [111]	0.324	0.484	0.057
Si [100] 1 × 1	−0.761	−0.082	0.0
Si [100] 1 × 2	−0.665	−0.243	0.038

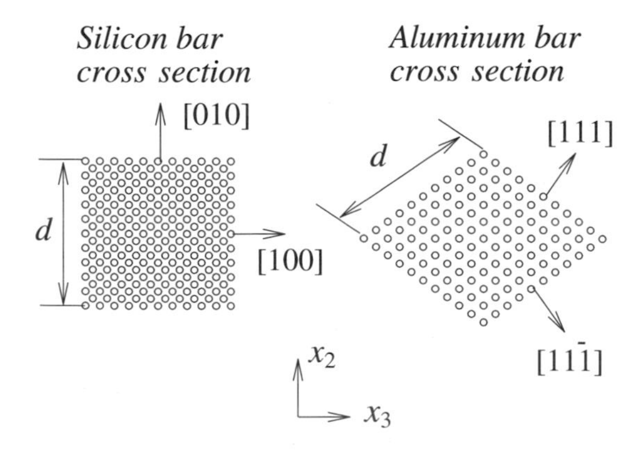

Figure 7.28 Schematic cross sections of the bars considered in augmented continuum theory. (After Miller and Shenoy.[64])

where the elastic modulus of the bulk can be shown to be

$$E = C_{11}\left[1 - \frac{2C_{12}^2}{C_{11}(C_{11}+C_{12})}\right]. \tag{7.109}$$

For the aluminum bar with the diamond cross section, the same expression holds except that the cross-sectional area is now $A = (2\sqrt{2}d^2)/3$ and the expression for the modulus E is considerably more complicated due to the crystal orientation.

When surface effects are included, the expression for the bar force P becomes

$$P = 4d\tau^0 + \underbrace{(EA + 4Sd)}_{D}\,\varepsilon, \tag{7.110}$$

where S is the surface elastic modulus, which is computed assuming that the Poisson contraction of the surface is equal to that of the bulk. The nondimensional difference between the true tensile modulus and that predicted by standard continuum theory is

$$\frac{D - D_c}{D_c} = \begin{cases} 4\dfrac{S}{E}\dfrac{1}{d} - \dfrac{4d_0}{d} & \text{for the silicon bar,} \\ 3\sqrt{2}\dfrac{S}{E}\dfrac{1}{d} = \frac{3\sqrt{2}d_0}{d} & \text{for the aluminum bar.} \end{cases} \tag{7.111}$$

Thus the theory identifies the intrinsic length scale to be the ratio of the surface modulus to the bulk elastic modulus. Figure 7.29 shows plots of $(D - D_c)/D_c$ as a function of d for the two bars. The Al bar with the (111) free surfaces represents an important test of the model. In this case, the model predicts a positive value for h_0, and thus an *increase* in stiffness with decreasing size. Indeed, this is borne out by the atomistic simulations and the model prediction remains correct.

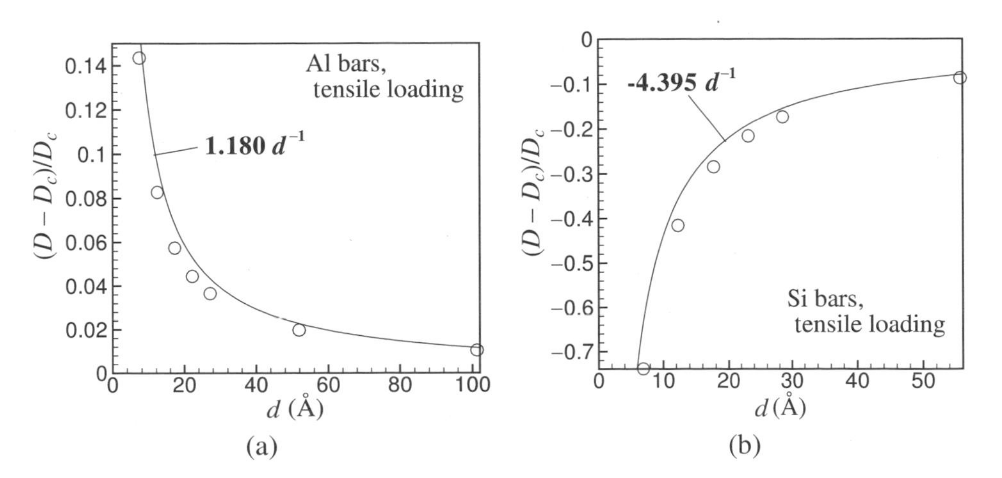

Figure 7.29 Nondimensional difference between bar modulus computed atomistically and that predicted by continuum theory: (a) Al bars and (b) Si bars. (After Miller and Shenoy.[64])

Similar models were developed for bending of plates and bars[65] (see Fig. 7.30). In all cases

$$\frac{D - D_c}{D_c} = A\frac{d_0}{d}, \tag{7.112}$$

where A is a constant that is determined with the augmented continuum theory, d is the size scale of the structure, and $d_0 = S/E$, where S is a surface elastic constant and E is a bulk elastic constant. The use of this theory is envisaged as follows. The bulk elastic constants and the surface elastic constants (for various surfaces) of materials of interest can be calculated and tabulated. Then the expressions for the constant A can be worked out for a host of cross-sectional shapes once and for all. A collection of such information will be useful for the designers of nanomechanical systems in that the need for direct atomistic simulations of nanoscale structures is obviated. Further improvements of this model are possible by considering nonlocal elastic effects (especially in the bulk) discussed earlier in this section.

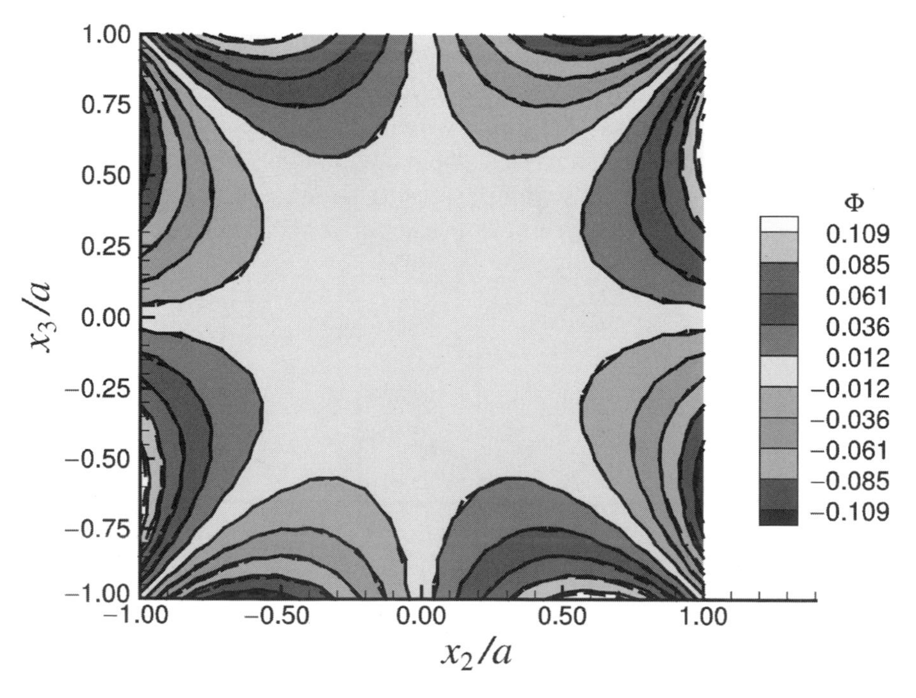

Figure 7.30 Comparison of the atomistically simulated and theoretical warping functions based on augmented continuum theory. The solid lines are the contours of the atomistic result while the dashed lines correspond to the theoretical calculation. The bar is made of aluminum with a width $2a$ of 10 lattice constants. (After Shenoy.[65])

7.5 Concluding remarks

This chapter presents a survey of methods used to study mechanics at the atomic scales. A detailed exposition of these methods can be found in the book by Phillips.[70] While atomistic models presented here are the most accurate methods for studying problems of nanomechanics, mixed models and augmented continuum theories are more suitable for conceptual clarity.

Much future work is required to elevate the present theoretical models to the status of predictive tools. An important point to be noted is that theoretical work in nanoscience will have to be necessarily interdisciplinary. While this chapter has not stressed this point, it is well recognized that at the atomic scale all phenomena are coupled. As an example, it is possible to change the electronic properties of a carbon nanotube by a simple process of straining—a nanotube can show drastic changes in its electronic band structure when subjected to strain.[71] It is the understanding of the *coupling of properties* and the exploitation of such phenomena that can likely produce useful devices. Thus, a concerted interdisciplinary effort in gaining theoretical understanding is required if nanotechnology is to live up to its promise.

Acknowledgments

Support for part of the work by the Indian National Science Academy under the Young Scientist Scheme is gratefully acknowledged. The author expresses his heartfelt thanks to Rob Phillips (teacher, adviser, and friend) of Caltech from whom he learned much of what is presented here. Rob's words on matters academic and nonacademic have proved invaluable to the author. Suggestions from N. Ravishankar and Biswaroop Mukherjee on the manuscript are thankfully acknowledged.

References

1. R. P. Feynman, "There's plenty of room at the bottom" (1960); Caltech's *Engineering and Science*; available on the Web at http://nono.xerox.com/nanotech/feynman.html.
2. K. E. Drexler, *Engines of Creation, The Coming Era of Nanotechnology*, Anchor Books/Doubleday, New York (1986).
3. D. M. Eigler and E. K. Schweizer, "Positioning of single atoms with a scanning tunneling microscope," *Nature* **344**, 524–526 (1990).
4. M. L. Roukes, "Nanoelectromechanical systems," Technical Digest, Solid-State Sensors and Actuator Workshop, Transducers Res. Found., Cleveland, OH, 367–376 (2000).
5. R. Gao, Z. L. Wang, Z. Bai, W. A. de Heer, L. Dai, and M. Gao, "Nanomechanics of individual carbon nanotubes from pyrolitically grown arrays," *Phys. Rev. Lett.* **85**, 622–625 (2000).

6. E. B. Tadmor, M. Ortiz, and R. Phillips, "Quasicontinuum analysis of defects in solids," *Phil. Mag. A* **73**, 1529–1563 (1996).
7. P. Chadwick, *Continuum Mechanics*, Wiley, New York (1976).
8. R. W. Ogden, *Non-linear Elastic Deformations*, Dover, New York (1997).
9. L. D. Landau and E. M. Lifshitz, *Theory of Elasticity*, Pergamon Books, Oxford (1989).
10. W. Noll, "The foundations of classical mechanics in the light recent advances in continuum mechanics," in *The Axiomatic Method with Special Reference to Geometry and Physics, Proceedings of an International Symposium*, L. Henkin, P. Suppes, and A. Tarski (Eds.), 266–281, North-Holland, Amsterdam (1959).
11. T. J. R. Hughes, *The Finite Element Method: Linear Static and Dynamic Finite Element Analysis*, Prentice-Hall, Englewood Cliffs, NJ (1987).
12. O. C. Zienkiewicz, *The Finite Element Method*, Vols. 1 and 2, 4th ed., McGraw-Hill, London (1991).
13. S. W. Sloan, "A fast algorithm for generating constrained Delaunay triangulations," *Comput. Struct.* **47**, 441–450 (1993).
14. L. I. Schiff, *Quantum Mechanics*, McGraw-Hill, New York (1968).
15. J. J. Sakurai, *Modern Quantum Mechanics*, Addison-Wesley, New York (1994).
16. A. Szabo and N. S. Ostlund, *Modern Quantum Chemistry*, Dover, New York (1996).
17. N. W. Ashcroft and N. D. Mermin, *Solid State Physics*, Saunders College, Philadelphia (1976).
18. O. Madelung, *Introduction to Solid State Theory*, Springer, Berlin (1978).
19. P. Hohenberg and W. Kohn, "Inhomogeneous electron gas," *Phys. Rev.* **136**, B864–B871 (1964).
20. W. Kohn and L. J. Sham, "Self-consistent equations including exchange and correlation effects," *Phys. Rev.* **140** A1133–A1138 (1965).
21. M. C. Payne, M. P. Teter, D. C. Allan, T. A. Arias, and J. D. Joannopoulos, "Iterative minimization techniques for *ab initio* total-energy calculations: molecular dynamics and conjugate gradients," *Rev. Mod. Phys.* **64**, 1045–1097 (1992).
22. K. Capelle, "A bird's-eye view of density-functional theory," Lectures given at the VIIIth Summer School on Electronic Structure of the Brazilian Physical Society (2002).
23. M. H. F. Sluiter, V. Kumar, and Y. Kawazoe, "Symmetry-driven phase transformations in single-wall carbon-nanotube bundles under hydrostatic pressure," *Phys. Rev. B* **65**, 161402R/1–4 (2002).
24. F. J. Ribeiro, D. J. Roundy, and M. L. Cohen, "Electronic properties and ideal tensile strength of MoSe nanowires," *Phys. Rev. B* **65**, 153401/1–4 (2002).
25. J. C. Slater and G. F. Koster, "Simplified LCAO method for the periodic potential problem," *Phys. Rev.* **94**, 1498–1524 (1954).

26. S. Itoh, P. Ordejón, D. A. Drabold, and R. M. Martin, "Structure and energetics of giant fullerenes: an order-N molecular-dynamics study," *Phys. Rev. B* **53**, 2132–2140 (1996).
27. C. M. Goringe, D. R. Bowler, and Hernández, "Tight-binding modelling of materials," *Rep. Prog. Phys.* **60**, 1447–1512 (1997).
28. T. Kawai, Y. Miyamoto, O. Sugino, and Y. Koga, "Nanotube and nanohorn nucleation from graphitic patches: tight-binding molecular-dynamics simulations," *Phys. Rev. B* **66**, 033404/1–4 (2002).
29. A. F. Voter, "Guest editor," *MRS Bull.* (Feb. 1996).
30. A. Y. Toukmaji and J. A. Board, "Ewald summation techniques in perspective: a survey," *Comput. Phys. Commun.* **95**, 73–92 (1996).
31. W. A. Harrison, *Electronic Structure and the Properties of Solids: The Physics of the Chemical Bond*, Freeman, San Fransisco, CA (1980).
32. V. Vitek, "Computer simulation of the screw dislocation motion in bcc metals under the effect of the external shear and uniaxial stress," *Proc. R. Soc. Lond. A* **352**, 109–124 (1976).
33. V. Vitek, "Structure of dislocation cores in metallic materials and its impact on their plastic behavior," *Prog. Mater. Sci.* **36**, 1–27 (1992).
34. M. S. Daw and M. I. Baskes, "Embedded-atom method: derivation and application to impurities, surfaces, and other defects in metals," *Phys. Rev. B* **29**, 6443–6453 (1984).
35. M. W. Finnis and J. E. Sinclair, "A simple empirical N-body potential for transtion metals," *Phil. Mag. A* **50**, 45–55 (1984).
36. S. M. Foiles, M. I. Baskes, and M. S. Daw, "Embedded-atom-method functions for the fcc metals Cu, Ag, Au, Ni, Pd, Pt and their alloys," *Phys. Rev. B* **33**, 7983–7991 (1986).
37. J. H. Rose, J. Ferrante, and J. R. Smith, "Universal binding energy curves for metals and bimetallic interaces," *Phys. Rev. Lett.* **47**, 675–678 (1981).
38. F. Ercolessi and J. B. Adams, "Interatomic potentials from first-principles calculations—the force-matching method," *Europhys. Lett.* **26**, 583 (1994).
39. A. F. Voter, *Intermetallic Compounds*, Vol. 1, *Principles*, 77–90, Wiley, New York (1994).
40. A. E. Carlsson, "Beyond pair potentials in elemental transition metals and semiconductors," *SSP* **43**, 1–91 (1990).
41. F. H. Stillinger and T. A. Weber, "Computer-simulation of local order in condensed phases of silicon," *Phys. Rev. B* **31**, 5262–5271 (1985).
42. J. Tersoff, "Empirical interatomic potential for silicon with improved elastic properties," *Phys. Rev. B* **38**, 9902–9905 (1988).
43. D. W. Brenner, "Empirical potential for hydrocarbons for use in simulating the chemical vapor deposition of diamond films," *Phys. Rev. B* **42**, 9458–9471 (1990).
44. M. B. Nardelli, B. I. Yakobson, and J. Bernholc, "Mechanism of strain release in carbon nanotubes," *Phys. Rev. B* **57**, R4277–R4280 (1998).
45. K. Huang, *Statistical Mechanics*, Wiley, New York (1987).

46. R. K. Pathria, *Statistical Mechanics*, Butterworth-Heinemann, Oxford (1996).
47. N. Metropolis and S. Ulam, "The Monte Carlo method," *J. Am. Stat. Assoc.* **44**, 335–341 (1949).
48. M. P. Allen and D. J. Tildesley, *Computer Simulation of Liquids*, Oxford Univerisity Press, New York (1987).
49. S. Grigoras, A. A. Gusev, S. Santos, and U. W. Suter, "Evaluation of the elastic constants of nanoparticles from atomistic simulations," *Polymer* **43**, 489–494 (2002).
50. L. Verlet, "Computer experiments on classical fluids. I. Thermodynamical properties of Lennard-Jones molecules," *Phys. Rev.* **159**, 98–103 (1967).
51. L. Verlet, "Computer experiments on classical fluids. II. Equilibrium correlation functions," *Phys. Rev.* **165**, 201–214 (1967).
52. D. Frenkel and B. Smit, *Understanding Molecular Dynamics: From Algorithms to Applications*, Academic Press, San Diego, CA (1996).
53. S. Nosé, "Constant temperature molecular dynamics methods," *Prog. Theo. Phys.* **103**, 1–46 (1991).
54. W. G. Hoover, *Computational Statistical Mechanics*, Elsevier, Amsterdam (1991).
55. S. Nose, "A unified formulation of the constant temprature molecular dynamics methods," *J. Chem. Phys.* **81**, 511–519 (1984).
56. F. F. Abraham, "Very large scale simulations of materials failure," *Phil. Trans. R. Soc. Lond. A* **360**, 367–382 (2002).
57. V. B. Shenoy, R. M. Miller, E. B. Tadmor, R. Phillips, and M. Ortiz, "Quasicontinuum models of interfacial deformation," *Phys. Rev. Lett.* **80**, 742–745 (1998).
58. V. B. Shenoy, R. M. Miller, E. B. Tadmor, D. Rodney, R. Phillips, and M. Ortiz, "An adaptive finite element approach to atomic-scale mechanics—the quasicontinuum method," *J. Mechan. Phys. Solids* **47**, 611–642 (1998).
59. E. Isaacson and H. B. Keller, *Analysis of Numerical Methods*, Dover, New York (1994).
60. D. Rodney and R. Phillips, "Structure and strength of dislocation junctions: an atomic level analysis," *Phys. Rev. Lett.* **82**, 1704–1707 (1999).
61. V. B. Shenoy, "Quasicontinuum models of atomic-scale mechanics," PhD Thesis, Brown University, Providence, RI (1998).
62. V. B. Shenoy, "Multi-scale modeling strategies in materials science—the quasi-continuum method," *Bull. Mater. Sci.* **26**, 53–62 (2003).
63. A. C. Eringen, "Continuum mechanics at the atomic scale," *Crys. Latt. Defect.* **7**, 109–130 (1977).
64. R. E. Miller and V. B. Shenoy, "Size-dependent elastic propertis of nanosized structual elements," *Nanotechnology* **11**, 139–147 (2000).
65. V. B. Shenoy, "Size-dependent rigidities of nanosized torsional elements," *Int. J. Solids Struct.* **39**, 4039–4052 (2002).
66. M. E. Gurtin and A. I. Murdoch, "A continuum theory of elastic material surfaces," *Arch. Rat. Mechan. Anal.* **57**, 291–323 (1975).

67. J. R. Rice and T.-Z. Chuang, "Energy variations in diffusive cavity growth," *J. Am. Ceram. Soc.* **64**, 46–53 (1981).
68. R. C. Cammarata, "Surface and interface stress effects in thin films," *Prog. Surf. Sci.* **46**, 1–38 (1994).
69. R. Shuttleworth, "The surface tension of solids," *Proc. Phys. Soc. Lond. A* **63**, 444–457 (1950).
70. R. Phillips, *Crystals, Defects and Microstructures–Modeling Across Scales*, Cambridge University Press, Cambridge (2001).
71. E. D. Minot, Y. Yaish, V. Sazonova, J.-Y. Park, M. Brink, and P. L. McEuen, "Tuning carbon nanotube band gaps with strain," http://arXiv.org/abs/cond-mat/0211152 (2002).

List of symbols

β	$1/k_B T$
γ	surface energy
ϵ_0	permittivity of free space
$\boldsymbol{\epsilon}$	small-strain tensor
$\kappa_{\alpha\beta}$	components of surface curvature tensor
ρ	electron density
$\boldsymbol{\sigma}$	stress tensor
$\tau_{\alpha\beta}$	components of surface stress tensor
ψ	electronic wave function
Ψ	many-electron wave function
$\boldsymbol{\omega}$	small-rotation tensor
∇	gradient operator
$\boldsymbol{a}_i$	acceleration of ith atom
$\boldsymbol{b}$	body force
$[\boldsymbol{B}]$	strain-displacement matrix
C_{ijkl}	components of elastic modulus tensor
d	size-scale of nanostructure
D	stiffness of nanostructure determined from atomistics
D_c	stiffness of nanostructure determined from continuum theory
E_g	ground state energy of many-electron system
E_{tot}	total energy of atomistic system
$\boldsymbol{E}$	Green-Lagrange strain tensor
$\boldsymbol{f}$	surface force
$F(\rho)$	embedding energy
$\boldsymbol{F}$	deformation gradient tensor
h	Planck's constant
$\hbar$	$h/2\pi$
$\mathcal{H}$	Hamiltonian operator
$H_{i\alpha j\beta}$	tight-binding Hamiltonian matrix elements
$\mathbf{i}$	$\sqrt{-1}$
$\boldsymbol{I}$	unit tensor
$J_{i\alpha j\beta}$	tight-binding overlap matrix elements
k_B	Boltzmann constant
m	mass of atom
m_e	mass of electron
$n(\boldsymbol{r})$	electron density
$\boldsymbol{n}$	normal vector
N_α	finite element shape function associated with node α
$\{\boldsymbol{N}\}$	shape function matrix
$\boldsymbol{p}_i$	momentum vector of ith atom
Q	configurational integral
$\boldsymbol{r}_j$	position vector of jth electron

r_{ij}	distance between atom i and atom j
$\mathcal{S}$	surface enclosing a body
t	time
$\boldsymbol{t}$	traction vector
T	temperature
$\boldsymbol{u}$	displacement vector
$\mathcal{V}$	region defining a body
W	strain energy density
$\boldsymbol{x}$	position vector in reference configuration
$\boldsymbol{x}_i$	position vector of atom i
$\boldsymbol{y}$	position vector in deformed configuration
Z	partition function

Vijay B. Shenoy received his BTech. degree in mechanical engineering from the Indian Institute of Technology, Madras, in 1992. In 1994, he received his MS degree in computational mechanics from Georgia Institute of Technology, Atlanta. He received his PhD degree from Brown University, Providence, Rhode Island, in 1998, for his work on the quasi-continuum method. He was an assistant professor with the Department of Mechanical Engineering, Indian Institute of Technology, Kanpur from 1999 to 2002. Currently, he is an assistant professor with the Materials Research Centre, Indian Institute of Science, Bangalore. His areas of interest are multiscale materials modeling, properties of nanostructures, and physics of soft thin films.

Chapter 8

Nanoscale Fluid Mechanics

Petros Koumoutsakos, Urs Zimmerli, Thomas Werder, and Jens H. Walther

8.1. Introduction 320
8.2. Computational nanoscale fluid mechanics 322
8.2.1. Quantum mechanical calculations 323
8.2.2. *Ab initio* calculations of water aromatic interaction 324
8.2.3. Atomistic computations 327
8.2.4. Multiscaling: linking macroscopic to atomistic scales 334
8.3. Experiments in nanoscale fluid mechanics 339
8.3.1. Diagnostic techniques for the nanoscale 339
8.3.2. Atomic force microscopy for fluids at the nanoscale 344
8.4. Fluid-solid interfaces at the nanoscale 347
8.4.1. Hydrophobicity and wetting 347
8.4.2. Slip flow boundary conditions 350
8.5. Fluids in confined geometries 355
8.5.1. Flow motion in nanoscale channels 355
8.5.2. Phase transitions of water in confined geometries 360
8.6. Nanofluidic devices 362
8.6.1. Solubilization 363
8.6.2. Nanofluids 363
8.6.3. CNT as sensors and AFM tips 364
8.6.4. Carbon nanotubes as storage devices—adsorption 365
8.6.5. Nanofluidics for microscale technologies 366
8.7. Outlook—go with the flow 370
Acknowledgments 371
References 371
List of symbols 391

8.1 Introduction

Nanoscale fluid mechanics (NFM) is the study of fluid (gas, liquid) flow around and inside nanoscale configurations. As we are increasingly enabled to study nanoscale systems, through advanced computations and innovative experiments, it becomes apparent that the ancient saying $\tau\alpha\ \pi\alpha\nu\tau\alpha\ \rho\epsilon\iota$ ("everything flows") remains valid in the era of nanotechnology. Nanoscale flow phenomena are ubiquitous!

As a start, biology evolves in an environment that is mostly water. While the percentage of water in human bodies is about 65%, it is generally higher in plants (about 90%), and even more so in week-old human embryos (up to 97%)! Where is the water? In human beings, 1/3 of it can be found in the extracellular medium, while 2/3 of it lies within the intracellular medium, a confined environment that is typically a few microns in diameter. From the words of Alberts et al., "Water accounts for about 70% of a cell's weight, and most intracellular reactions occur in an aqueous environment. Life on Earth began in the ocean, and the conditions in that primeval environment put a permanent stamp on the chemistry of living things. Life therefore hinges on the properties of water."[1]

As scientists and engineers develop nanoscale sensor and actuator devices for the study of biomolecular systems, NFM will play an increasingly important role. The study of fundamental nanoscale flow processes is a key aspect of our effort to understand and interact with biological systems. Many biomolecular processes such as the transport of DNA and proteins are carried out in aqueous environments, and aerobic organisms depend on gas exchange for survival. The development of envisioned nanoscale biomedical devices such as nanoexplorers and cell manipulators will require understanding of natural and forced transport processes of flows in the nanoscale. In addition, it will be important to understand transport processes around biomolecular sensing devices to increase the probability of finding target molecules and identifying important biological processes in the cellular and subcellular level in isolated or high background noise environments.

While there can be a large variety of nanoscale systems (from the individual molecules themselves to the assembly of those molecules into complex structures such as cellular membranes), it would be a formidable task to try to understand the essential physics of these systems by peering at every known device. For more than a century, engineering fluid mechanics has taught us that simple, canonical experiments, such as the flow around a circular cylinder, can provide us with all of the fundamental physics needed to understand the flow dynamics of much more complex systems, such as the aerodynamics of airplanes or the hydrodynamics of ships. Following this conjecture, one may consider that the study of fundamental nanoscale flow physics of prototypical configurations will enable further advances in the development of complex scientific and engineering devices. At the same time, we are reminded that thousands of airplanes had been flying without the engineers having understood every minute detail about turbulent flow.

Nanoscale flow physics also affects flows at larger scales in an inherently multiscaling way. For instance, phenomena such as wall turbulence and aircraft aerodynamics are dependent on the behavior of fluids in the near-wall region of aero/hydrodynamic structures. This near-wall region, the so-called boundary layer,[2] has often been modeled using the highly debated "no-slip condition." This condition hinges on nanoscale fluid flow phenomena, which only now have become amenable to experimental and theoretical investigations.

NFM is a complex and still pristine research subject, mainly because it cannot be tackled with conventional experimental means and also because no universally accepted model equation has ever been laid down for such flows. However, research on these frontiers is expected to bring advances that will largely enhance our understanding and will enable us to develop better engineering devices.

Currently open research issues in computational and experimental NFM can be categorized into four major tasks, namely,

1. In computational studies, it is important to develop suitable models and efficient computational tools for the systems that are being simulated. Key aspects include the development of suitable interaction potentials for molecular dynamics simulations based on experiments and *ab initio* calculations, the development of hybrid computational methods such as QM/MM methods, combining classical molecular mechanics with quantum mechanical calculations, e.g., Car-Parrinello molecular dynamics,[3] and the development of efficient multiscaling techniques. Specific algorithmic developments involve the treatment of the long-range forces between molecules and the development of efficient computational techniques to extend the time-scale of the simulation as well as to expand the range of solutes and solid substrates that can be studied.
2. Experimental diagnostic techniques need to be developed to provide quantitative information for phenomena that take place in the nanoscale. Techniques and instruments that are able to explore atomistic structures are invaluable on this front. The adoption of innovative and interdisciplinary approaches is necessary to face the challenges of this task.
3. As the third task, we consider the study of prototypical flows to identify key physical mechanisms such as the degrees of slip and sticking at the solid-liquid interfaces or to determine the changes in liquid viscosity and surface tension near the surfaces and inside small pores. Particular flows of interest involve flows inside nanopores and nanoscale flows as influencing the interface of nanoscale flows with larger-scale flow phenomena. A suitable synergy of experimental and computational techniques will benefit the problems at hand and the techniques themselves.
4. The fourth task involves the continuous exploration of fundamental and novel concepts for nanofluidic devices. Through an interdisciplinary approach and in a combined experimental and computational setting, we can consider preliminary designs using molecular simulations that need to be subsequently verified via appropriate experiments. In particular, the large-scale manufacturing of nanoscale flow devices needs to be addressed as well

as their interface with microscale devices. Included in these concepts are biomolecular sieves, nanopores, nanocilia, and nanopumps. These studies will provide the basis for a rational design of nanoscale biomolecular sensors and actuators.

Several comprehensive review articles have appeared in the area of nanoscale fluid mechanics, a nonexhaustive list of which is given here. Koplik and Banavar[4] presented one of the first reviews discussing the study of phenomena of macroscale systems from atomistic simulations while Micci et al.[5] have reviewed research of nanoscale phenomena related to atomization and sprays. In recent articles, Maruyama[6] and Poulikakos et al.[7] have reviewed molecular dynamics simulations of micro- and nanoscale thermodynamic phenomena. Moving up to mesoscales, Gad-el Hak[8] and Ho and Tai[9,10] presented reviews of the flow in microdevices and microelectromechanical systems (MEMS) devices. Vinogradova[11] and Churaev[12] reviewed the slippage of water over hydrophobic surfaces, including general properties of thin liquid layers.

However, nanotechnology is a very dynamic field and new information is constantly becoming available from improved computational models and experimental diagnostics. For example, much has changed since the review of Koplik and Banavar[4] on slip boundary conditions: the presence of slip has been demonstrated in experiments at hydrophobic[13] and at hydrophilic surfaces,[14] thus casting doubts on the validity of the no-slip condition. These words of caution must be kept in mind as well when assessing the works discussed in this review.

The chapter is structured as follows: Sec. 8.2 discusses computational aspects of NFM. We emphasize that practitioners understand the ramifications of seemingly benign tasks such as the choice of the molecular interaction potentials and simulation boundary conditions. The simulated physics critically depend on such choices.

Section 8.3 discusses experimental diagnostics techniques for nanoscale flow phenomena. The interdisciplinary and innovative approaches of scientists and engineers when probing flows at the nanoscale is exemplified in this topic. Section 8.4 discusses the flow phenomena at the interface of fluids and solids from the NFM perspective, while in Sec. 8.5 the effects of confinement to fluid mechanics are discussed. Finally, Sec. 8.6 discusses a selective list of applications where nanoscale flow phenomena play a critical role.

8.2 Computational nanoscale fluid mechanics

The difficulty of carrying out controlled experiments on nanoscale systems makes computational studies potent alternatives for characterizing their properties. This fact has led to several computational studies of nanoscale phenomena using molecular simulations, and many of the advances to date in nanotechnology have come from theoretical or computational predictions that were later confirmed by experiment (e.g., the metallic and semiconducting nature of carbon nanotubes[15]).

The goal of computational studies in NFM is to characterize prototypical nanofluidic systems as well as to explore specific nanoscale flow phenomena that may facilitate the development of nanoscale flow sensors and actuators, nanodevices capable of manipulating biomolecules in the form of molecular sieves, etc.

The development of efficient solvers for quantum mechanical (QM) and molecular mechanical (MM) simulations has enabled reliable simulation of phenomena involving up to a few thousand atoms. For larger systems, the method of molecular dynamics (MD) is used to simulate systems that can be described with up to a few million atoms. However, as nanoscale devices are often embedded in micro- and macroscale systems, the computation of such flows requires a proper integration of atomistic simulations with computational methods suitable for larger scales. One of the great challenges in computational NFM is the development of efficient computational methods to tackle the large number of time and space scales associated with NFM. Multiscaling techniques bridging nano and micro/macroscale flow phenomena may well be very fruitful areas of research in the near future.

8.2.1 Quantum mechanical calculations

Quantum mechanical phenomena are described by wavelike particles, which are mathematically represented by a wave function Ψ. The differential equation that describes their evolution in time was developed in 1925 by Schrödinger:

$$i\hbar\frac{\partial}{\partial t}\Psi = \hat{E}\Psi = \hat{H}\Psi = -\frac{\hbar^2}{2m}\nabla^2\Psi + V(r,t)\Psi, \tag{8.1}$$

where i is the imaginary unit, $\hbar$ is Planck's constant divided by 2π, the time is represented by t, and the energy operator is $\hat{E}$. The Hamiltonian operator $\hat{H}$ is the sum of the potential energy operator $V(r,t)$ and the kinetic energy operator, and m denotes the mass of the particle. The Schrödinger Eq. (8.1), though mostly used in its time-independent form, is the basis for the solution of atomistic systems that

- Involve the determining of structural problems, for example, questions regarding conformation and configuration of molecular systems as well as geometry optimizations; and
- Require finding energies under given conditions, for example, heat of formation, conformational stability, chemical reactivity, and spectral properties.

Analytic solutions of the Schrödinger equations are known only for special cases, where the potential energy contribution to the Hamilton operator is particularly simple. For example, this is the case if there is no potential energy contribution (free particle) or in the case of a single electron in the field of a nucleus (hydrogen atom).

In more complex situations, the Schrödinger equation has to be solved approximately. The approximation methods can be categorized as either *ab initio* or semi-empirical. While *ab initio* calculations tackle the full form of the equations, semi-empirical methods replace some of the time-consuming expressions and terms by empirical approximations. The parameters for semiempirical methods are usually either derived from experimental measurements or from *ab initio* calculations on model systems. For a detailed introduction into different methods and the according approximations, the reader is referred to quantum chemistry text books (e.g., Ref. 16).

Note that *ab initio* methods also depend on appoximations such as the Born-Oppenheimer approximation[17] or the choice of underlying basis sets[16] to model the wave function. While *ab initio* calculations are independent of fitted parameters and enable us to calculate properties of interest fully deterministically, the main advantage of semiempirical methods lies in the reduced computational cost, which enables the simulation of larger systems of one to two orders of magnitude. Therefore phenomena can be studied on different scales, and size restrictions of *ab initio* methods can be overcome.

In the following two sections, we review second-order Møller–Plesset and density functional theory (DFT) calculations on water interacting with aromatic systems and we focus the extrapolation of these results to the water graphite interaction. The water graphite interaction is reviewed here as it is of particular interest in the field of hydrophobic interactions. It provides a prototypical system to study hydrophobic interactions, which are important in various areas of NFM such as flow in nanopores and protein folding in aqueous environments.

8.2.2 *Ab initio* calculations of water aromatic interaction

Feller and Jordan[18] used an approach based on second-order Møller–Plesset perturbation theory[19] to calculate the interaction energy between a water molecule and a sequence of centrosymmetric, aromatic systems, consisting of up to 37 aromatic rings. An extrapolation of the results yields an estimated electronic binding energy of $-24.3\ \mathrm{kJ\,mol^{-1}}$ for a single water molecule interacting with a monolayer of graphite. In these calculations, the largest sources of uncertainty are the basis set superposition error, the incompleteness of the basis set, and the assumptions regarding the extrapolation from the clusters to the graphite sheet.[18]

The aforementioned estimate of the binding energy of a water molecule to a graphite sheet is appreciably larger than an experimentally determined estimate[18] of $-15\ \mathrm{kJ\,mol^{-1}}$. Nevertheless, this result, along with data presented in the next section allows to parametrize classical force fields.

The estimate of the water-graphite binding energy from Feller and Jordan[18] is slightly larger than the interaction between two water molecules but still significantly lower than the average electronic binding energy of a fully solvated water molecule, where hydrogen bonding provides a network leading to high binding energies.

Feller and Jordan[18] identified that the most important attractive interactions are the dipole-quadrupole, dipole-induced dipole (induction), and dispersion contributions in their study of the water-benzene complex. From the underlying data, they concluded that the dispersion interaction is critical, contributing of the order of -25 kJ mol^{-1} to the binding energy. This issue is further discussed in the following section.

8.2.2.1 Density functional theory calculations

High-order QM calculations, such as the second-order Møller–Plesset[19] approach, reproduce the interaction energy of weakly bound molecular systems reasonably. However, the systems that can be investigated with these methods are limited in size due to the high computational cost.

DFT provides an intermediate accuracy at lower computational cost by basing the calculation of system properties on the electron density. For a detailed introduction to DFT the reader is referred to Ref. 20. For the calibration of interaction potentials, a DFT study of larger weakly bound systems is of highest interest

DFT describes hydrogen bonds with reasonable accuracy,[21] whereas the description of weak interactions, generally denoted as dispersion interactions, is not correctly reproduced. The dispersion energy results from correlated fluctuations in the charge density, which contribute to the interaction energy even at distances where electron density overlap is negligible. Since all current DFT energy functionals are approximations based on expressions for local electron density, its gradient, and the local kinetic-energy density,[22] they fail to reproduce the dispersion contribution to the interaction energy.

Anderson and Rydberg[23] and Hult et al.[24,25] presented an approach to extend DFT calculations with local or semilocal approximations to include the dispersion contribution, and Rydberg et al. applied it to graphite.[26] Although their model depends on a cutoff to ensure finite polarizabilities at all electron densities,[25,27] their approach is promising with regard to a unified treatement within DFT.

Alternative approaches have been presented by Wu et al.[28] and Elstner et al.[29] Wu et al.[28] concentrated on the interaction between small molecules and presented a systematic search for a possible simplified representation of the weak interaction in DFT. In Ref. 30, this approach was extended to deal with the interaction between a flat semiconductor surface and a small molecule. Two distinct models are discussed that serve to calculate lower and upper bounds to the interaction energy. The model assumptions are then validated for a water benzene system and the method is applied to the water graphite case to obtain the lower and upper bound to the water graphite interaction.

In Wu et al.[28] a correction term ΔE_{disp} is proposed to account for the contribution of dispersion energy in the total interaction energy

$$\Delta E_{\text{tot}} = \Delta E_{\text{DFT}} + \Delta E_{\text{disp}}, \tag{8.2}$$

where ΔE_{DFT} is the DFT interaction energy and ΔE_{disp} is a damped correction term based on the first term of the dispersion energy expansion.[31] The dispersion energy expansion has the following form:

$$\Delta E_{\text{disp}} = \frac{C_n}{r^n} g_n(r), \tag{8.3}$$

where C_n denotes the dispersion coefficient, r is the distance between the two centers of mass, and n is a geometry-specific integer resulting from theoretical considerations.[32,33] In the asymptotic limit, at long distances, it can be shown that the coefficents for different geometries map onto each other.[34] Additionally, the dispersion energy correction has to be damped by a geometry-specific damping function $g_n(r)$, which is necessary as the dispersion correction diverges at short range instead of reaching saturation.[31,35]

The interaction energy between water and graphite can be bound (Fig. 8.1) as described in Ref. 30. The minimum interaction energy can be computed when considering graphite as a collection of isolated molecules,[33] while an upper bound can be computed when considering the graphite sheet as an ideal metal.[36] To model an ideal metal, the Jellium model was used, in which the electrons are free to move while only subject to a homogeneous background charge.

Grujicic et al.[39] carried out DFT calculations to analyze the effect on the ionization potential of carbon nanotubes due to the absorbtion of molecules with high dipole moments as well as clusters of water molecules at the tip of capped (5,5) metallic armchair nanotubes. The results obtained show that the adsorption energies of both single- and multimolecule clusters are quite low (typically less than

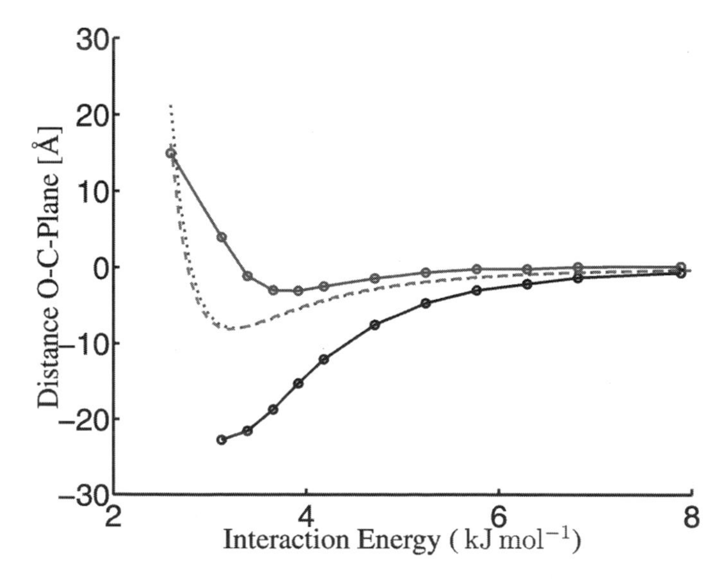

Figure 8.1 Upper and lower bounds (solid line) to the interaction energy compared with two force field expressions GROMOS (dotted)[37] and Werder et al.[38] (−−). The upper bound is obtained through the assumption of a noninteracting plane of atoms or molecules, whereas the lower bound is obtained through the assumption of an ideal metal.

2.9 kJ mol^{-1}). This suggests that the studied adsorbates are not stable and would most likely desorb quickly. In the same work, in sharp contrast, under a typical field-emission electric field the adsorbtion energy was found to be substantially higher making the adsorbates stable.

8.2.3 Atomistic computations

The computational cost of quantum mechanical calculations does not permit simulations of systems containing more than a few hundred atoms. In this case, the behavior of the system is modeled using MD simulations. MD involves computing the trajectories of particles that model the atoms of the system, as they result from relatively simplified interaction force fields.

MD has been used extensively in the past to model the structural and dynamic properties of complex fluids. The first MD simulations date back to the mid-1950s in works of Fermi et al.[40] Then in 1957 in Alder and Wainwright,[41] the phase diagram of a hard sphere system was investigated. A few years later, Aneesur Rahman at Argonne National Laboratory published his seminal work on correlations in the motion of atoms in liquid argon.[42] In 1967 Loup Verlet calculated the phase diagram of argon using the Lennard–Jones potential and computed correlation functions to test theories of the liquid state,[43,44] and two years later phase transitions in the same system were investigated by Hansen and Verlet.[45] In 1971 Rahman and Stillinger reported the first simulations of liquid water.[46] Since then, MD simulations have provided a key computational element in physical chemistry, material science, and NFM for the study of pure bulk liquids,[47] solutions, polymer melts,[48] and multiphase and thermal transport.[49–52] The motion of an ensemble of atoms in MD simulations is governed by interatomic forces obtained from the gradient of a potential energy function. This so-called force field is an approximation of the true interatomic forces arising from the interaction of electrons and nuclei. Thus, the qualitative and quantitative result of MD simulations is intimately related to the ability of the potential energy function to represent the underlying system.

Several "generic" force fields have been developed, ranging from general purpose force fields capable of describing a wide range of molecules, such as the universal force field,[53] to specialized force fields designed for graphitic and diamond forms of carbon,[54] for covalent systems,[55] and models for liquid water.[56–59] Several classes of force fields have been developed to account for specific types of molecules or chemical systems, e.g., for zeolites,[60] for biomolecules such as AMBER[61] and GROMOS,[37] and CHARMM for proteins,[62] or for organic molecules.[63]

With an abundance of potentials and parameters to account for interatomic forces, the user may wish to consider the following criteria for choosing a potential:

- Accuracy: the simulation should reproduce the properties of interest as closely as possible.

- Generalization/transferability: the force field expressions should be applicable to situations for which it was not explicitly fitted.
- Efficiency: force calculations are generally the most time-consuming part of a simulation and they should be as efficient as possible

The proper balance between these criteria depends to a large extent on the system to be investigated. Thus, for NFM studies that involve chemical reactions, the classical representation is usually not sufficient and a quantum or a hybrid quantum-classical technique is required[64,65] to capture the breaking and formation of chemical bonds. On the other hand, in large-scale simulations of nonreactive systems, computational efficiency is essential and simple expressions for the forces will suffice.

Force fields are generally empirical in the sense that a specific mathematical form is chosen and parameters are adjusted to reproduce available experimental data such as bond lengths, energies, vibrational frequencies, and density.[59,66] Generic force fields are developed to be suitable for a wide range of molecules. One should be aware of this fact when considering these generic force fields for the study of a specific system. In this case, it is not uncommon to conduct QM calculations for a small system in order to calibrate MD potentials for the system under consideration.

A complementary route to experimental results in developing interaction potentials involves their calibration using simulations from first principles. We exemplify this process by considering the problem of water-graphite interactions.[18,38] This can be seen as a model problem for more complex water-carbon interactions such as those involved when considering carbon nanotubes as biosensors and fullerenes as chemical reaction chambers or nanoreactors.[67] An added complexity to this problem is that the behavior of water in confined geometries is drastically different than in bulk systems.[68] Using MD simulations to reliably understand and analyze such systems, it is important to develop suitable models for the simulation of water in such environments.

While water-water potentials are well established in the literature,[46,56–58] there are no reliable water-nanotube potentials at the moment. In addition, one may need to reconsider the water-water potentials when considering its drastic change in behavior in confined geometries. The starting point for the development of such potentials is the quantification of the interaction of a single water molecule with a single layer of graphite. The reliablity of existing estimates for the interaction energy is questionable as they exhibit large variations ranging from $-5.07\ \mathrm{kJ\,mol^{-1}}$(Ref. 69) to $-24.3\ \mathrm{kJ\,mol^{-1}}$(Ref. 18), leaving a great uncertainty about predicted behavior. Furthermore, there exists surprisingly little experimental data, with a reported experimentally determined interaction energy[18] of $15\ \mathrm{kJ\,mol^{-1}}$. Werder et al.[38] presented a review of recently used interaction potentials for the water-graphite interaction and a linear relationship between the interaction energy and the contact angle of water on graphite could be determined. As there are, however, contradictory measurements of water graphite contact angles,[38] the actual interaction still remains an open question.

8.2.3.1 Molecular dynamics: force fields and potentials

The potential energy function or force field provides a description of the relative energy or forces of the ensemble for any geometric arrangement of its constituent atoms. This description includes energy for bending, stretching, and vibrations of the molecules and interaction energies between the molecules. Classical force fields are usually built up as composite potentials, i.e., as sums over many rather simple potential energy expressions. Mostly pair potentials $V(r_{ij})$ are used, but in the case of systems where bonds are determining the structure, multibody contributions $V(r_{ij}, r_{ik})$ and $V(r_{ij}, r_{ik}, r_{il})$ can also enter the expression, thus

$$U = \sum_{i,j} V(r_{ij}) + \sum_{i,j,k} V(r_{ij}, r_{ik}) + \sum_{i,j,k,l} V(r_{ij}, r_{ik}, r_{il}), \tag{8.4}$$

where $r_{ij} = |\mathbf{r}_i - \mathbf{r}_j|$ is the distance between ith and jth atoms. The contribution to the interaction potential can be ordered in two classes: intramolecular and intermolecular contributions. While the former describe interactions that arise in bonded systems, the latter are usually pair terms between distant atoms.

8.2.3.2 Intramolecular forces

Various intramolecular potentials are used to describe the dynamics of chemical bonds. The potential

$$V(r_{ij}) = \frac{1}{2} K_h (r_{ij} - r_0)^2 \tag{8.5}$$

is developed from a consideration of simple harmonic oscillators,[40] where r_{ij} and r_0 denote the bond length and the equilibrium bond distance, respectively. The force constant of the bond is given by K_h. Alternatively, the Morse potential,[70]

$$V(r_{ij}) = K_M \left(e^{-\beta(r_{ij}-r_0)} - 1\right)^2, \tag{8.6}$$

is used, allowing for bond breaking. Here K_M and β are the strength and distance related parameters of the potential.

For coordination centers, i.e., atoms where several bonds come together, usually bond angle terms are applied including harmonic bending via

$$V(\theta_{ijk}) = \frac{1}{2} K_\theta (\theta_{ijk} - \theta_c)^2, \tag{8.7}$$

or the harmonic cosine bending via

$$V(\theta_{ijk}) = \frac{1}{2} K_\theta (\cos\theta_{ijk} - \cos\theta_c)^2, \tag{8.8}$$

where θ_{ijk} is the angle formed by the bonds extending between the ith, jth, and kth atoms, and θ_c is the equilibrium angle. Dihedral angle potentials are often employed for systems involving chains of bonded atoms to ensure a consistent representation over several centers[71,72]

$$V(\phi_{ijkl}) = \frac{1}{2}\sum_{m=0}^{n} K_m \cos(m\phi_{ijkl}), \tag{8.9}$$

where the sum can contain up to 12 terms.

As an example, a single-walled carbon nanotube immersed in water was recently described using the Morse, harmonic cosine, and torsion potentials by Walther et al.[73] The torsion potential was fitted to quantum chemistry calculations of tetracene ($C_{18}H_{12}$) using density functional theory.[74]

An alternative to the direct modeling of bonded interactions and intramolecular forces is to constrain the bond length or bond angle.[75] As an example, most water models consider rigid molecules.[76] The high-frequency oscillation of the O–H bonds in water formally requires a quantum mechanical description, and removing these intramolecular degrees of freedom alleviates the problem. The computational efficiency is furthermore significantly improved by allowing a 5 to 10 times larger time step than the flexible models.[56] The constraints are imposed using iterative procedures such as SHAKE,[77–79] SETTLE,[80] or direct methods.[81]

8.2.3.3 Intermolecular forces

Commonly applied intermolecular force terms are van der Waals forces described through a Lennard–Jones 12–6 potential[82]

$$V(r_{ij}) = 4\epsilon\left[\left(\frac{\sigma}{r_{ij}}\right)^{12} - \left(\frac{\sigma}{r_{ij}}\right)^{6}\right], \tag{8.10}$$

where ϵ is the depth of the potential well and σ is related to the equilibrium distance between the atoms. The parameters are usually obtained through fitting to experimental data and/or theoretical considerations. For multiatomic fluids such as gaseous fluids, the Lorentz-Berthelot mixing rules are often used,[47] thus,

$$\epsilon_{IJ} = \sqrt{\epsilon_I \epsilon_J}, \qquad \sigma_{IJ} = \frac{1}{2}(\sigma_I + \sigma_J), \tag{8.11}$$

where I and J denote the Ith and Jth atomic species. However, recent work[83] has shown this approach to be inadequate for accurate liquid simulations, as quantities like liquid mass density are sensitive to the choice of parameters.

For large surfaces an average 10–4 Lennard–Jones potential may be obtained by integrating the 12–6 Lennard–Jones over the surface as

$$V(z) = 4\epsilon\sigma^2\left[\left(\frac{\sigma\pi}{z}\right)^{10} - \left(\frac{\sigma\pi}{z}\right)^{4}\right], \tag{8.12}$$

where z is the wall normal distance (cf., e.g., Ref. 84). The fast decay of the Lennard–Jones potential usually enables a spherical truncation of the potential at a cutoff distance r_c. Typical cutoff values are 1.5σ for purely repulsive interactions, and 2.5σ and 10σ for homogeneous and inhomogeneous systems.

The long-range electrostatic interactions are described through the Coulomb potential

$$V(r_{ij}) = \frac{q_i q_j}{4\pi\epsilon_0 r_{ij}}, \tag{8.13}$$

where q_i and q_j refer to the electric charges of the ith and jth atoms, and ϵ_0 is the permittivity of vacuum. Fractional charges are used for polar molecules, and integral values are used for monatomic ions. The long-range interaction implied by the electrostatics requires fast summation techniques, see Sec. 8.2.3.4. To accelerate the algorithmic development and computational time for homogeneous systems, the Coulomb potential can be truncated using a smooth tapering of the potential energy function,[85]

$$V(r_{ij}) \approx \frac{q_i q_j}{4\pi\epsilon_0 r_{ij}} S(r_{ij}), \tag{8.14}$$

where $S(r)$ is a smoothing function, e.g.,

$$S(r_{ij}) = \begin{cases} \left[1 - (r_{ij}/r_c)^2\right]^2 & r_{ij} < r_c, \\ 0 & r_{ij} \geqslant r_c. \end{cases} \tag{8.15}$$

Note, however, that the results obtained from MD simulations using a truncation may be significantly different from results using Ewald summation, in particular for systems with inhomogeneous charge distributions and for ionic solutions.[86] On the other hand, fast summation techniques may introduce artifically strong correlations in small systems,[87] and when employed with potentials calibrated with truncation, the results using Ewald summation techniques may be less accurate than using truncation.[59,88]

8.2.3.4 Computational issues in MD

Molecular dynamics simulations of heterogeneous nanoscale flows may involve the computation of the interaction of millions of atoms. For example, a cube of water with an edge length of 20 nm contains approximately one million atoms. The most time-consuming aspect of MD simulations of large systems is the accurate evaluation of the long-range interactions, which include electrostatic and dispersion interactions. Without an explicit cutoff, the computational cost scales as $\mathcal{O}(N^2)$ for N particles. Efficient algorithms have been devised to reduce the computational cost, ranging from simple sorting already provided by Verlet[44] to accurate fast summation techniques such as Ewald summation,[89–91] the Particle-Mesh

Ewald (PME) method,[92,93] and the particle-particle particle-mesh technique (P^3M) by Hockney and Eastwood and their colleagues.[94–97] While Ewald summation requires $\mathcal{O}(N^{1.5})$ operations, the PME and P^3M techniques scale as $\mathcal{O}(N \log N)$.

To achieve this computational efficiency, the P^3M method utilizes a grid to solve for the potential field (Φ)

$$\nabla^2 \Phi = -\frac{\rho}{\epsilon_0}, \tag{8.16}$$

where ρ is the charge density field reconstructed from the charges onto a regular mesh ($\mathbf{x}_m$) by a smooth projection

$$\rho(\mathbf{x}_m) \approx \frac{1}{h^3} \sum_i W(\mathbf{r}_i - \mathbf{x}_m) q_i, \tag{8.17}$$

and h denotes the mesh spacing. The Poisson equation [Eq. (8.16)] is solved on the mesh using fast Fourier transforms or efficient multigrid methods with an effective computational cost that scales as $\mathcal{O}(N \log N)$ or $\mathcal{O}(N)$, depending on the specific Poisson solver. The electrostatic field is computed from the potential on the mesh ($\mathbf{E} = -\nabla \Phi$) and interpolated onto the particles to allow the calculation of the electrostatic interaction

$$\mathbf{f}_i \approx q_i \sum_m W(\mathbf{r}_m - \mathbf{x}_i) \mathbf{E}_m. \tag{8.18}$$

The P^3M algorithm furthermore involves a particle-particle correction term for particles in close proximity (in terms of the grid spacing) to resolve subgrid scales.

Computations of potential forces employing a grid often involve simulations of periodic systems in order to take advantage of fast potential calculation algorithms such as fast Fourier transforms and multigrid methods. In addition, special care needs to be exercised in grid-particle interpolations so as not to induce spurious dissipation.

In the last 25 years, a number of mesh-free techniques based on the concept of multipole expansions have been developed that circumvent the need for simulating periodic systems and have minimal numerical dissipation. Examples of such methods involve the Barnes-Hut algorithm,[98] the fast multipole method (FMM),[99,100] and the Poisson integral method (PIM).[101,102] The methods employ clustering of particles and use expansions of the potentials around the cluster centers with a limited number of terms to calculate their far-field influence onto other particles. The savings are proportional to the ratio of the number of terms used in the expansions versus the number of particles in the cluster and scale nominally as $\mathcal{O}(N \log N)$. By allowing groups of particles to interact with each other by translating the multipole expansion into a local Taylor expansion, the algorithm achieves an $\mathcal{O}(N)$ scaling. It has been argued that the 3D version of the Greengard-Rokhlin algorithm is not efficient, as it adds nominally a computational cost of $\mathcal{O}(N \times P^4)$, where P is the number of terms retained in the truncated multipole expansion representation of the potential field. However, this issue has been resolved by suitable implementation of fast Fourier transforms.[103] Summarizing, these techniques rely on tree-data structures to achieve computational efficiency. The tree enables a spatial grouping of the particles, and the interactions of well-separated particles is computed using their center of mass or multipole expansions for the Barnes Hut and FMM algorithm, respectively.

Another advantage in tree-data structures is that they enable us to incorporate variable time steps and techniques. For example, in hierarchical internal coordinates,[104] some regions may be treated as rigid while only a subset or all degrees of freedom are considered for others. The Newton–Euler inverse mass operator method was developed for fast internal coordinate dynamics on a million atoms.[104,105] For a recent review of the treatment of long-range electrostatics in molecular dynamics simulations we refer the reader to Ref. 106.

8.2.3.5 Boundary conditions for MD

For situations involving the simulation of a solvent, the small volume of the computational box in which solvent and other molecules of interest are contained can introduce undesirable boundary effects if the boundaries are modeled as simple walls. To circumvent this problem, either the system can be placed in vacuum[47] or a periodic system can be assumed. In this approach, the original computational box containing the molecular system subject to investigation is surrounded with identical images of itself. Commonly, a cubic or rectangular parallelepiped box is used, but generally all space-filling shapes (e.g., truncated octahedron) are possible.[47] However, periodic boundary conditions imposed on small systems may introduce artifacts in systems that are not inherently periodic.[87]

Stochastic boundary conditions enable us to reduce the size of the system by partitioning the system into two zones with different functionality: a reaction zone and a reservoir zone. The reaction zone is the zone intended to be investigated, while the reservoir zone contains the portion that is of minor interest to the current study. The reservoir zone is excluded from MD calculations and is replaced by random forces whose mean corresponds to the temperature and pressure in the system. The reaction zone is further subdivided into a reaction zone and a buffer zone. The stochastic forces are only applied to atoms of the buffer zone. In Ref. 107, the application of stochastic boundary conditions to a water model is described and in Ref. 108, the method is derived.

8.2.3.6 Nonequilibrium molecular dynamics

To study nonequilibrium processes or dynamic problems, such as flows in capillaries and confined geometries, nonequilibrium MD (NEMD) is found to be a very efficient tool. It is based on the introduction of a flux in thermodynamic properties of the system.[47,109] In Ref. 110, NEMD is reviewed with regard to the computation of transport coefficients of fluids from the knowledge of pair interactions between molecules. In Ref. 111, rheological issues are addressed focusing on shear thinning and the ordering transition. Ryckaert et al.[112] compare the performance of NEMD with Green–Kubo approaches to evaluate the shear viscosity of simple fluids. In Ref. 113, a modified NEMD approach is presented to ensure energy conservation, and an elongated flow is studied in Ref. 114 with both spatial and temporal periodic boundary conditions. For detailed background about the underlying statistical mechanics of nonequilibrium systems, the reader is referred to Ref. 115.

Another form of NEMD is steered molecular dynamics (SMD), applied by Grubmüller et al.[116] to determine the rupture force of proteins. The principle of SMD is to superimpose a time-dependent force on selected atoms or molecules such that the molecules or the system are driven along certain degrees of freedom in order to investigate rare events. A short review is provided by Isralewitz et al.[117]

8.2.4 Multiscaling: linking macroscopic to atomistic scales

Nanoscale flows are often part of larger scale systems (as, for example, when nanofluidic channels are interfacing microfluidic domains) and in simulations we are confronted with an *inherently* multiscale problem when the nanoscale directly influences larger scales. The simulation of such flows is challenging, as one must suitably couple the nanoscale systems with larger spatial and time scales. The macroscale flows determine the external conditions that influence the nanoscale system, which in turn influences the larger scales by modifying its boundary conditions.

In the macroscale the state of a compressible, viscous, isothermal fluid can be described by its velocity field $\mathbf{u}$ and by its pressure P, temperature T, and density field ρ. The conservation of the system's mass, momentum and energy together with the continuum assumption lead to the compressible Navier–Stokes equations. The last 50 years have seen extensive research on the numerical simulation of these flows and a review is beyond the scope of this chapter. These equations inherently involve the computation of averaged quantities of the flow field. Hence, as in micro- and nanoscale flows, the continuum assumption and/or the associated constitutive relations eventually break down. Along with them the validity of the Navier–Stokes equations breaks down. To model a fluid at these scales, a computationally expensive atomistic description is required, such as direct simulation Monte Carlo (DSMC) for dilute gases or MD for liquids. Both methods are however subject to enormous CPU time requirements. An example of a recent MD study involving long simulation times (400 ns, 512 water molecules) is the one by Matsumoto et al.,[118] where they study the formation of ice.

To illustrate these limitations at the example of MD, consider that the time step δt in a molecular dynamics simulation is dictated by the fastest frequency one must resolve. For a simulation of pure water, $\delta t = 2$ fs when models with fixed O—H bonds and H—O—H angles are used; in other words, 500 million time steps are required for 1 μs of simulation time. With the optimistic assumption that the execution of single time step takes 0.1 s, a total of some 19 months of CPU time is required.

In this section, we review computational techniques that attempt to overcome these limitations either by combining the continuum and atomistic descriptions or through a mesoscopic model. First, we present the Navier–Stokes equations in conservation form, since in this form they are amenable to multiscale simulations. Then, hybrid algorithms are discussed that combine the continuum with atomistic descriptions. Finally, we discuss a mesoscopic model called dissipative particle dynamics (DPD).

8.2.4.1 Breakdown of the Navier–Stokes equation at small scales

The conservative form of the Navier–Stokes equations for a control volume V bounded by a surface S reads[2,119]

$$\frac{d}{dt}\int_V \rho\, dV + \int_S \rho\mathbf{u}\cdot\mathbf{n}\, dA = 0, \tag{8.19}$$

$$\frac{d}{dt}\int_V \rho\mathbf{u}\, dV + \int_S \rho\mathbf{u}\mathbf{u}\cdot\mathbf{n}\, dA - \int_S \boldsymbol{\sigma}\cdot\mathbf{n}\, dA, \tag{8.20}$$

with the stress tensor $\boldsymbol{\sigma}$ for a Newtonian fluid

$$\sigma_{ik} = -P\delta_{ik} + 2\mu\left(D_{ik} - \frac{1}{3}D_{mm}\delta_{ik}\right) + \lambda D_{mm}\delta_{ik}. \tag{8.21}$$

The rate-of-deformation tensor D is given by

$$D_{ik} = \frac{1}{2}\left(\frac{\partial u_i}{\partial x_k} + \frac{\partial u_k}{\partial x_i}\right). \tag{8.22}$$

The parameters μ and λ are the shear and bulk viscosities. To solve Eqs. (8.19) and (8.20) for a specific domain Ω, appropriate boundary conditions must be specified for $\partial\Omega$ with normal vector ($\mathbf{n}$). The equations must be complemented by boundary conditions such as solid, far-field, and porous boundary conditions. Here, we consider only the velocity boundary condition for solid surfaces. For macroscopic systems, it is a classical and widely used assumption that there is no relative motion between a flowing fluid and a solid boundary, i.e., $\mathbf{u} = 0$ at $\partial\Omega$. This postulate is called the *no-slip* boundary condition.

One of the fundamental questions in the context of micro- and nanofluidics is the range of validity of the Navier–Stokes equations and of the associated no-slip boundary condition. This range can be parametrized by the Knudsen number Kn, which is defined as the ratio between the mean free path and a characteristic length L of the system under consideration. The value of the Knudsen number determines the degree of *rarefaction* of the fluid and therefore the validity of the continuum flow assumption. Note that a *local* Knudsen number can be defined when L is taken to be the scale L of the macroscopic gradients[120] $L = \rho/(d\rho/dx)$. Until recently, noncontinuum (rarefied) gas flows were only encountered in low-density applications such as in the simulation of space shuttle reentries. However, in micro- and even more in nanofluidic applications, such as flows in nanopores or around nanoparticles, rarefaction effects are important at much higher pressures, due to the small characteristic length scales and the large gradients.[121] An empirical classification of gas flows is the following.[120] For $Kn < 0.01$, the flow is in the *continuum regime* and can be well described by the Navier–Stokes with no-slip boundary conditions. For $0.01 < Kn < 0.1$, the Navier–Stokes equations can still be used to describe the flow, provided that tangential slip-velocity boundary conditions are implemented along the walls of the flow domain. This is usually referred to as the *slip-flow* regime. In the *transition regime*, for $0.1 < Kn < 10$, the constitutive equation for the stress tensor of Eq. (8.21) starts to loose its validity. In this case, higher-order corrections to the constitutive equations are needed such as the Burnett or Woods equations, along with higher-order slip models at the boundary. At even larger Knudsen numbers ($Kn > 10$), the continuum assumption fails completely and atomistic descriptions of the gas flow are needed.[120] In Sec. 8.4.2, the slip-flow boundary conditions are discussed in more detail.

8.2.4.2 Hybrid atomistic-continuum computations

To maximize the effectivity of any hybrid scheme, the interface location must be chosen such that both schemes are valid around it, and such that the extent of the more expensive scheme is minimized. To locate this interface automatically, a variety of Navier–Stokes breakdown parameters have appeared in the literature.[122–125] These parameters are based on the coefficients of the higher-order terms of the Chapman–Enskog expansion of the solution of the Boltzmann equation. However, the validity and the cutoff value of these parameters are not yet very well understood.

An early attempt to extend the length scales accessible in MD simulations of fluids was undertaken by O'Connell and Thompson.[126] In their simulations, the particle (P) and continuum (C) regions were connected through an overlap region (X). The overlap region was used to ensure continuity of the momentum flux—or equivalently of the stress—across the interface between the P and the C regions. The average momentum of the overlap particles was adjusted through the application of constrained dynamics. The continuum boundary conditions at C were taken to be the spatially and temporally averaged particle velocities. O'Connell and Thompson[126] applied this algorithm to an impulsively started Couette flow where the P–C interface was chosen to be parallel to the walls. This ensured that there was no net mass flux across the MD-continuum interface.

As pointed out by Hadjiconstantinou and Patera,[127] the scheme proposed by O'Connell and Thompson decouples length scales, but not time scales. Hadjiconstantinou and Patera[127] and Hadjiconstantinou[128] therefore suggested to use the Schwarz alternating method for hybrid atomistic-continuum models. The continuum solution in C provides boundary conditions for a subsequent atomistic solution in P, which in turn results in boundary conditions for the continuum solution. The iteration is terminated when the solution in the overlap region X is identical for both the particle and the continuum descriptions. The usage of the Schwarz method avoids the imposition of fluxes in the overlap region, since flux continuity is automatically ensured if the transport coefficients in the two regions are consistent. The Schwarz method is inherently bound to steady-state problems. However, for cases in which the hydrodynamic time scale is much larger than the molecular time scale, a series of quasi-steady Schwarz iterations can be used to treat transient problems.[127]

Flekkøy et al.[129] presented a hybrid model that, in contrast to earlier hybrid schemes,[126,128] is explicitly based on direct flux exchange between the particle region and the continuum region. This scheme is robust in the sense that it does not rely on the use of the exact constitutive relations and equations of state to maintain mass, momentum, and energy conservation laws. The main difficulty in the approach of Flekkøy et al.[129] arises in the imposition of the flux boundary condition from the continuum region on the particle region. The scheme was tested for a 2D Lennard–Jones fluid coupled to a continuum region described by the compressible Navier–Stokes (NS) equations. To ensure consistency and to complement the NS equations, the viscosity ν and the equation of state $p = p(\rho, T)$ were measured in separate particle simulations. The first test was a Couette shear flow parallel to the P–C interface, and the second test involved a Poiseuille flow where the flow direction was perpendicular to the P–C interface. In both cases, good agreement between the observed and the expected velocity profiles was achieved. Wagner et al.[130] extended this work to include the energy equation and applied the technique to flow in a channel.

Flekkøy et al.[131] and Alexander et al.[132] studied how the continuum description plays the role of a statistical mechanical reservoir for the particle region in a hybrid computation. Both studies employed the example of a 1D diffusion process. Flekkøy et al.[131] used a finite difference (FD) discretization of the 1D (deterministic) diffusion equation coupled to a system of random walkers moving on a lattice. They found that the size of the particle fluctuations interpolates between those of an open system and those of a closed system depending on the ratio between the grid spacing of the FD discretization and the particle lattice constant. Alexander et al.[132] showed that a coupling of the deterministic diffusion equation to a system of random walkers does capture the mean of the density fluctuations across the particle-continuum interface, but that it fails in capturing the correct variance close to the interface. With a stochastic hybrid algorithm, where the fluctuating diffusion equation is solved in the continuum region, both the expected mean and variance of the density fluctuations are recovered.

Finally, Garcia et al.[125] have proposed a coupling of a DSMC solver embedded within an adaptive compressible Navier–Stokes solver. They have successfully tested their scheme on systems such as an impulsively started piston and flow past a sphere. The DSMC method is, however, restricted to dilute particle systems.

8.2.4.3 Mesoscopic models: dissipative particle dynamics

Coarse-grained models attempt to find a mesoscale description that enables the simulation of complex fluids such as colloidal suspensions, emulsions, polymers, and multiphase flows. The initial formulation of the DPD model was given by Hoogerbrugge and Koelman.[133] It is based on the notion of *fluid particles* representing a collection of atoms or molecules that constitute the fluid. These fluid particles interact pairwise through three types of forces, i.e., the force on particle i is given by

$$\mathbf{f}_i = \sum_{j \neq i} \left[\mathbf{F}^C(\mathbf{r}_{ij}) + \mathbf{F}^D(\mathbf{r}_{ij}, \mathbf{u}_{ij}) + \mathbf{F}^R(\mathbf{r}_{ij}) \right], \tag{8.23}$$

where

- $\mathbf{F}^C$ represents a conservative force that is derived from a soft repulsive potential, allowing for large time steps.
- The dissipative force $\mathbf{F}^D$ depends on the relative velocity $\mathbf{u}_{ij}$ of the particles to model friction

$$\mathbf{F}_{ij}^D = -\gamma \omega^D(r_{ij})(\mathbf{u}_{ij} \cdot \hat{\mathbf{r}}_{ij})\hat{\mathbf{r}}_{ij}, \tag{8.24}$$

  where $\hat{\mathbf{r}}_{ij}$ is a unit vector and γ is a scalar.
- Finally, a stochastic force $\mathbf{F}_{ij}^R$ models the effect of the suppressed degrees of freedom in the form of thermal fluctuations of amplitude σ

$$\mathbf{F}_{ij}^R = \sigma \omega^R(r_{ij})\xi_{ij}\hat{\mathbf{r}}_{ij}, \tag{8.25}$$

  where ξ_{ij} is a random variable.

Both $\mathbf{F}_{ij}^D$ and $\mathbf{F}_{ij}^R$ include r-dependent weight functions ω^D and ω^R, respectively. These weight functions and amplitudes σ and γ must satisfy the relations

$$w^D(r) = \left[w^R(r) \right]^2, \qquad \sigma^2 = 2\gamma k_B T \tag{8.26}$$

to simulate a canonical ensemble.[134] A review of DPD applied to complex fluids was given by Warren.[135] Although DPD has had considerable success in simulations of flows with polymers, its formulation has a conceptual difficulty.[136,137] First of all, its thermodynamic behavior is determined by the conservative forces and is therefore an output of the model and not (as desirable) an input. In addition, the physical scales that are simulated are not clearly defined. Recent reviews on mesoscale simulations of polymer materials can be found in articles by Glotzer and Paul[138] and Kremer and Müller–Plathe.[139]

Espanõl and Revenga[137] have recently introduced the smoothed dissipative particle dynamics method (SDPD), which combines elements of smoothed particle hydrodynamics (SPH) with DPD. SDPD emerges from a top-down approach, i.e., from a particle discretization of the Navier–Stokes equations in Lagrangian form similar to the SPH formulation. Every particle has an associated position, velocity, constant mass, and entropy. Two additional extensive variables, a volume and an internal energy, are associated with every particle. The particle volume enables us to give the conservative forces of the original DPD model in terms of pressure forces. Most importantly, the interpolant used in the SDPD formulation fulfills the second law of thermodynamics explicitly and thus enables the consistent introduction of thermal fluctuations through the use of the dissipation-fluctuation theorem. This will, for example, enable us to study the influence of thermal effects in the formation of bubbles.

8.3 Experiments in nanoscale fluid mechanics

The need for quantitative assessment of NFM has prompted the development of several novel experimental techniques and the adaptation of existing methodologies to the study of such phenomena in an interdisciplinary fashion.

As the noncontinuum, molecular structure of the fluids dominates the behavior of these systems, probing them requires diagnostics that can distinguish temporal and spatial scales at the atomistic level. Experimental techniques from diverse scientific fields are finding a "new life" in the area of nanoscale flow diagnostics. Techniques such as nuclear magnetic resonance (NMR) and devices such as the surface force apparatus (SFA) and the atomic force microscope (AFM) are invaluable tools in providing quantitative information that, along with computational results, probe the physics of NFM. At the same time, techniques such as molecular tagging that have been successfully implemented in biology are currently being adapted to monitor flow phenomena in the nanoscale.

8.3.1 Diagnostic techniques for the nanoscale

Nuclear magnetic resonance spectroscopy is increasingly being used to characterize microliter and smaller-volume samples. Substances at picomole levels have been identified using NMR spectrometers equipped with microcoil-based probes. These NMR probes that incorporate multiple sample chambers and the hyphenation of capillary-scale separations and microcoil NMR enable high-throughput experiments. The diagnostic capabilities of NMR spectroscopy have enabled the physico-chemical aspects of a capillary separation process to be characterized online.[140] Because of such advances, the application of NMR to smaller samples continues to grow. In particular, NMR techniques have been used extensively in the examination of diffusion, hydrodynamic dispersion, flow, and thermal convection under the influence of geometrical confinements and surface interactions in porous media.[141,142] The anomalous character of these phenomena is mostly characterized by the single-file diffusion behavior expected for atoms and molecules in 1D gas phases of nanochannels with transverse dimensions that do not allow for the particles to bypass each other. Although single-file diffusion may play an important role in a wide range of industrial catalytic, geologic, and biological processes, experimental evidence is scarce despite the fact that the dynamics differ substantially from ordinary diffusion.

Gas-phase NMR has great potential as a probe for a variety of interesting physical and biomedical problems that are not amenable to study by water or similar liquids. However, NMR of gases was largely neglected due to the low signal obtained from the thermally polarized gases with very low sample density. The advent of optical pumping techniques for enhancing the polarization of the noble gases He-3 and Xe-129 has bought new life to this field, especially in medical imaging where He-3 lung inhalation imaging is approaching a clinical application.[143] Meersmann et al.[144] and Ueda et al.[145] demonstrate the application of continuous-flow laser-polarized Xe-129 NMR spectroscopy for the study of gas transport into the effectively 1D channels of a microporous material. The novel methodology makes it possible to monitor diffusion over a time scale of tens of seconds, often inaccessible in conventional NMR experiments. The experimental observations indicate that single-file behavior for xenon in an organic nanochannel is persistent even at long diffusion times of over tens of seconds. In Kneller et al.,[146] the properties of the purified multiwalled carbon nanotubes are probed using C-13 and Xe-129 NMR spectroscopy under continuous-flow optical-pumping conditions. Xenon is shown to penetrate the interior of the nanotubes. A distribution of inner tube diameters gives rise to chemical shift dispersion. When the temperature is lowered, an increasing fraction of xenon resides inside the nanotubes and is not capable of exchanging with xenon in the interparticle space.

On a related front, recent years have seen an increase in the number of devices available to measure interaction forces between two surfaces separated by a thin film. One such device, the *surface forces apparatus*, measures static and dynamic forces (both normal and lateral) between optically transparent surfaces in a controlled environment, and is useful for studying interfacial and thin film phenomena at a molecular level. The SFA developed by the group of Israelachvilli at the University of California, Santa Barbara, in the late 1970s, is capable of measuring the forces between two molecularly smooth surfaces made of mica in vapors or liquids with a sensitivity of a few millidynes (10 nN) and a distance resolution of about 0.1 nm. The flat, smooth surfaces of mica can be covered to obtain the force between different materials. The basic instrument has a simple single-cantilever spring to which the lower silica disk is attached. The lower mica is brought near the upper mica by a piezoelectric device. If there is some interaction, the distance between the micas will not be the same as that given by the piezoelectric device. Therefore, the force is measured indirectly by the difference in the gap distance given, on one hand, by the piezoelectric device and, on the other hand, by that measured directly by interferometry (attractive forces make the micas closer and repulsive forces try to repell the micas). Interferometry is used in the SFA to measure the distance between the two surfaces of interest with high accuracy down to 1/1000 of a wavelength.[147]

Multiple-beam interferometry[148] uses intense white light, which is sent normally through the surfaces in the SFA. Each surface has a highly reflecting silver coating on one side, therefore, both surfaces form an optical cavity. The white light is reflected many times from these mirrors before it leaves the interferometer, each time interfering with the previously reflected beams. Some particular wavelengths fit an exact integral number of times inside the interferometer and lead to constructive interference. The light emerging from the interferometer is sent to a spectrograph, and it consists of well-defined wavelengths that fit an exact number of times into the optical resonator in the form of curved fringes.

The SFA has been used in many NFM studies.[149] The instrument relies on having a very low surface roughness over a very large interaction area. As the liquid is confined in spaces of 4 to 8 molecular diameters, between two macroscopic, molecularly smooth surfaces, it is forced into a discrete number of layers. SFA is the only apparatus thus far to demonstrate the continuous measurement of solvation forces in water as a function of surface-surface separation. However, most measurements have been limited to mica, a hydrophilic and chemically unreactive surface with no lateral characterization of the two surfaces possible or relevant.

The SFA is often used to investigate both short-range and long-range forces related to colloidal systems, adhesive interactions, and specific binding interactions. In addition, the SFA can be used to measure the refractive index of the medium between surfaces, adsorption isotherms, capillary condensation, surface deformations (due to surface forces), and dynamic interactions such as viscoelastic and frictional forces and the rheological properties of confined liquid films. Most notable studies involve the study of the no-slip boundary condition.[150,151] In these studies Newtonian alkane fluids (octane, dodecane, tetradecane) were placed between molecularly smooth surfaces that were either wetting (muscovite mica) or rendered partially wetted by adsorption of surfactants. The measured hydrodynamic forces agreed with predictions from the no-slip boundary condition when the flow rate was low but implied partial slip when it exceeded a critical level. A possible mechanism by which "friction modifiers" operate in oil and gasoline was identified. In a related study using the SFA, water confined between adjoining hydrophobic and hydrophilic surfaces (a Janus interface) is found to form stable films of nanometer thickness that have extraordinarily noisy responses to shear deformations.[152] From these studies the physical picture emerges whereas surface energetics encourage water to dewet the hydrophobic side of the interface, and the hydrophilic side constrains water to be present, resulting in a flickering, fluctuating complex. Difficulties with SFA measurements in thin films due to nanoparticles has been highlighted in Ref. 153. The authors propose that density anomalies in the thin liquid films are ultimately coupled to the presence of local-surface nonparalellism and the nanoparticles that are produced during the widely used mica-cutting procedure.

An optical technique for visualization in the nanoscale involves the measurement of the *index of refraction*. Kameoka and Craighead[154] reported fabrication and testing of a refractive index sensor based on photon tunneling in a nanofluidic system. The device comprises an extremely thin fluid chamber formed between two optically transparent layers. It can be used to detect changes in refractive index due to chemical composition changes of a fluid in the small test volume. Because the physical property measured is a refractive index change, no staining or labeling is required. The authors have tested the device with five samples, water and water with 1% ethanol, 2% ethanol, 5% ethanol, and 10% ethanol. The sensing was done by measuring the intensity of a reflected laser beam incident on the sensing element at around the critical angle.

Shadowgraphing techniques have been used to investigate the nonlinear optical properties of carbon nanotube suspensions in water and in chloroform for optical limiting.[155,156] Carbon nanotube suspensions are known to display interesting optical limiting properties as a result of the formation of solvent or carbon-vapor bubbles that scatter the laser beam. The main effect is nonlinear scattering, which is due to heat transfer from particles to solvent, leading to solvent bubble formation and to sublimation of carbon nanotubes. A clear correlation between the radius of the scattering centers and the evolution in transmittance of the sample has been observed. Also, the presence of compression waves that propagate parallel to the laser beam and can produce secondary cavitation phenomena after reflection on the cell walls has been observed.

Near-field scanning optical microscopy[157] helps extending measurements to the nanoscale for the optical characterization of thin films and surfaces. In this technique a light source or detector with dimensions less than the wavelength (λ) is placed in close proximity ($<\lambda/50$) to a sample to generate images with resolution better than the diffraction limit. Betzig et al.[158] developed a near-field probe that yields a resolution of approximately 12 nm ($\lambda/43$) and signals approximately 10^4- to 10^6-fold larger than those reported previously. In addition, image contrast is demonstrated to be highly polarization dependent. With these probes, near-field microscopy appears poised to fulfill its promise by combining the power of optical characterization methods with nanometric spatial resolution. In Reitz et al.,[159] a near-field scanning optical microscope system was implemented and adapted for nanoscale steady-state fluorescence anisotropy measurements. The system as implemented can resolve about 0.1-cP microviscosity variations with a resolution of 250 nm laterally in the near field, or approximately 10 μm when employed in a vertical scanning mode. The system was initially used to investigate the extent of microviscous vicinal water over surfaces of varying hydrophilicity.

Closed carbon nanotubes provide a unique opportunity for *in situ transmission electron microscope (TEM)* study of the chemical interactions between aqueous fluids and carbon. High-resolution *in situ* studies of an interface between fluid and carbon in TEM have been reported by Gogotsi et al.[160–162] and Megaridis et al.[163] (Fig. 8.2). Both groups reported that using hydrothermal synthesis produced closed hydrophilic multiwall carbon nanotubes filled with aqueous fluid. Strong interaction between the liquid and walls, intercalation of nanotubes with O—H species, and dissolution of walls on heating have been demonstrated.

After considerable success in biological systems, *molecular tagging* is being investigated as a means to probe flow phenomena in the nanoscale. In Gendrich et al.,[164] the development and applications of a new class of water-soluble compounds suitable for molecular tagging diagnostics are described. These molecular

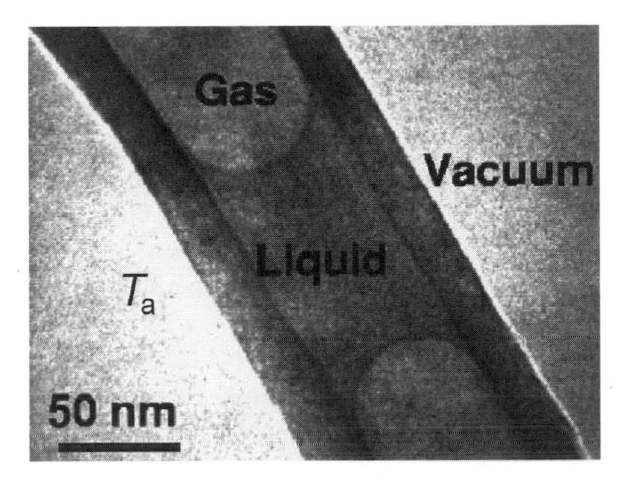

Figure 8.2 TEM micrograpahis showing an aqueous solution trapped in a closed multiwalled carbon nanotube. (From Gogotsi et al.[162])

complexes are formed by mixing a lumophore, an appropriate alcohol, and cyclodextrin. Using 1-BrNp as the lumophore, cyclohexanol is determined to be the most effective overall among the alcohols for which data are currently available. Information is provided for the design of experiments based on these complexes along with a less complex method for generating the grid patterns typically used for velocimetry. Implementation of a two-detector system is described that, in combination with a spatial correlation technique for determining velocities, relaxes the requirement that the initial tagging pattern be known *a priori*, eliminates errors in velocity estimates caused by variations in the grid pattern during an experiment, and makes it possible to study flows with nonuniform mixtures. *Fluorescent tagging*[165] has been used to track carbon nanotubes.[166] Modification of the surface of single-walled carbon nanotubes by using polymers enables the nanotubes to be distinctly visualized in solvents by fluorescence microscopy. Electrophoresis of the polymer-modified nanotubes under an alternating electric field was observed in real time, and a scanning electron microscopy image of the resultant nanotubes trapped on the electrodes revealed the consistency of the modification. This modification method will facilitate fabricating nanotube-based devices that can be detected with high sensitivity using simple light microscopes. A similar technique has been applied to fluorescent nanoparticles, enabling subnanometer precision by use of off-focus imaging.[167]

Information on the properties of liquids at surfaces at the nanoscale is also required to elucidate the mechanisms behind macroscopic observations. For example, there are few techniques for the study of contact angles at nanoscale resolution. The minimum drop size that can be accurately measured using a standard low-magnification goniometer is 1 mm. *Confocal laser scanning fluorescence microscopy (CLSM)* has recently been used to study the contact angles of thin oil films doped with a fluorescent dye.[168] Apart from an improvement in resolution over standard optical microscopy, CLSM has the ability to measure depth profiles and therefore can obtain accurate contact angles over the full range of angles. However, the liquid drops must be transparent and able to solvate the fluorescent dye.

8.3.2 Atomic force microscopy for fluids at the nanoscale

Oscillatory forces between two approaching surfaces in a solvent have long been the subject of study due to their possible influence on any surface-surface interactions mediated through a liquid or in the presence of a fluid film. Of particular interest is water, due to its omnipresence in all but the most stringently controlled environments and its role as the primary medium for biological interactions. The AFM was pioneered[169] to image the topography of surfaces, but is now becoming an important tool for investigating water-surface interactions.

As described in Ref. 170, both for the original (static) version of the SFA and for the AFM, the force is obtained from the deflection of a measuring spring or cantilever. However, in contrast to the SFA, wherein the cantilever deflection is detected interferometrically, the AFM uses electronic or digitally analyzed optical methods to sense deflections. Another difference is that various electronic techniques are used to control the motion of the surface. Moreover, in the case of the AFM the deflection versus the position of the piezo curves are much more sensitive at high speed than the SFA separation versus time curves. Hence, the AFM is much more convenient than the SFA for studying the highly dynamic phenomena in a thin gap.

The AFM has been developed and modified over recent years, utilizing different operating principles to explore a wide range of surface properties. Techniques such as tapping mode AFM (TMAFM), noncontact scanning force microscopy (NSFM) and scanning polarized force microscopy (SPFM), which were developed to image soft samples and weakly absorbed species are finding applications for the study of fluids in the nanoscale. Luna et al.[171] reported on a study of water droplets and films on graphite by NSFM. In a high-relative-humidity atmosphere (>90%), water adsorbs on the surface to form flat rounded islands of 5 nm in height that transform to 2-nm-high islands when the relative humidity stabilizes to 90%. This process is induced by the presence of the scanning tip. Desorption of the water present on the surface is achieved after the exposure of the sample to a dry atmosphere for several hours. The adsorption-desorption cycle is reversible. In addition to topography, TMAFM can probe the micromechanical behavior of a solid surface by analyzing the vibrational phase shift and amplitude of the probe as it interacts with the surface, providing information on viscoelastic and adhesive properties. In Attard et al.,[172] nanobubbles, whose existence on hydrophobic surfaces immersed in water has previously been inferred from measurements of long-ranged attractions between such surfaces, were directly imaged by TMAFM. Imaging of hydrophobic surfaces in water with TMAFM reveals[173] them to be covered with soft domains, apparently nanobubbles, that are close-packed and irregular in cross section, have a radius of curvature of the order of 100 nm, and have a height above the substrate of 20 to 30 nm. Complementary force measurements show features seen in previous measurements of the long-range hydrophobic attraction, including a jump into a soft contact and a prejump repulsion. The distance of the jump is correlated with the height of the images. The morphology of the nanobubbles and the time scale for their formation suggest the origin of their stability. TMAFM has also recently been employed to measure the contact angle of dewetted liquid bilayer thin films of two immiscible polymers, polystyrene and polybromostyrene.[174]

Ishida et al.[175] inserted an immersed silicon wafer hydrophobized with OTS into water to observe the surface *in situ* using a tapping-mode AFM. A large number of nanosize-domain images were found on the surface. Their shapes were characterized by the height image procedure of AFM, and the differences of the properties compared to those of the bare surface were analyzed using the phase image procedure and the interaction force curves. All of the results consistently implied that the domains represent the nanoscopic bubbles attached to the surface. The apparent contact angle of the bubbles was much smaller than that expected macroscopically, and this was postulated to be the reason bubbles were able to sit stably on the surface. Further studies of nanobubbles produced at liquid-solid interfaces using the AFM have been reported.[176,177] The atomic force microscope used to detect these nanobubbles showed that they can be seen on liquid-graphite and liquid-mica interfaces. The conformation of the bubbles was influenced by the atomic steps of the graphite substrate.

The AFM was employed in various complementary modes of operation to investigate the properties of nanometer-scale oil droplets existing on a polystyrene surface. Force curve mapping was used to gently probe the surface of the fluid droplets, and through automated analysis of the force curves the true topography and microscopic contact angle of the droplets were extracted. The interfacial tension of this oil-water junction was then measured using the AFM and again was found to be in close agreement with theory and macroscopic measurement. Using this information, the force exerted on the sample by a scanning tapping tip in fluid was derived and compared with forces experienced during tapping mode imaging in air. These results highlight the ability of AFM to both measure interfacial properties and investigate the topography of the underlying substrate at the nanometer scale.[174] Mugele et al.[178] used an AFM to image liquid droplets on solid substrates and to determine the contact line tension. Compared to conventional optical contact angle measurements, the AFM extends the range of accessible drop sizes by three orders of magnitude. By analyzing the global shape of the droplets and the local profiles in the vicinity of the contact line, it was shown that the optical measurement overestimates the line tension by approximately four orders of magnitude. Zitzler et al.[179] investigated the influence of the relative humidity on amplitude and phase of the cantilever oscillation while operating an AFM in the tapping mode. If the free-oscillation amplitude exceeds a certain critical amplitude $A(c)$, the amplitude- and phase-distance curves show a transition from a regime with a net attractive force between tip and sample to a net repulsive regime.

For hydrophilic tip and sample, $A(c)$ is found to increase with increasing relative humidity. In contrast, no such dependence was found for hydrophobic samples. Numerical simulations show that this behavior can be explained by assuming the intermittent formation and rupture of a capillary neck in each oscillation cycle of the AFM cantilever.

The use of the AFM is limited by the formation of *nanomenisci and nanobridges*. Colchero et al.[180] described a technique to measure the tip-sample interaction in a scanning force microscope setup with high precision. Essentially, the force exerted on the cantilever is acquired simultaneously with a spectrum of the cantilever. This technique is applied to study the behavior of the microscope setup as the tip approaches a sample surface in ambient conditions. The measured interaction can only be understood assuming the formation of a liquid neck and the presence of a thin liquid film on the tip as well as on the sample. Piner et al.[181] developed a direct-write "dip-pen" nanolithography (DPN) to deliver collections of molecules in a positive printing mode. An AFM tip is used to write alkanethiols with 30-nm-linewidth resolution on a gold thin film in a manner analogous to that of a dip pen. Molecules are delivered from the AFM tip to a solid substrate of interest via capillary transport, making DPN a potentially useful tool for creating and functionalizing nanoscale devices.

Calleja et al.[182] studied the dimensions of water capillaries formed by an applied electrical field between an atomic force microscope tip and a flat silicon surface. The lateral and vertical dimensions of the liquid meniscus are in the 5 to 30-nm range. The size depends on the duration and strength of the voltage pulse. It increases by increasing the voltage strength or the pulse duration. The meniscus size is deduced from the experimental measurement of the snap-off separation. In AFM studies of molecular thin films, a defined jump of the tip through the film is often observed once a certain threshold force has been exceeded. Butt and Franz[183] presented a theory to describe this film rupture and to relate microscopic parameters to measurable quantities. These models were later verified in Ref. 184.

Ahmed et al.[185] reported on studies aimed at employing AFM to measure the viscosity of aqueous solutions. At ambient temperature, the AFM cantilever undergoes thermal fluctuations that are highly sensitive to the local environment. The measurements revealed that variations in the resonant frequency of the cantilever in the different solutions are largely dependent on the viscosity of the medium. An application of this technique is to monitor the progression of a chemical reaction where a change in viscosity is expected to occur.

With magnetically activated AFM it has been possible to resolve molecular layers of large molecules such as octamethylcyclotetrasiloxane and *n*-dodecanol. With this method, magnetic material is deposited directly behind an AFM tip on the backside of the cantilever so that the tip position can be controlled by the addition of a magnetic field. The lever can be vibrated in an oscillating magnetic field in order to make dynamic measurements. One expected consequence of the success of this technique was a rapid exploitation of the experimental advantages over SFA such as various surface materials that can be studied and simultaneous lateral characterization. However, the current literature is restricted to measurements as a function of separation between a silicon tip and a mica or graphite surface. Further, using magnetically activated AFM it has not yet been possible to reproduce the solvation shell measurements in water measured by SFA. Jarvis et al.[186] attribute this to the long averaging times necessary to obtain a sufficiently sensitive signal-to-noise ratio using a lock-in amplifier and also because of the low aspect ratio tips commonly used.

One report of water layer and/or hydrated ion measurements using static AFM is that of Cleveland et al.[187] They show that with sufficiently long measurement times and sufficient stability it is possible even with static measurements to pinpoint different energy minima close to ionic crystals in water by using the thermal noise of the cantilever. Unfortunately, due to the long averaging times needed for this technique it is not readily applicable to location sensitive investigations. When the AFM is used for force measurements, the driving speed typically does not exceed a few microns per second. However, it is possible to perform the AFM force experiment at a much higher speed. In Ref. 149, theoretical calculations and experimental measurements are used to show that in such a dynamic regime the AFM cantilever can be significantly deflected due to viscous drag force. This suggests that in general the force balance used in a surface force apparatus does not apply to the dynamic force measurements with an AFM. Vinogradova et al.[149] also developed a number of models that can be used to estimate the deflection caused by viscous drag on a cantilever in various experimental situations. As a result, the conditions when this effect can be minimized or even suppressed are specified. This opens up a number of new possibilities to apply the standard AFM technique for studying dynamic phenomena in a thin gap.

8.4 Fluid-solid interfaces at the nanoscale

Hydrophobic effects and wetting phenomena have a long-standing history and open questions remain for both areas. The emphasis in this section is on the computational efforts to understand the molecular nature of wetting and hydrophobicity. For recent reviews on the general molecular theory of hydrophobic effects, the reader is directed to the works by Pratt[188] and Pratt and Pohorille.[189]

8.4.1 Hydrophobicity and wetting

The attribute *hydrophobic* (water-fearing) is commonly used to characterize substances like oil that do not mix with water. The classical interpretation of this phenomenon is that the interaction between the water molecules is so strong that it results in an effective oil-oil attraction. Interestingly, oil and water do in fact attract each other, but not nearly as much as water attracts itself. Lazaridis[190] performed a series of MD simulations with hypothetical solvents to identify the solvent characteristics that are necessary conditions for general solvophobic behavior. His findings support the classical view that solvophobicity is observed when the solvent-solvent interaction strength clearly exceeds the solvent-solute interaction. In the case of water, the large cohesive energy is mainly due to the strong hydrogen bond network. The importance of the hydrophobic effect as a source of protein stability was first identified by Kauzmann[191] and a review on dominant forces in protein folding is given by Dill.[192] At a certain solute size (around 1 nm), it becomes energetically more favorable to assemble hydrophobic units than to keep them apart by thermal agitation.[193,194]

The spreading and wetting of water on hydrophobic/hydrophilic surfaces is a related subject of great practical interest where substantial insight has been gained through the help of computation. The wetting behavior of a surface could be characterized through the contact angle that a liquid forms on it. One can distinguish at least two different states, namely, the wetting state, where a liquid spreads over the substrate to form a uniform film, and the partial wetting state, where the contact angle lies in between 0 and 90 deg. The microscopic contact angle θ for a droplet with base radius r is given by the modified Young's equation[195]

$$\gamma_{SV} = \gamma_{SL} + \gamma_{LV}\cos\theta + \frac{\tau}{r}, \tag{8.27}$$

where the γ's denote the surface tensions between the solid (S), liquid (L), and vapor (V) phases, respectively, and τ is the tension associated with the three-phase contact line. In the limit of macroscopic droplets, the effect of the line tension τ becomes insignificant, i.e., for $r \to \infty$, Eq. (8.27) reduces to the well-known Young's equation.[38] In the following, we review computational studies that aim at studying the validity of macroscopic concepts such as Young's or Laplace's equations at the nanoscale and at a molecular characterization (ordering, orientation, etc.) of a liquid at a hydrophobic or hydrophilic interface.

The wetting and drying of a liquid and a vapor phase enclosed between parallel walls was studied by Saville,[196] Sikkenk et al.,[197] and Nijmeijer et al.[198,199] The main difference in their simulations was the representation of the confining wall. The introduction of an "inert" wall[199] leads to good agreement between visually observed contact angles and the ones deduced from the surface tensions through Young's equation.

Hautman and Klein[200] have performed one of the first MD studies to investigate a liquid droplet on different solid substrates. They observed the equilibrium contact angle of water droplets containing merely 90 molecules on hydrophobic and hydrophilic surfaces that were formed by monolayers of long-chain molecules with terminal $-CH_3$ and $-OH$ groups, respectively.

Thompson et al.[201] tested and confirmed the validity of Young's and of Laplace's equation at microscopic scales for a fluid-fluid interface in a channel. The wetting properties of the fluids were controlled by setting different interaction strengths between the fluids and the wall; all interactions were modeled using the Lennard–Jones potential. Fan and Cağin[202] simulated the wetting of crystalline polymer surfaces by water droplets containing 216 water molecules. Furthermore, they introduced a different way to measure the contact angle between a liquid and a solid surface using the volume and contact area of the droplet instead of the droplet center-of-mass height above the surface. The dynamics of spreading at the molecular level were first studied by de Ruijter and De Coninck[203] and de Ruijter et al.[204] They monitored the relaxation of the contact angle for a fluid modeled by linear chain molecules and obtained good agreement with a molecular kinetic theory of wetting. MD studies of heat transfer at solid liquid interfaces has been reported.[51] Reviews of the dynamics of wetting are given in Refs. 205 and 206.

Bresme and Quirke[207,208] investigated by means of MD simulations the wetting and drying transitions of spherical particulates at a liquid-vapor interface as a function of the fluid-particulate interaction strengths and of the particulate size. They showed that the wetting transition for a small particulate occurs at a weaker interaction strength than for a large one. This suggests that a change in geometry of the particulate enhances its solubility. In a subsequent study, Bresme and Quirke[209] analyzed the dependence of the spreading of a lens in a liquid-liquid interface in terms of the liquid-lens surface tension. It was found that this dependence is well described by Neumann's construction, which is the analog to Young's equation when the three phases in contact are deformable. Werder et al.[69] studied the behavior of water droplets confined in pristine carbon nanotubes using molecular dynamics simulations (cf. Fig. 8.3). They found contact angles of 110 deg indicating a nonwetting behavior. Lundgren et al.[210] studied the wetting of water and water-ethanol droplets on graphite. For pure water droplets, they found contact angles that were in good agreement with the experimentally observed ones. On addition of ethanol, the contact angle decreased as expected and the ethanol molecules were concentrated close to the hydrophobic surface and at the water-vapor interface. Werder et al.[38] used the known wetting behavior of water on graphite to calibrate the water-graphite interaction in MD simulations (cf. Fig. 8.4). They found that water monomer binding energies of -6.33 and -9.37 kJ mol^{-1} are required to recover, in the macroscopic limit, contact angles of 86 deg (Ref. 211) and 42 deg (Ref. 212), respectively. Figure 8.5 shows micro-sized water droplets on a graphite

Figure 8.3 Molecular dynamics simulation of the contact angle of water droplets in single-walled carbon nanotubes.[69] The molecular structure (left) and the time-averaged isochor profiles (right) indicate a nonwetting behavior of the 5-nm-diameter water droplet. (Reprinted with permission from Ref. 69, © 2001 American Chemical Society.)

surface. These binding energies include a correction to account for the line tension that, through MD simulations of droplets of different sizes, is estimated to be positive and of the order of 2×10^{-10} J/m. For a simple Lennard–Jones interaction potential acting between the oxygen atoms of the water and the carbon atom sites, the corresponding interaction parameters to obtain the desired binding energies are $\sigma_{CO} = 3.19$ Å, $\epsilon_{CO} = 0.392$ kJ mol^{-1}, and $\epsilon_{CO} = 0.5643$ kJ mol^{-1}, respectively.

8.4.2 Slip flow boundary conditions

The conditions at the fluid-solid interface are of paramount interest to develop suitable computational models and to understand the governing physical mechanisms

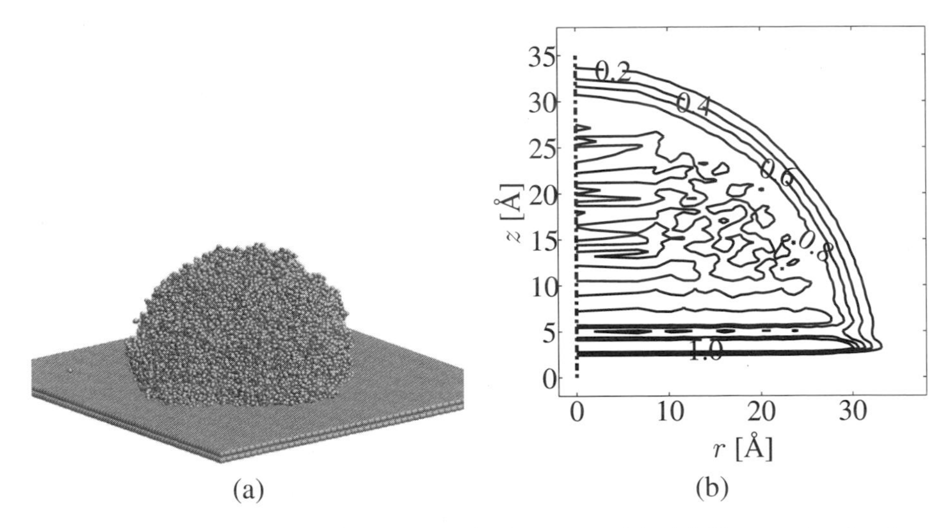

Figure 8.4 Side view of a 5-nm large water droplet on graphite (a). From molecular dynamics simulations by Werder et al.[69] The contact angle is extracted from the time-averaged water isochore profile (b). The isochore levels are 0.2, 0.4, 0.6, 0.8, and 1.0 g cm^{-3}. (Reprinted with permission from Ref. 69, © 2001 American Chemical Society.)

Figure 8.5 ESEM experiments of micron-sized water droplets condensed on a graphite surface showing contact angles of approximately 30 deg. (From Noca and Sansom.[213])

to design effective nanodevices. In the nanoscale the fluid-solid interfaces assume greater importance because the surface-to-volume ratio is larger than in macroscale flows and the flow length scale approaches the fluid molecule size. When solids are immersed in fluids, the boundary condition usually adopted in the modeling equations of the macroscale systems is a vanishing relative velocity between the fluid and the solid surface—the *no-slip* condition.[214,215] The validity of this condition is an active area of computational and experimental research.

8.4.2.1 Experimental evidence of no slip

Experimental evidence of the no-slip condition at wetting surfaces was provided by Whetham[216] and Bulkley.[217] On the other hand, slip is found to exist in narrow, hydrophobic capillaries, as demonstrated by Helmholtz and von Piotrowski,[218] and later confirmed by Schnell,[219] Churaev et al.,[220] and Baudry et al.[13] A thorough review of earlier works concerning the manifestation of slip can be found in Ref. 11. The existence of no-slip conditions for liquid flows in confined spaces is furthermore complicated by the unusual behavior of the fluid properties associated with phase changes of the fluid. For water, strong density fluctuations are furthermore observed within 1 nm of the solid surface,[221] and the water orientation and hydrogen bonding are perturbed.[222] Garnick[223] found that the viscosity attains a significantly higher value when the fluid is confined leading to a stick-slip behavior,[224,225] or solidification when the film thickness becomes sufficiently small.[226]

One important yet unresolved question in NFM is the amount of slip occurring at hydrophilic surfaces. Bonaccurso et al.[14] observed a persistent slip in measurements of water on mica and glass, whereas Vinogradova and Yakubov[170] recently found a no-slip condition in drainage experiments of thin films between silica surfaces.

The question remains if the transition from no-slip to slip follows the limit of zero to nonzero contact angle of the fluid solid interface, or if (weakly) hydrophobic surfaces can support a no-slip. The experimental evidence is strongly affected by uncertainties such as surface roughness, entrapped gas or vapor bubbles,[173,227] chemical impurities,[11] and the purity of the fluid.[228] Alternatively, molecular dynamics simulations free of such impurities may provide valuable insight into the nature of the no-slip condition. At the same time such conjectures rely on the existence of accurate interaction potentials that describe the fluid-solid interface.

To extend continuum fluid dynamics modeling to nanoscale flow systems, the liquid-solid boundary conditions must be determined and parameterized,[229] and the length scale where molecular-size effects become important should be known. Contrary to traditional continuum modeling, taking into account nanoscale flow phenomena implies that the conditions will depend on the specific molecular nature of the fluid and the surface.

The slip velocity Δu at a surface may be modeled according to Maxwell,[230] as

$$\Delta u = b\frac{\partial u}{\partial y}, \tag{8.28}$$

where b is the slip length, $\alpha = 1/b$ is the slip coefficient, and $\partial u/\partial y$ denotes the shear stress at the interface as shown in Fig. 8.6. The slip length is a function of the properties of the fluid-solid interface. At hydrophobic surfaces, Churaev et al.[220] and Baudry et al.[13] found slip lengths of the order of 30 to 40 nm. The slip observed by Bonaccurso et al.[14] at hydrophilic surfaces amounts to 8 to 9 nm.

A closed formula for the slip length was derived for dilute systems by Bocquet,[231] and an approximate formula for dense Lennard–Jones fluids was given by Barrat and Bocquet.[232] However, Richardson[233] showed that the dissipation of energy caused by the surface roughness (ϵ), and irrespective of the boundary condition imposed at the microscale (a no-slip or a zero shear boundary condition) results in an effective no-slip condition $b = \mathcal{O}(\epsilon)$. Recent measurements by Zhu et al.[150] confirmed that the effect of surface roughness dominates the local intermolecular interaction. The analysis of Richardson[233] is based on the separation of length scales; thus $l \ll \epsilon \ll L$, where l denotes the size of the molecules, and L is the bulk fluid length scales. This separation is not present in many nanoscale flows, such as the flow of waters ($l \approx 0.4\,\text{nm}$) passing a single-walled carbon nanotubes ($L \approx 1$ nm and $\epsilon \approx 0$ nm). As a consequence, the amount of slip found in nanoscale flows is expected to depend not only of the wetting properties of the fluid-solid interface, but also on the particular geometry.

8.4.2.2 MD simulations of slip

Molecular dynamics simulations provide a controlled environment for the study of slip in nanoscale systems free from impurities and surface roughness, but limited to studies of small systems, currently of the order of tens of nanometers and tens of nanoseconds. Also, most studies have been conducted for idealized systems such as Lennard–Jones fluids in simple geometries, often confined between smooth (Lennard–Jones type) solids. However, these studies have provided valuable insight into the fundamental mechanisms of slip. The following sections contain a short review of recent MD simulations of the internal flows such as the planar

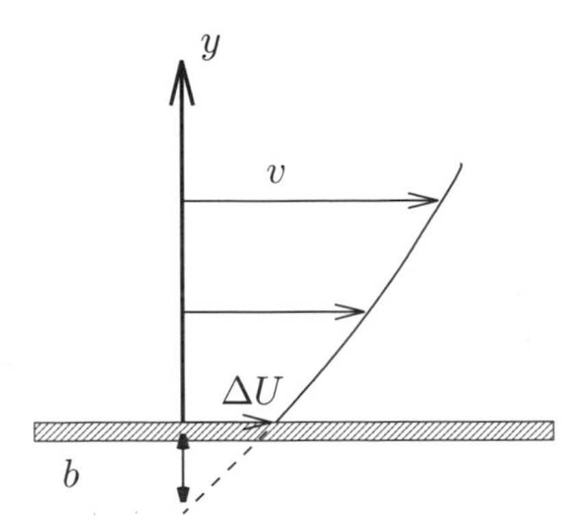

Figure 8.6 Slip at a fluid-solid interface is characterized by a finite velocity (ΔU) at the interface. This slip velocity is related to the slip length (b) through the shear rate at the interface: $\Delta U = b\partial u/\partial y$.

Poiseuille and Couette flows, and external flows, including flows past cylinders.

Koplik et al.[234] performed MD simulations of *Poiseuille flow* and moving contact lines. The no-slip condition was found to be satisfied for Lennard–Jones fluids confined between Lennard–Jones solids, and slip at the contact line. Bitsanis et al.[235] found velocity profiles with slip, but also a flatness of the velocity profile close to the reservoir walls used in their study.

In MD simulations of Poiseuille and Couette flows, Barrat and Bocquet[232] found the no-slip boundary condition to depend on the wetting properties of the fluid-solid interface. Both the fluid and solids were modeled as Lennard–Jones molecules using a modified Lennard–Jones potential

$$V_{ij}(r) = 4\epsilon\left[\left(\frac{\sigma}{r}\right)^{12} - c_{ij}\left(\frac{\sigma}{r}\right)^{6}\right], \tag{8.29}$$

where the parameter c_{ij} was used to adjust the relative strength of the interactions. Thus, the cohesion of the fluid was increased from the usual Lennard–Jones fluid using a value of $c_{FF} = 1.2$, and the fluid-solid interaction was varied between 0.5 and 1.0, corresponding to contact angles of 140 and 90 deg, respectively. The Poiseuille flow was driven by imposing an external (gravity) force, and the slip length was found to vary between 40σ and $\mathcal{O}(\sigma)$ for for contact angles of 90 and 140 deg, respectively. The slip length was found to decrease as a function of the pressure in the channel.

In a series of simulations of flows in narrow pores, Todd et al.[236] and Travis et al.[237,238] found the velocity profile to deviate significantly from the quadratic form predicted by the Navier–Stokes formalism. Both the solid and fluid atoms were modeled using a purely repulsive Lennard–Jones type (Weeks–Chander–Andersen) potential, or the full 12–6 Lennard–Jones potential. The density of the solid surface was approximately 80% of the fluid density, resulting in a high surface corrugation and a no-slip condition at the fluid-solid interface.

Mo and Rosenberger[239] modeled the surface corrugation explicitly in 2D simulations of a Lennard–Jones system. Both sinoidally and randomly roughened walls were considered with various amplitudes. The no-slip condition was found to hold when the molecular mean free path is comparable to the surface roughness. In the *planar Couette flow*, the fluid is confined between two solid planar walls. The flow is generated by moving one or both walls with constant (opposite) velocity and the imposed shear diffuses into the flow developing a linear velocity profile. Thompson and Robins[240] studied a Lennard–Jones fluid in a planar Couette flow and found slip, no-slip and locking depending on the amount of structure (corrugation) induced by the solid walls. Highly corrugated walls would result in a no-slip condition, whereas weak fluid-wall interaction would result in slip. At strong interactions, a epitaxial ordering was induced in the first fluid layers, effectively locking these to the wall. Thus the slip would occur within the fluid. For Couette flows driven by a constant force, this locking results in a stick-slip motion involving a periodic shear-melting transition and recrystallization of the film.[241,242]

The importance of the surface corrugation was later emphasized by Thompson and Troian,[243] who found that the slip length diverges at a critical shear rate ($\dot{\gamma}_c$) as

$$b = \frac{b^0}{\sqrt{1 - \dot{\gamma}/\dot{\gamma}_c}}, \tag{8.30}$$

where b^0 is the slip length in the limit of low-shear rate. The critical shear rate is reached while the fluid is still Newtonian, and depends on the corrugation of the surface energy. The importance of the corrugation of the surface was later confirmed by Cieplak et al.[244] for simple and chain-molecule fluids, and by Jabbarzadeh et al.[245] for alkenes confined between rough atomic sinusoidal walls. They found that the amount of slip is governed by the relative size of the molecular length to the wall roughness.

Sokhan et al.[246] considered methane modeled as spherical Lennard–Jones molecules confined between (high-density) graphite surfaces. They found a significant slip even in the strongly wetting case, and recovered the no-slip condition by artificially reducing the density of the wall. The constant gravity force imposed to drive the flow resulted in low-frequency oscillations of the mean flow with a time scale ranging from 10 ps to 2 ns. Both flexible or rigid walls were considered but the dynamics of the wall was found to have little influence of the slip length.

For water confined between hydrophobic graphite surfaces, Walther et al.[247] found slip lengths in the range of 31 to 63 nm for pressures between 1 and 1000 bar. Changing the wetting properties of the interface to hydrophilic reduced the slip length to 14 nm. Other confined flows include the *Hagen–Poiseuille (pipe) flow* as considered in Heinbuch and Fischer,[248] who found that two layers of molecules would stick to the wall for sufficiently strong fluid-wall interaction. Similar studies involve the flows of monoatomic fluids,[249] and methane[250] through single-walled carbon nanotubes. Similar to their study of methane flowing in a slit carbon nanotube pore, Sokhan et al.[250] found a large slip in the range of 5.4 to 7.8 nm, which is significantly less than the values found for the planar graphite surface, due to the high curvature and increased friction in the carbon nanotube.

Hirshfeld and Rapaport[251] conducted MD simulations of the Taylor–Couette flow. Using a purely repulsive Lennard–Jones potential and hard walls, they found good agreement with experiments and theory. In a recent study, Walther et al.[247] performed nonequilibrium molecular dynamics simulations of water flowing past an array of single-walled carbon nanotubes. For diameters of the carbon nanotube of 1.25 and 2.50 nm and onset flow speeds in the range of 50 to 200 $\mathrm{m\,s^{-1}}$, they found the no-slip condition to hold as demonstrated in Fig. 8.7. Application of the same model to the Couette flow resulted in significant slip, indicating an influence of the geometry on the slip.

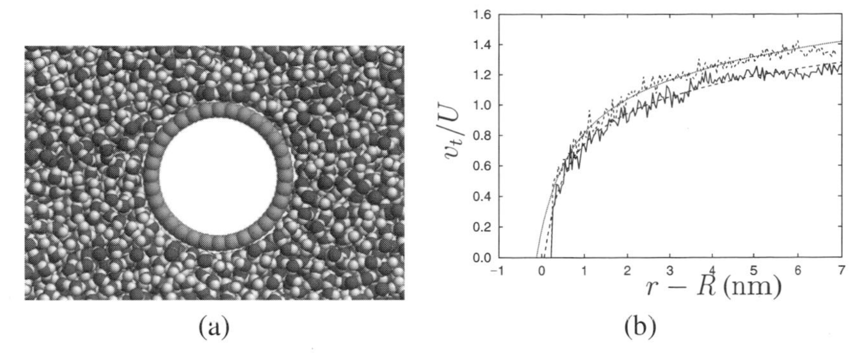

Figure 8.7 NEMD simulation for the study of hydrodynamic properties of carbon nanotubes.[252] The simulations involve water flowing past an array of 1.25- and 2.50-nm-diameter carbon nanotubes. A closeup of the systems is shown in (a), and the time-averaged tangential component of the velocity is shown in (b). The profiles are obtained for a 1.25-nm tube: ; measured; −−; fit, and 2.50-nm tube. - -, measured, · · ·, fit, and compared with the Stokes–Oseen solution. The slip length extracted from these simulations indicates that the continuum no-slip condition is valid.

8.5 Fluids in confined geometries

An understanding of the interaction of water-based liquids with carbon in confined nanoscale geometries at the nanoscale is very important for exploring the potential of devices such as carbon nanotubes (CNTs) in nanofluidic chips, probes, and capsules for drug delivery. The hollow interior of carbon nanotubes can serve as a nanometer-sized capillary. The nanotube cavities are weakly reacting with a large number of substances and, hence, may serve as nanosize test tubes. The small diameter of CNTs points to using their filled cavities as a mold or a template in material fabrication. Ugarte et al.[253] filled open carbon nanotubes with molten silver nitrate by capillary forces producing chains of silver nanobeads separated by high-pressure gas pockets.

Finally, the ability to encapsulate a material in a nanotube also offers new possibilities for investigating dimensionally confined phase transitions. In particular, water molecules in confinement exhibit several phase transitions as their network of hydrogen bonds is disrupted.

The prospect of controlled transport of picoliter volumes of fluid and single molecules requires addressing phenomena such as a local density increase of several orders of magnitude and layering of transported elements in confined nanoscale geometries.[254] This presents a unique set of concerns for transport and lubrication of films in the nanometer scale.

8.5.1 Flow motion in nanoscale channels

Nanoscale channels such as ion channels are one of the most important natural devices for the transport of molecules into and out of biological cells. The behavior of confined fluids in nanoscale geometries is an area that has been under study for some time in zeolites and ideal nanoporous systems. The understanding of such processes is of great interest for nanotechnology applications in biotechnology.

Experiments have demonstrated that fluid properties become drastically altered when the separation between solid surfaces approaches the atomic scale.[224,255] In the case of water, so-called drying transitions occur on this scale as a result of strong hydrogen bonding between water molecules, which can cause the liquid to recede from nonpolar surfaces and form distinct layers separating the bulk phase from the surface. In addition, changes such as increased effective shear viscosity as compared to the bulk, prolonged relaxation times and nonlinear responses set in at lower shear rates.[223] Computational studies of the behavior of molecules in nanoporous structures have played an important role in understanding the behavior of fluids in the nanometer scale, complementing experimental works. A detailed study regarding the behavior of a fluid in close confinment was reported by Thompson and Robbins,[240] who used molecular-dynamics simulations of Lennard–Jones liquids sheared between two solid walls. A broad spectrum of boundary conditions was observed including slip, no-slip, and locking. It was shown that the degree of slip is directly related to the amount of structure induced in the fluid by wall-fluid interaction potential. For weak wall-fluid interactions, there is little ordering and slip was observed. At large interactions, substantial epitaxial ordering was induced and the first one or two fluid layers became locked to the wall. The liquid density oscillations also induced oscillations in other microscopic quantities normal to the wall, such as the fluid velocity in the flow direction and the in-plane microscopic stress tensor, that are contrary to the predictions of the continuum Navier–Stokes equations. However, averaging the quantities over length scales that are larger than the molecular lengths produced smooth quantities that satisfied the Navier–Stokes equations.

Molecular dynamics and Monte Carlo simulations have been used to simulate systems that include films of spherical molecules, straight chain alkanes, and branched alkanes.[235,256–258] Bitsanis and his coworkers[235] have reported on the flow of fluids confined in molecularly narrow pores. They observed departure from the continuum as strong density variations across the pore rendered the usual dependence of the local viscosity on local density inappropriate. At separations greater than four molecular diameters flow can be described by a simple redefinition of local viscosity. In narrower pores, a dramatic increase of effective viscosities is observed and is due to the inability of fluid layers to undergo the gliding motion of planar flow. This effect is partially responsible for the strong viscosity increases observed experimentally in thin films that still maintain their fluidity. The simulations for Couette and Poiseulle types of flow yielded wall parallel velocity profiles that deviate from the shape predicted by continuous assumptions. Confinement also affects the electronic properties of the enclosed substances. Intermolecular dipole-dipole interactions were once thought to average to zero in gases and liquids as a result of rapid molecular motion that leads to sharp nuclear magnetic resonance lines. In Ref. 259, it is shown that a much larger, qualitatively different intermolecular dipolar interaction remains in nanogases and nanoliquids.

The dipolar coupling that characterizes such interactions is identical for all spin pairs and depends on the shape, orientation (with respect to the external magnetic field), and volume of the gas/liquid container. This nanoscale effect is useful in the determination of nanostructures.

Flows of argon, helium, and a buckyball and helium fluid inside carbon nanotubes have been reported using molecular dynamics simulations.[249,260] The fluid was started at some initial velocity; fluid particles were allowed to recycle axially through the tube via minimum image boundary conditions. Argon slowed down more quickly than helium. In addition, the behavior of the fluid strongly depended on the rigidity of the tube; a dynamic tube slowed down the fluid far more quickly than one in which the tube was held frozen. Another study[261] reports a molecular dynamics simulation to investigate the properties and design space of molecular gears fashioned from carbon nanotubes with teeth added via a benzyne reaction. A number of gear and gear-shaft configurations are simulated on parallel computers. One gear is powered by forcing the atoms near the end of the nanotube to rotate, and a second gear is allowed to rotate by keeping the atoms near the end of its nanotube constrained to a cylinder. The meshing aromatic gear teeth transfer angular momentum from the powered gear to the driven gear. Results suggest that these gears can operate at up to 50 to 100 GHz in a vacuum at room temperature. The failure mode involves tooth slip, not bond breaking, so failed gears can be returned to operation by lowering the temperature and/or rotation rate.

Manipulation of the geometry at the nanoscale may be readily utilized for controlled fluid transport. This was demonstrated[262] by fluidic control in lipid nanotubes 50 to 150 nm in radius, conjugated with surface-immobilized unilamellar lipid bilayer vesicles. Transport in nanotubes was induced by continuously increasing the surface tension of one of the conjugated vesicles, for example, by ellipsoidal shape deformation using a pair of carbon microfibers controlled by micromanipulators as tweezers. The shape deformation resulted in a flow of membrane lipids toward the vesicle with the higher membrane tension; this lipid flow in turn moved the liquid column inside the nanotube through viscous coupling. By control of the membrane tension difference between interconnected vesicle containers, fast and reversible membrane flow (moving walls) with coupled liquid flow in the connecting lipid nanotubes was achieved.

8.5.1.1 Biological nanochannels

Ion channels consist of a particular natural form of nanochannels with particular importance to biological systems. They belong to a class of proteins that forms nanoscopic aqueous tunnels acting as a route of communication between intra and extracellular compartments. Each ion channel consists of a chain of aminoacids carrying a strong and rapidly varying electric charge. Ion channels regulate cell internal ion composition, control electrical signaling in the nervous system and in muscle contraction, and are important for the delivery of many clinical drugs. Channels are usually "gated," i.e., they contain a region that can interrupt the flow of molecules (water, ions) that is often coupled to a sensor that controls the gate allosterically.[263] They exhibit selectivity on the types of ions that get transmitted and may exhibit switching properties similar to other electronic devices. At the same time, channels or pores for uncharged molecules mediate transport through the membrane by diffusion driven by the gradient of this substance. Pores or channels are known to exist for water and small molecules like urea, glycerol, and others. One particular nongated channel of interest is a water channel called an aquaporin. In aquaporins, the general belief was that water diffuses through the lipids of biological membranes. On the other hand, it has been known for many years that a large portion of water transport is protein-mediated. The question of how gating works at an atomic level is one of considerable complexity. A pattern is emerging for some channels in which the most constricted region of the pore (which is usually identified with the gate) is ringed by hydrophobic amino acid side chains, e.g., leucine or valine. So, is an effect other than steric occlusion able to close a channel, i.e., hydrophobic gating? Experimental evidence in favor of such a mechanism comes from studies of pores in modified Vycor glass, which showed that water failed to penetrate these pores once a threshold hydrophobicity of the pore walls was exceeded.[263]

Molecular dynamics simulations through atomistic models of nanopores embedded within a membrane mimetic have been used to identify whether a hydrophobic pore can act as a gate of the passage of water. Both the geometry of a nanopore and the hydrophilicity vs. hydrophobicity of its lining determine whether water enters the channel. For purely hydrophobic pores, there is an abrupt transition from a closed state (no water in the pore cavity) to an open state (cavity water at approximately bulk density) once a critical pore radius is exceeded. This critical radius depends on the length of the pore and the radius of the mouth region. Furthermore, a closed hydrophobic nanopore can be opened by adding dipoles to its lining.

The prospect of employing structures such as pure and doped carbon nanotubes for molecular transport has not been unnoticed. As a step in understanding the governing physical phenomena, in long (>50 ns) simulations of a carbon nanotube submerged in water, Hummer et al.[68] (Fig. 8.8) observed water flux through a pore occuring in a pulsatory fashion, with fluctuations in flux on a time scale of 4 ns.

Waghe et al.[264] have studied the kinetics of water filling and emptying the interior channel of carbon nanotubes using molecular dynamics simulations. Filling and emptying occur predominantly by sequential addition and/or removal of water to or from a single-file chain inside the nanotube. Advancing and receding water chains are orientationally ordered. This precludes simultaneous filling from both tube ends, and forces chain rupturing to occur at the tube end where a water molecule donates a hydrogen bond to the bulk fluid. They used transition path concepts and a Bayesian approach to identify a transition state ensemble that was characterized by its commitment probability distribution. At the transition state, the tube is filled with all but one water molecule. One important observation is that filling thermodynamics and kinetics depend on the strength of the attractive nanotube-water interactions that increases with the length of the tubes.

Figure 8.8 Pseudo-1D ordering of water molecules in a 8.5-Å-diameter carbon nanotube. (From unpublished simulations performed in our group based on work by Hummer et al.[68])

Computational requirements for the simulation of transport across nanoscale channels has been identified as a challenging multiscale problem due to the disparate scales that are present. In their review article, Aluru et al.[265] proposed the use of continuum simulation techniques for handling the complex geometries to resolve the drift-diffusion equation for charge flow. At the same time, ion traversal can be a rather rare event. Continuum models can then be parametrized to match current-voltage characteristics by specifying a suitable space and/or energy-dependent diffusion coefficient, which accounts for the ions' interactions with the local environment.

Particle methods can be implemented for the solution of such flows. A Brownian dynamics approach can be used for the description of the ion flow, in which ion trajectories evolve according to the Langevin equation. An N-body solver can be used to account for all of the pairwise ion interactions, while external forces induced by the potential can be computed from solving the potential equation for the externally computed potential fields. A frictional term is included to account for ion-water scattering, while a short-range repulsion term is used to account for ionic core repulsion. MD and Monte Carlo methods can be used to model water-ion interactions, while Monte Carlo methods offer an interesting alternative as water and protein are treated as background dielectric media and only the individual ion trajectories are resolved.[266]

Beckstein et al.[263] present simulations of a model comprised of a membrane-spanning channel of finite length allowing water molecules within the pore to equilibrate with those in the bulk phase, thus avoiding any prior assumptions about water density. Effectively, the interior of the pore is simulated in a grand canonical ensemble and entry or exit of water to or from an atomistic model of a nanopore is probed, while retaining control over its geometry and the charge pattern of its pore lining. In summary, hydrophobicity per se can close a sterically open channel to penetration by water and hence, by simple extension, to ions and small polar solutes. Such a channel can be opened by adding a relatively small number of dipoles to the lining of the pore or by a modest increase in radius. The critical gating radius depends on the geometry of the mouth region of the pore. Simulation studies of gramicidin suggested that 87% of overall channel resistance to water permeation comes from the energetic cost for a bulk water to enter the mouth. Thus, both overall dimensions and the extents of hydrophobic and hydrophilic regions in the lining provide a key to gating of nanopores.

8.5.2 Phase transitions of water in confined geometries

Encapsulation of a second phase inside carbon nanotubes offers a new avenue to investigate dimensionally confined phase transitions. When pure liquid water is encapsulated inside narrow carbon nanotubes, water molecules would be expected to line up into some quasi-1D structures, and on freezing, may exhibit quite different crystalline structures from bulk ice. Confinement may change not only resulting crystalline structures but also the way liquids freeze.[267]

Supercooled water and amorphous ice have a rich metastable phase behavior. In addition to transitions between high- and low-density amorphous solids and between high- and low-density liquids, a fragile-to-strong liquid transition has recently been proposed and supported by evidence from the behavior of deeply supercooled bilayer water confined in hydrophilic slit pores.[268] Evidence from molecular dynamics simulations suggests another type of first-order phase transition—a liquid-to-bilayer amorphous transition—above the freezing temperature of bulk water at atmospheric pressure as reported in Koga et al.[269] This transition occurs only when water is confined in a hydrophobic slit pore[270] with a width of less than 1 nm. On cooling, the confined water, which has an imperfect random hydrogen-bonded network, transforms into a bilayer amorphous phase with a perfect network (owing to the formation of various hydrogen-bonded polygons) but no long-range order.

Molecular dynamics simulations were performed in Noon et al.[271] at physiological conditions (300 K and 1 atm) using nanotube segments of various diameters submerged in water. The results show that water molecules can exist inside the nanotube segments and that the water molecules inside the tubes tend to organize themselves into a highly hydrogen-bonded network, i.e., solid-like wrapped-around ice sheets. The disorder-to-order transition of these ice sheets can be achieved purely by tuning the size of the tubes.

Particularly intriguing is the conjecture[272–274] that matter within the narrow confines of a carbon nanotube might exhibit a solid-liquid critical point beyond which the distinction between solid and liquid phases disappears. This unusual feature, which cannot occur in bulk material, would allow for the direct and continuous transformation of liquid matter into a solid. In Koga et al.[272] simulations of the behavior of water encapsulated in carbon nanotubes suggest the existence of a variety of new ice phases not seen in bulk ice, and of a solid-liquid critical point. Using carbon nanotubes with diameters ranging from 1.1 to 1.4 nm and applied axial pressures of 50 to 500 MPa, they found that water can exhibit a first-order freezing transition to hexagonal and heptagonal ice nanotubes, and a continuous phase transformation into solid-like square or pentagonal ice nanotubes (Fig. 8.9).

Slovak et al.[275] performed a series of MD simulations to examine in more detail the results of a water simulation, which shows that a thin film of water, when confined in a hydrophobic nanopore, freezes into a bilayer ice crystal composed of two layers of hexagonal rings. They found that only in one case the confined water completely freezes into perfect bilayer ice, whereas in two other cases, an imperfect crystalline structure consisting of hexagons of slightly different shapes is observed.

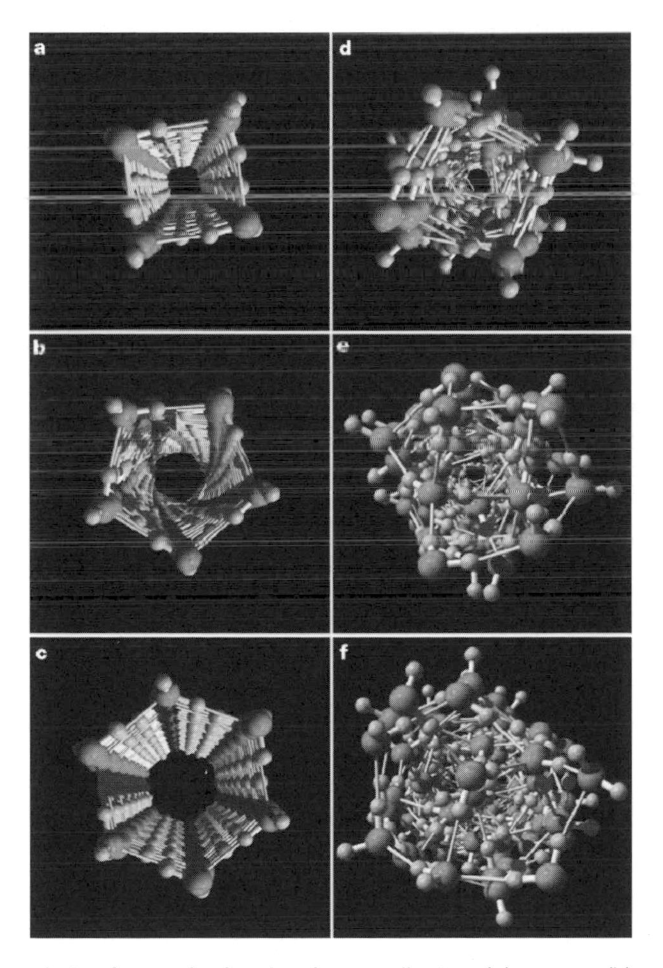

Figure 8.9 Snapshots of quenched molecular coordinates: (a) square; (b) pentagonal; and (c) hexagonal ice nanotubes in (14,14), (15,15), and (16,16) SWCNs; and (d) to (f), the corresponding liquid phases. The ice nanotubes were formed on cooling under an axial pressure of 50 MPa in molecular dynamics simulations. The nearest-neighbor distances in both ice nanotube and encapsulated liquid water are fairly constant, about 2.7 to 2.8 Å, and this is in part responsible for the novel phase behavior. (Reprinted with permission from Ref. 272, © 2001 The Nature Publishing Group.)

This imperfection apparently hinders the growth of a perfect bilayer crystal. After adjusting the area density to match spatial arrangements of molecules, the latter two systems are able to crystallize completely. As a result, we obtain three forms of bilayer crystals differing in the area density and hexagonal rings alignment.

The same group in a later study[267] considered simulations of phase behavior of quasi-1D water confined inside a carbon nanotube, in the thermodynamic space of temperature, pressure, and diameter of the cylindrical container. Four kinds of solid-like ordered structures—ice nanotubes—form spontaneously from liquid-like disordered phases at low temperatures. In the model system, the phase change occurs either discontinuously or continuously, depending on the path in the thermodynamic space.

Confinement of liquids such as water in nanoscales can also induce properties that correspond to water properties in *supercritical conditions*. While at room temperature, water is forming tetrahedral units of five molecules linked by hydrogen bonds. When temperature is raised and/or density is reduced, some of the hydrogen bonds are broken. Most of the dominant order is then lost and the remaining structures are linear and bifurcated chains of H-bonded water molecules, which can be regarded as parts of broken tetrahedrals. The destruction of the hydrogen bonds affects the water so that its compressibility and transport properties are intermediate between those of liquid and gas. However, increasing temperature and/or decreasing density are not the only means to achieve this effect. Recent MD simulations indicate that when water is introduced inside carbon nanotubes, its hydrogen bonding structure is also compromised[84,276] with an important decrease in the average number of hydrogen bonds with respect to bulk supercritical water. This reduction is greater than for water in standard conditions. The atomic density profiles are slightly smoother, but with the same general features than for water at lower temperatures.

8.6 Nanofluidic devices

The previous sections have discussed some of the fundamental issues of nanoscale fluid mechanics. Understanding the governing principles of these flows through novel computational and experimental techniques will lead to the development of devices that are able to exploit the unique characteristics of these flows. In parallel, engineers are developing nanofluidic devices by ingeniously adopting concepts from areas such as biology and chemistry. Several key issues remain unresolved such as the manufacturing of nanoscale devices either by self-assembly or by controlled interaction with microscale devices. The stage is set for new and inventive engineering concepts to continue to feed fundamental research in NFM, while the envelope of what can be accomplished by exploiting nanoscale fluid mechanics is pushed. In the following sections, we review a partial list of nanofluidic concepts and devices as they are linked to the flow physics addressed in the previous sections.

8.6.1 Solubilization

Most applications employing the unique electronic, thermal, optical, and mechanical properties of individual single-wall carbon nanotubes will require the large-scale manipulation of stable suspensions at a high weight fraction. Tube solubilization provides access to solution-phase separation methodologies[277] and facilitates chemical derivatization, controlled dispersion and deposition,[278] microfluidics, fabrication of nanotube-based fibers and composites,[279] and optical diagnostics. Unfortunately, nanotubes aggregate easily and are difficult to suspend as a result of substantial van der Waals attractions between tubes.[280]

Thus far, some progress has been made toward the solubilization of single-walled carbon nanotubes (SWNTs) in both organic and aqueous media. Dissolution in organic solvents has been reported with bare SWNT fragments (100 to 300 nm in length) and with chemically modified SWNTs.[281] Dissolution in water, which is important because of potential biomedical applications and biophysical processing schemes, has been facilitated by surfactants and polymers,[282,283] by polymer wrapping,[284] and by attaching glucosamine, which has both an amine group that can easily form an amide bond with the SWNT and high water solubility.[285] In the method reported by O'Connell et al.[286] the formation of any chemical bond was avoided by wrapping the SWNT in macromolecules such as poly(vinylpyrrolidone) PVP and polystyrene sulfonate PSS. Sano et al.[287] functionalized the SWNTs with monoamine-terminated poly(ethylene oxide) PEO using a preparation method via acyl chloride. High-weight-fraction suspensions of surfactant-stabilized SWNTs in water are reported in Islam et al.,[288] with a large fraction of single tubes. A single-step solubilization scheme was developed by the nonspecific physical adsorption of sodium dodecylbenzene sulfonate. The diameter distribution of nanotubes in the dispersion, measured by atomic force microscopy, showed that even at 20 mg/mL, about 65% of single-wall carbon nanotube bundles exfoliated into single tubes. In Riggs et al.,[283] solubilization of the shortened carbon nanotubes was achieved by attaching the nanotubes to highly soluble polyethylenimine or by functionalizing the nanotubes with octadecylamine. The soluble carbon nanotube samples formed homogeneous solutions in room-temperature chloroform. Optical limiting properties of these solutions were also determined for 532-nm pulsed-laser irradiation, and the results indicate that the carbon nanotubes exhibit significantly weaker optical limiting responses in homogeneous solutions than in suspensions.

8.6.2 Nanofluids

Common fluids with particles of the order of nanometers in size are termed nanofluids. These nanofluids have created considerable recent interest for their improved heat transfer capabilities. With a very small volume fraction of such particles, the thermal conductivity and convective heat transfer capability of these suspensions are significantly enhanced without the problems encountered in common slurries such as clogging, erosion, sedimentation, and increase in pressure drop.

Heating or cooling fluids is important for many industrial sectors, including energy supply and production, transportation, and electronics. The thermal conductivity of these fluids plays a vital role in the development of energy-efficient heat transfer equipment. However, conventional heat transfer fluids have poor thermal transfer properties compared to most solids. To improve the thermal conductivity of these fluids numerous theoretical and experimental studies of the effective thermal conductivity of liquids containing suspended milli- or microsized solid particles have been conducted. A number of procedures have been proposed for the development of nanofluids. In Xuan and Li,[289] a procedure is presented for preparing a nanofluid by a suspension of copper nanophase powder and a base liquid. Wilson et al.[290] used colloidal metal particles as probes of nanoscale thermal transport in fluids. They investigated suspensions of 3- to 10-nm-diameter Au, Pt, and AuPd nanoparticles as probes of thermal transport in fluids and determined approximate values for the thermal conductance G of the particle/fluid interfaces. The measured G are within a factor of 2 of theoretical estimates based on the diffuse-mismatch model. Thermal transport in nanofluids has also been considered through experimental study of pool boiling in water-Al_2O_3 nanofluids.[291] The results indicate that the nanoparticles have pronounced and significant influence on the boiling process deteriorating the boiling characteristics of the fluid. This effect is attributed to the change of surface characteristics during boiling by particles trapped on the surface.

Nanofluids consisting of CuO or Al_2O_3 nanoparticles in water or ethylene glycol exhibit enhanced thermal conductivity. A maximum increase in thermal conductivity of approximately 20% was observed in Lee et al.[292] for 4 vol% CuO nanoparticles with an average diameter of 35 nm dispersed in ethylene glycol. A similar behavior has been observed in a Al_2O_3/ethylene glycol nanofluid.[293] Furthermore, the effective thermal conductivity has shown to be increased by up to 40% for the nanofluid consisting of ethylene glycol containing approximately 0.3 vol% Cu nanoparticles of mean diameter <10 nm, and the effective thermal conductivity of a nanofluid consisting of carbon nanotubes (1 vol%) in oil exhibits 160% enhancement.[294]

8.6.3 CNT as sensors and AFM tips

The low bending force constants of carbon nanotubes make them ideal candidates for gentle imaging of soft samples. Moreover, due to their small (5- to 20-nm) diameter and cylindrical shape, they provide excellent lateral resolution and are ideal for scanning high-aspect-ratio objects.

Dai et al.[295] first suggested mounting a CNT on silicon as a probe for tapping-mode AFM to image the structure of nanoscale liquid samples. They attached individual nanotubes several microns in length to the Si cantilevers of conventional atomic force microscopes. Because of their flexibility, the tips are resistant to damage from tip crashes, while their slenderness enables imaging of sharp recesses in surface topology. The authors were also able to exploit the electric conductivity of nanotubes by using them for scanning tunneling microscopy. These developments open up the possibility of investigating water layers under a variety of experimental conditions and as a function of precise lateral position on any surface including biological membranes and macromolecules. Among the many and varied roles of water layers are effects on biomolecular adhesion, colloid dispersion, and tribology, which can now be investigated with nanometer lateral resolution and with a wider range of materials than that previously provided by a surface force apparatus.

Building on this work, Moloni et al.[296] proposed an improved technique for obtaining tapping mode scanning force microscopy (TMSFM) images of soft samples submerged in water. This technique makes use of a carbon nanotube several microns in length mounted on a conventional silicon cantilever as the TMSFM probe. The sample is covered by a shallow water layer and only a portion of the nanotube is submerged during imaging. This mode of operation largely eliminates the undesirable effects of hydrodynamic damping and acoustic excitation that are present during conventional tapping mode operation in liquids and leads to high-quality TMSFM images. A limitation of probes based on open-ended MWNT is due to their limited lateral resolution as the tips of these probes have a flat cylindrical endform of 5 nm or more in diameter. Implementation of a SWNT with tips of about 1 nm may be the next step in perfecting scanning force microscopy.

The combination of a carbon nanotube probe and a highly sensitive dynamic measurement scheme enabled the use of an AFM to measure oscillatory forces in water approaching a surface that has been laterally characterized on a nanometer scale. One important aspect of these results, in particular for colloidal systems, is that forces appear to scale with the surface dimensions from the mesoscopic, as measured by the surface forces apparatus, to the nanoscale.[186] Also of importance is the observation of solvation shells on a nonrigid surface. Application of these techniques may help elucidate phenomena associated with the detailed mechanism of hydrophobic drying of surfaces in aqueous environments (Fig. 8.10).

8.6.4 Carbon nanotubes as storage devices—adsorption

Carbon nanotubes have been envisioned as suitable storage devices for hydrogen and hydrogen-based fuels. Hydrogen-based fuels are considered a promising prospect for the ever-growing demand for energy. Hydrogen's byproduct is water, and it can be easily regenerated, thus meeting the rising concern of environmental pollution and the call for new and clean fuels. Unfortunately, owing to the lack of a suitable storage system satisfying a combination of both volume and weight limitations, the use of hydrogen energy technology has been restricted from automobile application. Therefore, to implement hydrogen energy for electrical vehicles, the first step is to look for an economical and safe hydrogen-storage medium. Recent reports on very high and reversible adsorption of hydrogen in nanostructured

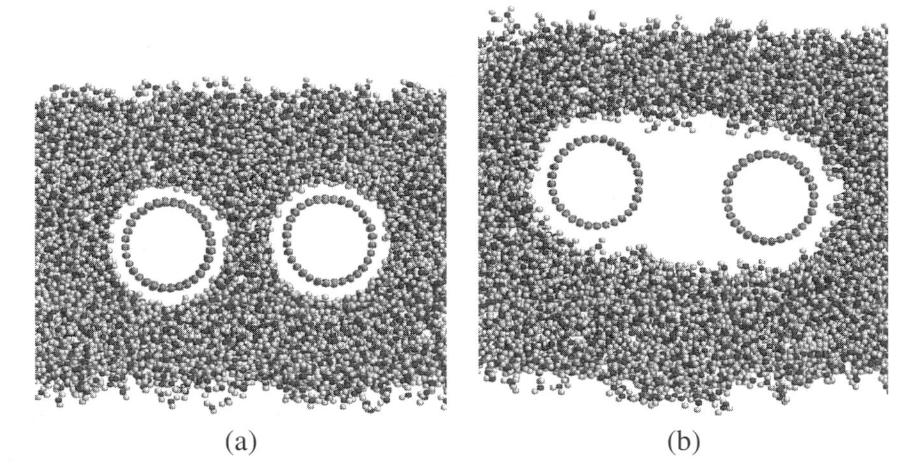

Figure 8.10 Drying of carbon nanotubes immersed in water. The range of the drying behavior is strongly dependent on the wetting properties of the interface. A hydrophobic, but partially wetting surface (a) displays a persistent wetting behavior for tube spacing exceeding two layers of water molecules, whereas a purely repulsive interface (b) shows an extended drying behavior. (From molecular dynamics simulations by Walther et al.[297])

carbon materials such as carbon nanotubes,[298] graphite nanofibers,[299] and alkali-doped nanotubes,[300] have stimulated many experimental works[301] and computational studies.[302–304] Experimental results demonstrate that nanostructured carbon materials have relatively high gravimetric hydrogen storage capacity. This capacity is dependent on the purity of the carbon nanotubes with increased capacity observed of the purified carbon nanotubes compared with that of the as-prepared counterparts.[305] This improvement is attributed to the removal of the impurities, oxygen-containing functionalities, and adsorbed species in the MWNTs.

To investigate the capabilities and the specific mechanisms of gas adsorption by CNTs, a number of computational studies have been performed. Such simulations have examined gas molecules (NO_2, O_2, NH_3, N_2, CO_2, CH_4, H_2O, H_2, Ar) on SWNTs and bundles using molecular dynamics[303] and first principles methods.[306] The adsorption and desorption energy of hydrogen atoms depend on the hydrogen coverage and the diameter of the SWNTs. The adsorption energy decreases with the increasing diameter of the armchair tubes. Most molecules adsorb weakly on SWNTs and can be either charge donors or acceptors to the nanotubes. Zhao et al.[303] found that the gas adsorption on the bundle interstitial and groove sites is stronger than that on individual nanotubes. The electronic properties of SWNTs are sensitive to the adsorption of certain gases such as NO_2 and O_2. Charge transfer and gas-induced charge fluctuation might significantly affect the transport properties[307] of SWNTs.

8.6.5 Nanofluidics for microscale technologies

Almost a decade after the first miniaturized gas chromatography system was successfully fabricated on a silicon wafer,[308] the first liquid-phase separation was demonstrated, thereby catalyzing the development of micrototal analysis systems. Since that time there has been an enormous amount of research devoted to developing miniaturized systems for separations, chemical and biological sensing.[309] Simultaneously, a number of technological factors have driven the development of fluidic architectures toward the nanometer length scale. However, nanostructures proposed to date for chemical and biological applications rely on self-assembling and self-organizational processes.[310] A technical challenge is to construct such units into integrated 3D systems. The ultimate nanofluidic device is one that can handle single molecules and colloid particles. Such devices require unprecedented control over transport and mixing behaviors, and to advance current fluidics into the single-molecule regime, we must develop systems having physical dimensions in the nanometer scale. To create such devices, we can draw much knowledge from biological systems. For example, the Golgi-endoplasmic reticulum network in eukaryotic cells has many attractive features for sorting and routing of single molecules, such as transport control and the capability to recognize different molecular species, and for performing chemical transformations in nanometer-sized compartments with minimal dilution. It is, however, extremely difficult to mimic these biological systems by using traditional microfabrication technologies and materials because of their small scale, complex geometries, and advanced topologies. Advanced nanofabrication techniques are necessary to construct such devices, and a number of devices, such as nanochannels and nanomembranes are currently being implemented.

The key characteristic feature of nanofluidic channels is that fluid flow occurs in structures of the same size as the physical parameters that govern the flow. Another factor that favors the development of nanoscale interconnects is the enhanced surface area-to-volume ratio characteristic of the nanochannels in these membranes. The ability to interface nanochannels with conventional microfluidics alleviates the need for nanofabrication techniques, and yet still enables a number of important applications that use the unique characteristics of the nanopores. For instance, the small pore size system can be used to concentrate dilute analytes, or clean up analyte solutions. This latter point is especially important for biological separations where often the major components (whether salts or proteins) in a mixture obscure the ability to separate and collect the desired trace level components. While a simple transfer of a band is demonstrated here from one microfluidic channel to the other, this concept can be extended to chemical manipulation in the receiving channel with derivatizing reagents. Besides the chemical manipulations possible between isolated microchannels, the high surface-to-volume ratio of the nanochannels offers additional opportunities. For example, by including molecular recognition elements on the interior of the nanopores, it should be possible to effect intelligent fluidic switching in which certain elements of the fluidic stream being transported through the nanopores are retained, reacted, degraded or otherwise chemically processed before being released into the next microfluidic channel.

8.6.5.1 Nanofluidic networks, sieves, and arrays

Networks of nanofluidic tubes have been manufactured by using a heat-depolymerizable polycarbonate (HDP) as a sacrificial layer.[311] A patterned HDP film is used as a temporary support for another film that is stable at the depolymerization temperature. Heating the structure removes the HDP, leaving a network of nanofluidic tubes without the use of solvents or other chemicals as required in most other sacrificial layer processes. Tube dimensions of 140-nm height, 1-μm width, and 1-mm length are reported, and fabrication of other structures is discussed. Nanoimprint lithography has been used[312] to manufacture channels with a cross section as small as 10 by 50 nm, which can be of great importance for confining biological molecules into ultrasmall spaces. To avoid entropic traps in introducing biological molecules such as DNA in fluidic channels directly from the macroscale, diffraction gradient lithography techniques have been used to fabricate continuous spatial gradient structures that smoothly narrow the cross section of a volume from the micron to the nanometer length scale.[313]

Nanofluidic devices are gaining popularity as DNA separation devices thus replace the standard electrophoresis techniques. When passing through such nanoscale sieves, ordinarily a long chain DNA molecule in liquid will clump into a roughly spherical shape, and to move through a sieve it must uncoil and slide in lengthwise. This movement involves an entropic force that causes DNA molecules only partially within a sieve to withdraw when the force pulling them in is removed. The effect results from the motion of segments in the chain molecule as they interact with the beginning of the barrier. The force is called "entropic" because the molecule moves out of the restricted space of the sieve into an open area where it can be more disordered. A nanofluidic channel device,[309] consisting of many *entropic traps*, was designed and fabricated for the separation of long DNA molecules. The channel comprises narrow constrictions and wider regions that cause size-dependent trapping of DNA at the onset of a constriction. This process creates electrophoretic mobility differences, thus enabling efficient separation without the use of a gel matrix or pulsed electric fields. Samples of long DNA molecules (5000 to similar to 160,000 base pairs) were efficiently separated into bands in 15-mm-long channels. Multiple-channel devices operating in parallel were demonstrated. The efficiency, compactness, and ease of fabrication of the device suggest the possibility of more practical integrated DNA analysis systems.

An alternative device involves nanosphere arrays[314] prepared by colloidal templating, which traps the macromolecules within a 2D array of spherical cavities interconnected by circular holes. Across a broad DNA size range, diffusion does not proceed by the familiar mechanisms of reptation or sieving. Rather, because of their inherent flexibility, DNA molecules strongly localize in cavities and only sporadically jump through holes. By reducing DNA's configurational freedom, the holes act as molecular weight-dependent entropic barriers.

Fluidic control in nanometer-size channels using a moving wall provides plug-like liquid flows, offers a means for efficient routing and trapping of small molecules, polymers, and colloids, and offers new opportunities to study chemistry in confined spaces. Networks of nanotubes and vesicles might serve as a platform to build nanofluidic devices operating with single molecules and nanoparticles. Soft microfabrication technologies for processing of fluid-state liquid crystalline bilayer membranes have been presented in Karlsson et al.[262] They have developed a micro-electrofusion method for construction of fluid-state lipid bilayer networks of high geometrical complexity up to fully connected networks with genus = 3 topology. Within networks, self-organizing branching nanotube architectures could be produced where intersections spontaneously arrange themselves into three-way junctions. It is also demonstrated that materials can be injected into specific containers within a network by nanotube-mediated transport of satellite vesicles having defined contents. Using a combination of microelectrofusion, spontaneous nanotube pattern formation, and satellite-vesicle injection, complex networks of containers and nanotubes can be produced for a range of applications in, for example, nanofluidics and artificial cell design. In addition, this electrofusion method enables integration of biological cells into lipid nanotube-vesicle networks.

8.6.5.2 Nanoporous membranes

Nanoporous membranes containing monodisperse distributions of nanometer diameter channels have been proposed as an effective medium for controlled molecular transport.[315] The facility with which molecular manipulations may be accomplished at the nanometer scale suggests their use for integrating multilevel microfluidic systems. The use of commercially available nanoporous membranes enables quick and economical fabrication of nanochannel architectures to provide fluidic communication between microfluidic layers. By incorporating these nanoporous membranes into microfluidic systems, a variety of novel flow control concepts can be implemented. The cylindrical nanochannels ($10 \text{ nm} < d < 200 \text{ nm}$) of the membranes can be used as nanofluidic interconnects to establish controllable fluidic communication between micron-scale channels operating in different planes. Kuo et al. initially investigated the ability to manipulate macroscopic transport using these nanochannels,[316] and recently reported on interfacing the nanoporous membranes with microfluidic channels.[317] More importantly, these nanoporous membranes add functionality to the system as gateable interconnects. These nanofluidic interconnects enable control of net fluid flow based on a number of different physical characteristics of the sample stream, the microfluidic channels and the nanochannels, leading to hybrid fluidic architectures of considerable versatility. Because the nanofluidic membrane can have surfaces with excess charge of either polarity, the net flow direction inside the microdevices is principally controlled by two factors: the magnitude of the electrical and physical flow impedance of the nanoporous membrane relative to that of the microchannels and the surface chemical functionalities, which determine the polarity of the excess charge in the nanochannels. The nanochannel impedance can be manipulated by varying membrane pore size. Flow control is investigated by monitoring electrokinetic transport of both neutral and negatively charged fluorescent probes, by means of laser-induced fluorescence and fluorescence microscopy, while varying solution and nanochannel properties.

Sun and Crooks[319] used multiwall carbon nanotubes as templates to fabricate single-pore membranes. These membranes are better experimental models for testing specific predictions of mass transport theories than arrays of nanopores because they require fewer adjustable parameters and they have well-defined geometry and chemical structures. Using polystyrene particles as probes, they demonstrated that quantitative information about fundamental modes of transport, such as hydrodynamic and electrophoretic flow, can be obtained using these single-pore membranes.

Miller et al.[320] and Miller and Martin[321] prepared carbon nanotube membranes (CNMs) using chemical vapor deposition of graphitic carbon into the pores of microporous alumina template membranes. This approach yields a freestanding membrane containing a parallel array of carbon nanotubes (with the outside diameter similar to 200 nm, and a wall thickness similar to 40 nm) that spans the complete thickness of the membrane (60 μm). The electro-osmotic flow (EOF) can be driven across these CNMs by allowing the membrane to separate two electrolyte solutions and using an electrode in each solution to pass a constant ionic current through the nanotubes. The as-synthesized CNM has anionic surface charge and as a result, the EOF is in the direction of cation migration across the membrane. In Lee et al.[322] synthetic bionanotube membranes were developed and used to separate two enantiomers of a chiral drug. These membranes are based on alumina films that have cylindrical pores with monodisperse nanoscopic diameters (for example, 20 nm). Silica nanotubes were chemically synthesized within the pores of these films, and an antibody that selectively binds one of the enantiomers of the drug was attached to the inner walls of the silica nanotubes. These membranes selectively transport the enantiomer that specifically binds to the antibody, relative to the enantiomer that has lower affinity for the antibody. The solvent dimethyl sulfoxide was used to tune the antibody binding affinity. The enantiomeric selectivity coefficient increases as the inside diameter of the silica nanotubes decreases.

Melechko et al.[323] report a method to fabricate nanoscale pipes ("nanopipes") suitable for fluidic transport. Vertically aligned carbon nanofibers grown by plasma-enhanced chemical vapor deposition are used as sacrificial templates for nanopipes with internal diameters as small as 30 nm and lengths up to several micrometers that are oriented perpendicular to the substrate. This method provides a high level of control over the nanopipe location, number, length, and diameter, permitting them to be deterministically positioned on a substrate and arranged into arrays.

8.7 Outlook—go with the flow

As the promise of nanotechnology is beginning to be realized, the new scientific frontiers for this field are outlined. In particular, the interface of nanotechnology with biology seems to emerge as a rich ground for fundamental scientific research and engineering applications. The close link between life and aqueous environments will continue to be explored. While visions of nanomedicine may continue to be controversial, understanding of nanoscale fluid mechanics will continue to offer tools for the exploration of molecular-level drug delivery and on-site interfacing with biological cells.

Fluid mechanics at the nanoscale is an emerging field in need of powerful computational tools and innovative experimental diagnostic techniques aimed at better understanding these phenomena. In computation there is much need for the development of multiscale computational techniques linking the atomistic to the nano, meso, and continuum scales. In parallel, the development of new techniques for experimental diagnosis and manipulation of fluids at the nanoscale will have a significant impact in the coming decades. These experiments and simulations will certainly enable new understandings and findings for the underlying flow physics. The exploitation of these findings to areas ranging from new computer architectures to disease fighting methods will be a breeding ground for further fluid mechanics research at the nanoscale in the near future.

Acknowledgments

Our research in nanoscale fluid mechanics has benefited tremendously from direct and indirect contacts with several research groups working in the area of nanotechnology. The references include a partial list of people whose work has served us as guidance and motivation. We wish to particularly acknowledge many inspirational discussions and fruitful collaborations with Dr. Richard Jaffe (NASA Ames), Dr. Timur Halcioglou (NASA Ames), and Dr. Flavio Noca (JPL/NASA). Finally, we wish to acknowledge the invaluable help of Daniela Wiesli with the typesetting of this report.

References

1. B. Alberts, D. Bray, A. Johnson, J. Lewis, M. Raff, K. Roberts, and P. Walter, *Essential Cell Biology*, Garland Publication, Inc., New York (1997).
2. G. K. Batchelor, *An Introduction to Fluid Dynamics*, 1st ed., Cambridge University Press, Cambridge (1967).
3. R. Car and M. Parinello, "Unified approach for molecular dynamics and density-functional theory," *Phys. Rev. Lett.* **55**, 2471–2475 (1985).
4. J. Koplik and J. R. Banavar, "Corner flow in the sliding plate problem," *Phys. Fluids* **7**, 3118–3125 (1995).
5. M. M. Micci, T. L. Kaltz, and L. N. Long, "Molecular dynamics simulations of atomization and spray phenomena," *Atom. Sprays* **11**, 351–363 (2001).
6. S. Maruyama, "Molecular dynamics methods for microscale heat transfer," in *Advances in Numerical Heat Transfer*, W. J. Minkowycz and E. M. Sparrow, Eds., Vol. II, 189–226, Taylor and Francis, New York (2000).

7. D. Poulikakos, S. Archidiacono, and S. Maruyama, "Molecular dynamics simulations in nanoscale heat transfer: a review," *Micro. Therm. Eng.* **7**, 181–206 (2003).
8. M. Gad-el Hak, "The fluid mechanics of microdevices—the Freeman Scolar lecture," *J. Fluids Eng.* **121**, 5–33 (1999).
9. C.-M. Ho and Y.-C. Tai, "Review: MEMS and its application for flow control," *J. Fluids Eng.* **118**, 437–447 (1996).
10. C.-M. Ho and Y.-C. Tai, "Micro-electro-mechanical-systems (mems) and fluid flow," *Annu. Rev. Fluid Mech.* **30**, 579–612 (1998).
11. O. I. Vinogradova, "Slippage of water over hydrophobic surfaces," *Int. J. Miner. Process.* **56**, 31–60 (1999).
12. N. V. Churaev, "Thin liquid layers," *Colloid J.* **58**, 681–693 (1996).
13. J. Baudry, E. Charlaix, A. Tonck, and D. Mazuyer, "Experimental evidence for a large slip effect at a nonwetting fluid-solid interface," *Langmuir* **17**, 5232–5236 (2001).
14. E. Bonaccurso, M. Kappl, and H.-J. Butt, "Hydrodynamic force measurements: boundary slip of water on hydrophilic surfaces and electrokinetic effects," *Phys. Rev. Lett.* **88**, 076103 (2002).
15. P. Harris, *Carbon Nanotubes and Related Structures*, Cambridge University Press, New York (1999).
16. A. Szabo and N. S. Ostlund, *Modern Quantum Chemistry*, rev. ed., McGraw-Hill, New York (1989).
17. M. Born and R. Oppenheimer, "Zur Quantentheorie der Molekeln," *Ann. Phys.* **84**, 457–484 (1927).
18. D. Feller and K. D. Jordan, "Estimating the strength of the water/single-layer graphite interaction," *J. Phys. Chem. A* **104**, 9971–9975 (2000).
19. Møller, C. and M. S. Plesset, "Note on an approximation treatment for many-electron systems," *Phys. Rev.* **46**, 618–622 (1934).
20. W. Koch and M. C. Holthausen, *A Chemist's Guide to Density Functional Theory*, 2nd ed., Wiley, Weinheim, Germany (2001).
21. F. Sim, A. St-Amant, I. Papai, and D. R. Salahub, "Gaussian density functional calculations on hydrogen-bonded systems," *J. Am. Chem. Soc.* **114**, 4391–4400 (1992).
22. J. P. Perdew, S. Kurth, A. Zupan, and P. Blaha, "Accurate density functional with correct formal properties: a step beyond the generalized gradient approximation," *Phys. Rev. Lett.* **82**, 2544–2547 (1999).
23. Y. Andersson and H. Rydberg, "Dispersion coefficients for van der Waals complexes, including C_{60}–C_{60}," *Physica Scripta* **60**, 211–216 (1999).
24. E. Hult, Y. Andersson, B. I. Lundqvist, and D. C. Langreth, "Density functional for van der Waals forces at surfaces," *Phys. Rev. Lett.* **77**, 2029–2032 (1996).
25. E. Hult, H. Rydberg, B. I. Lundqvist, and D. C. Langreth, "Unified treatment of asymptotic van der Waals forces," *Phys. Rev. B* **59**, 4708–4713 (1999).

26. H. Rydberg, N. Jacobson, P. Hyldgaard, S. Simak, B. I. Lundqvist, and D. C. Langreth, "Hard numbers on soft matter," *Surf. Sci.* **532**, 606–610 (2003).
27. Y. Andersson, D. C. Langreth, and B. I. Lundqvist, "Van der Waals Interactions in density-functional theory," *Phys. Rev. Lett.* **76**, 102–105 (1996).
28. X. Wu, M. C. Vargas, S. Nayak, V. Lotrich, and G. Scoles, "Towards extending the applicability of density functional theory to weakly bound systems," *J. Chem. Phys.* **115**, 8748–8757 (2001).
29. M. Elstner, P. Hobza, T. Frauenheim, S. Suhai, and E. Kaxiras, "Hydrogen bonding and stacking interactions of nucleic acid base pairs: a density-functional-theory based treatment," *J. Chem. Phys.* **114**, 5149–5155 (2001).
30. U. Zimmerli, "On the water graphite interaction," Internal report, Institute of Computational Science, ETH Zürich (2003).
31. C. Douketis, G. Scoles, S. Marchetti, M. Zen, and A. J. Thakkar, "Intermolecular forces via hybrid Hartree-Fock-SCF plus damped dispersion (HFD) energy calculations. An improved spherical model," *J. Chem. Phys.* **76**, 3057–3063 (1982).
32. E. Zaremba and W. Kohn, "Van der Waals interaction between an atom and a solid surface," *Phys. Rev. B* **13**, 2270–2285 (1976).
33. C. Mavroyannis and M. J. Stephen, "Dispersion forces," *Molec. Phys.* **5**, 629–638 (1962).
34. J. N. Israelachvili, *Intermolecular and Surface Forces. With Applications to Colloidal and Biological Systems*, 2nd ed., Academic Press, London (1992).
35. K. T. Tang and J. P. Toennies, "An improved simple model for the van der Waals potential based on universal damping functions for the dispersion coefficients," *J. Comput. Phys.* **80**, 3726–3741 (1984).
36. T. H. Boyer, "Unretarded London-van der Waals forces derived from classical electrodynamics with classical electromagnetic zero-point radiation," *Phys. Rev. A* **6**, 314–319 (1972).
37. W. F. van Gunsteren, S. R. Billeter, S. R. Eising, P. H. Hünenberger, P. Krüger, A. E. Mark, W. R. P. Scott, and I. G. Tironi, *Biomolecular Simulation: The GROMOS96 Manual and User Guide*, Vdf Hochschulverlag AG, Zürich (1996).
38. T. Werder, J. H. Walther, R. L. Jaffe, T. Halicioglu, and P. Koumoutsakos, "On the water-graphite interaction for use in MD simulations of graphite and carbon nanotubes," *J. Phys. Chem. B* **107**, 1345–1352 (2003).
39. M. Grujicic, G. Caoa, and B. Gerstenb, "Enhancement of field emission in carbon nanotubes next term through adsorption of polar molecules," *Appl. Surf. Sci.* **206**, 167–177 (2003).
40. E. Fermi, J. Pasta, and S. Ulam, "Studies in nonlinear problems," Los Alamos report LA-1940 (1955).
41. B. J. Alder and T. E. Wainwright, "Phase transition for a hard sphere system," *J. Chem. Phys.* **27**, 1208–1209 (1957).
42. A. Rahman, "Correlations in the motion of atoms in liquid argon," *Phys. Rev.* **136**, 405–411 (1964).

43. L. Verlet, "Computer experiments on classical fluids. I. Thermodynamical properties of Lennard-Jones molecules," *Phys. Rev.* **159**, 98–103 (1967).
44. L. Verlet, "Computer experiments on classical fluids. II. Equilibrium correlation functions," *Phys. Rev.* **165**, 201–214 (1968).
45. J.-P. Hansen and L. Verlet, "Phase transitions of the Lennard-Jones system," *Phys. Rev.* **184**, 151–162 (1969).
46. A. Rahman and F. H. Stillinger, "Molecular dynamics study of liquid water," *J. Chem. Phys.* **55**, 3336–3359 (1971).
47. M. P. Allen and D. J. Tildesley, *Computer Simulation of Liquids*, Clarendon Press Oxford, Oxford (1987).
48. D. N. Theodorou and U. W. Suter, "Atomistic modeling of mechanical properties of polymeric glasses," *Macromolecules* **19**, 139–154 (1986).
49. D. G. Cahill, K. E. Ford, G. D. Mahan, A. Majumdar, H. J. Maris, R. Merlin, and S. R. Phillpot, "Nanoscale thermal transport," *J. Appl. Phys.* **93**, 793–818 (2003).
50. T. L. Kaltz, L. N. Long, M. M. Micci, and J. K. Little, "Supercritical vaporization of liquid oxygen droplets using molecular dynamics," *Combust. Sci. Techol.* **136**, 279–301 (1998).
51. S. Maruyama and T. Kimura, "A study of thermal resistance over a solid-liquid interface by the molecular dynamics method," *Thermal Sci. Eng.* **7**, 63–68 (1999).
52. J. H. Walther and P. Koumoutsakos, "Molecular dynamics simulation of nanodroplet evaporation," *J. Heat Transfer* **123**, 741–748 (2001).
53. A. K. Rappé, C. J. Casewit, K. S. Colwell, W. A. Goddard III, and W. M. Skiff, "UFF, a full periodic table force field for molecular mechanics and molecular dynamics simulations," *J. Am. Chem. Soc.* **114**, 10024–10035 (1992).
54. D. W. Brenner, "Empirical potential for hydrocarbons for use in simulating the chemical vapor deposition of diamond films," *Phys. Rev. B* **42**, 9458–9471 (1990).
55. J. Tersoff, "New empirical approach for the structure and energy of covalent systems," *Phys. Rev. B* **37**, 6991–7000 (1988).
56. O. Teleman, B. Jönsson, and S. Engström, "A molecular dynamics simulation of a water model with intramolecular degrees of freedom," *Molec. Phys.* **60**, 193–203 (1987).
57. H. J. C. Berendsen, J. R. Grigera, and T. P. Straatsma, "The missing term in effective pair potentials," *J. Phys. Chem.* **91**, 6269–6271 (1987).
58. W. L. Jorgensen, "Revised TiPS for simulations of liquid water and aqueous solutions," *J. Comput. Phys.* **77**, 4156–4163 (1982).
59. M. W. Mahoney and W. L. Jorgensen, "A five-site model for liquid water and the reproduction of the density anomaly by rigid, nonpolarizable potential functions," *J. Chem. Phys.* **112**, 8910–8922 (2000).
60. E. D. Burchart, V. A. Verheij, H. van Bekkum, and B. de Graaf, "A consistent molecular mechanics force-field for all-silica zeolites," *Zeolites* **12**, 183–189 (1992).

61. W. D. Cornell, P. Cieplak, C. I. Bayly, I. R. Gould, K. M. Merz, Jr., D. M. Ferguson, D. C. Spellmeyer, T. Fox, J. W. Caldwell, and P. A. Kollman, "A second generation force field for the simulation of proteins, nucleic acids, and organic molecules," *J. Am. Chem. Soc.* **117**, 5179–5197 (1995).
62. B. R. Brooks, R. E. Bruccoleri, B. D. Olafson, D. J. States, S. Swaminathan, and M. Karplus, "CHARMM—a program for macromolecular energy, minimization, and dynamics calculations," *J. Comput. Chem.* **4**, 187–217 (1983).
63. S. L. Mayo, B. D. Olafson, and W. A. Goddard III, "Dreiding—a generic force-field for molecular simulations," *J. Phys. Chem.* **94**, 8897–8909 (1990).
64. J. Gao and M. A. Thompson, *Combined Quantum Mechanical and Molecular Mechanical Methods*, ACS Symposium Series 712, American Chemical Society, Washington, DC (1998).
65. L. Guidoni, P. Maurer, S. Piana, and U. Röthlisberger, "Hybrid Car-Parrinello/molecular mechanics modelling of transition metal complexes: structure, dynamics and reactivity," *Quant. Struct. Act. Relat.* **21**, 119 127 (2002).
66. A. Glättli, X. Daura, and W. F. van Gunsteren, "Derivation of an improved simple point charge model for liquid water: SPC/A and SPC/L," *J. Chem. Phys.* **116**, 9811–9828 (2002).
67. J. Kong, N. R. Franklin, C. Zhou, M. G. Chapline, S. Peng, C. Kyeongjae, and H. Dai, "Nanotube molecular wires as chemical sensors," *Science* **287**, 622–625 (2000).
68. G. Hummer, J. C. Rasaiah, and J. P. Noworyta, "Water conduction through the hydrophobic channel of a carbon nanotube," *Nature* **414**, 188–190 (2001).
69. T. Werder, J. H. Walther, R. Jaffe, T. Halicioglu, F. Noca, and P. Koumoutsakos, "Molecular dynamics simulations of contact angles of water droplets in carbon nanotubes," *Nano Lett.* **1**, 697–702 (2001).
70. P. M. Morse, "Diatomic molecules according to the wave mechanics. II. Vibrational levels," *Phys. Rev.* **34**, 57–64 (1929).
71. G. Marechal and J.-P. Ryckaert, "Atomic versus molecular description of transport-properties in polyatomic fluids. n-butane as an illustration," *Chem. Phys. Lett.* **101**, 548–554 (1983).
72. J.-P. Ryckaert and A. Bellemans, "Molecular dynamics of liquid n-butane near its boiling point," *Chem. Phys. Lett.* **30**, 123–125 (1975).
73. J. H. Walther, R. Jaffe, T. Halicioglu, and P. Koumoutsakos, "Carbon nanotubes in water: structural characteristics and energetics," *J. Phys. Chem. B* **105**, 9980–9987 (2001).
74. M. J. Frisch, G. W. Trucks, H. B. Schlegel, G. E. Scuseria, M. A. Robb, J. R. Cheeseman, V. G. Zakrzewski, J. A. Montgomery, Jr., R. E. Stratmann, J. C. Burant, S. Dapprich, J. M. Millam, A. D. Daniels, K. N. Kudin, M. C. Strain, O. Farkas, J. Tomasi, V. Barone, M. Cossi, R. Cammi, B. Mennucci, C. Pomelli, C. Adamo, S. Clifford, J. Ochterski, G. A. Petersson, P. Y. Ayala, Q. Cui, K. Morokuma, D. K. Malick, A. D. Rabuck, K. Raghavachari, J. B. Foresman, J. Cioslowski, J. V. Ortiz, A. G. Baboul,

B. B. Stefanov, G. Liu, A. Liashenko, P. Piskorz, I. Komaromi, R. Gomperts, R. L. Martin, D. J. Fox, T. Keith, M. A. Al-Laham, C. Y. Peng, A. Nanayakkara, C. Gonzalez, M. Challacombe, P. M. W. Gill, B. Johnson, W. Chen, M. W. Wong, J. L. Andres, C. Gonzalez, M. Head-Gordon, E. S. Replogle, and J. A. Pople, "Gaussian 98, revision a.7," Technical report, Gaussian, Inc., Pittsburgh, PA (1998).

75. J.-P. Ryckaert, G. Cicotti, and H. J. C. Berendsen, "Numerical integration of the cartesian equations of motion of a system with constraints: molecular dynamics of n-alkanes," *J. Comput. Phys.* **23**, 327–341 (1977).
76. W. L. Jorgensen, J. Chandrasekhar, J. D. Madura, R. W. Impey, and M. L. Klein, "Comparison of simple potential functions for simulating liquid water," *J. Chem. Phys.* **79**, 926–935 (1983).
77. W. F. van Gunsteren and H. J. C. Berendsen, "Algorithms for macromolecular dynamics and constraint dynamics," *Molec. Phys.* **37**, 1311–1327 (1977).
78. T. R. Forester and W. Smith, "SHAKE, rattle, and roll: efficient constaint algorithms for linked rigid bodies," *J. Comput. Chem.* **19**, 102–111 (1998).
79. V. Kräutler, W. F. van Gunsteren, and P. H. Hünenberger, "A fast SHAKE algorithm to solve distance constraint equations for small molecules in molecular dynamics simulations," *J. Comput. Chem.* **22**, 501–508 (2001).
80. S. Miyamoto and P. A. Kollman, "SETTLE: an analytical version of the SHAKE and RATTLE algorithm for rigid water models," *J. Comput. Chem.* **13**, 952–962 (1992).
81. M. Yoneya, "A generalized non-iterative matrix method for constraint molecular dynamics simulations," *J. Comput. Phys.* **172**, 188–197 (2001).
82. J. E. Lennard–Jones and J. Corner, "The calculation of surface tension from intermolecular forces," *Trans. Faraday Soc.* **36**, 1156–1162 (1940).
83. G. D. Smith and R. L. Jaffe, "Comparative study of force fields for benzene," *J. Phys. Chem.* **100**, 9624–9630 (1996).
84. M. C. Gordillo and J. Martí, "Molecular dynamics description of a layer of water molecules on a hydrophobic surface," *J. Chem. Phys.* **117**, 3425–3430 (2002).
85. P. J. Steinbach and B. R. Brooks, "New spherical-cutoff methods for long-range forces in macromolecular simulations," *J. Comput. Chem.* **15**, 667–683 (1994).
86. D. S. Vieira and L. Degrève, "Molecular simulation of a concentrated aqueous KCl solution," *J. Molec. Struct.* **580**, 127–135 (2002).
87. P. Hünenberger and J. A. McCammon, "Ewald artifacts in computer simulations of ionic solvation and ion-ion interaction: a continuum electrostatic study," *J. Chem. Phys.* **110**, 1856–1872 (1999).
88. M. Lísal, J. Kolafa, and I. Nezbeda, "An examination of the five-site potential (TIP5P) for water," *J. Chem. Phys.* **117**, 8892–8897 (2002).
89. P. P. Ewald, "Die Berechnung Optischer und Elektrostatische Gitterpotentiale," *Ann. Phys.* **64**, 253–287 (1921).

90. M. J. L. Sangster and M. Dixon, "Interionic potentials in alkali halides and their use in simulations of the molten salts," *Adv. Phys.* **25**, 247–342 (1976).
91. H. G. Petersen, "Accuracy and efficiency of the particle mesh Ewald method," *J. Chem. Phys.* **103**, 3668–3679 (1995).
92. T. Darden, D. York, and L. Pedersen, "Particle mesh Ewald: an $N \cdot \log N$ method for Ewald sums in large systems," *J. Chem. Phys.* **98**, 10089–10092 (1993).
93. U. Essmann, L. Perera, M. L. Berkowitz, T. Darden, H. Lee, and L. G. Pedersen, "A smooth particle mesh Ewald method," *J. Chem. Phys.* **103**, 8577–8593 (1995).
94. R. W. Hockney, S. P. Goel, and J. W. Eastwood, "A 10000 particle molecular dynamics model with long-range forces," *Chem. Phys. Lett.* **21**, 589–591 (1973).
95. R. W. Hockney and J. W. Eastwood, *Computer Simulation Using Particles*, 2nd ed., IOP, Bristol (1988).
96. B. A. Luty, M. E. Davis, I. G. Tironi, and W. F. van Gunsteren, "A comparison of particle-particle, particle-mesh and Ewald methods for calculating electrostatic interactions in periodic molecular systems," *Molec. Sim.* **14**, 11–20 (1994).
97. J. H. Walther, "An influence matrix particle-particle particle-mesh algorithm with exact particle-particle correction," *J. Comput. Phys.* **184**, 670–678 (2003).
98. J. Barnes and P. Hut, "A hierarchical $O(N \log N)$ force-calculation algorithm," *Nature* **324**, 446–449 (1986).
99. L. Greengard and V. Rokhlin, "The rapid evaluation of potential fields in three dimensions," *Lect. Notes Math.* **1360**, 121–141 (1988).
100. K. E. Schmidt and M. A. Lee, "Implementing the fast multipole method in three dimensions," *J. Stat. Phys.* **63**, 1223–1235 (1991).
101. C. R. Anderson, "An implementation of the fast multipole method without multipoles," *SIAM J. Sci. Stat. Comput.* **13**, 923–947 (1992).
102. Y. Hu and S. L. Johnsson, "A data-parallel implementation of hierarchical N-body methods," *Int. J. Supercomput. Appl.* **10**, 3–40 (1996).
103. W. D. Elliott and J. A. Board, Jr., "Fast Fourier transform accelerated fast multipole algorithm," *SIAM J. Sci. Stat. Comput.* **17**, 398–415 (1996).
104. A. M. Mathiowetz, A. Jain, N. Karasawa, and W. A. Goddard III, "Protein simulations using techniques for very large systems—the cell multipole method for nonbond interactions and the Newton-Euler inverse mass operator method for internal coordinate dynamics," *Proteins* **20**, 227–247 (1994).
105. N. Vaidehi, A. Jain, and W. A. Goddard III, "Constant temperature constrained molecular dynamics: the Newton-Euler inverse mass operator method," *J. Phys. Chem.* **100**, 10508–10517 (1996).
106. C. Sagui and T. D. Darden, "Molecular dynamics simulations of biomolecules: long-range electrostatic effects," *Annu. Rev. Biophys. Biomol. Struc.* **28**, 155–179 (1999).

107. A. Brünger, C. L. Brooks, and M. Karplus, "Stochastic boundary conditions for molecular dynamics simulations of ST2 water," *Chem. Phys. Lett.* **105**, 495–500 (1984).
108. M. Berkowitz and J. A. McCammon, "Molecular-dynamics with stochastic boundary-conditions," *Chem. Phys. Lett.* **90**, 215–217 (1982).
109. G. Ciccotti, G. J. Martyna, S. Melchionna, and M. E. Tuckerman, "Constrained isothermal-isobaric molecular dynamics with full atomic virial," *J. Phys. Chem. B* **105**, 6710–6715 (2001).
110. P. T. Cummings and D. J. Evans, "Nonequilibrium molecular dynamics approaches to transport properties and non-Newtonian fluidy rheology," *Ind. Eng. Chem. Res.* **31**, 1237–1252 (1992).
111. W. Loose and S. Hess, "Rheology of dense model fluids via nonequilibrium molecular dynamics: shear thinning and ordering transition," *Rheol. Acta* **28**, 91–101 (1989).
112. J.-P. Ryckaert, A. Bellemans, G. Ciccotti, and G. V. Paolini, "Evaluation of transport coefficients of simple fluids by molecular dynamics: comparison of Green-Kubo and nonequilibrium approaches for shear viscosity," *Phys. Rev. A* **39**, 259–267 (1989).
113. M. E. Tuckerman, C. J. Mundy, S. Balasubramanian, and M. L. Klein, "Modified nonequilibrium molecular dynamics for fluid flows with energy conservation," *J. Chem. Phys.* **106**, 5615–5621 (1997).
114. B. D. Todd and P. J. Daivis, "Nonequilibrium molecular dynamics simulations of planar elongational flow with spatially and temporally periodic boundary conditions," *Phys. Rev. Lett.* **81**, 1118–1120 (1998).
115. J. R. Dorfman, *An Introduction to Chaos in Nonequilibrium Statistical Mechanics*, Vol. 1, Cambridge University Press, Cambridge (1999).
116. H. Grubmüller, B. Heymann, and P. Tavan, "Ligand binding: molecular mechanics calculation of the streptavidin-biotin rupture force," *Science* **271**, 997–999 (1996).
117. B. Isralewitz, M. Gao, and K. Schulten, "Steered molecular dynamics and mechanical functions of proteins," *Curr. Opin. Struct. Biol.* **11**, 224–230 (2001).
118. M. Matsumoto, S. Saito, and I. Ohmine, "Molecular dynamics simulation of the ice nucleation and growth process leading to water freezing," *Nature* **416**, 409–413 (2002).
119. L. D. Landau and E. M. Lifshitz, *Fluid Mechanics*, Vol. 6, 2nd ed., Pergamon Press, New York (1987).
120. G. A. Bird, *Molecular Gas Dynamics and the Direct Simulation of Gas Flows*, Clarendon Press Oxford, Oxford (1994).
121. G. E. Karniadakis and A. Beskok, *Micro Flows. Fundamentals and Simulation*, Springer, New York (2002).
122. G. A. Bird, "Breakdown of translational and rotational equilibrium in gaseous expansions," *AIAA J.* **8**, 1998–2003 (1970).

123. I. D. Boyd and G. Chen, "Predicting failure of the continuum fluid equations in transitional hypersonic flows," *Phys. Fluids* **7**, 210–219 (1995).
124. S. Tiwari, "Coupling of the Boltzmann and Euler equations with automatic domain decomposition," *J. Comput. Phys.* **144**, 710–726 (1998).
125. A. L. Garcia, J. B. Bell, W. Y. Crutchfield, and B. J. Alder, "Adaptive mesh and algorithm refinement using direct simulation Monte Carlo," *J. Comput. Phys.* **154**, 134–155 (1999).
126. S. T. O'Connell and P. A. Thompson, "Molecular dynamics-continuum hybrid computations: a tool for studying complex fluid flow," *Phys. Rev. E* **52**, R5792–R5795 (1995).
127. N. G. Hadjiconstantinou and A. T. Patera, "Heterogeneous atomistic-continuum representations for dense fluid systems," *Int. J. Mod. Phy. C* **8**, 967–976 (1997).
128. N. G. Hadjiconstantinou, "Hybrid atomistic-continuum formulations and the moving contact-line problem," *J. Comput. Phys.* **154**, 245–265 (1999).
129. E. G. Flekkøy, G. Wagner, and J. Feder, "Hybrid model for combined particle and continuum dynamics," *Europhys. Lett.* **52**, 271–276 (2000).
130. G. Wagner, E. Flekkøy, J. Feder, and T. Jossang, "Coupling molecular dynamics and continuum dynamics," *Comp. Phys. Commun.* **147**, 670–673 (2002).
131. E. G. Flekkøy, J. Feder, and G. Wagner, "Coupling particles and fields in a diffusive hybrid model," *Phys. Rev. E* **64**, 066302-1–066302-7 (2001).
132. F. J. Alexander, A. L. Garcia, and D. M. Tartakovsky, "Algorithm refinement for stochastic partial differential equations. I. Linear diffusion," *J. Comput. Phys.* **182**, 47–66 (2002).
133. P. J. Hoogerbrugge and J. M. V. A. Koelman, "Simulating microscopic hydrodynamics phenomena with dissipative particle dynamics," *Europhys. Lett.* **19**, 155–160 (1992).
134. P. Español and P. Warren, "Statistical-mechanics of dissipative particle dynamics," *Europhys. Lett.* **30**, 191–196 (1995).
135. P. B. Warren, "Dissipative particle dynamics," *Curr. Opin. Colloid Inter. Sci.* **3**, 620–629 (1998).
136. M. Serrano and P. Español, "Thermodynamically consistent mesoscopic fluid particle model," *Phys. Rev. E* **64**, 046115-1–046115-18 (2001).
137. P. Español and M. Revenga, "Smoothed dissipative particle dynamics," *Phys. Rev. E* **67**, 026705-1–026705-12 (2003).
138. S. C. Glotzer and W. Paul, "Molecular and mesoscale simulation methods for polymer materials," *Annu. Rev. Mater. Res.* **32**, 401–436 (2002).
139. K. Kremer and F. Müller–Plathe, "Multiscale simulation in polymer science," *Molec. Sim.* **28**, 729–750 (2002).
140. A. M. Wolters, D. A. Jayawickrama, and J. V. Sweedler, "Microscale NMR," *Curr. Opin. Chem. Biol.* **6**, 711–716 (2002).
141. R. Kimmich, "Strange kinetics, porous media, and NMR," *Chem. Phys.* **284**, 253–285 (2002).

142. M. Weber and R. Kimmich, "Rayleigh-Benard percolation transition of thermal convection in porous media: computational fluid dynamics, NMR velocity mapping, NMR temperature mapping," *Phys. Rev. E* **66**, 026306 (2002).
143. R. W. Mair and R. L. Walsworth, "Novel MRI applications of laser-polarized noble gases," *Appl. Mag. Res.* **22**, 159–173 (2002).
144. T. Meersmann, J. W. Logan, R. Simonutti, S. Caldarelli, A. Comotti, P. Sozzani, L. G. Kaiser, and A. Pines, "Exploring single-file diffusion in 1D nanochannels by laser-polarized Xe-129 NMR spectroscopy," *J. Phys. Chem.* **104**, 11665–11670 (2000).
145. T. Ueda, T. Eguchi, N. Nakamura, and R. E. Wasylishen, "High-pressure Xe-129 NMR study of xenon confined in the nanochannels of solid (+/−)-[co(en)(3)]cl-3," *J. Phys. Chem. B* **107**, 180–185 (2003).
146. J. M. Kneller, R. J. Soto, S. E. Surber, J. F. Colomer, A. Fonseca, J. B. Nagy, G. van Tendeloo, and T. Pietrass, "TEM and laser-polarized Xe-129 NMR characterization of oxidatively purified carbon nanotubes," *J. Am. Chem. Soc.* **122**, 10591–10597 (2000).
147. M. Heuberger, "The surface forces apparatus," in *Encyclopedia of Chemical Physics and Physical Chemistry*, J. H. Moore and N. D. Spencer, Eds., Vol. II, 1517–1536, Institute of Physics Publishing, Bristol (2001).
148. H. Komatsu and S. Miyashita, "Comparison of atomic force microscopy and nanoscale optical microscopy for measuring step hieghts," *Jpn. J. Appl. Phys.* **32**, 1478–1479 (1993).
149. O. I. Vinogradova, H. J. Butt, G. E. Yakubov, and F. Feuillebois, "Dynamic effects on force measurements. I. Viscous drag on the atomic force microscope cantilever," *Rev. Sci. Instrum.* **72**, 2330–2339 (2001).
150. H. W. Zhu, C. L. Xu, D. H. Wu, B. Q. Wei, R. Vajtai, and P. M. Ajayan, "Direct synthesis of long single-walled carbon nanotube strands," *Science* **296**, 884–886 (2002).
151. Y. Zhu and S. Granick, "Limits of the hydrodynamic no-slip boundary condition," *Phys. Rev. Lett.* **88**, 106102-1–106102-4 (2002).
152. L. T. Zhang, G. J. Wagner, and W. K. Liu, "A parallelized meshfree method with boundary enrichment for large-scale CFD," *J. Comput. Phys.* **176**, 483–506 (2002).
153. M. Heuberger and M. Zäch, "Nanofluidics: structural forces, density anomalies, and the pivotal role of nanoparticles," *Langmuir* **19**, 1943–1947 (2003).
154. J. Kameoka and H. G. Craighead, "Nanofabricated refractive index sensor based on photon tunneling in nanofluidic channel," *Sens. Actuat. B Chem.* **77**, 632–637 (2001).
155. L. Vivien, J. Moreau, D. Riehl, P. A. Alloncle, M. Autric, F. Hache, and E. Anglaret, "Shadowgrapic imaging of carbon nanotube suspensions in water and in chloroform," *J. Opt. Soc. Am. B* **19**, 2665–2672 (2002).
156. L. Vivien, D. Riehl, F. Hache, and E. Anglaret, "Optical limiting properties of carbon nanotubes," *Physica B* **323**, 233–234 (2002).

157. D. W. Pohl, "Scanning near-field optical microscopy," Chap. 12, in *Advances in Optical and Electron Microscopy*, C. J. R. Sheppard and T. Mulvey, Eds., Academic Press, London (1990).
158. E. Betzig, J. K. Trautman, T. D. Harris, J. S. Weiner, and R. L. Kostelak, "Breaking the diffraction barrier—optical microscopy on a nanometric scale," *Science* **251**, 1468–1470 (1991).
159. F. B. Reitz, M. E. Fauver, and G. H. Pollack, "Fluorescence anisotropy near-field scanning optical microscopy FANSOM: a new technique for nanoscale microviscometry," *Ultramicroscopy* **90**, 259–264 (2002).
160. Y. Gogotsi, J. A. Libera, and M. Yoshimura, "Hydrothermal synthesis of multiwall carbon nanotubes," *J. Mater. Res.* **15**, 2591–2594 (2000).
161. Y. Gogotsi, J. A. Libera, A. Güvenç-Yazicioglu, and C. M. Megaridis, "In situ multiphase fluid experiments in hydrothermal carbon nanotubes," *Appl. Phys. Lett.* **79**, 1021–1023 (2001).
162. Y. Gogotsi, N. Naguib, and J. A. Libera, "In situ chemical experiments in carbon nanotubes," *Chem. Phys. Lett.* **365**, 354–360 (2002).
163. C. M. Megaridis, A. Güvenç-Yazicioglu, and J. A. Libera, "Attoliter fluid experiments in individual closed-end carbon nanotubes: liquid film and fluid interface dynamics," *Phys. Fluids* **14**, L5–L8 (2002).
164. C. P. Gendrich, M. M. Koochesfahani, and D. G. Nocera, "Molecular tagging velocimetry and other novel applications of a new phosphorescent supramolecule," *Exper. Fluids* **23**, 361–372 (1997).
165. M. J. Saxton and K. Jacobson, "Single-particle tracking: applications to membrane dynamics," *Annu. Rev. Biophys. Biomol. Struct.* **26**, 373–399 (1997).
166. K. Otobe, H. Nakao, H. Hayashi, F. Nihey, M. Yudasaka, and S. Iijima, "Fluorescence visualization of carbon nanotubes by modification with silicon-based polymer," *Nano Lett.* **2**, 1157–1160 (2002).
167. M. Speidel, A. Jonas, and E. L. Florin, "Three-dimensional tracking of fluorescent nanoparticles with subnanometer precision by use of off-focus imaging," *Opt. Lett.* **28**, 69–71 (2003).
168. J. P. S. Farinha, M. A. Winnik, and K. G. Hahn, "Characterization of oil droplets under a polymer film by laser scanning confocal fluorescence microscopy," *Langmuir* **16**, 3391–3400 (2000).
169. G. Binnig, C. F. Quate, and C. Gerber, "Atomic force microscope," *Phys. Rev. Lett.* **56**, 930–933 (1986).
170. O. I. Vinogradova and G. E. Yakubov, "Dynamic effects on force measurements. 2. Lubrication and the atomic force microscope," *Langmuir* **19**, 1227–1234 (2003).
171. M. Luna, J. Colchero, and A. M. Baró, "Study of water droplets and films on graphite by noncontact scanning force microscopy," *J. Phys. Chem. B* **103**, 9576–9581 (1999).
172. P. Attard, M. P. Moody, and J. W. G. Tyrrell, "Nanobubbles: the big picture," *Physica A* **314**, 696–705 (2002).

173. J. W. G. Tyrrell and P. Attard, "Images of nanobubbles on hydrophobic surfaces and their interaction," *Phys. Rev. Lett.* **87**, 176104-1–176104-4 (2001).
174. S. D. A. Connell, S. Allen, C. J. Roberts, J. Davies, M. C. Davies, S. J. B. Tendler, and P. M. Williams, "Investigating the interfacial properties of single-liquid nanodroplets by atomic force microscopy," *Langmuir* **18**, 1719–1728 (2002).
175. N. Ishida, T. Inoue, M. Miyahara, and K. Higashitani, "Nano bubbles on a hydrophobic surface in water observed by tapping-mode atomic force microscopy," *Langmuir* **16**, 6377–6380 (2000).
176. S.-T. Lou, Z.-Q. Ouyang, Y. Zhang, X.-J. Li, J. Hu, M.-Q. Li, and F.-J. Yang, "Nanobubbles on solid surface imaged by atomic force microscopy," *J. Vac. Sci. Technol. B* **18**, 2573–2575 (2000).
177. S. T. Lou, J. X. Gao, X. D. Xiao, X. J. Li, G. L. Li, Y. Zhang, M. Q. Li, J. L. Sun, X. H. Li, and J. Hu, "Studies of nanobubbles produced at liquid/solid interfaces," *Mater. Char.* **48**, 211–214 (2002).
178. F. Mugele, T. Becker, R. Nikopoulos, M. Kohonen, and S. Herminghaus, "Capillarity at the nanoscale: an AFM view," *J. Adhesion Sci. Technol.* **16**, 951–964 (2002).
179. L. Zitzler, S. Herminghaus, and F. Mugele, "Capillary forces in tapping mode atomic force microscopy," *Phys. Rev. B* **66**, 155436 (2002).
180. J. Colchero, A. Storch, M. Luna, J. G. Herrero, and A. M. Baró, "Observation of liquid neck formation with scanning force microscopy techniques," *Langmuir* **14**, 2230–2234 (1998).
181. R. D. Piner, J. Zhu, F. Xu, S. H. Hong, and C. A. Mirkin, "Dip-pen nanolitography," *Science* **283**, 661–663 (1999).
182. M. Calleja, M. Tello, and R. Garcia, "Size determination of field-induced water menisci in noncontact atomic force microscopy," *J. Appl. Phys.* **92**, 5539–5542 (2002).
183. H. J. Butt and V. Franz, "Rupture of molecular thin films observed in atomic force microscopy. I. Theory," *Phys. Rev. E* **66**, 031601 (2002).
184. S. Loi, G. Sun, V. Franz, and H. J. Butt, "Rupture of molecular thin films observed in atomic force microscopy ii. Experiment," *Phys. Rev. E* **66**, 031602-1–031602-7 (2002).
185. N. Ahmed, D. F. Nino, and V. T. Moy, "Measurement of solution viscosity by atomic force microscopy," *Rev. Sci. Instrum.* **72**, 2731–2734 (2001).
186. S. P. Jarvis, T. Uchihashi, T. Ishida, H. Tokumoto, and Y. Nakayama, "Local solvation shell measurement in water using a carbon nanotube probe," *J. Phys. Chem. B* **104**, 6091–6094 (2000).
187. J. P. Cleveland, T. E. Schaeffer, and P. K. Hansma, "Probing oscillatory hydration potentials using thermal-mechanical noise in an atomic-force microscope," *Phys. Rev. B* **52**, R8692–R8695 (1995).
188. L. R. Pratt, "Molecular theory of hydrophobic effects: 'She is too mean to have her name repeated'," *Annu. Rev. Phys. Chem.* **53**, 409–436 (2002).

189. L. R. Pratt and A. Pohorille, "Hydrophobic effects and modeling of biophysical aqueous solution interfaces," *Chem. Rev.* **102**, 2671–2692 (2002).
190. T. Lazaridis, "Solvent size vs cohesive energy as the origin of hydrophobicity," *Acc. Chem. Res.* **34**, 931–937 (2001).
191. W. Kauzmann, "Some factors in the interpretation of protein denaturation," *Adv. Prot. Chem.* **14**, 1–63 (1959).
192. K. A. Dill, "Dominant forces in protein folding," *Biochemistry* **29**, 7133–7155 (1990).
193. D. Chandler, "Two faces of water," *Nature* **417**, 491 (2002).
194. K. Lum, D. Chandler, and J. D. Weeks, "Hydrophobicity at small and large length scales," *J. Phys. Chem. B* **103**, 4570–4577 (1999).
195. L. Boruvka and A. W. Neumann, "Generalization of the classical theory of capillarity," *J. Chem. Phys.* **66**, 5464–5476 (1977).
196. G. Saville, "Computer simulation of the liquid-solid-vapour contact angle," *J. Chem. Soc. Faraday Trans.* **5**, 1122–1132 (1977).
197. J. H. Sikkenk, J. O. Indekeu, J. M. J. van Leeuwen, E. O. Vossnack, and A. F. Bakker, "Simulation of wetting and drying at solid fluid interfaces on the Delft molecular-dynamics processor," *J. Stat. Phys.* **52**, 23–44 (1988).
198. M. Nijmeijer, C. Bruin, and A. Bakker, "A visual measurement of contact angles in a molecular-dynamics simulation," *Physica A* **160**, 166–180 (1989).
199. M. J. P. Nijmeijer, C. Bruin, A. F. Bakker, and J. M. J. van Leeuwen, "Wetting and drying on an inert wall by a fluid in a molecular dynamics simulation," *Phys. Rev. A* **42**, 6052–6059 (1990).
200. J. Hautman and M. L. Klein, "Microscopic wetting phenomena," *Phys. Rev. Lett.* **67**, 1763–1766 (1991).
201. P. A. Thompson, W. B. Brinckerhoff, and M. O. Robbins, "Microscopic studies of static and dynamic contact angles," *J. Adhesion Sci. Technol.* **7**, 535–554 (1993).
202. C. F. Fan and T. Cağin, "Wetting of crystalline polymer surfaces: a molecular dynamics simulation," *J. Chem. Phys.* **103**, 9053–9061 (1995).
203. M. J. de Ruijter and J. de Coninck, "Contact angle relaxation during the spreading of partially wetting drops," *Langmuir* **13**, 7293–7298 (1997).
204. M. J. de Ruijter, T. D. Blake, and J. de Coninck, "Dynamic wetting studies by molecular modeling simulations of droplet spreading," *Langmuir* **15**, 7836–7847 (1999).
205. M. Voué and J. de Coninck, "Spreading and wetting at the microscopic scale: recent developments and perspectives," *Acta Mater.* **48**, 4405–4417 (2000).
206. J. de Coninck, M. J. de Ruijter, and M. Voué, "Dynamics of wetting," *Curr. Opin. Colloid Inter. Sci.* **6**, 49–53 (2001).
207. F. Bresme and N. Quirke, "Computer simulation study of the wetting behavior and line tensions of nanometer size particulates at a liquid-vapor interface," *Phys. Rev. Lett.* **80**, 3791–3794 (1998).
208. F. Bresme and N. Quirke, "Computer simulation of wetting and drying of spherical particulates at a liquid-vapor interface," *J. Chem. Phys.* **110**, 3536–3547 (1999).

209. F. Bresme and N. Quirke, "Computer simulation studies of liquid lenses at a liquid-liquid interface," *J. Chem. Phys.* **112**, 5985–5990 (2000).
210. M. Lundgren, N. L. Allan, T. Cosgrove, and N. George, "Wetting of water and water/ethanol droplets on a non-polar surface: a molecular dynamics study," *Langmuir* **18**, 10462–10466 (2002).
211. F. M. Fowkes and W. D. Harkins, "The state of monolayers adsorbed at the interface solid-aqueous solution," *J. Am. Chem. Soc.* **62**, 3377–3377 (1940).
212. M. E. Schrader, "Ultrahigh-vacuum techniques in the measurement of contact angles. 5. LEED study of the effect of structure on the wettability of graphite," *J. Phys. Chem.* **84**, 2774–2779 (1980).
213. F. Noca and E. Sansom, Private communication (2003).
214. C. L. M. H. Navier, "Memoire sur les lois du mouvement des fluides," *Mem. Acad. R. Sci. Inst. Fr.* **6**, 389 (1827).
215. G. G. Stokes, "On the effect of the internal friction of fluids on the motion of pendulums," *Trans. Cambridge Phil. Soc.* **9**, 8 (1851).
216. W. C. D. Whetham, "On the alleged slipping at the boundary of a liquid in motion," *Phil. Trans. R. Soc. London A* **181**, 559–582 (1890).
217. R. Bulkley, "Viscous flow and surface films," *Bur. Stand. J. Res.* **6**, 89–112 (1931).
218. H. Helmholtz and G. von Piotrowski, "Über reibung tropfbarer flussigkeiten," *Sitz. Kaiserlich Akad. Wissen.* **40**, 607–658 (1860).
219. E. Schnell, "Slippage of water over nonwettable surfaces," *J. Appl. Phys.* **27**, 1149–1152 (1956).
220. N. V. Churaev, V. D. Sobolev, and A. N. Somov, "Slippage of liquids over lyophobic solid surfaces," *J. Coll. Interface Sci.* **97**, 574–581 (1984).
221. C. Y. Lee, J. A. McCammon, and P. J. Rossky, "The structure of liquid water at an extended hydrophobic surface," *J. Chem. Phys.* **80**, 4448–4455 (1984).
222. L. F. Scatena, M. G. Brown, and G. L. Richmond, "Water at hydrophobic surfaces: weak hydrogen bonding and strong orientation effects," *Science* **292**, 908–912 (2001).
223. S. Granick, "Motion and relaxations of confined liquids," *Science* **253**, 1374–1379 (1991).
224. D. Y. C. Chan and R. G. Horn, "The drainage of thin liquid films between solid surfaces," *J. Chem. Phys.* **83**, 5311–5324 (1985).
225. T. D. Blake "Slip between a liquid and a solid: D. M. Tolstoi's (1952) theory reconsidered," *Coll. Surf.* **47**, 135–145 (1990).
226. J. Klein and E. Kumacheva, "Simple liquids confined to molecularly thin layers. i. Confinement-induced liquid-to-solid phase transitions," *J. Chem. Phys.* **108**, 6996–7009 (1998).
227. P. Attard, "Bridging bubbles between hydrophobic surfaces," *Langmuir* **12**, 1693–1695 (1996).
228. U. Raviv and J. Klein, "Fluidity of bound hydration layers," *Science* **297**, 1540–1543 (2002).

229. J. Koplik and J. R. Banavar, "Continuum deductions from molecular hydrodynamics," *Annu. Rev. Fluid Mech.* **27**, 257–292 (1995).
230. J. C. Maxwell, "On stress in rarefied gases arising from inqualities of temperature," *Phil. Trans. R. Soc. Lond.* **170**, 231–256 (1879).
231. L. Bocquet, "Glissement d'un fluide sur une surface de rugosité modèle," *C. R. Acad. Sci. II* **316**, 7–12 (1993).
232. J.-L. Barrat and L. Bocquet, "Large slip effect at a nonwetting fluid-solid interface," *Phys. Rev. Lett.* **82**, 4671–4674 (1999).
233. S. Richardson, "On the no-slip boundary condition," *J. Fluid Mech.* **59**, 707–719 (1973).
234. J. Koplik, J. R. Banavar, and J. F. Willemsen, "Molecular dynamics of Poiseuille flow and moving contact lines," *Phys. Rev. Lett.* **60**, 1282–1285 (1988).
235. I. Bitsanis, S. A. Somers, H. T. Davis, and M. Tirrell, "Microscopic dynamics of flow in molecularly narrow pores," *J. Chem. Phys.* **93**, 3427–3431 (1990).
236. B. D. Todd, P. J. Daivis, and D. J. Evans, "Pressure tensor for inhomogeneous fluids," *Phys. Rev. E* **52**, 1627–1638 (1995).
237. K. P. Travis, B. D. Todd, and D. J. Evans, "Departure from Navier-Stokes hydrodynamics in confined liquids," *Phys. Rev. E* **55**, 4288–4295 (1997).
238. K. P. Travis and K. E. Gubbins, "Poiseuille flow of Lennard-Jones fluids in narrow slit pores," *J. Chem. Phys.* **112**, 1984–1994 (2000).
239. G. Mo and F. Rosenberger, "Molecular-dynamics simulation of flow in a 2D channel with atomically rough walls," *Phys. Rev. A* **42**, 4688–4692 (1990).
240. P. A. Thompson and M. O. Robbins, "Shear flow near solids: epitaxial order and flow boundary conditions," *Phys. Rev. A* **41**, 6830–6841 (1990).
241. P. A. Thompson and M. O. Robbins, "Origin of stick-slip motion in boundary lubrication," *Science* **250**, 792–794 (1990).
242. M. O. Robbins and P. A. Thompson, "Critical velocity of stick-slip motion," *Science* **253**, 916 (1991).
243. P. A. Thompson and S. M. Troian, "A general boundary condition for liquid flow at solid surfaces," *Nature* **389**, 360–362 (1997).
244. M. Cieplak, J. Koplik, and J. R. Banavar, "Boundary conditions at a fluid-solid interface," *Phys. Rev. Lett.* **86**, 803–806 (2001).
245. A. Jabbarzadeh, J. D. Atkinson, and R. I. Tanner, "Effect of the wall roughness on slip and rheological properties of hexadecane in molecular dynamics simulation of Couette shear flow between two sinusoidal walls," *Phys. Rev. E* **61**, 690–699 (2000).
246. V. P. Sokhan, D. Nicholson, and N. Quirke, "Fluid flow in nanopores: An examination of hydrodynamic boundary conditions," *J. Chem. Phys.* **115**, 3878–3887 (2001).
247. J. H. Walther, R. Jaffe, T. Werder, T. Halicioglu, and P. Koumoutsakos, "On the boundary condition for water at a hydrophobic surface," *Proceedings of the summer program 2002*, Center for Turbulence Research, Stanford Univ. and NASA Ames, 317–329 (2002).

248. U. Heinbuch and J. Fischer, "Liquid flow in pores: slip, no-slip or multilayer sticking," *Phys. Rev. A* **40**, 1144–1146 (1989).
249. R. E. Tuzun, D. W. Noid, B. G. Sumpter, and R. C. Merkle, "Dynamics of fluid flow inside carbon nanotubes," *Nanotechnology* **7**, 241–248 (1996).
250. V. P. Sokhan, D. Nicholson, and N. Quirke, "Fluid flow in nanopores: accurate boundary conditions for carbon nanotubes," *J. Chem. Phys.* **117**, 8531–8539 (2002).
251. D. Hirshfeld and D. C. Rapaport, "Molecular dynamics simulation of Taylor-Couette vortex formation," *Phys. Rev. Lett.* **80**, 5337–5340 (1998).
252. J. H. Walther, T. Werder, R. L. Jaffe, and P. Koumoutsakos, "Hydrodynamic properties of carbon nanotubes," *Phys. Rev. E*, accepted for publication.
253. D. Ugarte, A. Châtelain, and W. A. de Heer, "Nanocapillarity and chemistry in carbon nanotubes," *Science* **274**, 1897–1899 (1996).
254. J. Israelachvili, M. Gee, P. McGuiggan, and A. Homola, "Dynamic properties of molecularly thin liquid-films," *Abstracts of Papers of the American Chemical Society*, Vol. 196, p. 277 (1988).
255. M. Gee, P. McGuiggan, J. Israelachvili, and A. Homola, "Liquid to solidlike transitions of molecularly thin films under shear," *J. Chem. Phys.* **93**, 1895–1906 (1990).
256. M. Schoen, D. J. Diestler, and J. H. Cushman, "Fluids in micropores: I. Structure of a simple classical fluid in a slit-pore," *J. Chem. Phys.* **87**, 5464–5476 (1987).
257. I. Bitsanis, J. J. Magda, M. Tirrell, and H. T. Davis, "Molecular dynamics of flow in micropores," *J. Chem. Phys.* **87**, 1733–1750 (1987).
258. P. E. Sokol, W. J. Ma, K. W. Herwig, W. M. Snow, Y. Wang, J. Koplik, and J. R. Banavar, "Freezing in confined geometries," *Appl. Phys. Lett.* **61**, 777–779 (1992).
259. J. Baugh, A. Kleinhammes, D. X. Han, Q. Wang, and Y. Wu, "Confinement effect on dipole-dipole interactions in nanofluids," *Science* **294**, 1505–1507 (2001).
260. R. E. Tuzun, D. W. Noid, B. G. Sumpter, and R. C. Merkle, "Dynamics of He/C_{66} flow inside carbon nanotubes," *Nanotechnology* **8**, 112–118 (1997).
261. J. Han, A. Globus, R. Jaffe, and G. Deardorff, "Molecular dynamics simulations of carbon nanotube-based gears," *Nanotechnology* **8**, 95–102 (1997).
262. R. Karlsson, M. Karlsson, A. Karlsson, A.-S. Cans, J. Bergenholtz, B. Åkerman, A. G. Ewing, M. Voinova, and O. Orwar, "Moving-wall-driven flows in nanofluidic systems," *Langmuir* **18**, 4186–4190 (2002).
263. O. Beckstein, P. C. Biggin, and M. S. P. Sansom, "A hydrophobic gating mechanism for nanopores," *J. Phys. Chem. B* **105**, 12902–12905 (2001).
264. A. Waghe, J. C. Rasaiah, and G. Hummer, "Filling and emptying kinetics of carbon nanotubes in water," *J. Chem. Phys.* **117**, 10789–10795 (2002).
265. N. R. Aluru, J.-P. Leburton, W. McMahon, U. Ravaioli, S. Rotkin, M. Stedele, T. van der Straaten, B. R. Tuttle, and K. Hess, "Modeling electronics at the nanoscale," in *Handbook of Nanoscience, Engineering and Technology*,

W. A. I. Goddard, D. W. Brenner, S. E. Lyshevski, and G. J. Iafrate, Eds., pp. 11.1–11.32, CRC Press, Boca Raton, FL (2003).
266. M. Amini, S. K. Mitra, and R. W. Hockney, "Molecular dynamics study of boron trioxide glass," *J. Phys. C* **14**, 3689–3700 (1981).
267. K. Koga, G. T. Gao, H. Tanaka, and X. C. Zeng, "How does water freeze inside carbon nanotubes?" *Physica A* **314**, 462–469 (2002).
268. O. Mishima and H. E. Stanley, "The relationship between liquid, supercooled and glassy water," *Nature* **396**, 329–335 (1998).
269. K. Koga, H. Tanaka, and X. C. Zeng, "First-order transition in confined water between high-density liquid and low-density amorphous phases," *Nature* **408**, 564–567 (2000).
270. K. Koga, X. C. Zeng, and H. Tanaka, "Freezing of confined water: a bilayer ice phase in hydrophobic nanopores," *Phys. Rev. Lett.* **79**, 5262–5265 (1997).
271. W. H. Noon, K. D. Ausman, R. E. Smalley, and J. Ma, "Helical ice-sheets inside carbon nanotubes in the physiological condition," *Chem. Phys. Lett.* **355**, 445–448 (2002).
272. K. Koga, G. T. Gao, H. Tanaka, and X. C. Zeng, "Formation of ordered ice nanotubes inside carbon nanotubes," *Nature* **412**, 802–805 (2001).
273. E. Dujardin, T. W. Ebbesen, H. Hiura, and K. Tanigaki, "Capillarity and wetting of carbon nanotubes," *Science* **265**, 1850–1852 (1994).
274. P. M. Ajayan and S. Iijima, "Capillarity-induced filling of carbon nanotubes," *Nature* **361**, 333–334 (1993).
275. J. Slovák, H. Tanaka, K. Koga, and X. C. Zeng, "Computer simulation of water-ice transition in hydrophobic nanopores," *Physica A* **292**, 87–101 (2001).
276. M. C. Gordillo and J. Martí, "Hydrogen bonding in supercritical water confined in carbon nanotubes," *Chem. Phys. Lett.* **341**, 250–254 (2001).
277. S. K. Doorn, R. E. Fields, H. Hu, M. A. Hamon, R. C. Haddon, J. P. Selegue, and V. Majidi, "High resolution capillary electrophorensis of carbon nanotubes," *J. Am. Chem. Soc.* **124**, 3169–3174 (2002).
278. K. H. Choi, J. P. Bourgoin, S. Auvray, D. Esteve, G. S. Duesberg, S. Roth, and M. Burghard, "Controlled deposition of carbon nanotubes on a patterned substrate," *Surf. Sci.* **462**, 195–202 (2000).
279. M. J. Biercuk, M. C. Llaguno, M. Radosavljevic, J. K. Hyun, A. T. Johnson, and J. E. Fischer, "Carbon nanotube composites for thermal management," *Appl. Phys. Lett.* **80**, 2767–2769 (2002).
280. L. A. Girifalco, M. Hodak, and R. S. Lee, "Carbon nanotubes, buckyballs, ropes, and a universal graphitic potential," *Phys. Rev. B* **62**, 13104–13110 (2000).
281. M. G. C. Kahn, S. Banerjee, and S. S. Wong, "Solubilization of oxidized single-walled carbon nanotubes in organic and aqueous solvents through organic derivatization," *Nano Lett.* **2**, 1215–1218 (2002).
282. K. B. Shelimov, R. O. Esenaliev, A. G. Rinzler, C. B. Huffman, and R. E. Smalley, "Purification of single-wall carbon nanotubes by ultrasonically assisted filtration," *Chem. Phys. Lett.* **282**, 429–434 (1998).

283. J. E. Riggs, D. B. Walker, D. L. Carroll, and Y.-P. Sun, "Optical limiting properties of suspended and solubilized carbon nanotubes," *J. Phys. Chem. B* **104**, 7071–7076 (2000).
284. A. Star, D. W. Steuerman, J. R. Heath, and J. F. Stoddart, "Starched carbon nanotubes," *Angew. Chem.* **41**, 2508–2512 (2002).
285. F. Pompeo and D. E. Resasco, "Water solubilization of single-walled carbon nanotubes by functionalization with glucosamine," *Nano Lett.* **2**, 369–373 (2002).
286. M. J. O'Connell, S. M. Bachilo, C. B. Huffman, V. C. Moore, M. S. Strano, E. H. Haroz, K. L. Rialon, P. J. Boul, W. H. Noon, C. Kittrell, J. Ma, R. H. Hauge, R. B. Weisman, and R. E. Smalley, "Band gap fluorescence from individual single-walled carbon nanotubes," *Science* **297**, 593–596 (2002).
287. N. Sano, H. Wang, M. Chhowalla, I. Alexandrou, and G. A. J. Amaratunga, "Synthesis of carbon 'onions' in water," *Nature* **414**, 506–507 (2001).
288. M. F. Islam, E. Rojas, D. M. Bergey, and A. T. Johnson, "High weight fraction surfactant solubilization of single-wall carbon nanotubes in water," *Nano Lett.* **3**, 269–273 (2003).
289. Y. M. Xuan and Q. Li, "Heat transfer enhancement of nanofluids," *Int. J. Heat Fluid Flow* **21**, 58–64 (2000).
290. K. R. Wilson, R. D. Schaller, D. T. Co, R. J. Saykally, B. S. Rude, T. Catalano, and J. D. Bozek, "Surface relaxation in liquid water and methanol studies by x-ray adsorption spectroscopy," *J. Chem. Phys.* **117**, 7738–7744 (2002).
291. S. K. Das, N. Putra, and W. Roetzel, "Pool boiling characteristics of nanofluids," *Int. J. Heat Mass Trans.* **46**, 851–862 (2003).
292. S. Lee, S. U. S. Choi, S. Li, and J. A. Eastman, "Measuring thermal conductivity of fluids containing oxide nanoparticles," *J. Heat Trans.* **121**, 280–289 (1999).
293. H. Q. Xie, J. C. Wang, T. G. Xi, Y. Liu, F. Ai, and Q. R. Wu, "Thermal conductivity enhancement of suspensions contraining nanosize alumina particles," *J. Appl. Phys.* **91**, 4568–4572 (2002).
294. S. U. S. Choi, Z. G. Zhang, W. Yu, F. E. Lockwood, and E. A. Grulke, "Anomalous thermal conductivity enhancement in nanotube suspensions," *Appl. Phys. Lett.* **79**, 2252–2254 (2001).
295. H. Dai, J. H. Hafner, A. G. Rinzler, D. T. Colbert, and R. E. Smalley, "Nanotubes as nanoprobes in scanning probe microscopy," *Nature* **384**, 147–151 (1996).
296. K. Moloni, M. R. Buss, and R. P. Andres, "Tapping mode scaling force microscopy in water using a carbon nanotube probe," *Ultramicroscopy* **80**, 237–246 (1999).
297. J. H. Walther, R. L. Jaffe, E. Kotsalie, T. Werder, T. Halicioglu, and P. Koumoutsakos, "Hydrophobic hydration of C_{60} and carbon nanotubes in water," submitted for publication.
298. A. C. Dillon, K. M. Jones, T. A. Bekkedahl, C. H. Kiang, D. S. Bethune, and M. J. Heben, "Storage of hydrogen in single-walled carbon nanotubes," *Nature* **386**, 377–379 (1997).

299. Y. Y. Fan, B. Liao, M. Liu, Y. L. Wei, M. Q. Lu, and H. M. Cheng, "Hydrogen uptake in vapor-grown carbon nanofibers," *Carbon* **37**, 1649–1652 (1999).
300. P. Chen, X. Wu, J. Lin, and K. L. Tan, "High H_2 uptake by alkali-doped carbon nanotubes under ambient pressure and moderate temperatures," *Science* **285**, 91–93 (1999).
301. H. W. Zhu, A. Chen, Z. Q. Mao, C. L. Xu, X. Xiao, B. Q. Wei, J. Liang, and D. H. Wu, "The effect of surface treatments on hydrogen storage of carbon nanotubes," *J. Mater. Sci. Lett.* **19**, 1237–1239 (2000).
302. S. M. Lee and Y. H. Lee, "Hydrogen storage in single-walled carbon nanotubes," *Appl. Phys. Lett.* **76**, 2877–2879 (2000).
303. M. W. Zhao, Y. Y. Xia, Y. C. Ma, M. J. Ying, X. D. Liu, and L. M. Mei, "Tunable adsorption and desorption of hydrogen atoms on single-walled carbon nanotubes," *Chin. Phys. Lett.* **19**, 1498–1500 (2002).
304. G. E. Froudakis, "Hydrogen interaction with carbon nanotubes: a review of *ab initio* studies," *J. Phys. Condens. Matter.* **14**, R453–R465 (2002).
305. P. X. Hou, Q. H. Yang, S. Bai, S. T. Xu, M. Liu, and H. M. Cheng, "Bulk storage capacity of hydrogen in purified multiwalled carbon nanotubes," *J. Chem. Phys.* **106**, 963–966 (2002).
306. J. Zhao, A. Buldum, J. Han, and J. P. Lu, "Gas molecule adsorption in carbon nanotubes and nanotube bundles," *Nanotechnology* **13**, 195–200 (2002).
307. L.-M. Peng, Z. L. Zhang, Z. Q. Xue, Q. D. Wu, Z. N. Gu, and D. G. Pettifor, "Stability of carbon nanotubes: how small can they be?" *Phys. Rev. Lett.* **85**, 3249–3252 (2000).
308. S. C. Terry, J. H. Jerman, and J. B. Angell, "A gas chromatographic air analyzer fabricated on a silicon wafer," *IEEE Trans. Electron Devices* **26**, 1880–1886 (1979).
309. J. Han and H. G. Craighead, "Separation of long DNA molecules in a microfabricated entropic trap array," *Science* **288**, 1026–1029 (2000).
310. Y. Chen, J.-G. Weng, J. R. Lukes, A. Majumdar, and C.-L. Tien, "Molecular dynamics simulation of the meniscus formation between two surfaces," *Appl. Phys. Lett.* **79**, 1267–1269 (2001).
311. C. K. Harnett, G. W. Coates, and H. G. Craighead, "Heat-depolymerizable polycarbonates as electron beam patternable sacrificial layers for nanofluidics," *J. Vac. Sci. Technol. B* **19**, 2842–2845 (2001).
312. H. Cao, Z. N. Yu, J. Wang, J. O. Tegenfeldt, R. H. Austin, E. Chen, W. Wu, and S. Y. Chou, "Fabrication of 10 nm enclosed nanofluidic channels," *Appl. Phys. Lett.* **81**, 174–176 (2002).
313. H. Cao, J. O. Tegenfeldt, R. H. Austin, and S. Y. Chou, "Gradient nanostructures for interfacing microfluidics and nanofluidics," *Appl. Phys. Lett.* **81**, 3058–3060 (2002).
314. D. Nykypanchuk, H. H. Strey, and D. A. Hoagland, "Brownian motion of DNA confined within a 2D array," *Science* **297**, 987–990 (2002).
315. P. J. Kemery, J. K. Steehler, and P. W. Bohn, "Electric field mediated transport in nanometer diameter channels," *Langmuir* **14**, 2884–2889 (1998).

316. T.-C. Kuo, L. A. Sloan, J. V. Sweedler, and P. W. Bohn, "Manipulating molecular transport through nanoporous membranes by control of electrokinetic flow: effect of surface charge density and Debye length," *Langmuir* **17**, 6298–6303 (2001).
317. T.-C. Kuo, D. M. Cannon, Jr., W. Feng, M. A. Shannon, J. V. Sweedler, and P. W. Bohn, "Three-dimensional fluidic architectures using nanofluidic diodes to control transport between microfluidic channels in microelectromechanical devices," *Proc. of the mTAS Symposium*, Monterey, 60–62 (2001).
318. T. C. Kuo, D. M. Cannon, M. A. Shannon, P. W. Bohn, and J. Sweedler, "Hybrid 3D nanofluidic/microfluidic devices using molecular gates," *Sens. Actuat. A Phys.* **102**, 223–233 (2003).
319. L. Sun and R. M. Crooks, "Single carbon nanotube membranes: a well-defined model for studying mass transport through nanoporous materials," *J. Am. Chem. Soc.* **122**, 12340–12345 (2000).
320. S. A. Miller, V. Y. Young, and C. R. Martin, "Electroosmotic flow in template-prepared carbon nanotube membranes," *J. Am. Chem. Soc.* **123**, 12335–12342 (2001).
321. S. A. Miller and C. R. Martin, "Controlling the rate and direction of electroosmotic flow in template-prepared carbon nanotube membranes," *J. Elec. Chem.* **522**, 66–69 (2002).
322. S. B. Lee, D. T. Mitchell, L. Trofin, T. K. Nevanen, H. Soderlund, and C. R. Martin, "Antibody-based bio-nanotube membranes for enantiomeric drug separations," *Science* **296**, 2198–2200 (2002).
323. A. V. Melechko, T. E. McKnight, M. A. Guillorn, V. I. Merkulov, B. Ilic, M. J. Doktycz, D. H. Lowndes, and M. L. Simpson, "Vertically aligned carbon nanofibers as sacrificial templates for nanofluidic structures," *Appl. Phys. Lett.* **82**, 976–978 (2003).

List of symbols

C_n	dispersion coefficient of order n
E	energy
K_h	force constant for a harmonic oscillator
K_M	force constant for the Morse bond potential
K_Θ	force constant for a bond angle potential
U	internal energy
V	potential function
b	slip length
b^0	slip length in the low shear rate limit
$c_{i,j}$	adjustment parameter for Lennard–Jones potential
d	diameter
q_i	charge associated to atom i
r	droplet radius
r_0	equilibrium distance between two centers
r_c	cutoff radius
$r_{i,j}$	distance between two centers i and j
z	distance from a plane
Δu	slip velocity
$\Phi_{i,j,k,l}$	dihedral bond angle over four centers i, j, k, and l
Θ	contact angle
Θ_c	equilibrium bond angle
$\Theta_{i,j,k}$	bond angle between three centers i, j, and k
α	slip coefficient
β	parameter for the Morse bond potential
γ	surface tension
$\dot{\gamma}$	shear rate
$\dot{\gamma}_c$	critical shear rate
ϵ	surface roughness
ϵ_0	free-space permittivity
$\epsilon_{I,J}$	Lennard–Jones energy parameter for interaction between atoms I and J
λ	wavelength
$\sigma_{I,J}$	Lennard–Jones distance parameter for interaction between atoms I and J
τ	line tension

Petros Koumoutsakos has been full Professor of Computational Science at ETH Zurich since July 2000.

Petros Koumoutsakos, a Greek citizen, was born in Gythion, Laconia, Greece in 1963. He studied at the National Technical University of Athens (1981–1986) and received his Diploma in Naval Architecture and Mechanical Engineering. He received a master's degree (1987) in Naval Architecture from the University of Michigan, Ann Arbor. He continued his graduate studies at the California Institute of Technology, where he received a master's degree in Aeronautics (1988) and a PhD in Aeronautics and Applied Mathematics (1992). During 1992–1994, he was a National Science Foundation postdoctoral fellow in parallel supercomputing at Caltech. Since 1994, he has been a senior research associate and maintains an active affiliation with the Center for Turbulence Research (CTR) at NASA Ames/Stanford University. From September 1997 to June 2000, he was an assistant professor in Computational Fluid Dynamics at ETH Zurich. Since October 1999, he is a member of the Center for Computational Astrobiology at NASA Ames/Stanford University. He is the Director of the Institute of Computational Science (www.icos.ethz.ch) and of the ETHZ Computational Laboratory (www.colab.ethz.ch)

His research activities are in the areas of multiscale particle methods, machine learning and biologically inspired computation and the application of these techniques to problems of interest in the areas of Engineering and Life Sciences.

Urs Zimmerli is a PhD student at the Institute of Computational Science at ETH Zurich.

Urs Zimmerli was born in San Pedro Sula, Honduras, on 4 June 1977. From 1997–2001, he studied Chemical Engineering at the Department of Chemistry at ETH Zurich and at the Department for Chemical Engineering and Chemical Technology at Imperial College in London. In 2001, he received his degree in Chemical Engineering from ETH Zurich.

His research activities are focused on the derivation of interaction potentials from first principles and the development of simulation tools for biomolecular flows.

Thomas Werder is a PhD student at the Institute of Computational Science at ETH Zurich.

Thomas Werder was born in Baden, Switzerland, on 12 December 1974. From 1995–1997, he studied Mechanical Engineering at EPFL Lausanne, and in 2000, he received his degree in Computational Science from ETH Zurich. Since 2000, Thomas Werder has been a PhD student at the Institute of Computational Science at ETH Zurich.

His research activities are focused on Molecular Dynamics Simulations and Multiscale Physics.

Jens Walther is a Senior Research Associate at the Institute of Computational Science at ETH Zurich.

Jens Walther was born in Aalborg, Denmark, on 13 February 1966. He studied at the Aalborg University, Denmark, where he received his MSc in Mechanical Engineering in 1991. In 1994, he achieved his PhD in Mechanical Engineering at the Technical University of Denmark. From 1994–1996, he was working as a scientist at the Danish Meteorological Institute, Denmark, and from 1996–1999, he was employed as a project manager in Industrial Fluid Dynamics at the Danish Maritime Institute, Denmark. From 1997–2000, Jens Walther was a postdoctoral fellow at the Institute of Fluid Dynamics at ETH Zurich, and since 2000, he has been working as a research associate at the Institute of Computational Science at ETH Zurich.

His research activities are focused on the development and application of new algorithms for particle methods such as vortex particle methods and smoothed particle hydrodynamics for macro-scale fluid dynamics and molecular dynamics simulations for nanofluidics and material science.

Chapter 9

Introduction to Quantum Information Theory

Mary Beth Ruskai

9.1. Overview 397
9.1.1. Introduction 397
9.1.2. Encoding information 397
9.1.3. Effective parallelism 398
9.1.4. Choosing a basis 400
9.1.5. Perspective 403
9.2. Basic quantum principles 405
9.2.1. Isolated systems 405
9.2.2. Quantum measurement 406
9.2.3. Mixed states 407
9.2.4. Open systems 409
9.2.5. Notation and Pauli matrixes 412
9.2.6. No-cloning principle 413
9.3. Entanglement 414
9.3.1. Bell states and correlations 414
9.3.2. An experiment 415
9.3.3. Bell inequalities and locality 416
9.3.4. An important identity 417
9.3.5. More on entanglement 418
9.4. Quantum computation algorithms 420
9.4.1. The Deutsch–Jozsa problem 420
9.4.2. Grover's algorithm 422
9.4.3. Period finding via the QFT 425
9.4.4. Implementing the quantum Fourier transform 429
9.5. Other types of quantum information processing 430
9.5.1. Quantum key distribution 430
9.5.2. Quantum cryptography 432
9.5.3. Dense coding 433
9.5.4. Quantum teleportation 434
9.5.5. Quantum communication 435

9.6. Dealing with noise 436
9.6.1. Accessible information 436
9.6.2. Channel capacity 439
9.6.3. Quantum error correction 441
9.6.4. Fault-tolerant computation 444
9.6.5. DFS encoding 445
9.7. Conclusion 446
9.7.1. Remarks 446
9.7.2. Recommendations for further reading 447
Appendix 9.A. Dirac notation 449
Appendix 9.B. Trace and partial trace 450
Appendix 9.C. Singular value and Schmidt decompositions 451
Appendix 9.D. A more complete description 453
9.D.1. Continuous variables 453
9.D.2. The hidden spatial wave function 453
9.D.3. The Pauli principle 454
Acknowledgment 454
References 455
List of acronyms 464

9.1 Overview

9.1.1 Introduction

In nanotechnology, one is dealing with physical systems at a scale so small that quantum effects are important. As the size of computer chips decreases, one eventually reaches the point where quantum effects, wanted or not, occur. If these quantum effects can be controlled, they can be exploited to build computers that can do some tasks more effectively than classical ones.

Moreover, quantum particles can be be used advantageously for other purposes in information technology, most notably for communication and cryptography. In fact, experiments demonstrating the feasibility of the process known as quantum key distribution (QKD) are quite impressive. QKD, which is described in Sec. 9.5.1, is likely to be practical long before a full-fledged quantum computer (QC) is built.

The area of quantum information theory (QIT) encompasses quantum computation, quantum communication, and quantum cryptography. A device for implementing any of these is called a quantum information processor. We begin by considering methods of encoding information in such devices.

9.1.2 Encoding information

In classical situations, information is encoded in strings of 0's and 1's; the basic unit of information is a bit, which is in one of two mutually exclusive physical states, e.g., "on" or "off," which are interpreted as 0 and 1. Thus the state of a classical information processor can be identified as an element of $\mathbf{Z}_2^{\otimes n}$, i.e., a binary n-tuple.

When quantum particles are used to process information, the basic unit is a "qubit," which can be identified with a normalized vector in the two-dimensional complex vector space $\mathbf{C}_2$. For example, one can represent 0 and 1 as

$$|0\rangle = \begin{pmatrix} 1 \\ 0 \end{pmatrix}, \quad |1\rangle = \begin{pmatrix} 0 \\ 1 \end{pmatrix}. \tag{9.1}$$

When these states are realized using the spin components of spin-$\frac{1}{2}$ particles, they correspond to spin "up" and "down," respectively; alternatively they can be realized using vertical and horizontal polarization of single photons. One can then use products to represent classical strings, e.g.,

$$|1001\rangle = \begin{pmatrix} 0 \\ 1 \end{pmatrix} \otimes \begin{pmatrix} 1 \\ 0 \end{pmatrix} \otimes \begin{pmatrix} 1 \\ 0 \end{pmatrix} \otimes \begin{pmatrix} 0 \\ 1 \end{pmatrix}, \tag{9.2}$$

as vectors in the vector space $\mathbf{C}_2^{\otimes n}$. However, there are many more vectors in $\mathbf{C}_2$. For example, one could use the vectors

$$|0\rangle_x = \frac{1}{\sqrt{2}}\begin{pmatrix}1\\1\end{pmatrix} = \frac{1}{\sqrt{2}}(|0\rangle_z + |1\rangle_z), \tag{9.3a}$$

$$|1\rangle_x = \frac{1}{\sqrt{2}}\begin{pmatrix}1\\-1\end{pmatrix} = \frac{1}{\sqrt{2}}(|0\rangle_z - |1\rangle_z), \tag{9.3b}$$

as an alternative way of encoding 0 and 1, choosing the direction of a magnetic field to quantize the spin so that its eigenvectors are characterized as "right" and "left." However, this is not the only possible interpretation of the vectors in Eqs. (9.3). They can also be regarded as representing *both* 0 and 1, each with probability $\frac{1}{2}$.

More generally, the state $\binom{a}{b} = a|0\rangle + b|1\rangle$ (with $|a|^2 + |b|^2 = 1$) can be interpreted as containing 0 and 1 with probabilities $|a|^2$ and $|b|^2$, respectively. This is explained in Sec. 9.2.2, when the quantum measurement process is discussed. For now, note only that the probabilities associated with a superposition via the squared amplitudes of coefficients in this way are nonclassical, and behave differently than mixtures. (The term superposition is used to describe a linear combination of vectors when the result is constrained to have norm 1.) The n-qubit state

$$|0\rangle_x \otimes |0\rangle_x \otimes \ldots |0\rangle_x = 2^{-n/2} \sum_{i_1 i_2 \ldots i_n} |i_1 i_2 \ldots i_n\rangle, \tag{9.4}$$

where $i_k \in \{0, 1\}$ is thus a superposition of all possible n-bit strings of 0 and 1, each of which occurs with probability 2^{-n}. Any action on this vector can then be regarded as effectively acting in parallel on all possible 2^n classical n-bit strings. However, the usefulness of this parallelism is restricted by the measurement process used to extract information. This is discussed in more detail in Sec. 9.2.2. For now, note only that we are restricted to making one measurement, which enables us to extract one piece of information, equivalent to the identification of a classical string or binary n-tuple.

9.1.3 Effective parallelism

We describe the situation schematically as follows. In an ordinary sequential computer, one has a physical device on which only one operation can be performed at a time. Information processing then requires a long sequence of operations, as shown schematically in Fig. 9.1. In a parallel processor machine, operations can be performed simultaneously on n physical devices, as shown in Fig. 9.2.

For some algorithms, the length of the sequence of operations may be decreased at a cost of employing more physical devices. Moreover, the use of n processors yields n outputs, which can then be extracted and analyzed or combined further. In a QC, we have only one physical device with a state that can be described by a superposition, as shown in Fig. 9.3. The logical operations, or gates, in a QC are implemented by unitary operators that act on the vector, not on the individual pieces in the superposition. The result is a single vector $\sum_K y_K|K\rangle$ from which one can extract only the information equivalent to that in n classical bits.

There is really nothing mysterious about the effective parallelism. It is an immediate consequence of the fact that gates are implemented via unitary operators that act *linearly* on vectors in $\mathbf{C}^{2^n}$. What is difficult is the extraction of useful information after the operation. Unlike a classical parallel processor, the accessible information is limited by the principles of quantum measurement.

To explain this further, consider the well-known example of computing the fast Fourier transform (FFT), which takes a vector with components x_K to one with components $y_K = \sum_J e^{(2\pi i)JK/N} x_J$. On a classical computer, this can be done on a vector of size $N = 2^n$ in $O(N \log N)$ steps. We can view this process using N classical parallel processors schematically as in Fig. 9.4. The vertical lines between blocks in different processors reflect the fact that the reduction to $\log N = n$ requires some swapping between processors. Nevertheless, the total combined resources needed for the FFT is still Nn in the form of N physical devices and time n.

By contrast, the quantum Fourier transform (QFT) can be viewed schematically as in Fig. 9.5. It requires only $n = \log N$ steps, but the character of the output is quite different. To apply the QFT, one must first encode the information in the vector $\boldsymbol{x}$ by using its components x_K as the amplitudes of the vector $|\phi\rangle = \sum_{K=1}^N x_K|K\rangle$. It can then be shown that the QFT can be performed in $O(dn)$ steps, where d is the number of binary digits. Indeed, if one accepts that the QFT is really the FFT acting in parallel, this is almost obvious. However, the information

Figure 9.1 Schematic representation of a classical sequential computer.

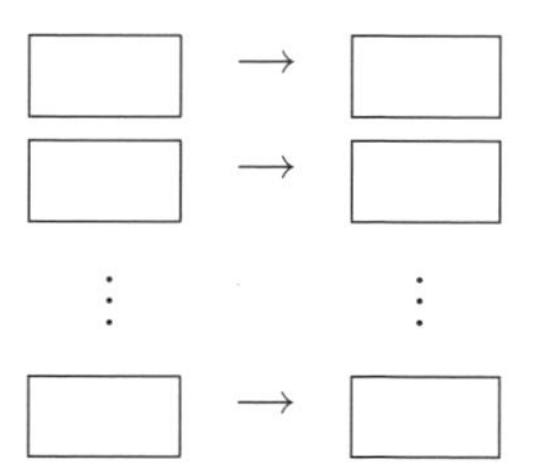

Figure 9.2 Schematic representation of a classical parallel computer.

$$|\phi\rangle = \sum_K x_K|K\rangle \rightarrow U|\phi\rangle = \sum_K x_K\, U|K\rangle$$

Figure 9.3 Schematic representation of a quantum computer.

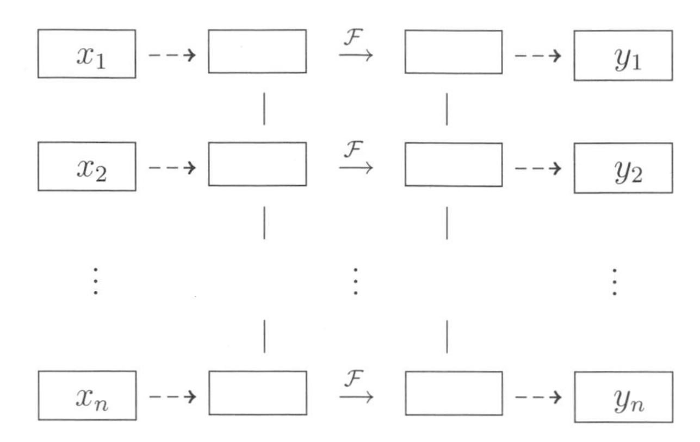

Figure 9.4 FFT with N parallel processors.

$$|\phi\rangle = \sum_{K=1}^{N} x_K |K\rangle \quad \xrightarrow{\mathcal{F}} \quad \mathcal{F}|\phi\rangle = \sum_{K=1}^{N} y_K |K\rangle$$

Figure 9.5 Quantum Fourier transform.

that can be extracted from the QC is quite different. Indeed, one can not obtain any information* about the Fourier coefficients y_K encoded in the final state!

One might wonder if the effective parallelism has been achieved without any mechanism for using the information. In fact, the purposes for which the QFT is used are necessarily quite different from those of the FFT. The most common use of the QFT occurs in situations for which (as in the period-finding algorithms discussed in Sec. 9.4.3) the state of the QC is such that most of the $y_k \approx 0$ in the final state. A measurement then yields one of the states with $y_k \neq 0$, leading to the identification of the set of k with nonzero y_k. This is useful when the set of nonzero y_k has a particular property, such as denoting multiples of a single integer.

9.1.4 Choosing a basis

9.1.4.1 The computational basis

Since $\mathbf{C}_2^{\otimes n}$ is isomorphic to $\mathbf{C}^{2^n}$, the state of a quantum information processor can be defined as a 1D subspace of $\mathbf{C}^{2^n}$, typically described by a vector $|\phi\rangle$. (As explained in Secs. 9.2.1 and 9.2.3, a state described by a vector is more properly termed a *pure state*. Even if the normalization is chosen so that $\|\phi\| = 1$, the representative vector is only defined up to an overall phase factor.) It is customary to represent 0 and 1 as a pair of orthonormal vectors in $\mathbf{C}_2$, as in Eq. (9.1). Taking tensor products as in Eq. (9.2) then yields an orthonormal basis for $\mathbf{C}^{2^n}$ of the form

*Actually, one could obtain estimates of $|y_K|^2$ by repeating the entire process—encoding in $|\phi\rangle$, application of the QFT and measurement—many times. However, this defeats the purpose of using the QC and still yields only information about $|y_K|$.

$|j_1 j_2 \ldots j_n\rangle = |j_1\rangle \otimes |j_2\rangle \otimes \ldots \otimes |j_n\rangle$. This is referred to as the "computational basis." The elements of the computational basis can be identified with binary n-tuples or elements of $\mathbf{Z}_2^n$. As in classical information processing, a state corresponding to a binary n-tuple $(j_1 j_2 \ldots j_n)$, can be interpreted in various ways, of which the most common is as the binary representation of an integer.

An arbitrary state or vector in $\mathbf{C}^{2^n}$ can always be written as a superposition of elements of the computational basis

$$|\phi\rangle = \sum_{j_1 j_2 \cdots j_n} c_{j_1 j_2 \cdots j_n} |j_1 j_2 \cdots j_n\rangle, \tag{9.5}$$

with

$$c_{j_1 j_2 \cdots j_n} = \langle j_1 j_2 \ldots j_n, \phi\rangle. \tag{9.6}$$

Now, a linear operator can be defined by specifying its action on a set of basis vectors, such as those in the computational basis. It can then be extended to arbitrary vectors in $\mathbf{C}^{2^n}$ by linearity. The so-called "effective parallelism" is actually an artifact of the convention of defining gates on a basis that can be identified with classical, as well as quantum, states but letting them act on arbitrary states. For example, the action of a rotation on a vector in three dimensions can be specified by an axis and an angle, or by a 3×3 matrix (with elements given by Euler angles) in a particular basis. However, one need not decompose a vector into components in that basis to implement the rotation. The rotation operation is independent of its description in a particular basis.

The implementation of a gate requires that one find a physical operation that has the desired effect, regardless of how it is defined. This is not at all trivial and is the essence of the construction of a QC or quantum information processor. Although discussion of practical implementation is beyond the scope of this chapter, the important point is that in any successful implementation, the state of the system can be described by any vector in $\mathbf{C}^{2^n}$ and the gates affect the state of the system and not its basis vectors. Indeed, the state of a physical system is entirely independent of the basis in which one chooses to represent it.

In the computational basis, a measurement can be regarded as a mechanism for identifying one of the basis vectors, Even when the system is in a superposition of basis vectors, the outcome of the measurement always yields one of the computational basis states, as explained in Sec. 9.2.2.

9.1.4.2 Nonorthogonal bases

Thus far, we have considered only orthogonal bases for encoding binary strings; in fact, as explained in Sec. 9.2.6, only orthogonal states can be reliably distinguished. Nevertheless there are situations in which it is advantageous to use nonorthogonal bases. One of these occurs when dealing with noise. For some types of noise it is actually possible to more reliably distinguish the corrupted outputs of a noisy channel when nonorthogonal inputs are used.[43,79,130]

Some procedures in quantum cryptography also use nonorthogonal encodings. For example, one might use

$$|0\rangle_A = \begin{pmatrix} 1 \\ 0 \end{pmatrix} \qquad |1\rangle_A = \frac{1}{\sqrt{2}} \begin{pmatrix} 1 \\ 1 \end{pmatrix}, \tag{9.7a}$$

or

$$|0\rangle_B = \frac{1}{\sqrt{2}} \begin{pmatrix} 1 \\ -1 \end{pmatrix} \qquad |1\rangle_B = \begin{pmatrix} 0 \\ 1 \end{pmatrix}. \tag{9.7b}$$

An application using such encodings for QKD is considered in Sec. 9.5.1.

One might also wonder if one could use a single qubit to encode more than just 0 and 1. For example, could one choose to represent 0, 1, 2, 3 as

$$\begin{array}{cccc} |0\rangle = \begin{pmatrix} 1 \\ 0 \end{pmatrix} & |1\rangle = \frac{1}{\sqrt{2}} \begin{pmatrix} 1 \\ 1 \end{pmatrix} & |2\rangle = \begin{pmatrix} 1 \\ 0 \end{pmatrix} & |3\rangle = \frac{1}{\sqrt{2}} \begin{pmatrix} 1 \\ -1 \end{pmatrix}, \\ \uparrow & \rightarrow & \downarrow & \leftarrow \end{array} \tag{9.8}$$

corresponding to, say, spin up, right, down, and left? One might even note that the most general state of a qubit can be written as

$$\sqrt{1-c^2}|0\rangle + c|1\rangle \quad \text{or} \quad \begin{pmatrix} \sin\theta \\ e^{i\varphi}\cos\theta \end{pmatrix}, \tag{9.9}$$

suggesting that one could represent any real number in the interval $[-1, 1]$ or $[0, 2\pi]$ or even a pair of real numbers corresponding to the angles θ and φ. One might expect that there would be a practical limit to the accuracy with which one could distinguish between such encodings of real numbers. However, there is a more significant constraint known as the *Holevo bound on the accessible information*. This bound, stated precisely (and proved) in Sec. 9.6.1, implies that one can not extract more information from n qubits than for n classical bits. Thus one must pay a price to use encodings of the form of Eq. (9.8). This price might be a high error rate, as in the B92 protocol for QKD, or the need to provide supplementary information, as in the BB84 protocol. In these cryptographic procedures (explained in Sec. 9.5.1), one is willing to pay this price because the use of nonorthogonal states provides protection against eavesdroppers that is not available when orthogonal encodings are used.

9.1.4.3 Physical implementations

Writing states as vectors in $\mathbf{C}^{2^n}$ would serve little purpose unless they can be realized in a physical system. There are two common, and rather natural, ways of implementing the qubit states already described. One uses states of spin-$\frac{1}{2}$ particles, such as electrons or protons. The other uses the polarization of single photons. Although the geometric properties of the latter are quite different from the former;

Table 9.1 Some states of a two-level quantum system.

			Spin		Polarization	
$\|0\rangle_z$	$\|0\rangle$	$\begin{pmatrix}1\\0\end{pmatrix}$	up	↑	vertical	↑
$\|1\rangle_z$	$\|1\rangle$	$\begin{pmatrix}0\\1\end{pmatrix}$	down	↓	horizontal	→
$\|0\rangle_x$	$\frac{1}{\sqrt{2}}(\|0\rangle + \|1\rangle)$	$\frac{1}{\sqrt{2}}\begin{pmatrix}1\\1\end{pmatrix}$	right	→		↗
$\|1\rangle_x$	$\frac{1}{\sqrt{2}}(\|0\rangle - \|1\rangle)$	$\frac{1}{\sqrt{2}}\begin{pmatrix}1\\-1\end{pmatrix}$	left	↓		↘
$\|0\rangle_y$	$\frac{1}{\sqrt{2}}(\|0\rangle + i\|1\rangle)$	$\frac{1}{\sqrt{2}}\begin{pmatrix}1\\i\end{pmatrix}$	out		right circular	↻
$\|1\rangle_y$	$\frac{1}{\sqrt{2}}(i\|0\rangle + \|1\rangle)$	$\frac{1}{\sqrt{2}}\begin{pmatrix}i\\1\end{pmatrix}$	in		left circular	↺

their algebraic representations are equivalent. This correspondence is summarized in Table 9.1.

In addition to forming states of qubits, one also needs to control them, i.e., to implement gates. This poses a greater challenge. In trying to meet it, physicists have proposed a number of other possible implementations. In fact, any two-dimensional quantum system, such as two low-level energy states of an atom, will suffice.

One could, in principle, use a quantum information processor whose fundamental units (called "qutrits" or "qudits") are described by states in $\mathbf{C}_3$ or $\mathbf{C}_d$, respectively. For example, one might implement a "qutrit" using a spin-1 particle, or three low-lying states of an atom. Although the feasibility and utility of this have yet to be established, examining the properties of such systems is an active area of research. There are some significant differences between $d = 2$ and $d > 2$, and understanding these provides additional insight into the special properties of qubit systems. Moreover, because multiqubit systems correspond to $d = 2^n$ one must understand at least some facets of this situation.

The actual implementation of a full-fledged QC is an extremely challenging problem in nanotechnology that is beyond the scope of this chapter. For an excellent overview, see DiVincenzo.[36] Some other types of quantum information processors have been implemented successfully, most notably QKD, which is discussed in Sec. 9.5.1.

9.1.5 Perspective

9.1.5.1 Reversibility

Historically, quantum computation grew out of questions raised by Landauer[88] about the reversibility of information processing. It was subsequently shown that classical computation could be done reversibly if suitable gates were used. While considering this question, Benioff[9–11] developed an early quantum model of a computer. Manin,[97] independently, and Feynman[41,42] (probably influenced[89] by Benioff's work), speculated about the possibility that QCs might be able to simulate quantum systems in ways that can not be done on a classical computer. The explicit introduction of quantum parallelism seems to have first appeared in the fundamental paper of Deutsch.[34] Reversible models of quantum computation, which use unitary operators as gates, are quite natural, and the action of unitary operators on superpositions gives rise to effective parallelism.

As a result, reversibility is sometimes regarded as an essential component of quantum computation. However, a few years ago Nielsen[106] (see also Ref. 90) showed that measurements could also be used to generate gates. This development was followed by several proposals for using measurements to construct irreversible or "one-way" QCs of which the most extensively developed is that by Raussendorf and Briegel.[118–120]

9.1.5.2 Circuits and models

Although the earliest model for quantum computation was that of a quantum Turing machine, the most common model is that of a quantum circuit composed from a small set of unitary operations known as quantum gates. In fact, it can be shown that a rather short list of 1-bit gates, together with one type of nontrivial 2-bit gate* (e.g., CNOT or SWAP), suffice in the sense that any unitary operation on n bits can be approximated as a product of these basic gates. To analyze the computational complexity of an operation on n bits, we must know how many basic 1- and 2-bit gates are required to implement it. For this reason, some introductions to quantum computation focus on the quantum circuit model.

In a departure from this trend, this chapter does not use the quantum circuit model at all. Moreover, the only n-bit operator analyzed is the QFT. The main reason for this is my firm conviction that understanding the role of the quantum measurement process is essential to understanding quantum algorithms. Moreover, the key feature of a particular algorithm is the method it uses to convert the initial state of the QC to one in which a measurement can yield useful information. At this point, the obstacle to developing new algorithms does not seem to be finding efficient decompositions of n-bit unitary operators, but finding methods for changing the state of the computer to one from which a measurement can extract useful information.

Furthermore, a detailed description of a quantum circuit is not required to gauge the complexity of an algorithm. Indeed, most texts on the analysis of algorithms for classical computation describe them in a pseudo-language, using "if-then-else style" constructs, rather than machine language or some other decomposition into primitives. A similar approach to describing algorithms is used here, although the language is different.

*A CNOT gate takes $|j\rangle \otimes |k\rangle \mapsto |j\rangle \otimes |j+k\rangle$, and SWAP $|j\rangle \otimes |k\rangle \mapsto |k\rangle \otimes |j\rangle$. See also Eq. (9.121).

9.1.5.3 Outline

Sections 9.2.1 and 9.2.2 are critical for understanding the rest of the chapter. The reader primarily interested in algorithms can move to Sec. 9.4 after reading Secs. 9.2.1, 9.2.2, and 9.2.5. Section 9.2.4 can be skipped on first reading; this material, although important, is directly relevant only to Sec. 9.6. Section 9.3 contains material that is important for understanding quantum correlations, but is most relevant primarily to Sec. 9.5.

Section 9.4 describes several important algorithms for quantum computation. Section 9.5 describes some procedures used in quantum cryptography and quantum communication. Section 9.6 gives a brief description of a few issues associated with noise, namely, the fundamental Holevo bound on accessible information, channel capacity, the quantum error correction process, other issues in fault-tolerant computation, and other types of encodings.

This presentation follows the convention of using physicists' Dirac notation for vectors and projections; indeed, this has already been done implicitly. An explanation is given in Appendix 9.A and is essential to understanding the quantum information processing (QIP) literature, as well as this chapter. Appendixes 9.B and 9.C summarize some standard mathematical results that are needed, but may be unfamiliar to many readers. Appendix 9.D gives a brief overview of some issues regarding continuous variables and permutational symmetry.

9.2 Basic quantum principles

9.2.1 Isolated systems

As discussed at the start of Sec. 9.1.4, the state of an isolated quantum system can be described (up to an arbitrary phase factor) by a normalized vector in an appropriate Hilbert space $\mathcal{H}$. For most purposes in QIP, it suffices to assume $\mathcal{H} = \mathbf{C}^{2^n}$.

The time development of an isolated system is determined by a self-adjoint operator, H, known as the Hamiltonian. The time development of a system in the state $|\psi\rangle$ is then governed by the Schrödinger equation

$$i\hbar\frac{\partial}{\partial t}|\psi(t)\rangle = H|\psi(t)\rangle. \tag{9.10}$$

This implies that the time-evolution is unitary, i.e., that $|\psi(t)\rangle = U(t)|\psi(t_0)\rangle$, where $U(t)$ is a one parameter unitary group. When the Hamiltonian is independent of time, $U(t) = e^{-i\hbar t H}$ so that $|\psi(t)\rangle = e^{-i\hbar t H}|\psi(t_0)\rangle$.

A gate in quantum computation is usually regarded as a fixed unitary operator V. Any gate can be written in the form $V = U(t_1) - U(t_0)$ for some unitary group. In fact, given V, there is a self-adjoint operator A such that $V = e^{iA}$, in which case $V = U(1)$ with $U(t) = e^{itA}$. Thus, one might regard a sequence of unitary gates $V_n V_{n-1} \ldots V_2 V_1$ as arising from the dynamics of a Hamiltonian of the form $e^{it[H_0 + \sum_k \delta(t-t_j) A_j]}$ with $V_j = e^{it_j A_j}$. In general, the A_j do not commute;

therefore, this correspondence is not exact. However, it does demonstrate that the state of a QC is essentially governed by the dynamics of the Hamiltonian of the system.

9.2.2 Quantum measurement

9.2.2.1 The measurement postulate

One of the assumptions of quantum theory is that observables are represented by self-adjoint operators. Such operators always have a spectral decomposition,* which we write in the form

$$A = \sum_k a_k |\alpha_k\rangle\langle\alpha_k| = \sum_k a_k E_k, \tag{9.11}$$

where a_k is an eigenvalue of A, $|\alpha_k\rangle$ is the corresponding eigenvector, and $|\alpha_k\rangle\langle\alpha_k| = E_k$ is the projection onto its eigenspace. We have made the simplifying assumption that A has only discrete spectra, as this suffices for most purposes in QIP. One of the fundamental principles of quantum theory is that when a measurement is made on a system in the state $|\psi\rangle$ using the observable represented by A, then the following hold:

1. The only possible outcome is one of the eigenvalues a_k.
2. After the measurement the system is an eigenstate $|\alpha_k\rangle$ of a_k.
3. The probability of this outcome is $|\langle\psi, \alpha_k\rangle|^2 = \langle\psi, E_k\psi\rangle$.

If many measurements of A are made with the system in the state $|\psi\rangle$, then these postulates imply that the average value of the observable A is $\langle\psi, A\psi\rangle$.

The measurement process may seem quite remarkable. A system in one state $|\psi\rangle$ is changed rather suddenly (and irreversibly) into another state $|\alpha_k\rangle$. Moreover, information about the initial state has been lost; only the state of the system after the measurement is known. We shall not even attempt to explain how this happens; it has been the subject of extensive debate[150] since the dawn of quantum theory. However mysterious one may find this description of the measurement process, its validity as a physical model has been verified experimentally far beyond any reasonable doubt.

It may help to think of the special case of measuring the polarization of a single photon by using a filter and a detector. Suppose the filter is designed so that only vertically polarized photons go through. If a photon is polarized at a 45-deg angle (or if it is circularly polarized), it may or may not go through the filter, with

*For more information on the postulates of quantum theory, see Sec. 2.2 of Ref. 107 or a text on quantum mechanics, e.g., Landau and Lifschitz.[87] Jordan[157] gives a very readable summary of the spectral theory of operators needed for quantum theory; unifortunately, it is not readily available. The many excellent mathematics texts that discuss spectral theory include Halmos,[56] Horn and Johnson,[65] and Naylor and Sell.[105]

probability $\frac{1}{2}$. But if the photon passes through the filter, what emerges is always a vertically polarized photon. Thus, if a photon is detected, one cannot conclude that a vertically polarized photon was sent. The possible polarizations of the photon before it passed through the filter might have been at a 30-deg angle, a 45-deg angle, or right circular. The only thing that can be said with certainty is that the photon was not horizontally polarized; no photon in that state could go through the filter.

9.2.2.2 von Neumann versus POVM measurements

For simplicity, we explained the measurement process as if all eigenvalues are nondegenerate. In the case of degenerate eigenvalues, one should use the expression $\sum_k a_k E_k$ with distinct eigenvalues a_k and projections E_k onto eigenspaces whose dimension is the degeneracy. A measurement can then be identified with a set of orthogonal projections, i.e., a set of self-adjoint operators $\{E_k\}$ satisfying $E_j E_k = \delta_{jk} E_k$ and $\sum_k E_k = I$ (where δ_{jk} denotes the Kronecker delta for which $\delta_{jj} = 1$ and $\delta_{jk} = 0$ when $j \neq k$). This is called a von Neumann measurement.

Many treatments of quantum theory consider only von Neumann measurements, and these suffice for identifying the computational basis as well as for quantum error correction. The actual measurement process will, in general, use a set of commuting self-adjoint operators sufficient to distinguish between a set of subspaces. For example, the identification of the state of a hydrogen atom with quantum numbers n, ℓ, m is the result of measuring the energy, angular momentum, and the so-called z component of angular momentum. Similarly, the operators $\{Z_k\}$, $k = 1 \ldots n$, where

$$Z_k = I \otimes \ldots I \otimes \sigma_z \otimes I \ldots \otimes I, \tag{9.12}$$

with σ_z in the kth position, suffice to make measurements in the computational basis. Indeed, the eigenvalues ± 1 of Z_k correspond to $i_k - 0, 1$ since $Z_k |i_1 \ldots i_k \ldots i_n\rangle = (-1)^{i_k} |i_1 \ldots i_k \ldots i_n\rangle$.

In certain situations in QIP, one needs a more general type of measurement known as a *positive operator valued measure* (POVM). In this case, the requirement $E_j E_k = \delta_{jk} E_k$ (which implies that each E_k has eigenvalues 0, 1) is dropped and replaced by the weaker requirement that E_k is positive semi-definite. Thus, we define a POVM as a set of operators $\{F_b\}$ such that each $F_b > 0$ and $\sum_b F_b = I$. Unlike a von Neumann measurement, the result of a POVM depends on the order in which the operations are performed. Nevertheless, a POVM can always be represented as a von Neumann measurement on a larger Hilbert space involving an auxiliary space in much the same way as in Sec. 9.2.4.

9.2.3 Mixed states

Representing states of an isolated quantum system, also known as *pure states*, by vectors is not entirely satisfactory. A state can also be represented by the projection $|\psi\rangle\langle\psi|$ onto the subspace spanned by $|\psi\rangle$. This has the advantage of avoiding the ambiguity associated with overall phase factors. Thus, every pure state can be described uniquely by a rank-one projection.

A mixed state ρ can then be defined as a convex combination of pure states, i.e.,

$$\rho = \sum_k p_k |\psi_k\rangle\langle\psi_k| \quad \text{with} \quad p_k > 0, \ \sum_k p_k = 1. \tag{9.13}$$

There is a sense in which a mixed state can be regarded as a (classical) probability distribution over quantum states. The pure states in Eq. (9.13) need not be orthogonal; when they are, Eq. (9.13) is simply the spectral decomposition of ρ. If a measurement is made on a system in a mixed state, the average outcome is

$$\langle A\rangle = \sum_k p_k \langle\psi_k, A\psi_k\rangle = \sum_k p_k \mathrm{Tr} A(|\psi_k\rangle\langle\psi_k|) = \mathrm{Tr} A\rho, \tag{9.14}$$

where Tr denotes the trace as defined in Appendix 9.B.

A density matrix is a positive semidefinite matrix ρ such that $\mathrm{Tr}\rho = 1$. Thus, there is a one-to-one correspondence between mixed states and density matrixes. A density matrix ρ describes a pure state if and only if $\rho^2 = \rho$.

Mixed states arise in various contexts, including quantum statistical mechanics, subsystems of larger systems, and noisy quantum systems. Roughly speaking, a mixed quantum state can be thought of as having two types of probabilities. That given by the p_k in Eq. (9.13) behaves very much like classical probability describing a distribution over a set of (pure) quantum states. However, these quantum states also have the nonclassical probabilistic properties associated with superpositions. Consider the three density matrixes

$$|0\rangle\langle 0|_x = \frac{1}{2}\begin{pmatrix} 1 & 1 \\ 1 & 1 \end{pmatrix}, \quad |1\rangle\langle 1|_x = \frac{1}{2}\begin{pmatrix} 1 & -1 \\ -1 & 1 \end{pmatrix}, \quad \frac{1}{2}\begin{pmatrix} 1 & 0 \\ 0 & 1 \end{pmatrix}. \tag{9.15}$$

All three yield $+1$ with probability $\frac{1}{2}$ and -1 with probability $\frac{1}{2}$ when a measurement is made using σ_z. However, the first two are pure states that give very different results when a measurement is made using σ_x; the first would yield $+1$ with probability 1 and the second -1 with probability 1; but the third would give ± 1, each with probability $\frac{1}{2}$. Although all three matrixes have the same diagonal, they have different off-diagonal elements, which express quantum correlations.

It is useful to have a quantitative measure of the extent to which a state is pure or mixed. Although there are many possibilities, we consider only the von Neumann entropy, which is the most important and widely used. It is defined as

$$S(\rho) = -\mathrm{Tr}\rho \log \rho = -\sum_k \lambda_k \log \lambda_k, \tag{9.16}$$

where λ_k are the eigenvalues of ρ and we use the convention that $0 \log 0 = 0$. Note that the von Neumann entropy of a given density matrix could be regarded as the Shannon entropy* of its eigenvalues. However, we prefer to consider the von Neumann entropy as more fundamental. Indeed, von Neumann defined his entropy‡ in 1927, more than 20 years before Shannon put forth his theory in 1948. Moreover, the von Neumann entropy includes the Shannon entropy as a special case.

In fact, one can embed classical discrete probability within the formalism associated with mixed quantum states. A discrete probability vector with elements p_k can be written as a diagonal matrix with elements $p_k \delta_{jk}$. It is sometimes useful to let $\mathcal{D}(\mathbf{C}^d)$ denote the diagonal $d \times d$ matrixes. The positive semidefinite matrixes in $\mathcal{D}(\mathbf{C}^d)$ then correspond to classical probability vectors.

9.2.4 Open systems

9.2.4.1 A basic model of noise

Noise arises because no system is really isolated, but interacts with its environment. Denote the Hilbert space of the system (typically a QC or quantum communication channel) by $\mathcal{H}_C$ and that of the environment by $\mathcal{H}_E$. The combined system is described by a Hamiltonian acting on $\mathcal{H}_C \otimes \mathcal{H}_E$ with the form

$$H_{CE} = H_C \otimes I_E + I_C \otimes H_E + V_{CE}, \tag{9.17}$$

where H_C is the Hamiltonian of the QC, H_E is that of the environment, and V_{CE} describes the interaction between the QC and its environment.

The statement that the system is in the pure state ψ_C^0 can carry hidden assumptions; in this case, that the total system is in a product state of the form $\psi_C^0 \otimes \psi_E^0 \equiv \Psi_{CE}^0$, which evolves in time according to

$$|\Psi_{CE}(t)\rangle = U_{CE}(t)|\psi_C^0 \otimes \psi_E^0\rangle = \sum_k c_k |\psi_C^k \otimes \psi_E^k\rangle, \tag{9.18}$$

where $U_{CE}(t)$ is the unitary group determined by Eqs. (9.10) and (9.17) and we have rewritten $|\Psi_{CE}(t)\rangle$ as a superposition of products of two sets of basis vectors whose first elements are $|\psi_C^0\rangle$ and $|\psi_E^0\rangle$, respectively. However, Eq. (9.18) is somewhat unwieldy as a description of the system C. One can obtain a more compact description by taking the partial trace (explained in Appendix 9.B) over the Hilbert space $\mathcal{H}_E$. The result

$$\mathrm{Tr}_E(|\Psi_{CE}\rangle\langle\Psi_{CE}|) = \sum_k |c_k|^2 |\psi_C^k\rangle\langle\psi_C^k| \tag{9.19}$$

*Those unfamiliar with Shannon entropy can regard it as a special case of the von Neumann entropy in which the density matrix is diagonal.

‡Admittedly, von Neumann was motivated by very different considerations than Shannon. He wanted to extend classical statistical mechanics to the quantum setting, and his definition was a natural generalization of Gibbs' approach. It is remarkable that two such different contexts led to very similar mathematical structures.

is a mixed state on $\mathcal{H}_C$. Having motivated this definition, we now extend it to arbitrary (mixed) states as

$$\rho_C \mapsto \Phi(\rho_C) \equiv \mathrm{Tr}_E[U_{CE}(t_0)\rho_C \otimes \gamma_E U_{CE}^\dagger(t_0)], \tag{9.20}$$

with t_0 fixed and with γ_E a fixed reference state on $\mathcal{H}_E$. The map Φ gives a snapshot of the effect of noise at time t_0. Maps of this form are known as completely positive and trace-preserving (CPT) maps.

At a minimum, one would expect such a map to take density matrixes to density matrixes. This implies that it should preserve the trace, i.e., $\mathrm{Tr}\,\Phi(\rho) = \mathrm{Tr}\,\rho$, and that it should be positivity preserving in the sense that it takes positive semidefinite matrixes to positive semidefinite matrixes. In fact, maps of the form of Eq. (9.20) satisfy a stronger condition known as "complete positivity," which means that $I \otimes \Phi$ is also positivity preserving on any space of the form $\mathbf{C}^n \otimes \mathcal{H}_C$. There is an extensive literature on completely positive maps in various contexts, some quite abstract. We note here only that complete positivity implies that Φ is also positivity preserving when extended to include any system with which the QC is entangled.

At first glance, this model of noise may seem rather different from the common classical one of convolution with another signal. However, one can show that the action of a CPT map restricted to diagonal matrixes, $\Phi : \mathcal{D}(\mathbf{C}^d) \mapsto \mathcal{D}(\mathbf{C}^d)$, is equivalent to the action of a column stochastic matrix on the probability vectors given by the diagonals, and a convolution is equivalent to multiplication by a cyclic stochastic matrix. Thus, the noise model of CPT maps includes classical noise as a special case.

9.2.4.2 Terminology

A variety of names have been used for these CPT maps in the QIP literature. In the 1990s, they were often referred to as "superoperators" (because they are linear operators acting on operators); however, this is somewhat unsatisfactory since the same term is used in other contexts for maps that do not satisfy the special requirement of complete positivity. The term "quantum operation," employed in the influential book of Nielsen and Chuang,[107] is frequently used. The term "stochastic map" is sometimes used to reflect the fact that CPT maps can be regarded as the noncommutative analogue of the action of a column stochastic matrix on a probability vector. This terminology has the merit that a *unital* CPT map, i.e., one for which $\Phi(I) = I$ as well, is naturally called "bistochastic," but this does not seem to have caught on. Because of the role played by CPT maps in the study of noisy quantum communication, they are sometimes referred to simply as "channels."

9.2.4.3 Equivalent descriptions

There are several important equivalent ways of describing CPT maps.

- Choi[31] showed that a linear map on $\mathcal{B}(\mathbf{C}^d)$ (the $d \times d$ matrixes acting on $\mathbf{C}^d$) is completely positive if and only if the matrix

$$\Gamma_\Phi \equiv (I \otimes \Phi)(|\beta\rangle\langle\beta|) \tag{9.21}$$

is positive semidefinite, where

$$\beta = d^{-1/2} \sum_k |k\rangle \otimes |k\rangle. \tag{9.22}$$

Moreover, there is[72] a one-to-one correspondence between CP maps on $\mathcal{B}(\mathbf{C}^d)$ and positive semidefinite matrixes on $\mathbf{C}^{d^2}$. By restricting to matrixes that also satisfy the condition $\text{Tr}_B \Gamma = I_A$ (or $\text{Tr}_A \Gamma = I_B$), one can extend this correspondence to CPT maps (or to unital CP maps).

- In QIP, a CPT map is often given by its Kraus representation,[85,86] which is a set of operators A_k (often called *Kraus operators*) satisfying

$$\Phi(P) = \sum_k A_k P A_k^\dagger \quad \text{with} \quad \sum_k A_k^\dagger A_k = I. \tag{9.23}$$

This representation is not unique; indeed, if $\widetilde{A}_j = \sum_k u_{jk} A_k$ with U unitary, then $\{\widetilde{A}_j\}$ also forms a set of Kraus operators for the map Φ. However, Choi[31] gave a canonical prescription for A_k in terms of the eigenvectors of the matrix Γ_Φ in (9.21). (For a very nice exposition of this construction see [Ref. 91].) Finally, note that $\Phi(I) = I \Leftrightarrow \sum_k A_k A_k^\dagger = I$.

- Stinespring[145] showed that given a CPT map Φ one can always find an auxiliary space $\mathcal{H}_B$, a reference state Q_B, and a unitary operator U on $\mathcal{H} \otimes \mathcal{H}_B$ such that

$$\Phi(P) = \text{Tr}_B[U P \otimes Q_B U^\dagger]. \tag{9.24}$$

Stinespring's fundamental work[145] considerably predated that of Kraus.[85,86] Nevertheless, the Kraus representation does provide a convenient mechanism for constructing Eq. (9.24), as explained* in Sec. III.D of Ref. 123.

Thus, Stinespring showed that any CPT map can be represented as if it arose from the noise model in Sec. 9.2.4.1. The Kraus and Stinespring representations are really equivalent, but the former is more commonly used in QIP.

*In Ref. 123 this result was attributed to Lindblad,[94] who obtained it using another variant of the Stinespring representation and showed its utility in his work on entropy inequalities.

9.2.5 Notation and Pauli matrixes

9.2.5.1 Pauli matrixes

The three Pauli matrixes,

$$\sigma_x = \begin{pmatrix} 0 & 1 \\ 1 & 0 \end{pmatrix}, \quad \sigma_y = \begin{pmatrix} 0 & -i \\ i & 0 \end{pmatrix}, \quad \sigma_z = \begin{pmatrix} 1 & 0 \\ 0 & -1 \end{pmatrix}, \tag{9.25}$$

are extremely important in QIP where they play a number of different roles. For example, the standard choice for encoding 0 and 1, as given in Eq. (9.1), uses the eigenstates of σ_z; the alternative given in Eq. (9.3) uses eigenstates of σ_x.

It is often convenient to replace the subscripts x, y, and z by 1, 2, and 3. One can then write the anticommutation relations as

$$\sigma_j \sigma_k + \sigma_k \sigma_j = 2I\delta_{jk}, \tag{9.26}$$

and note that Eq. (9.26) follows from the fundamental property

$$\sigma_j \sigma_k = i\sigma_\ell, \tag{9.27}$$

with j, k, and ℓ cyclic. The Pauli matrixes also represent some of the basic single-qubit operations used in QIP. In particular $\sigma_x|j\rangle = |j+1\rangle$, where $j = 0$ or 1 denotes a vector in the basis (9.1) and addition is mod 2; and $\sigma_z|j\rangle = (-1)^j|j\rangle$. When these actions are desired and implemented in a controlled way, they are regarded as gates; when they arise unwanted as the result of noise, the Pauli matrixes represent fundamental errors.

Moreover, any single-bit error can be written as a linear combination of the identity and Pauli matrixes, and any multibit error as a linear combination of products of Pauli matrixes and/or the identity. This has important implications for quantum error correction, as discussed in Sec. 9.6.3.3.

9.2.5.2 Bloch sphere representation

The Pauli matrixes, together with the identity I on $\mathbf{C}_2$, form an orthonormal basis with respect to the inner product defined by Eq. (9.126) for the vector space of $\mathcal{B}(\mathbf{C}_2)$ of 2×2 matrixes. When using them in this context, it is convenient to define $\sigma_0 = I$. Then one can write any 2×2 matrix A in the form $A = \sum_{k=0}^{3} a_k \sigma_k$. Moreover, $A = A^\dagger \Leftrightarrow$ all a_k are real; $\mathrm{Tr} A = 2a_0$; and A is positive semi-definite $\Leftrightarrow \sqrt{\sum_{k=1}^{3} a_k^2} \le a_0$. Thus, one can write the density matrix for a mixed state on $\mathbf{C}_2$ in the form

$$\rho = \frac{1}{2}[I + \boldsymbol{w} \cdot \sigma], \tag{9.28}$$

where $\boldsymbol{w} \in \mathbf{R}^3$ and $|\boldsymbol{w}| \le 1$. Moreover, ρ is a pure state if and only if $|\boldsymbol{w}| = 1$ This gives a one-to-one correspondence between pure states in $\mathbf{C}_2$ and unit vectors in $\mathbf{R}^3$, called the Bloch sphere representation, shown in Fig. 9.6. This picture is extremely useful. For example, the image of the Bloch sphere under a CPT map is an ellipsoid. However, not every ellipsoid is the image of a CPT map. For further discussion, see Ref. 125.

9.2.5.3 Rotations and the Hadamard transform

There is one-to-one correspondence between rotations in $\mathbf{R}^3$ and unitary matrixes on $\mathbf{C}^2$ with $\det U = 1$. In particular, if $\rho = \frac{1}{2}[I + \boldsymbol{w} \cdot \sigma] \mapsto U^\dagger \rho U$, with U unitary, then there is a rotation R such that $\boldsymbol{w} \mapsto R\boldsymbol{w}$ and $U^\dagger \rho U = \frac{1}{2}[I + R\boldsymbol{w} \cdot \sigma]$. Alternatively, $\sigma \mapsto R^T \sigma$ corresponds to a rotation of the Bloch sphere, and the relations in Eqs. (9.26) and (9.27) are invariant under rotations. Since $\rho = \frac{1}{2}[I + R\boldsymbol{w} \cdot R^T \sigma]$, the density matrix ρ corresponds to the point $R\boldsymbol{w}$ on the Bloch sphere in the rotated basis $R^T \sigma$.

The unitary matrix $H = \frac{1}{2}\begin{pmatrix} 1 & 1 \\ 1 & -1 \end{pmatrix}$, known as the Hadamard transform or Hadamard gate, plays an important role in QIP. It is self-adjoint as well as unitary, i.e., $H = H^\dagger = H^{-1}$; and $H\sigma_x H = \sigma_z$ and vice versa. Thus, H maps the basis (9.1) to (9.3) and *vice versa*. However, since $\det H = -1$, the Hadamard transform does not correspond to a rotation.

9.2.6 No-cloning principle

The standard formulation of the quantum measurement process, which leaves a system in an eigenstate of the observable measured, suggests that one cannot obtain sufficient information about a state to duplicate it. In 1982, Wootters and Zurek[154] put this into precise form and gave a proof of what is now called the "no-cloning theorem," although I prefer the term "principle." Their argument is predicated on the assumption that controlled quantum processes can only be carried out by unitary operations. Hence, the question is, given a Hilbert space $\mathcal{H}$ can one find a

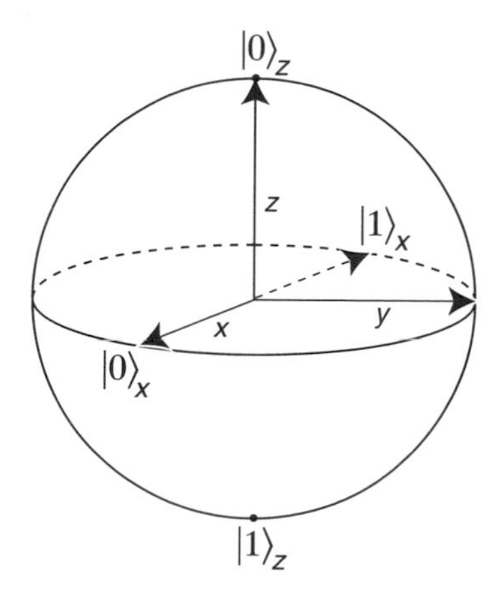

Figure 9.6 Bloch sphere representation.

unitary operator U (on the larger space $\mathcal{H} \otimes \mathcal{H} \otimes \mathcal{H}$), which can convert an arbitrary vector $\psi \in \mathcal{H}$ into a vector containing two copies of ψ. In other words, can one find a unitary U that satisfies

$$U(\psi \otimes \alpha \otimes \beta) = \psi \otimes \psi \otimes \gamma, \tag{9.29}$$

where α and β are fixed vectors that may be chosen explicitly if desired. Since ψ is arbitrary, it is also true that

$$U(\phi \otimes \alpha \otimes \beta) = \phi \otimes \phi \otimes \delta, \tag{9.30}$$

for any ϕ, and there is no loss of generality in assuming that all of the vectors are normalized to 1. Then the fact that a unitary operator preserves inner products implies

$$\langle \phi, \psi \rangle = \langle \phi, \psi \rangle^2 \langle \gamma, \delta \rangle. \tag{9.31}$$

From this (and the conditions for equality in the Schwarz inequality) one obtains a contradiction unless $\phi = \psi$ or $\langle \phi, \psi \rangle = 0$, i.e., one cannot clone arbitrary vectors, but only those in a fixed orthonormal basis. Now one might well ask whether this is not belaboring the obvious since a unitary operator is linear and a mapping of the form $\psi \mapsto \psi \otimes \psi$ is manifestly nonlinear. Clearly, the impossibility result is implicit in the restriction to *linear* unitary operations. However, one gains something more, namely, that one *can* clone a fixed set of orthonormal vectors. In the context of QIT, this means that one also recovers the result that one can clone the subset of vectors corresponding to classical bits.

Moreover, we have proved only that cloning is impossible within a certain framework or model. The argument does not apply in alternative theories, such as Bohmian mechanics, which include "hidden variables" with nonlinear behavior. If such theories cannot exclude cloning on other grounds, there are practical implications. For example, cloning would threaten the security of the QKD described in Sec. 9.5.1. Advocates of alternative theories have focused on recovering the predictions of conventional quantum theory. We are now in a position to consider consequences that are not only experimentally testable, but may have practical applications. Those who want some alternative theory to be taken seriously ought to propose such experiments.

9.3 Entanglement

9.3.1 Bell states and correlations

We begin by describing a special set of states on $\mathbf{C}_2 \otimes \mathbf{C}_2 = \mathbf{C}^4$ known as the maximally entangled Bell states. These can be defined so that, if expanded in the form

$$|\beta_k\rangle = \sum_{j,k} a_{jk}|jk\rangle, \tag{9.32}$$

the coefficient matrix a_{jk} is exactly that of the Pauli matrix σ_k (with the convention that $\sigma_0 = I$). They are given explicitly in Table 9.2, from which one can see that they also satisfy $|\beta_k\rangle = (\sigma_k \otimes I)|\beta_0\rangle$. The Bell states are simultaneous eigenvectors to the commuting operators $\sigma_x \otimes \sigma_x$ and $\sigma_z \otimes \sigma_z$ with the eigenvalues shown in Table 9.2. Therefore, these states can be identified by a measurement, known as a "Bell measurement" with this pair of observables.

9.3.2 An experiment

Suppose that every morning when you log onto your computer the screen shows three boxes

A B C

flashing on and off with the words "CLICK ME." You cannot proceed to check your e-mail (or do anything else) until you choose one of the boxes. As soon as you click, the other two boxes disappear and the remaining box changes to either Win or Lose indicating that your "frequent web buyer" account has won or lost 500 points.

You choose at random, but, in the hope of finding a better strategy, keep careful notes of your choice and the result. The game appears fair, in the sense that you win 50% of the time; however, no strategy appears. After some months, you attend an SPIE conference where you meet a colleague from the opposite coast who uses the same Internet provider and has kept similar records. You compare notes and discover an amazing coincidence. On those days when you both choose the same box, one wins and the other loses. Further investigation reveals that other engineers using this particular Internet provider seem to be paired up in a similar way. When

Table 9.2 Bell states and eigenvalues.

	$\sigma_x \otimes \sigma_x$	$\sigma_z \otimes \sigma_z$	
$\lvert\beta_0\rangle = \frac{1}{\sqrt{2}}(\lvert 00\rangle + \lvert 11\rangle)$	+1	+1	(9.33a)
$\lvert\beta_1\rangle = \frac{1}{\sqrt{2}}(\lvert 01\rangle + \lvert 10\rangle) = (\sigma_x \otimes I)\lvert\beta_0\rangle$	+1	−1	(9.33b)
$\lvert\beta_2\rangle = \frac{i}{\sqrt{2}}(\lvert 10\rangle - \lvert 01\rangle) = (\sigma_y \otimes I)\lvert\beta_0\rangle$	−1	−1	(9.33c)
$\lvert\beta_3\rangle = \frac{1}{\sqrt{2}}(\lvert 00\rangle - \lvert 11\rangle) = (\sigma_z \otimes I)\lvert\beta_0\rangle$	−1	+1	(9.33d)

two members of a pair choose the same box, one, and *only one*, wins. It appears that the Internet entrepreneur is sending out paired messages programmed so that the boxes are complementary—e.g., if your boxes are coded W W L, your partner's are L L W.

However, one astute pair notices something curious. On the days when they choose *different* boxes, both win $\frac{3}{8}$ of the time and both lose $\frac{3}{8}$ of the time; only $\frac{1}{4}$ of the time does one win and the other lose with different boxes. Yet, an elementary calculation shows that with complementary pairs, both should win $\frac{1}{4}$ and both should lose $\frac{1}{4}$ of the time. This seems to eliminate the complementary box hypothesis; moreover, it would also imply uncorrelated probabilities when the parties choose different boxes.

What other explanations are possible? The Internet provider (located in Kansas) may be sending entangled pairs of polarized photons. Clicking on box A, B, or C selects one of three polarization filters set at 120° angles. A "win" occurs when the photon passes through the filter and hits the detector.

Why is this explanation consistent? Let

$$|\phi\rangle = i|\beta_2\rangle = \frac{1}{\sqrt{2}}(|01\rangle - |10\rangle), \tag{9.34}$$

and consider an alternative description of the system in a rotated basis for which

$$|0\rangle = \cos\theta|\hat{0}\rangle + \sin\theta|\hat{1}\rangle \tag{9.35a}$$

$$|1\rangle = -\sin\theta|\hat{0}\rangle + \cos\theta|\hat{1}\rangle. \tag{9.35b}$$

It is not hard to see that the state $|\phi\rangle$ (known as a "singlet") has the same form in any rotated basis, i.e., $|\phi\rangle = 2^{-1/2}(|\hat{0}\hat{1}\rangle - |\hat{1}\hat{0}\rangle)$. Thus, whenever the two players choose the same rotation, one, and only one, wins. Choosing different boxes is equivalent to one party leaving the basis unchanged and the other choosing a rotation by $\pm\frac{2\pi}{3}$. Writing $|\phi\rangle$ accordingly, one finds

$$|\phi\rangle = \frac{1}{\sqrt{2}}(-\sin\theta|0\hat{0}\rangle + \cos\theta|0\hat{1}\rangle + \cos\theta|1\hat{0}\rangle + \sin\theta|1\hat{1}\rangle). \tag{9.36}$$

Thus, the probability of win-win is the amplitude squared of the coefficient of $|0\hat{0}\rangle$, which is $\frac{1}{2}\sin^2\theta = \frac{3}{8}$, when $\theta = \pm\frac{2\pi}{3}$. Similarly the probabilities of win-lose, lose-win, and lose-lose are determined by the coefficients of $|0\hat{1}\rangle$, $|1\hat{0}\rangle$ and $|1\hat{1}\rangle$, respectively; for $\theta = \pm\frac{2\pi}{3}$, these are $\frac{1}{4}$, $\frac{1}{4}$, and $\frac{3}{8}$.

9.3.3 Bell inequalities and locality

The analysis in Sec. 9.3.2 shows how conventional quantum theory explains the results of an idealized experiment. However, it does not rule out other explanations, i.e., it does not resolve the question of whether some alternative theory might also give a satisfactory explanation.

Moreover, because these correlations involve only the spin components, they should exist even for particles widely separated in spatial distance. This raises questions about locality, e.g., whether quantum correlations can be exploited for superluminal (faster than light) communication.

Bell considered this question and showed that if a theory is both local and has hidden variables, the correlations must satisfy certain inequalities, known as "Bell inequalities." The proofs of these inequalities require only elementary classical probability; the subtlety comes in a careful definition of what is meant by "local" and "hidden variables." The actual experiments designed to test such inequalities have found that they are violated and that the correlations do not depend on distance. Thus far, quantum mechanics is the only theory which, albeit nonlocal, can explain these correlations without permitting superluminal communication. (This requires a distinction between "passive" and "active" nonlocality.) For further discussion see Bell,[8] Faris,[40] Mermin,[102] Werner and Wolf,[149] and Wick.[150]

Experiments of the type described in Sec. 9.3.2 are often referred to as "EPR" experiments because of their connection to a famous 1935 paper of Einstein, Podolosky, and Rosen[38] on the implications of quantum correlations at long distances. However, the original EPR proposal was for an experiment using the continuous variables of position and momentum. A variant using discrete two level systems was first discussed* by Bohm[21] in 1951. However, experiments with angles other than 90° were neither considered nor performed, until after the work of Bell[8] in the 1960s.

9.3.4 An important identity

The identity in Eq. (9.37) plays a key role in the quantum process known as "teleportation," described in Sec. 9.5.4. Let $|\phi\rangle = a|0\rangle + b|1\rangle$ be the state of a qubit and β_0 be the Bell state of Eq. (9.33a). Then

$$|\phi\rangle_R \otimes |\beta_0\rangle_{ST} = \frac{1}{4}\sum_{k=0}^{3} |\beta_k\rangle_{RS} \otimes \sigma_k|\phi\rangle_T, \tag{9.37}$$

where we have introduced subscripts to emphasize that we are now working on a tensor product of three Hilbert space $\mathcal{H}_R \otimes \mathcal{H}_S \otimes \mathcal{H}_T$ and enable us to keep track of states on the various subspaces.

As discussed in Sec. 9.5.4, this identity implies that when two parties share an entangled pair of states, the information encoded in a third qubit can be transmitted

*The first explicit discussion of EPR correlations in spin systems appeared in Bohm's[21] 1951 book on quantum theory. Curiously, the first experiment of this type had been performed earlier (in 1949) by Wu and Shaknov[155] for a somewhat different purpose, following a 1946 proposal by Wheeler.[153] The analysis of EPR correlations in the Wu-Shaknov experiment was given by Bohm and Aharanov[22] in 1957. As far as I am aware, the original EPR experiment has not been done. However, a similar experiment (i.e., one that uses the continuous variables of position and momentum), that was proposed by Popper, has now been performed by Kim and Shih.[78]

using a pair of classical bits. It is is straightforward, but tedious, to verify Eq. (9.37) by simply writing out all terms on both sides. The following proof (which the reader may prefer to postpone until after seeing its application in Sec. 9.5.4) may gave more insight.

The Bell states form an orthonormal basis on $\mathbf{C}_2 \otimes \mathbf{C}_2$ so that any vector $|\chi\rangle$ can be expanded in the form

$$|\chi\rangle = \sum_{k=0}^{3} c_k |\beta_k\rangle, \tag{9.38}$$

with $c_k = \langle \beta_k, \chi \rangle$. We can generalize this to an expansion of the form

$$|\Psi\rangle_{RST} = \sum_{k=0}^{3} |\beta_k\rangle_{RS} \otimes |\gamma_k\rangle_T, \tag{9.39}$$

with

$$|\gamma_k\rangle_T = \langle \beta_k, \Psi \rangle_{RS}, \tag{9.40}$$

where the inner product is taken only over the subspace $\mathcal{H}_R \otimes \mathcal{H}_S$, yielding a vector $|\gamma_k\rangle$ on $\mathcal{H}_T$ rather than a constant c_k. We now apply this to $|\Psi\rangle_{RST} = |\phi\rangle_R \otimes |\beta_0\rangle_{ST}$. Then

$$\begin{aligned}
|\gamma_k\rangle_T &= \langle \beta_k, \phi \otimes \beta_0 \rangle_{RS} = \langle (\sigma_k \otimes I)_{RS} \beta_0, \phi \otimes \beta_0 \rangle_{RS} && (9.41)\\
&= \langle \beta_0, \sigma_k \phi \otimes \beta_0 \rangle_{RS} && (9.42)\\
&= \frac{1}{4} \big(\langle 00|_{RS} + \langle 11|_{RS} \big) \big(\sigma_k |\phi\rangle_R \otimes |00\rangle_{ST} + \sigma_k |\phi\rangle_R \otimes |11\rangle_{ST} \big) && (9.43)\\
&= \langle 0, \sigma_k \phi \rangle_R |0\rangle_T + \langle 1, \sigma_k \phi \rangle_R |1\rangle_T && (9.44)\\
&= \langle 0, \sigma_k \phi \rangle_T |0\rangle_T + \langle 1, \sigma_k \phi \rangle_T |1\rangle_T && (9.45)\\
&= (|0\rangle\langle 0| + |1\rangle\langle 1|) \sigma_k |\phi\rangle_T = \sigma_k |\phi\rangle_T, && (9.46)
\end{aligned}$$

where Eqs. (9.44) and (9.45) exploit the fact that the value of an inner product is the same in $\mathcal{H}_T$ and $\mathcal{H}_R$ since both are equal to $\mathbf{C}_2$.

9.3.5 More on entanglement

Most states on $\mathbf{C}^{2^n}$ are neither product states nor maximally entangled. A pure state is said to be entangled if it *cannot* be written as a product in *any* basis. Thus, one has a continuum of possible degrees of entanglement.

If one takes a superposition of entangled states it need not become more entangled. Indeed, $2^{-1/2}(|\beta_0\rangle + |\beta_3\rangle) = |00\rangle$ is easily seen to yield a product. With a bit more effort, one can see that $2^{-1/2}(|\beta_0\rangle - |\beta_1\rangle) = |0\rangle_x \otimes |1\rangle_x$ is also a product.

However, $2^{-1/2}(|\beta_0\rangle + i|\beta_1\rangle)$ cannot be written as a product; indeed, it is maximally entangled. What does this mean? How does one know if a state can be written as a product in some basis? More generally, given a superposition such as $\frac{3}{5}|00\rangle + \frac{4}{5}|00\rangle$ or $(\sqrt{3}/2)|\beta_0\rangle + \frac{1}{2}|\beta_3\rangle$, can one quantify the extent to which it is entangled?

In the case of pure states, there is a very simple criterion for answering this question. The entanglement of a pure state $|\psi\rangle$ on the space $\mathcal{H}_A \otimes \mathcal{H}_B$ is the von Neumann entropy of its reduced density matrix, i.e.,

$$S(\mathrm{Tr}_B|\psi_{AB}\rangle\langle\psi_{AB}|) = -\mathrm{Tr}_B|\psi_{AB}\rangle\langle\psi_{AB}|\log(\mathrm{Tr}_B|\psi_{AB}\rangle\langle\psi_{AB}|). \tag{9.47}$$

There is no ambiguity in this definition because the entropies of the two reduced density matrixes $\rho_A = \mathrm{Tr}_B|\psi_{AB}\rangle\langle\psi_{AB}|)$ and $\rho_B = \mathrm{Tr}_A|\psi_{AB}\rangle\langle\psi_{AB}|)$ are equal. In fact, ρ_A and ρ_B have the same nonzero eigenvalues. However, this is *only* true for pure states. It is a direct consequence of the so-called "Schmidt decomposition," which states that any bipartite pure state can be written in the form

$$|\psi_{AB}\rangle = \sum_k \mu_k|\phi_k\rangle \otimes |\chi_k\rangle, \tag{9.48}$$

with $\{\phi_k\}$ and $\{\chi_k\}$ orthogonal. This result, which plays an important role in QIP, is really just a special case of the singular value decomposition, as discussed in Appendix 9.C. (See also Appendix A of Ref. 80.) Moreover, it follows from Eq. (9.48) that the nonzero eigenvalues of both ρ_A and ρ_B are given by $|\mu_k|^2$ so that the entanglement of $|\psi_{AB}\rangle$ is $-\sum_k |\mu_k|^2 \log |\mu_k|^2$.

The preceding discussion applies to any bipartite composite system, i.e., in any situation for which one can write the underlying Hilbert space as $\mathcal{H} = \mathcal{H}_A \otimes \mathcal{H}_B$. However, it can not be extended to multipartite states, e.g., $\mathcal{H}_A \otimes \mathcal{H}_B \otimes \mathcal{H}_C$. On the other hand, it does apply to n-qubit systems in situations in which the qubits can be divided into two sets, one with k qubits and the other with $n-k$ using the isomorphism between $\mathbf{C}^{2^n}$ and $\mathbf{C}^{2^k} \otimes \mathbf{C}^{2^{n-k}}$.

The question of measuring the entanglement of bipartite mixed states is quite complex; and, in general, the entropy of one of the reduced density matrixes does not suffice. To see why, observe that one can construct a density matrix that is a convex combination of (nonorthogonal) products, but whose eigenvectors are not product states. Neither reduced density matrix will have entropy zero, although this state is not entangled in the sense that one has a mixture of products. There are several inequivalent definitions of entanglement for mixed states, corresponding to different physical situations. See Refs. 17, 26, 67 for a summary.

The classification of multipartite entanglement, even for pure states, is far from straightforward and seems to require a large number of invariants. However, a few special classes are worth a brief mention. A state of the form

$$\frac{1}{\sqrt{2}}(|00\ldots0\rangle + |11\ldots1\rangle) \tag{9.49}$$

is known as a GHZ or "cat" state. It is sometimes regarded as highly entangled because there is a sense in which all the particles are entangled with one another. However, the measurement of σ_z on a single qubit would destroy the entanglement, leaving the system in a product state. Recently, n-qubit states known as "cluster states" were found,[25] which are characterized by persistent entanglement in the sense that a minimum of $\frac{n}{2}$ single-qubit measurements are required before all entanglement is destroyed. Such states occur naturally in spin lattice models, such as the Ising model, and play an essential role in the Raussendorf–Briegel model on one-way computation.[117,118] They were recently shown[126] to be related to the stabilizer groups (mentioned in Sec. 9.6.4), which arise in quantum error correction.

The classification and quantification of entanglement is an active area of current research well beyond the scope of this chapter.

9.4 Quantum computation algorithms

9.4.1 The Deutsch–Jozsa problem

In the simple 2-qubit version of the Deutsch–Jozsa problem,[34,35] one has a function $f : \{0, 1\} \mapsto \{0, 1\}$ and an associated unitary operator U_f whose action on a basis of product vectors $|j, k\rangle$ is given by

$$U_f|j, k\rangle = |j, k + f(j)\rangle, \tag{9.50}$$

with addition mod 2. There are four possible functions; however, we are only interested in learning whether f is one of the two constant functions, or one of the other two, known as "balanced." We do not give details of the operator U_f; it is assumed that it can be carried out by what is known as an "oracle."

We first consider the effect of U_f on a product when the second vector has the form $H|1\rangle = 2^{-1/2}(|0\rangle - |1\rangle)$. Then,

$$U_f\Big(|j\rangle \otimes H|1\rangle\Big) = \frac{1}{\sqrt{2}} U_f|j, 0\rangle - U_f|j, 1\rangle \tag{9.51a}$$

$$= \frac{1}{\sqrt{2}}|j, f(j)\rangle - |j, 1 \oplus f(j)\rangle$$

$$= \frac{1}{\sqrt{2}} \begin{cases} |j, 0\rangle - |j, 1\rangle & \text{if } f(j) = 0 \\ |j, 1\rangle - |j, 0\rangle & \text{if } f(j) = 1 \end{cases}$$

$$= (-1)^{f(j)} \frac{1}{\sqrt{2}}(|j, 0\rangle - |j, 1\rangle) \tag{9.51b}$$

$$= (-1)^{f(j)}|j\rangle \otimes H|1\rangle. \tag{9.51c}$$

Thus, the effect of U_f on this special product is simply to multiply by a phase factor $(-1)^{f(j)}$. Although overall phase factors are not physically observable, the action of U_f is not restricted to products and the phase factor becomes extremely significant when U_f acts on a superposition.

Note that no special properties of $|j\rangle$ have been used; what is important is only that the range of f is $\{0, 1\}$. We could replace $|j\rangle$ by a multiqubit state $|J\rangle = |j_1 j_2 \ldots j_n\rangle$ as long as the function $f : J \mapsto \{0, 1\}$ has range $\{0, 1\}$.

Now consider the effect of U_f on the product state

$$(H \otimes H)|0, 1\rangle = \frac{1}{2}(|0\rangle + |1\rangle) \otimes (|0\rangle - |1\rangle). \tag{9.52}$$

It is not necessary to explicitly consider the effect of U_f on all four products. Since

$$U_f(H \otimes H)|0, 1\rangle = \frac{1}{\sqrt{2}}(U_f|0\rangle \otimes H|1\rangle + U_f|1\rangle H|1\rangle) \tag{9.53a}$$

$$= \frac{1}{\sqrt{2}}((-1)^{f(0)}|0\rangle \otimes H|1\rangle + (-1)^{f(1)}|1\rangle \otimes H|1\rangle$$

$$= \begin{cases} (-1)^{f(0)}\dfrac{1}{\sqrt{2}}(|0\rangle + |1\rangle) \otimes H|1\rangle & \text{if } f(0) = f(1) \\ (-1)^{f(0)}\dfrac{1}{\sqrt{2}}(|0\rangle - |1\rangle) \otimes H|1\rangle & \text{if } f(0) \neq f(1) \end{cases}$$

$$= \begin{cases} (-1)^{f(0)}(H \otimes H)|0, 1\rangle & \text{if } f(0) = f(1) \\ (-1)^{f(0)}(H \otimes H)|1, 1\rangle & \text{if } f(0) \neq f(1) \end{cases}. \tag{9.53b}$$

Thus, a measurement on the first qubit suffices to distinguish the two cases. One can summarize the algorithm as follows:

1. With the QC initialized in the state $|00\rangle$, act with $I \otimes \sigma_x$ to convert it to the state $|01\rangle$.
2. Act on both bits with the Hadamard transform $H \otimes H$.
3. Effectively evaluate the function in parallel with a single call to the "oracle" by acting with the unitary operator U_f.
4. Apply the Hadamard transform to both bits of the result. This leaves the QC in the state $\begin{cases} (-1)^{f(0)}|0, 1\rangle & \text{if } f(0) = f(1) \\ (-1)^{f(0)}|1, 1\rangle & \text{if } f(0) \neq f(1) \end{cases}$.
5. Make a measurement to determine if the first bit is 0 or 1.

Thus, a quantum computation can distinguish between the two types of functions with only one application of U_f, while a classical algorithm requires two-function evaluations. Although this may seem a rather small advantage in an artificial problem, it does establish that there is something that a QC can do more efficiently. Extensions and modifications of the Deutsch–Jozsa problem have been considered and used to provide additional demonstrations of the potential power of quantum computation. (See, e.g, Bernstein and Vazirani.[20]) More recently, Nathanson[104] showed that the Deutsch–Jozsa algorithm could be adapted to solve a problem that arises in models of the Internet.

9.4.2 Grover's algorithm

9.4.2.1 Introduction

Grover's algorithm[52,53] performs an unsorted search for a target state $|J_T\rangle$. It addresses the measurement question head-on by constructing an operation whose action incrementally increases the amplitude of the coefficient of the target state $|J_T\rangle$ in a superposition $\sum_J a_J|J\rangle$. This operation is performed until the QC is in a state with $|a_{J_T}|^2 > \frac{1}{2}$ so that a measurement has more than a 50% chance of identifying the target state.

For the purpose of explaining the algorithm, we assume the QC is initially in a superposition of the form

$$|\Psi\rangle = \frac{1}{\sqrt{M}} \sum_{J \in S} |J\rangle, \tag{9.54}$$

where S is a subset of the binary n-tuples with $M \leq 2^n$ elements. We might interpret the state $|J\rangle = |j_1 j_2 \ldots j_n\rangle$ as representing (or encoding)

1. A tag, such as a license plate or phone number, in m bits and the associated name in the remaining $n - m$, or
2. A candidate solution to a problem, such as factoring, whose validity can easily be checked.

9.4.2.2 The Grover oracle

We assume that an efficient process for determining whether J satisfies the requisite condition can be constructed; that this process yields an output $f(J)$, which is 1 or 0, depending on whether the condition is satisfied; and that the QC has an additional register bit whose state is changed from $|k\rangle$ to $|k \oplus f(J)\rangle$. The net result of this process is called the Grover oracle, G, and its action is equivalent to the unitary operation

$$G(|J\rangle \otimes H|1\rangle) = (-1)^{f(J)}|J\rangle \otimes H|1\rangle. \tag{9.55}$$

The analysis showing that G has this effect on states of the form $|J\rangle \otimes H|1\rangle$ is virtually identical to that used in Eq. (9.51) for the Deutsch–Jozsa algorithm. Because the register bit does not play an explicit role in what follows, it will be omitted. Now, $(-1)^{f(J)} = -1$ if and only if J is the target state. Therefore, when G acts on a superposition,

$$G \sum_J a_J|J\rangle = \sum_J (-1)^{f(J)} a_J|J\rangle, \tag{9.56}$$

its effect is to mark the target state by changing the sign of its coefficient so that $a_{J_T} \mapsto -a_{J_T}$. Thus G can be written as $G = I - 2|J_T\rangle\langle J_T|$. However, it is *not necessary to know* J_T *in advance* to construct G. Nor is it necessary to check each J individually. If a sequence of unitary operations (as described) yields Eq. (9.55) on a set of basis vectors, then the same operations on an arbitrary superposition have the effect of Eq. (9.56).

9.4.2.3 The algorithm

For any vector $|\phi\rangle$, $U_\phi \equiv I - 2|\phi\rangle\langle\phi|$ is the unitary operator that multiplies $|\phi\rangle$ by -1 and acts like the identity on its orthogonal complement. Geometrically, this corresponds to reflecting an arbitrary vector across the hyperplane orthogonal to $|\phi\rangle$. Grover's algorithm uses the repeated application of the product $U_{\Psi^\perp}G = -U_\Psi G$, with Ψ given by Eq. (9.54). Since this is a product of two reflections, the result is a rotation. Thus, we can restrict attention to the plane orthogonal to the rotation axis. This is spanned by $|J_T\rangle$ and $|\Psi\rangle$ (and the vector $|\Psi^\perp\rangle$ is uniquely defined by taking the orthogonal complement in this subspace).

We now write

$$|\Psi\rangle = \frac{1}{\sqrt{M}}|J_T\rangle + \sqrt{\frac{M-1}{M}}|J_T^\perp\rangle, \tag{9.57}$$

where $|J_T^\perp\rangle = \frac{1}{\sqrt{M-1}}\sum_{J\neq J_T}|J\rangle$, and note that

$$-U_\Psi = 2|\Psi\rangle\langle\Psi| - I = I - 2(I - |\Psi\rangle\langle\Psi|) = I - 2|\Psi^\perp\rangle\langle\Psi^\perp| = U_{\Psi^\perp}. \tag{9.58}$$

Thus, the actions of $-U_\Psi = 2|\Psi\rangle\langle\Psi| - I$ and G can then be described as reflections across $|\Psi\rangle$ and $|J_T^\perp\rangle$, respectively, as shown in Fig. 9.7. Let θ denote the angle between $|\Psi\rangle$ and $|J_T^\perp\rangle$. Then the net effect of the composite operation $-U_\Psi G = U_{\Psi^\perp}G$ is a rotation in the plane by an angle 2θ. Thus, L applications take

$$|\Psi\rangle \mapsto (-U_\Psi G)^L|\Psi\rangle = \cos(2L\theta)|J_T\rangle + \sin(2L\theta)|J_T^\perp\rangle. \tag{9.59}$$

When $\frac{\pi}{4} < 2L\theta < \frac{3\pi}{4}$, a measurement would yield J_T with probability greater than $\frac{1}{2}$. It follows from Eq. (9.57) that $\theta = \tan^{-1}(1/\sqrt{M-1}) \approx 1/\sqrt{M}$ so that $L \approx \frac{3\pi}{8\sqrt{M}}$ applications of $-U_\Psi G$ suffice.

9.4.2.4 Caveats

Preparing a superposition of the form of Eq. (9.54), which incorporates the correlations between tags and names, requires considerable resources, defeating the purpose of the algorithm. Therefore, the algorithm is used in a somewhat different manner. We chose the description in Sec. 9.4.2.1 because it is easy to envision a sequence of unitary operations that have the desired effect.

In practice, one expects to begin with the QC in the state $|00\ldots0\rangle$ and act on it with the Hadamard transform $H^{\otimes n}$ to obtain the evenly weighted superposition

$$\Psi = \sum_{J=j_1 j_2 \ldots j_n} |J\rangle = \sum_{J=0}^{2^n-1} |J\rangle \tag{9.60}$$

of all states in the computational basis. For applications of type 2, the only effect is to change the size of the set from M to 2^n. For applications of type 1, we would now have all possible names associated with every tag, which would be quite useless. Instead, we assume that J now denotes *only* the name and not the tag. Identification requires a more complex oracle process. The tag is stored in an m-qubit register, and the correlated list of tags and names read into another register, as shown in Fig. 9.8. The action of the oracle is now to output 1 if (and only if) the tags match and the effect is to multiply the coefficient of the corresponding name (encoded in the state $|J\rangle$) by -1. It might appear that implementing such an oracle would require that the list be stored in a quantum state of the form of Eq. (9.56). However, it turns out that this oracle process can be implemented using a classical memory for the list, provided that a quantum addressing procedure is available.

For a discussion of how this might be done, see Sec. 6.5 of Ref. 107. The complexity of the oracle process does raise questions about the practicality of the Grover algorithm for actually performing searches. It may be more useful for applications of type 2.

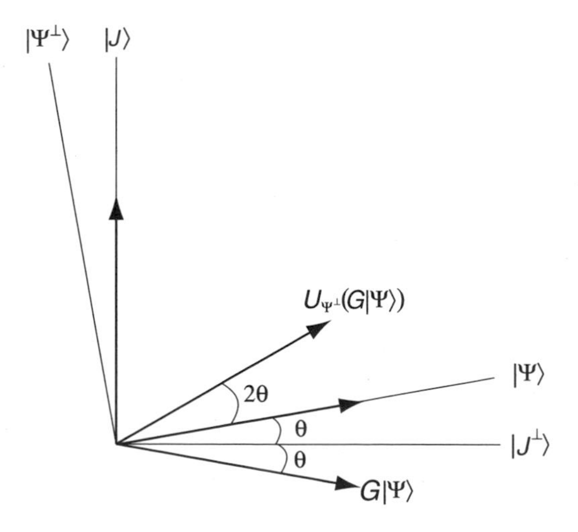

Figure 9.7 Grover diagram.

name	tag	tag-k	name-k
n	m	$m+n$	

Figure 9.8 Register structure for Grover search.

9.4.3 Period finding via the QFT

9.4.3.1 Introduction

Shor[131,132] gave an algorithm for factoring large numbers by reducing this problem to one of finding the period of a function. Since then quantum algorithms for other number theoretic problems have been found by reducing them to period finding. Some, most notably Hallgren's algorithm[54] for solving what is known as Pell's equations, require extending Shor's result to functions with an irrational period.[55] Because quantum computation plays a role only in the period-finding part of these algorithms, we focus on that and omit number theoretic considerations.

From the standpoint of computational complexity, neither Shor's algorithm nor the related number theory problems prove that a QC can solve a problem exponentially faster than a classical one, because it has not been demonstrated that any of these problems require exponential time on a classical computer. However, the ability of a QC to provide exponential speed-up was shown earlier by Simon[139] for the closely related problem of period finding for a function $f : \mathbf{Z}_2^n \mapsto \mathbf{Z}_2^n$. Indeed, the structure of Simon's and Shor's algorithms are virtually identical except for the interpretation of a binary n-tuple as the binary representation of an integer in the latter.

The brief presentation that follows draws heavily on that in Jozsa's review[73] to which the reader is referred for more details. Indeed, this review is highly recommended for its progressive treatment that begins with the Deutsch Jozsa algorithm, builds on it to explain Simon's and Shor's algorithms, and explains how the last two fit into the common framework of the Abelian hidden subgroup problem. Jozsa recently wrote another article[74] on Hallgren's algorithm, and the extension of Shor's algorithm to functions over **R** with an irrational period.

9.4.3.2 Shor's algorithm

Now let $N = 2^n \gg M$ and suppose that we have a function $f : \mathbf{Z}_N \mapsto \mathbf{Z}_M$ such that $f(x + r) = f(x)$. [An example of such a function is $f(x) = y^x \bmod M$ with $y < M$ and coprime to M. In this case, r is the smallest integer for which $y^r = 1 \bmod M$.] For simplicity, we also assume that N is a multiple of r. Now suppose that $2^{m-1} \leq M < 2^m$ and define the following operator on $\mathbf{C}^{2^n} \otimes \mathbf{C}^{2^m}$

$$U_f : |J\rangle \otimes |K\rangle \mapsto |J\rangle \otimes |K + f(J)\rangle, \tag{9.61}$$

where $|J\rangle = |j_1 j_2 \ldots j_n\rangle$; $|K\rangle = |k_1 k_2 \ldots k_m\rangle$; addition is mod M; and J and K are interpreted as integers in $\mathbf{Z}_N$ and $\mathbf{Z}_M$, respectively.

If this operator is applied to a QC in the state $H^{\otimes n}|0\rangle \otimes |0\rangle$,

$$U_f : H^{\otimes n}|0\rangle \otimes |0\rangle = 2^{-n/2} \sum_J |J\rangle \otimes |0\rangle \mapsto 2^{-n/2} \sum_J |J\rangle \otimes |f(J)\rangle. \tag{9.62}$$

If one then makes a measurement on the last m qubits, the process essentially selects a state $|K\rangle$ in the last m bits and leaves the computer in a state that is a superposition corresponding to all integers J mapped to K, i.e., those for which $f(J) = K$. The QC is then in the state

$$\sum_{J:f(J)=K} |J\rangle \otimes |K\rangle = \sum_{\lambda} |x + \lambda r\rangle \otimes |f(x)\rangle. \tag{9.63}$$

Unfortunately, a measurement on the first n bits in this state would simply select one of the integers $J = x + \lambda r$, which map to $f(K)$, without revealing any information about x, λ, or r. Because the QC is left in a state of the form $|J\rangle \otimes |K\rangle$, further measurements will give the same result. If one repeats the entire process, one may get a different integer K'. Then a measurement on the first n bits gives a number of the form $J' = x' + \lambda' r$. This provides no more information than a pair of random integers in $\mathbf{Z}_N$.

To extract additional information, one applies the QFT, given by Eq. (9.76), to the first n bits in Eq. (9.63) before doing the next measurement.

$$\mathcal{F}\Big(\sum_{\lambda} |x + \lambda r\rangle\Big) \otimes |f(x)\rangle = \sum_{\lambda}\sum_{L} e^{(2\pi i)(x+\lambda r)(L/N)}|L\rangle \otimes |f(x)\rangle \tag{9.64}$$

$$= e^{(2\pi i)\frac{x}{N}} \sum_{L}\Big\{\sum_{\lambda} e^{[2\pi(L/\mu)i]\lambda}\Big\}|L\rangle \otimes |f(x)\rangle$$

$$= e^{(2\pi i)x/N} \sum_{t} |t\,\mu\rangle \otimes |f(x)\rangle. \tag{9.65}$$

The analysis above used the assumption $N = r\mu$, but the key point is that

$$\sum_{\lambda} e^{[2\pi(L/\mu)i]\lambda} = \begin{cases} 0 & \text{if } \dfrac{L}{\mu} \text{ is not an integer} \\ \nu & \text{if } L = t\mu \text{ for some integer } t \end{cases}. \tag{9.66}$$

Thus, the QFT changes the state of the QC from a superposition of states of the form $|x + \lambda r\rangle$, with x and r fixed, to a superposition of states of the form $|t\mu\rangle$. With the QC in the state of Eq. (9.65), a measurement on the first n bits yields a multiple of μ.

Moreover, applying the QFT to the superposition of Eq. (9.63) always yields the state $\sum_t |t\mu\rangle$, i.e., the output is independent of x. A measurement on the first n bits is now guaranteed to yield a multiple of μ. If the process is repeated, one again obtains a multiple of μ. From this, one can eventually determine $r = \frac{N}{\mu}$. In fact, it can be shown that $O(n)$ repetitions suffice.

In general, it is not true that $N = 2^n = r\mu$. However, it can be shown that this analysis is approximately correct when $n > 2\log M$.

It is not actually necessary to perform the first measurement that generates Eq. (9.63). It suffices to observe that

$$\sum_J |J\rangle \otimes |f(J)\rangle = \sum_K \sum_{J:f(J)=K} |J\rangle \otimes |K\rangle, \tag{9.67}$$

so that

$$\mathcal{F}\sum_J |J\rangle \otimes |f(J)\rangle = \left(\sum_t |t\mu\rangle\right) \otimes \left(\sum_{x_K} e^{(2\pi i)(x_k/N)} |f(x_K)\rangle\right), \tag{9.68}$$

where the notation reflects the fact that one can associate an $x \equiv x_K$ with each K in the range of f.

9.4.3.3 Abelian hidden subgroup problem

A group theoretic interpretation of Shor's algorithm has led to similar algorithms for other problems in algebra and number theory. For a given f with period r the set

$$S_f = \{J : f(J) = f(0)\} = \{\lambda r : \lambda \in \mathbf{Z}\} \tag{9.69}$$

is a subgroup of $\mathbf{Z}_N$, the integers mod N, whose cosets have the form

$$S_K = \{J : f(J) = K\} = \{x_K + \lambda r : \lambda \in \mathbf{Z}\}. \tag{9.70}$$

Thus, the two sums in Eq. (9.67) can be interpreted as a sum over cosets, and a sum of elements within the coset, respectively; and Eq. (9.65) implies that the QFT acting on an evenly weighted superposition of elements in a coset is independent of the coset. Indeed, with $\mu = \frac{1}{r}$, the set $\{t\mu : t \in \mathbf{Z}\}$ can be interpreted as the factor group $\frac{G}{S_f}$. Finding the period of f is equivalent to determining the subgroup S_f. After the QFT has been applied, a measurement on the QC yields an element of the factor group $\frac{G}{S_f}$. This is repeated until enough elements of $\frac{G}{S_f}$ are known to enable one to determine S_f.

Shor's procedure can be extended to other instances of the Abelian hidden subgroup problem with the QFT replaced by an unitary map equivalent to the Fourier transform on groups. See Refs. 73 and 107 for more information.

One can summarize the Abelian hidden subgroup algorithm as follows:

1. Initialize the computer in the state $|0\ldots0\rangle \otimes |0\ldots0\rangle$ where the first product has n bits and the second m.
2. Apply the Hadamard transform to the first n bits to put the QC in the state $H^{\otimes n}|0\rangle \otimes |0\rangle = 2^{-n/2}\sum_J |J\rangle \otimes |0\rangle$.
3. Apply a unitary operator associated with the subgroup S to convert the QC to a state of the form $(2^{-n/2}\sum_J |J\rangle) \otimes |f(J)\rangle$.

4. Convert the QC to a state in which the first n bits have the form $\sum_{J \in S} |K_0 + J\rangle$. [This can be done either by a measurement on the last m bits, or by rewriting, as in Eq. (9.70). In the latter case, the QC is actually in a superposition of states of the desired from.]
5. Apply a suitable quantum Fourier transform to change the state of the first n qubits to the form $\sum_{K \in G/S} |K\rangle$.
6. Make a measurement in the computational basis to identify an element of $\frac{G}{S}$.
7. Repeat the entire process until one has enough elements to identify $\frac{G}{S}$.

9.4.3.4 Simon's algorithm

We now illustrate the group theoretic view by describing Simon's algorithm as a special case. In this case, the n-qubit state $|j_1 j_2 \dots j_n\rangle$ is interpreted as an element of $\mathbf{Z}_2^n \equiv Z_2^{\otimes n}$ rather than as an element of Z^{2^n}. Then $f : \mathbf{Z}_2^n \mapsto \mathbf{Z}_2^n$ is a $2:1$ function and the period $(u_1 u_2 \dots u_n)$ is an element of $\mathbf{Z}_2^n$, i.e.,

$$f[(j_1 j_2 \dots j_n) \oplus (u_1 u_2 \dots u_n)] = f(j_1 j_2 \dots j_n), \qquad (9.71)$$

where $\oplus$ denotes pointwise binary addition. The subgroup

$$S_f = \{(00\dots0),\ (u_1 u_2 \dots u_n)\} \qquad (9.72)$$

has two elements, and its cosets have the form

$$\{(j_1 j_2 \dots j_n),\ (j_1 j_2 \dots j_n) \oplus (u_1 u_2 \dots u_n)\}. \qquad (9.73)$$

The Hadamard transform $H^{\otimes n}$ now plays the role of the QFT. When applied to a superposition corresponding to elements in the set (9.73), it yields

$$\sum_{\mathbf{u}^{\perp}} |k_1 k_2 \dots k_n\rangle, \qquad (9.74)$$

where, in this case, the factor group G/S_f is the orthogonal complement of $(u_1 u_2 \dots u_n)$ as a vector in $\mathbf{Z}_2^n$, i.e.,

$$\mathbf{u}^{\perp} = \Big\{ (j_1 j_2 \dots j_n) : (j_1 j_2 \dots j_n) \cdot (u_1 u_2 \dots u_n) = \sum_i j_i u_i = 0 \bmod 2 \Big\}. \qquad (9.75)$$

Then, making a measurement yields an element of $\mathbf{u}^{\perp}$. If one repeats this process [which typically takes $O(n^2)$ times] until one has $n-1$ linearly independent vectors in $\mathbf{u}^{\perp}$, one has sufficient information to find $\mathbf{u}$.

9.4.4 Implementing the quantum Fourier transform

The QFT can be defined on basis vectors as $\mathcal{F}|K\rangle = \sum_J e^{JK(2\pi i)/2^n}|J\rangle$. It follows that for an arbitrary vector of the form $\sum_K x_K|K\rangle$

$$\mathcal{F}\left(\sum_K x_K|K\rangle\right) = \sum_J (\mathcal{F}x)_J|J\rangle, \tag{9.76}$$

where $(\mathcal{F}x)_J = \sum_K e^{JK(2\pi i)/2^n} x_K$ denotes the usual DFT on a vector x_K of length $N = 2^n$.

The key to implementing both the QFT and the FFT is the identity

$$\sum_{k_1 \dots k_n} e^{2\pi i[(j_1 2^{n-1}+\dots+2j_{n-1}+j_n)(k_1 2^{n-1}+\dots+2k_{n-1}+k_n)]/2^n}|k_1 \dots k_n\rangle \tag{9.77}$$
$$= [|0\rangle + e^{0.j_n(2\pi i)}|1\rangle][|0\rangle + e^{0.j_{n-1}j_n(2\pi i)}|1\rangle]\dots[|0\rangle + e^{(2\pi i)0.j_1 j_2 \dots j_n}|1\rangle],$$

where

$$0.j_k j_{k+1} \dots j_n = j_k 2^{-1} + j_{k+1} 2^{-2} + \dots + j_n 2^{-n+k-1} \tag{9.78}$$
$$= \frac{2^{n-1} j_k + 2^{n-2} j_{k+1} + \dots + j_n}{2^n}.$$

We first consider implementing the action

$$|j_1 j_2 \dots j_m\rangle \mapsto |0\rangle + e^{0.j_1 j_2 \dots j_m(2\pi i)}|1\rangle, \tag{9.79}$$

which can be done using the Hadamard transform and the controlled phase gate Q_j, which takes

$$|0\rangle \otimes |j\rangle \mapsto \frac{1}{\sqrt{2}}(|0\rangle + |1\rangle) \otimes |j\rangle, \tag{9.80a}$$

$$|1\rangle \otimes |j\rangle \mapsto \frac{1}{\sqrt{2}}(|0\rangle + e^{2\pi i/2^j}|1\rangle) \otimes |j\rangle. \tag{9.80b}$$

One can then verify that

$$H \otimes I^{m-1}|j_1 j_2 \dots j_m\rangle = \frac{1}{\sqrt{2}}[|0\rangle + e^{(2\pi i)0.j_1}|1\rangle] \otimes |j_2 \dots j_m\rangle \tag{9.81a}$$

$$Q_2 H \otimes I^{m-1}|j_1 j_2 \dots j_m\rangle = Q_2 \frac{1}{\sqrt{2}}[|0\rangle + e^{(2\pi i)0.j_1}|1\rangle] \otimes |j_2 \dots j_m\rangle\rangle \tag{9.81b}$$

$$= \frac{1}{\sqrt{2}}[|0\rangle + e^{(2\pi i)0.j_1 j_2}|1\rangle] \otimes |j_2 \dots j_m\rangle \tag{9.81c}$$

$$Q_m \dots Q_2 H \otimes I^{m-1} |j_1 j_2 \dots j_m\rangle$$
$$= \frac{1}{\sqrt{2}}(|0\rangle + e^{(2\pi i)0.j_1 j_2 \dots j_m}|1\rangle) \otimes |j_2 \dots j_m\rangle, \tag{9.81d}$$

where it is understood that Q_j acts on the first bit (the control bit) and the jth bit. Similarly,

$$Q_m \dots Q_3 H \otimes I^{m-2} |j_2 \dots j_m\rangle = \frac{1}{\sqrt{2}}[|0\rangle + e^{(2\pi i)0.j_2 \dots j_m}|1\rangle] \otimes |j_3 \dots j_m\rangle, \tag{9.82}$$

where the second bit is now the control bit. Thus, the QFT can be implemented by first using swapping operations to convert $|j_1 j_2 \dots j_n\rangle \mapsto |j_n j_{n-1} \dots j_1\rangle$ and then applying the process above to the first m bits with $m = 1 \dots n$. Each factor in Eq. (9.77) requires m gates, one Hadamard, and $m - 1$ controlled phase gates. Therefore, the QFT can be implemented using a total of $\sum_{m=1}^{n} m = \frac{n(n+1)}{2}$ gates plus $\frac{n}{2}$ swap gates, which yields $O(n^2)$ or $O(\log N)^2$ operations.

Although it may appear that the QFT requires $O(\log N)^2$ steps for the part of the classical FFT, which requires $O(\log N)$, the usual estimates for the FFT requires $O(N \log N)$, and hides the fact that the number of operations also depends on the accuracy. If d is the number of binary digits, one could say it requires $O(d \log N)$ operations. In the period-finding algorithms to which the QFT is applied, d is not constant, but $O(\log N)$. Thus the net result $O(\log N)^2$ can be interpreted as $O(d \log N)$, consistent with the classical FFT.

9.5 Other types of quantum information processing

9.5.1 Quantum key distribution

In QKD, one uses quantum particles to generate a secret code in the form of a string of 0s and 1s that can then be used as a classical one-time key pad. One approach[39] could be described using a variant of the EPR experiment in Sec. 9.3.2. In this setup, the two parties (traditionally known as Alice and Bob, but here called Sue and Tom to avoid confusion with the box labels) exchange e-mail telling which box A, B, or C each clicked on. They then discard all data for which they clicked on different boxes, and use the data from the remaining times. Because they retain only data from times when they chose the same box, their results are perfectly correlated. They can then apply an agreed upon procedure (e.g., Sue's W-L record corresponds to $0, 1$ and Tom's to $1, 0$) to their data to obtain a classical binary string. Even if their e-mail is intercepted, the eavesdropper (traditionally known as Eve, but here called Irv) only learns which boxes they chose; not whether they obtained a W or L.

It is not, however, at all necessary to use pairs of entangled particles for QKD. Several single-qubit protocols exist, and experiments based on them have given impressive results. We describe two of these before discussing some of the additional ingredients needed for security in both procedures.

Table 9.3 B92.

Encoding			Probabilities		
	0	1	Tom\Sue	0	1
Sue	$\uparrow$	$\nearrow$	0	$\frac{1}{2}$	0
Tom	$\nwarrow$	$\rightarrow$	1	0	$\frac{1}{2}$

In the B92 protocol,[12] Sue uses polarized photons to encode 0 and 1 as in Eq. (9.7a), while Tom measures the polarization using the encoding in Eq. (9.7b). The left part of Table 9.3 summarizes the encoding scheme; the right part the probability that Tom gets a signal when Sue sends as indicated in the top row and Tom measures as in the left column. If Sue sends a random string of 0s and 1s and Tom measures randomly, he can expect to get a signal about 25% of the time. What is certain is that he can never get a signal when Sue sends a 0 and he measures a 1 in their respective encodings. Thus, Tom receives a signal only when he and Sue both choose 0 or both choose 1. Tom then uses a public channel to tell Sue which measurements yielded signals. They retain the data corresponding to those measurements and discard the rest. Assuming that they have an accurate mechanism for labeling and recording their data, this yields a suitable secret key. An eavesdropper on the public channel can learn which signals are being used, but only Sue knows what was sent and Tom what was measured.

In the BB84 protocol,[13] orthogonal bases are used, but Sue randomly chooses between the bases Eqs. (9.1) and (9.3). Tom also randomly chooses between these same bases when making measurements. After publicly exchanging information about which bases they used, they retain the data from times in which they chose the same basis and discard the rest.

Security in the BB84 and B92 protocols requires some additional ingredients. First, one must suppose that they start with a short shared private key that can be used to authenticate any messages they exchange (e.g., to preclude an impersonator). Thus, these are more correctly termed procedures for "privacy amplification" rather than key distribution.

Next, they must check for the presence of an eavesdropper. For this, it is important that nonorthogonal states were involved. Suppose that BB84 is used. Irv can measure the signal, but does not know which basis Sue used. Therefore, Irv cannot reliably transmit the signal to Tom in the same basis. This will introduce errors into the supposedly perfectly correlated results when Sue and Tom use the same bases. Now if Sue and Tom sacrifice part of their shared key to perform an error detection procedure, they will be able to learn if significant eavesdropping has occurred.

Actually proving that these protocols can be used for what is called "unconditional security" in an idealized setting is not at all trivial. Nevertheless, this has been done for the BB84 and, more recently, the B92 protocol. Although the first arguments were rather complex, Shor and Preskill[137] used ideas from quantum error correction to give a simple argument for BB84, which has since been generalized. (Unconditional security does not mean absolute in the sense that no information can be obtained by an eavesdropper. It means that acceptable bounds on the eavesdropper's information assume that the adversary has access to any device permitted by physical principles, whether or not such devices have been built.)

It is an indication of how successful experiments on QKD have been that much current research is now concerned with imperfections in the experiments. In addition to efforts to minimize these, analyzing the impact on security of imperfect scenarios is an extremely active area of research. Recently, some proofs of unconditional security have been extended to nonideal settings.[51,70]

A number of experimental groups have now demonstrated the feasibility of QKD in various circumstances. For example, Hughes et al.[69] have demonstrated single photon protocol QKD in free space in daylight over 10 km under conditions that indicate that free-space QKD will be practical over much longer ranges. In particular, their work suggests that ground-to-satellite implementation of QKD are quite promising. For a detailed survey of both theory and experiment through 2001, see Ref. 44. More recently, several groups have described QKD protocols that use continuous variables. See Ref. 138 for a brief discussion and references to earlier work.

9.5.2 Quantum cryptography

QIP has the potential to provide both new methods for breaking codes and new methods of protecting data. For example, Shor's algorithm for factoring large numbers is a potential threat to the security of the RSA (Rivest, Shamir, Adelman) system currently in use. On the other hand, QKD provides new methods for generating secure one-time keypads. Although quantum information processors are potentially more powerful than classical ones, quantum cryptographic procedures are neither more nor less powerful than classical ones. Quantum theory offers new methods for breaking codes, eavesdropping, and interfering with messages, as well as new methods of encryption. The study of new cryptographic protocols is an active area of research for which we briefly mention only a few examples.

The first proposals for quantum cryptography, including a procedure for money that could not be counterfeit, seem to have been made by Wiesner about 1970, in work that was not published. Its eventual publication[151] in 1983 was facilitated by Bennett, whose appreciation for these ideas led to the first QKD proposal with Brassard.[13]

The next area to be actively explored was quantum bit commitment. In this process, Sue (one of two mutually untrusting parties) encodes information in a qubit that is sent to the other party, Tom, but cannot be read by him. When the bit is subsequently revealed to Tom, Sue is also required to prove that it has not been tampered with. Early in the development of QIP, a number of quantum bit commitment schemes were proposed. However, in 1997 it was shown by Mayers[100] that unconditionally secure quantum bit commitment is impossible. (See also, the independent proof in Ref. 96, and Ref. 23 for an overview.) Using ideas first proposed by Crépeau and Kilian,[33] some progress has been made on procedures for secure quantum bit commitment in the presence of noise.[152] Since cheating in quantum bit commitment generally exploits entanglement, and sufficient noise can be shown to break entanglement, it is not unexpected that noise can be used to enhance quantum cryptography.

Secret sharing is a method for distributing information among M parties so that no unauthorized subgroup (typically, $M-1$ of them) can use it. In a variant known as data hiding, $M-1$ parties cannot use the information even if classical communication among them is permitted. There is now extensive literature on various quantum protocols for secret sharing and related topics, such as secure distributed computing. See, e.g., Refs. 32, 37, and 49.

Another area with considerable practical interest is the development of quantum methods for authentication[50] or "digital signatures." It was shown in Ref. 6 that any scheme to authenticate quantum messages must also encrypt them. (In contrast, one can authenticate a classical message while leaving it publicly readable.)

QIT also offers the possibility of cryptographic methods that have no classical counterpart. For example, Gottesman[49] proposed using quantum particles to encrypt classical data, after which it could not be cloned.

In addition to generating new quantum methods for cryptography, QIT has also given new insights into classical procedures. Recently, Kerenidis and de Wolf[77] even used quantum methods to prove something new about classical codes. They use a quantum argument to show that what are known as locally decodable codes (i.e., codes from which information can be extracted from small pieces) must be exponentially long when only two classical queries are permitted. Their argument is based on an equivalence between two classical bits and one quantum bit in certain contexts. This equivalence also plays a role in Secs. 9.5.3 and 9.5.4.

9.5.3 Dense coding

We describe this process using a fictitious scenario in which Sue is spying behind enemy lines and wants to let Tom know the direction from which to expect the next attack—N, S, E, or W. Two classical bits would be needed to transmit this information. Now Sue wants to minimize the number of signals she transmits to avoid detection. Moreover, Sue and Tom share a pair of photons in an entangled Bell state, say $|\beta_0\rangle$, and have agreed on a correspondence between the four Bell states and the four directions N, S, E, and W. When Sue wants to send information, she applies one of I, σ_x, σ_y, or σ_z to her photon. Sue's operation converts the state of the entangled pair to $|\beta_k\rangle$, as in Eq. (9.33). If Sue then sends her photon to Tom, he will have a pair of entangled photons on which he can make a Bell measurement to learn the direction of the attack.

Sue has used a single qubit to encode and transmit information that would require two classical bits. This process is known as "dense coding." It does not contradict the Holevo bound on accessible information because Tom's measurement requires a pair of photons—one sent by Sue, and another that Tom has from the start.

9.5.4 Quantum teleportation

Sue and Tom again share the entangled Bell state $|\beta_0\rangle$. However, Sue now wants to transmit the quantum information encoded in the state $|\phi\rangle_R = a|0\rangle + b|1\rangle$ to Tom. This process uses three particles, two, initially located in Sue's lab, in the Hilbert spaces $\mathcal{H}_R$ and $\mathcal{H}_S$ and another, in the Hilbert spaces $\mathcal{H}_T$, with Tom. The full system is described by the Hilbert space $\mathcal{H}_R \otimes \mathcal{H}_S \otimes \mathcal{H}_T$ and the initial state of the system is $|\phi\rangle_R \otimes |\beta_0\rangle_{ST}$, where subscripts indicate subsystems.

Sue and Tom now use the following procedure:

- Sue makes a Bell measurement (i.e., one that can distinguish between the four states $|\beta_k\rangle$ in Table 9.2) on the composite system $\mathcal{H}_{RS}$. By the identity Eq. (9.37) this will leave the system in one of the four states $|\beta_k\rangle_{RS} \otimes \sigma_k|\phi\rangle_S$ and Sue will learn the value of k.
- Sue uses a classical communication channel to transmit the value of k to Tom. This requires the transmission of two classical bits to distinguish among 0, 1, 2, and 3.
- Having learned k, Tom performs the operation σ_k on the qubit $\sigma_k|\phi\rangle_S$ in his possession, converting it to the state $|\phi\rangle_S$ since $\sigma_k^2 = I$.

The net result is that the information encoded in $|\phi\rangle$ has been transmitted from Sue to Tom, without actually sending the qubit $|\phi\rangle_R$. The information originally encoded in a qubit in subspace $\mathcal{H}_R$ is now encoded in a qubit in the subspace $\mathcal{H}_T$ which could, in principle, be quite far away. It should be emphasized that *only* information has been teleported; not the physical qubits (or any form of matter).

Note that the state $|\phi\rangle$ has not been cloned. Sue's measurement destroys the qubit $|\phi\rangle_R$ before Tom has the information needed to construct $|\phi\rangle_T$. Moreover, information is not transmitted instantaneously. Sue must communicate two classical bits that cannot reach Tom faster than the speed of light.

What application might there be for this procedure? The answer depends on the various purposes for which quantum information might be used. Procedures, such as the single-photon QKD protocols described in Sec. 9.5.1, require the transmission of quantum information. However, quantum states are easily corrupted, making faithful transmission over long distances difficult. One way to overcome this is to use teleportation to construct "quantum repeaters"[24] along classical channels.

What is important in both the physical protocol and the mathematical argument is that the Hilbert spaces $\mathcal{H}_S$, $\mathcal{H}_S$, and $\mathcal{H}_R$ are all isomorphic to $\mathbf{C}_2$. It is *not* necessary to use identical particles. The procedure could, in principle, be applied with R representing the spin of a proton, S, T the spin of electrons (or even a positron-electron pair). In such a case, the information encoded in the spin of a proton would be transferred to the spin of an electron. Although this might be experimentally*

*Although progress on achieving entanglement between an electron and a spin-$\frac{1}{2}$ nucleus was recently reported.[101]

difficult, it does demonstrate that the procedure transmits only information, not matter.

9.5.5 Quantum communication

Both quantum teleportation and dense coding are special cases of quantum communication, i.e., the use of quantum particles to transmit information. In the former, the information to be transmitted is quantum, i.e., one transmits the information encoded in the state of a qubit; in the latter, the information is classical and equivalent to that encoded in two classical bits. In both cases, the communication is assisted in the sense that a resource, a pair of shared entangled particles, is used in addition to a communication channel. In fact, with the additional resource of a shared entangled EPR pair, two parties can transmit either one qubit of quantum information by sending two classical bits or two bits of classical information by sending one qubit.

There are many other ways to use quantum particles to transmit both classical and quantum information, with and without additional resources, such as shared entanglement or additional classical communication channels. Even protocols whose practicality for the direct transmission of messages may seem doubtful may have important applications within the full spectrum of QIP. In addition to using teleportation as an ingredient in the construction of quantum repeaters, one might use it to transfer information within a large QC.

It is natural to ask questions about quantum communication similar to those raised by Shannon for classical communication. In the study of quantum information, the concept of a typical sequence is replaced by that of a typical subspace, which was introduced by Schumacher.[127] Generalizations of Shannon's so-called "noiseless" coding theorem have been proved in a variety of circumstances. See, e.g., Refs. 68, 76, and 127 and the discussion in Chapter 12 of Ref. 107.

One can also ask for the maximum rate at which a noisy quantum channel can be used to transmit information. Because of the greater variety of protocols, the theory of quantum channel capacity is much richer than its classical counterpart. This topic requires notation introduced in Sec. 9.6.1, and also comes under the general category of noise. Therefore, it is discussed in Sec. 9.6.2.

It is worth pointing out that entangled particles can be transmitted using a channel capable of sending only one particle at a time. As explained in Appendix 9.D.2, the description of qubits we have been using is incomplete. Suppose one has an entangled pair of particles in the Bell state

$$\frac{1}{\sqrt{2}}[|f(\boldsymbol{x},t),0\rangle \otimes |g(\boldsymbol{x},t),1\rangle + |f(\boldsymbol{x},t),1\rangle \otimes |g(\boldsymbol{x},t),0\rangle], \tag{9.83}$$

where $f(\boldsymbol{x},t)$ describes the (spatial) probability distribution at time t. For example, a situation in which Sue initially has both particles occurs when both $f(\boldsymbol{x},t_0)$ and $g(\boldsymbol{x},t_0)$ have support in Sue's lab. Sending half of the entangled pair to Tom corresponds to a physical process that modifies $f(\boldsymbol{x}, t)$ until, at time t_1, the function $f(\boldsymbol{x}, t_1)$ has support in Tom's lab. The second particle can then be sent (using the same channel) by modifying $g(\boldsymbol{x}, t)$ so that $g(\boldsymbol{x}, t_2)$ also has support in Tom's lab; thus, at time t_2, Tom has the entire entangled pair. Alternatively, after the first step, Tom could perform an action such as $\sigma_k \otimes I$ and then again modify $f(\boldsymbol{x}, t)$ to send the particle back to Sue, who might then make a joint measurement as in the dense coding protocol.

9.6 Dealing with noise

9.6.1 Accessible information

9.6.1.1 The Holevo bound

We now consider the question alluded to after Eq. (9.8). What is the maximum amount of information that can be extracted from n qubits under ideal circumstances? Can a sufficiently clever encoding and measurement permit the extraction of more information than could be encoded in n classical bits? To answer this precisely, we must formalize the process of obtaining information from quantum systems.

Let $\{\rho_i\}$ denote the (possibly mixed) state of a quantum system in dimension d, and let π_j denote the probability that the system is in the state ρ_j. The average state of the system is $\rho = \sum_j \pi_j \rho_j$ and the set $\mathcal{E} = \{\pi_j, \rho_j\}$ is referred to as an ensemble. We are primarily interested in the case of n qubits for which $d = 2^n$.

The most general type of measurement one can make is a POVM (defined at the end of Sec. 9.2.2) of the form $\mathcal{M} = \{F_b\}$ with $F_b > 0$ and $\sum_b F_b = I$. The accessible information associated with a given measurement and ensemble can then be defined as the classical mutual information associated with the (discrete classical) probability distribution $p(j, b) = \pi_j \mathrm{Tr}\, \rho_j F_b$, which can be written as

$$I(\mathcal{E}, \mathcal{M}) = S[\mathrm{Tr} \rho E_b] - \sum_j \pi_j S[\mathrm{Tr} \rho_j E_b], \tag{9.84}$$

where we have used $S[y_b]$ to denote the entropy associated with a classical probability distribution with $y_b = \mathrm{Tr}\, \rho_j E_b$. The *Holevo bound*[59] states that Eq. (9.84) is bounded above by an analogous quantity involving the von Neumann entropy of Eq. (9.16), i.e.,

$$I(\mathcal{E}, \mathcal{M}) \leq S(\rho) - \sum_j \pi_j S(\rho_j) \equiv \chi(\pi_j, \rho_j). \tag{9.85}$$

Since $0 \leq S(\gamma) \leq \log d$, it follows immediately from Eq. (9.85) that

$$I(\mathcal{E}, \mathcal{M}) \leq S(\rho) \leq \log d = \log 2^n = n, \tag{9.86}$$

for n qubits. Thus, one can not obtain more information from n qubits than from n classical bits.

The expression for the quantum mutual information in Eq. (9.85) is known as the "Holevo χ quantity." Equality can be attained in Eq. (9.85) if and only if all of the ρ_j commute.[59,109,123]

9.6.1.2 Relative entropy and mutual information

The Holevo bound can be proved in a variety of ways. We use an approach based on the quantum relative entropy, which is defined as

$$H(\rho, \gamma) \equiv \mathrm{Tr}\rho(\log \rho - \log \gamma). \tag{9.87}$$

One can show that $H(\rho, \gamma) \geq 0$ with equality if and only if $\rho = \gamma$. Although the relative entropy is not a true distance, it is sometimes used as a measure of how different two states are. One expects that noise should make two states harder to distinguish so that

$$H[\Phi(\rho), \Phi(\gamma)] \leq H(\rho, \gamma), \tag{9.88}$$

where Φ is a CPT map, as discussed in Sec. 9.2.4. The inequality (9.88) is a deep property known as the "monotonicity of of relative entropy," and it is closely related to a property of quantum entropy known as strong subadditivity. The reader is referred to Refs. 108, 123, and 147 for a proof and discussion of the properties of entropy and relative entropy for quantum systems.

The mutual information in a mixed state ρ_{AB} on a tensor product space $\mathcal{H}_A \otimes \mathcal{H}_B$ can be defined using the relative entropy as

$$H(\rho_{AB}, \rho_A \otimes \rho_B) = -S(\rho_{AB}) + S(\rho_A) + S(\rho_B), \tag{9.89}$$

where $\rho_A = \mathrm{Tr}_B\, \rho_{AB}$ and $\rho_B = \mathrm{Tr}_A\, \rho_{AB}$ are the reduced density matrixes defined via the partial trace. (See Appendix 9.B.) If we now identify $\mathcal{H}_A$ with $\mathcal{H}$ and let $\mathcal{H}_B = \mathbf{C}^m$, we can formally associate the ensemble $\mathcal{E}$ with the mixed state

$$\rho_{AB} = \sum_{j=1}^{m} \pi_j \rho_j \otimes |j\rangle\langle j| = \begin{pmatrix} \pi_1\rho_1 & 0 & \dots & 0 \\ 0 & \pi_2\rho_2 & \dots & 0 \\ \vdots & & \ddots & \vdots \\ 0 & 0 & \dots & \pi_m\rho_m \end{pmatrix}. \tag{9.90}$$

Moreover, $\rho_A = \sum_j \pi_j \rho_j = \rho$ and $\rho_B = \sum_{j=1}^{m} \pi_j |j\rangle\langle j|$ is the diagonal matrix with nonzero elements π_j. Thus one finds

$$H(\rho_{AB}, \rho_A \otimes \rho_B) = \sum_j \mathrm{Tr}\,(\pi_j \rho_j) \log(\pi_j \rho_j) + S(\rho) + S[\pi_j] \tag{9.91}$$

$$= \sum_j \pi_j \mathrm{Tr}\,\rho_j \log \rho_j + \sum_j \pi_j \log \pi_j + S(\rho) + S[\pi_j]$$

$$= S(\rho) - \sum_j \pi_j S(\rho_j) = \chi(\pi_j, \rho_j). \tag{9.92}$$

9.6.1.3 Proof of Holevo bound

The proof of the Holevo bound requires one more ingredient, the recognition that the result of a POVM can be expressed as a special case of a CPT map.[62,123,156] For the POVM $\mathcal{M} = \{F_b\}$, with $d_{\mathcal{M}}$ elements define the map $\Phi_{\mathcal{M}} : \mathcal{B}(\mathcal{H}) \mapsto \mathcal{D}(\mathbf{C}_{d_{\mathcal{M}}})$ by

$$\Phi_{\mathcal{M}}(\gamma) = \sum_b \mathrm{Tr}\,(\gamma F_b) |b\rangle\langle b|. \tag{9.93}$$

Thus, $\Phi_{\mathcal{M}}$ maps the density matrix γ to the diagonal matrix with elements $\delta_{bc} \mathrm{Tr}\,(\gamma F_b)$. It is a CPT map [which one can verify by observing that $\Phi_{\mathcal{M}}$ can be written in the form of Eq. (9.23) with $A_{kb} = |b\rangle\langle k\sqrt{F_b}|$]; in fact, it is a special type[62] known as "quantum-classical," since it maps mixed quantum states to classical ones.

To prove the Holevo bound of Eq. (9.85), it suffices to observe that

$$I(\mathcal{E}, \mathcal{M}) = S[\mathrm{Tr}\,\rho E_b] - \sum_j \pi_j S[\mathrm{Tr}\,\rho_j E_b] \tag{9.94}$$

$$= H[(\Phi_{\mathcal{M}} \otimes I)(\rho_{AB}), (\Phi_{\mathcal{M}} \otimes I)(\rho_A \otimes \rho_B)] \tag{9.95}$$

$$\leq H(\rho_{AB}, \rho_A \otimes \rho_B) \tag{9.96}$$

$$= S(\rho) - \sum_j \pi_j S(\rho_j) = \chi(\pi_j, \rho_j). \tag{9.97}$$

It is important that the representation of the measurement operation $\Phi_{\mathcal{M}}$ as a CPT map is as a map on $\mathcal{B}(\mathcal{H}_A)$ or, equivalently, as a map of the form $\Phi_{\mathcal{M}} \otimes I_B$ on $\mathcal{B}(\mathcal{H}_A \otimes \mathcal{H}_B)$.

There is another way[156] of using the monotonicity of relative entropy to prove the Holevo bound. It uses the observation that

$$\chi(\pi_j, \rho_j) = \sum_j \pi_j H(\rho_j, \rho), \tag{9.98}$$

from which it follows immediately that

$$I(\mathcal{E},\mathcal{M}) = \sum_j \pi_j H[\Phi_{\mathcal{M}}(\rho_j), \Phi_{\mathcal{M}}(\rho)] \tag{9.99}$$

$$\leq \sum_j \pi_j H(\rho_j, \rho) = \chi(\pi_j, \rho_j). \tag{9.100}$$

Despite the brevity of this argument, we prefer the first because it demonstrates the role of mutual information. On the other hand, the identity of Eq. (9.98) is important because it leads to a very useful characterization[109,130] of the optimal inputs for noisy channels. For additional discussion of the properties of entropy used in this argument, see Ref. 123. A proof based on the strong subadditivity property of entropy was given in Ref. 128 and presented in Ref. 107.

9.6.2 Channel capacity

9.6.2.1 Background

We now briefly mention a few results concerning channel capacity in the case of a memoryless channel. In this model, the noise associated with a single use of the channel is given by the CPT map Φ that for n uses is simply the tensor product $\Phi^{\otimes n}$. This is a realistic model, even for entangled particles, assuming that they are sent one at a time, as described at the end of Sec. 9.5.5. Shannon's classical noisy coding theorem says that the optimal asymptotic transmission rate for a memoryless channel is given by a "one-shot" formula corresponding to a single use of the channel. In QIP this is not always true. Indeed, one of the features of quantum communication is the possibility of using entanglement to enhance communication.

9.6.2.2 Classical information

In the simplest type of communication, classical information is encoded in quantum particles and a POVM made on the information received, with no additional resources. With the input ensemble $\{\pi_, \rho_j\}$, the maximum information that can be obtained from a single use of the channel is the accessible information in the output ensemble $\Phi(\mathcal{E}) = \{\pi_j \Phi(\rho)_j\}$ or

$$\sup_{\mathcal{M}} I(\Phi(\mathcal{E}),\mathcal{M}) = \sup_{\mathcal{M}} \left(S[\mathrm{Tr}\, E_b \Phi(\rho)] - \sum_j \pi_j S[\mathrm{Tr}\, E_b \Phi(\rho_j)] \right). \tag{9.101}$$

The asymptotic capacity is then

$$C_{EE}(\Phi) \equiv \lim_{n\to\infty} \frac{1}{n} \sup_{\mathcal{E}_n \mathcal{M}_n} I[\Phi^{\otimes n}(\mathcal{E}_n), \mathcal{M}_n], \tag{9.102}$$

where the subscripts indicate that at the nth level in the supremum in Eq. (9.102), the allowed ensemble $\mathcal{E}_n$ and POVM $\mathcal{M}_n$ in $\mathbf{C}^{2^n}$ may include entangled states. Perhaps surprising, a closed-form expression for Eq. (9.102) is not known; however, it is known that Eq. (9.102) can be strictly greater than the supremum of Eq. (9.101) over all possible (one-shot) input ensembles. One can define capacities restricted to product inputs $C_{PE}(\Phi)$ and/or product measurements $C_{EP}(\Phi)$. It is known that

$$C_{EP}(\Phi) = C_{PP}(\Phi) = \sup_{\mathcal{E},\mathcal{M}} I[\Phi(\mathcal{E}), \mathcal{M}] \tag{9.103}$$

$$\leq \sup_{\mathcal{E}} \left\{ S[\Phi(\rho)] - \sum_j \pi_j S[\Phi(\rho_j)] \right\} \tag{9.104}$$

$$= C_{PE}(\Phi) \equiv C_{\mathrm{Holv}}(\Phi). \tag{9.105}$$

The first inequality is simply a special case of the Holevo bound in Eq. (9.85). The fact that entangled inputs do not increase the capacity if only product measurements are allowed, i.e., that $C_{EP}(\Phi) = C_{PP}(\Phi)$, was proved independently by King and Ruskai[81] and by Shor[134] and is implicit in Ref. 62. The fact that Eq. (9.104) can be achieved using product inputs and entangled measurements is a deep result, first considered in Ref. 60 and proved independently by Schumacher and Westmoreland[129] and by Holevo.[61] Moreover, Holevo[60,62] showed that $C_{\mathrm{Holv}}(\Phi)$ can be strictly greater than $C_{PP}(\Phi)$, i.e., that entangled measurements can increase the capacity of a memoryless channel.

What is still unresolved is whether using entangled input states can ever increase the capacity. This is closely related to the question of whether strict inequality ever holds in

$$C_{\mathrm{Holv}}(\Phi \otimes \Omega) \geq C_{\mathrm{Holv}}(\Phi \otimes \Omega). \tag{9.106}$$

If equality in Eq. (9.106) holds whenever $\Omega = \Phi^{\otimes k}$, then it would follow from Eq. (9.102) that $C_{EE}(\Phi) = C_{\mathrm{Holv}}(\Phi)$. Although the question is still open (and does not seem more difficult for general Ω), additivity has been shown in many special cases. Recently it was shown[2,3,99,136] to be equivalent to similar questions about the additivity of minimal entropy and other quantities characterizing the output state $\Phi(\rho)$ closest to a pure state; and to properties of a quantity called the "entanglement of formation."

More is actually known about the so-called "entanglement-assisted capacity" (EAC). This is the capacity of a memoryless channel when quantum particles are used to transmit classical information, but the sender and receiver have access to an unlimited amount of shared entanglement. As in the dense coding protocol, one expects this to enhance the capacity and this is, indeed, the case.[18,19,64] The capacity is the supremum over mutual information in states of the form $\gamma_{12} = (I \otimes \Phi)(|\Psi\rangle\langle\Psi|)$. Thus, the entanglement assisted capacity is given by

$$\mathrm{EAC}(\Phi) = \sup_{\Psi} H[\gamma_{12}, \gamma_1 \otimes \gamma_2]$$

$$= \sup_{\rho} \{S[\Phi(\rho)] + S[\rho] - S[(I \otimes \Phi)(|\Psi\rangle\langle\Psi|)]\}, \tag{9.107}$$

with $\rho = \mathrm{Tr}_A|\Psi\rangle\langle\Psi|$. The state $|\Psi\rangle$ is called a "purification" of ρ, and the quantity

$$S[\Phi, \rho] = S[(I \otimes \Phi)(|\Psi\rangle\langle\Psi|)] \quad \text{with} \quad \rho = \mathrm{Tr}_A|\Psi\rangle\langle\Psi| \tag{9.108}$$

is sometimes called the "entropy exchange."[63,107] Note that the entropy exchange is considered to be a function of the noise Φ and input ρ and can be shown to be independent of the purification $|\Psi\rangle$. (Purifications are discussed at the end of Appendix 9.C.)

9.6.2.3 Coherent information

When quantum information is transmitted, the capacity is associated with a quantity called "coherent information,"

$$\begin{aligned} I_{\mathrm{coh}}(\Phi) &= \sup\{S[\Phi(\rho)] - S[(I \otimes \Phi)(|\Psi\rangle\langle\Psi|)] : \rho = \mathrm{Tr}_A|\Psi\rangle\langle\Psi|\} \\ &= \sup_{\rho_{12}}(S[\Phi(\rho_2)] - S[(I \otimes \Phi)(\rho_{12})]) \\ &= \sup_{\rho_{12}}[H[(I \otimes \Phi)(\rho_{12}), (1/d)I \otimes \Phi(\rho_2)] - \log d, \end{aligned} \tag{9.109}$$

where the supremum, initially over reduced density matrixes of pure states, can be relaxed because Eq. (9.109) is a convex function of ρ_{12}.

If no additional resources are available, the asymptotic capacity for transmitting quantum information is $\lim_{n\to\infty} \frac{1}{n} I_{\mathrm{coh}}(\Phi^{\otimes n})$. The upper bound was proved by Barnum, Nielsen, and Schumacher[7] in 1997; the lower bound by Shor[135] in 2002. Furthermore, it is also known that one-way classical communication cannot increase the capacity.[16]

The most complex situation to analyze is the transmission of quantum information with two-way classical communication available. This enables one to apply a process known as "distillation" to optimize the use of entanglement in mixed states as described in Ref. 17. Although one expects this capacity to be less than any capacity for transmitting classical information, this has not been proved.

9.6.3 Quantum error correction

9.6.3.1 Basic error correction process

Quantum error correction poses several challenges. There are new nonclassical types of errors to correct. In addition, the fragility of a QC means that error correction is needed during the computation process. In general, one does not care if a message is destroyed while extracting the desired information. However, error correction during computation requires the ability to restore the QC to the correct quantum state. Moreover, the no-cloning theorem precludes copying the quantum state, while any attempt to discern it via measurements would seem to destroy it. Indeed, many scientists once thought that error correction in a QC would not be possible.

We illustrate the error correction process with the simple example of a 3-bit repetition code for correcting single bit flips. Let

$$|c_0\rangle = |000\rangle, \quad |c_1\rangle = |111\rangle, \tag{9.110}$$

give an encoding of 0 and 1 into three qubits. (This does not violate the no-cloning principle because the two states are orthogonal.) Then a general state can be written in the form

$$|\psi\rangle = a|c_0\rangle + b|c_1\rangle = a|000\rangle + b|111\rangle. \tag{9.111}$$

It is convenient to adopt the standard convention of using X_k and Z_k to represent the action of σ_x and σ_z on the kth bits, e.g., $X_3 = I \otimes I \otimes \sigma_x$, $Z_2 = I \otimes \sigma_z \otimes I$, and the product $Z_1 Z_2 = \sigma_z \otimes \sigma_z \otimes I$.

A single-bit flip is implemented by σ_x acting on one of these three bits, and takes $|\psi\rangle$ to a state in one of three orthogonal subspaces, which can be characterized by the eigenvectors of $Z_1 Z_2$ and $Z_2 Z_3$ as shown in Table 9.4.

The idealized error correction process is thus quite simple, and involves the following steps:

- Make a measurement with the commutating operators $Z_1 Z_2$ and $Z_2 Z_3$ to determine which subspace the state is in, according to Table 9.4.
- Apply σ_x to the corresponding bit, i.e., apply X_k. Since $\sigma_x^2 = I$ this will return the system to the original state $|\psi\rangle$.

Note that the error correction process does not require finding the parameters a and b, which determine the state $|\psi\rangle$. It works on "unknown" quantum states. This is because it uses a measurement process that distinguishes between four orthogonal two-dimensional subspaces, but does not distinguish between vectors within these subspaces.

9.6.3.2 Phase errors

To see this limitation of the code, consider the effect of a single Z_k. One finds

$$Z_k|\psi\rangle = Z_k(a|000\rangle + b|111\rangle) = a|000\rangle - b|111\rangle, \tag{9.113}$$

Table 9.4 Effect of bit flip errors.

	$Z_1 Z_2$	$Z_2 Z_3$	
$\lvert\psi\rangle = a\lvert 000\rangle + b\lvert 111\rangle$	$+1$	$+1$	(9.112a)
$X_1\lvert\psi\rangle = a\lvert 100\rangle + b\lvert 011\rangle$	-1	$+1$	(9.112b)
$X_2\lvert\psi\rangle = a\lvert 010\rangle + b\lvert 101\rangle$	-1	-1	(9.112c)
$X_3\lvert\psi\rangle = a\lvert 001\rangle + b\lvert 110\rangle$	$+1$	-1	(9.112d)

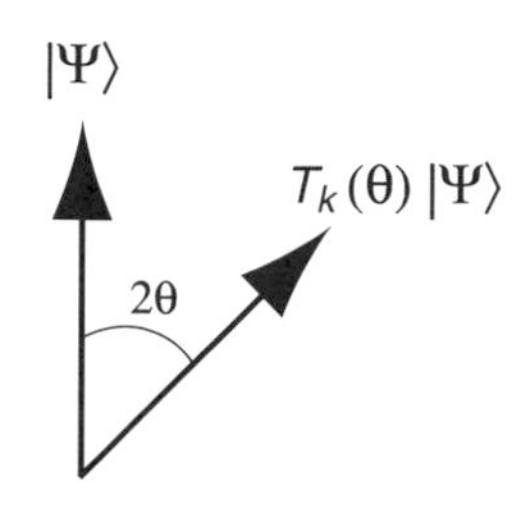

Figure 9.9 A bit tip of θ.

for $k = 1$, 2, or 3. The result lies in the space spanned by $|c_0\rangle$ and $|c_1\rangle$. Errors of this type, known as "phase errors" would not be detected. The code

$$|C_0\rangle = H^{\otimes 3}|c_0\rangle + H^{\otimes 3}|c_1\rangle = \frac{1}{2}(|000\rangle + |011\rangle + |101\rangle + |110\rangle), \tag{9.114a}$$

$$|C_1\rangle = H^{\otimes 3}|c_0\rangle - H^{\otimes 3}|c_1\rangle = \frac{1}{2}(|111\rangle + |100\rangle + |010\rangle + |001\rangle), \tag{9.114b}$$

can correct phase errors, but not bit flips. Shor[133] showed that concatenating the codes given by Eqs. (9.110) and (9.113) yields a 9-bit code that can correct all single-bit errors. Subsequently a 7-bit code, known as a CSS code[27,141] and related to the classical 7-bit Hamming code, and then a 5-bit code, which is essentially unique, were found. These can also correct all single-bit errors.

9.6.3.3 Linear combinations of errors

Although the code of Eq. (9.110) cannot correct phase errors (or σ_y errors), it can correct some additional errors, which one might regard as bit "tips." Explaining this gives considerable insight into quantum error correction. Let

$$T_k(\theta) = \begin{pmatrix} \cos\theta & \sin\theta \\ \sin\theta & \cos\theta \end{pmatrix} = \cos\theta I + \sin\theta X_k. \tag{9.115}$$

Then $T_k(\theta)$ has the effect of "tipping" the spin by an angle 2θ, as shown in Fig. 9.9, and

$$T_k(\theta)|\psi\rangle = \cos\theta|\psi\rangle + \sin\theta X_k|\psi\rangle. \tag{9.116}$$

If one now uses the error correction process described in Sec. 9.6.3.1, there are two possible outcomes:

- With probability $\cos^2\theta$, the procedure detects no error. However, this causes no problem, because the QC is left in the state $|\psi\rangle$, i.e., the measurement has effectively corrected the error.
- With probability $\sin^2\theta$, the procedure detects a flip in the kth bit. However, it also leaves the QC in the state $X_k|\psi\rangle$, from which the next step restores the QC to state $|\psi\rangle$.

Thus, the procedure for correcting bit flips, also corrects "tips" without even revealing whether they have occurred.

More generally, any linear combination of correctable errors is also correctable. This leads to the conclusion[17,83] that a set of errors $\{E_1, E_2, \ldots, E_t\}$ is correctable if and only if

$$\langle E_p c_j, E_q c_k \rangle = \delta_{jk} d_{pq}, \tag{9.117}$$

where the matrix d_{pq} does not depend on $j = 0, 1$. One might expect that a set of errors is correctable if and only if the subspaces $\{E_p(a|c_0\rangle + b|c_1\rangle)\}$ are mutually orthogonal, in which case the right side of Eq. (9.117) would be $\delta_{jk}\delta_{pq}$. However, the weaker condition, Eq. (9.117), suffices. If V is the matrix that diagonalizes D, then $\widetilde{E}_p = \sum_q u_{pq} E_q$ gives another set of errors for which the subspaces $\{\widetilde{E}_p(a|c_0\rangle + b|c_1\rangle)\}$ are orthogonal. Moreover, since the original errors $E_q = \sum_p \overline{u}_{pq} \widetilde{E}_p$ are linear combinations of the modified ones, they can be corrected in much the same way as the bit tips already described.

9.6.4 Fault-tolerant computation

There is a very elegant group-theoretic method for constructing quantum error correction codes, yielding what are known as "stabilizer" codes[28,29,45,46] or codes over $GF(4)$. Pollatsek[110] has given a nice exposition of this procedure. A subclass of these, known as CSS codes, were found earlier[27,140,141] by using a classical code, together with a dual code, to generate a quantum code. Several examples[115,116,124] of other codes, called "nonadditive," have also been found. However, there has been little systematic study of nonadditive codes and it is not yet known whether or not there are situations in which they may prove advantageous.

Finding error-correcting codes is but one aspect of fault-tolerant computation. One must also find a mechanism for implementing the basic gates on the encoded logical units as well as physical qubits. For stabilizer codes, Gottesman[46] has shown that this can always be done. In some situations, as discussed in the next section, encoding can actually facilitate the implementation of certain gates.

One must also correct errors faster than they propagate. It is certainly not practical to periodically stop the computer, make a measurement and go through the described process. One wants to incorporate error correction into the computation process, e.g., into the quantum circuit. This can be done using additional bits, known as "ancilla," to store the measurement outcome. Gates can be constructed whose effect is equivalent to storing the measurement outcome in the ancilla; then other gates can use this information to restore the QC to the correct state when necessary.

Finally, one must design the entire process to minimize propagation of errors. Whether or not this can be done depends on the actual error rates in the elements of the QC. Analyses of error thresholds for simple models have now been performed,[1,45,107] leading to threshold estimates of about 10^{-4} or 10^{-5}. When the probability of error in the gates is below this threshold, an arbitrarily long computation can be performed with only a polylogarithmic increase in the size of the circuit needed to achieve sufficiently small error. Most of these estimates have been made using the "depolarizing channel," which is equivalent to assuming that any of the three possible Pauli errors occurs with probability ϵ, i.e., the Kraus operators of Eq. (9.23) are $A_0 = \sqrt{1-3\epsilon}\sigma_k$ and $A_k = \sqrt{\epsilon}\sigma_k$ for $k = 1, 2, 3$. More realistic models, which correspond to specific physical implementations of a QC and include the possibility of correlated errors, must be studied.

One can relate the fundamental errors to the noise model in the form of Eq. (9.23) by thinking of the QC as being in the mixed state

$$\sum_k A_k|\Psi\rangle\langle\Psi|A_k^\dagger = \sum_k p_k|E_k\Psi\rangle\langle E_k\Psi|, \tag{9.118}$$

where $E_k\Psi = (A_k\Psi)/\|A_k\Psi\|$ and $p_k = \|A_k\Psi\|^2$. Then the process of replacing E_k by $\widetilde{E}_k = \sum_k v_{k\ell}E_\ell$ corresponds to making a linear transformation on the Kraus operators A_k. But, as pointed out after Eq. (9.23), the operators A_k are not unique; in fact, they are determined only up to a unitary transformation. Thus, selecting errors $\widetilde{E}_k$ that diagonalize Eq. (9.117) corresponds to making a choice of Kraus operators in the underlying noise model.

For further discussion of fault-tolerant computation and references to additional work, see Chapter 10 of Ref. 107 and the review articles cited in Sec. 9.7.2.

9.6.5 DFS encoding

There is another approach to dealing with noise[84,92] that is worth mentioning because these encodings have other important applications.

The interaction term of Eq. (9.17) can be written as $V_{CE} = \sum_k S_j \otimes T_k$ with S_j and T_j acting on $\mathcal{H}_C$ and $\mathcal{H}_E$, respectively. In general, one does not expect the eigenvectors V_{CE} or H_{CE} to be product states. One exception occurs when all of the operators S_j commute so that they have simultaneous eigenvectors. However, even when the S_j do not commute they may have a few simultaneous eigenvectors or, more generally, an invariant subspace $\mathcal{K}_C$ for which $S_j\mathcal{K}_C \subset \mathcal{H}_C$ for all j. If $\mathcal{K}_C$ is also an invariant subspace for H_C, it will be invariant under H_{CE} and the unitary group $U(t)$ determined by Eq. (9.10). Thus, a system initially in a state in $\mathcal{K}_C$ will remain there. This is called a decoherence free subsystem* (DFS).

It might seem that this situation is so special that it would rarely arise. However, there are physically realistic scenarios in which this does occur. In the most common, the operators S_j generate a group (or a Lie algebra) for which $\mathcal{K}_C$ transforms as an irreducible representation. One underlying physical model corresponds to a situation at low temperatures in which the errors are highly correlated.

*The acronym DFS is used for both decoherence free subsystem[84] and decoherence free subspace.[92] In the latter, only the trivial representation is allowed.

In considering the resilience of certain DFS codes against exchange errors, Bacon et al. realized[4] that the exchange interaction could actually be used for universal computation. To understand the underlying idea, suppose that 0 and 1 are encoded as

$$|c_0\rangle = |01\rangle, \quad |c_1\rangle = |10\rangle. \tag{9.119}$$

This code can detect, but not correct, single bit flips. The exchange operator,

$$E_{jk} = I + X_j X_k + Y_j Y_k + Z_j Z_k, \tag{9.120}$$

interchanges the values of bits j and k, i.e.,

$$E_{jk}|i_1 \ldots i_j \ldots i_k \ldots i_n\rangle = |i_1 \ldots i_k \ldots i_j \ldots i_n\rangle. \tag{9.121}$$

Exchange (also known as SWAP) is equivalent to a pair of bit flips if and only if $i_j \neq i_k$. Thus, $E_{12}|01\rangle = |10\rangle$ so that $E_{12}|c_0\rangle = |c_1\rangle$ and the exchange has the same effect on the encoded logical units $|c_0\rangle$ and $|c_1\rangle$ as σ_x. It was shown in Ref. 4 that certain 4-bit DFS encodings had the property that all the gates needed for universal quantum computation could be implemented using exchange on physical qubits. Subsequently, it was realized[5] that 3-bit encodings would suffice for universal computation with the exchange interaction.

These encodings may be quite useful in certain implementations, such as quantum dots. Implementing a σ_x or σ_z gate requires control of an anisotropic magnetic field. However, the exchange interaction in Eq. (9.120) can be implemented with an isotropic field. Multiplying the total number of qubits needed by a factor of 3 or 4 may be a small price to pay for efficient implementation of gates.

9.7 Conclusion

9.7.1 Remarks

9.7.1.1 Quantum theory

When quantum theory was first proposed, some aspects seemed so puzzling and contrary to ordinary experience that many were reluctant to accept it. However, its success in explaining physical phenomena and predicting the results of experiments were soon more than adequate to validate it as a physical theory. Since then, it has repeatedly been vindicated experimentally and its domain of applicability extended "from atoms to stars."[93] Nevertheless, some puzzling features continued to be debated for decades.

With the advent of QIP, a new attitude has emerged. Instead of expecting the physical world to conform to views shaped by experience with phenomena that can be explained by classical physics, we accept and try to understand the quantum world on its own terms. Rather than regarding quantum theory as full of paradoxes to be explained away, we look for new ways to exploit "quantum weirdness." This view has led to new advances in physics and in information theory and has shaped my exposition in this chapter.

9.7.1.2 Entanglement

A topic that has generated considerable discussion is the role of entanglement in quantum computation. Is this the key feature that makes quantum computation powerful? For two insightful discussions, see Jozsa and Linden[75] and Steane.[140] In my view, it is useful to distinguish between the explicit role of entanglement correlations, which are essential in the EPR experiment (Sec. 9.3.2) and such procedures as dense coding (Sec. 9.5.3) and teleportation (Sec. 9.5.4), and the implicit uses of entanglement. Although one can find bases for $\mathbf{C}^{2^n}$ composed entirely of product states, most states in $\mathbf{C}^{2^n}$ *cannot* be written as products. Any algorithm that requires access to arbitrary states in $\mathbf{C}^{2^n}$ uses entangled states, whether or not the explicit correlations play a role.

For example, Lloyd[95] showed that Grover's algorithm on a list of size M can be implemented without entanglement if M distinct states (say the M lowest levels of an oscillator) are used. However, this is not practical when M is large and implementations using tensor products lead to entangled states. Moreover, if an algorithm never uses entangled states, all gates necessarily take product states to product states, implying that only 1-bit gates are used. Thus, in some sense, quantum parallelism requires superpositions, but universality requires entanglement.

9.7.1.3 Physical implementation

Among the most commonly asked questions about QCs are "How realistic is this?" or "When will someone actually build a quantum computer?" The answer depends, to some extent, on exactly what one means by a QC. Rather than even attempting to answer this question, I refer the reader to the excellent article by DiVincenzo.[36] The following brief quotations from the introduction

> It does not require science fiction to envision a quantum computer ...

and conclusion,

> So, what is the "winning" technology going to be? I don't think that any living mortal has an answer to this question, and at this point it may be counterproductive even to ask it.

still give an accurate picture of the situation in 2003.

Although experiments demonstrating various facets of quantum computation have been performed, a full-fledged QC seems to be a long way off. However, some other types of QIP seem more feasible. Of these, QKD distribution seem the most likely to soon be realized at a practical level.

9.7.2 Recommendations for further reading

This chapter could provide only a very brief introduction to the many facets of QIT, and some important subtopics received little or no mention. I have endeavored to provide references to key papers and to a representative selection of recent work through which readers can find additional references on particular topics. However, the references are not comprehensive and many important papers are not cited. In this section, I try to provide some guidance to those who wish to learn more about various aspects of the fascinating field of QIT.

The best general reference on QIT is the text by Nielsen and Chuang.[107] It is comprehensive and thorough, yet begins each topic at an elementary level, requiring no background beyond linear algebra. The recent text by Kitaev et al.,[82] which has a rather different flavor and a focus on computational models and complexity, is also recommended. The lecture notes of Preskill[111] and the links on his website were invaluable before the publication of Ref. 107, and remain an important resource.

Wick[150] has given a very readable account of the historical development of the foundations of quantum theory and the experiments associated with Bell's inequalities. This book also contains a more mathematical appendix by Faris[40] that is highly recommended. David Mermin has written a number of insightful articles on related topics; many of these were published in *Physics Today* (e.g., Ref. 103) and are collected in a delightful volume of essays.[102] Many of Bell's papers, which are available in Ref. 8, are quite readable. Two recent reviews by Werner[148] and Werner and Wolf[149] contain useful insights from a perspective more directly connected to QIP.

There are a number of review articles that provide a good introduction to particular topics. The excellent pair of articles[73,74] by Jozsa were already mentioned in Sec. 9.4.3. For a nice account of the development of Grover's algorithm, see Ref. 53. For valuable reflections on the nature of the power of quantum computation, see Steane.[142] Those who want an introduction to quantum computation that includes the quantum circuit model might consult Refs. 30 and 122.

Pollatsek[110] has given a nice description of the construction of stabilizer codes. For additional introductory treatments of quantum codes and other aspects of fault-tolerant computation, Gottesman,[47] Preskill,[113,114] and Steane[143] are all highly recommended. For more detailed accounts, see Gottesman[45,46] or Preskill.[112] Lidar and Whaley have recently written a survey[92] of another approach, the DFS method alluded to in Sec. 9.6.5.

The detection and quantification of entanglement are extremely active areas of current research. The reviews by Bruss[26] and by Horodecki et al.[67] give good overviews of this complex subject.

To learn more about quantum entropy, the best place to begin is Wehrl's review article.[147] One can also consult the monograph by Ohya and Petz[108] and the recent review.[123] The extension of Shannon's information theory to quantum systems is an active area of research. See Bennett and Shor[16] for an introduction to the different types of capacities. For a more advanced account of many results and related topics, the monograph by Holevo[63] is recommended. For the most recent results, one should see the references cited in Sec. 9.6.2.

Finally, some websites are worth mentioning. Most people working in QIP post preprints at arxiv.org/quant-ph and check it regularly for the latest developments.

To find the most recent results that have been published in refereed archival journals, one can consult the *Virtual Journal of Quantum Information*,[146] edited by D. DiVincenzo. In fall 2002, a series of workshops were held at the Mathematical Sciences Research Institute (MSRI) at Berkeley. The talks, including some excellent tutorials, are available as streaming video (together with pdf files of the notes) at the MSRI website.[98] Many talks from workshops at the Institute for Theoretical Physics at the University of California at Santa Barbara in fall 2001 are also available on the Internet.[71]

Appendix 9.A Dirac notation

Most of the literature in QIP uses the physicist's convention of writing vectors and projections using Dirac's bra and ket notation, which is explained here.

If we let u represent a column vector in $\mathbb{C}^m$ and $u^\dagger$ its conjugate transpose, then (except for placement of the complex conjugate) the usual Hermitian form can be written as

$$\langle v, u\rangle = v^\dagger u = (\overline{v}_1, \ldots, \overline{v}_m) \begin{pmatrix} u_1 \\ \vdots \\ u_m \end{pmatrix}. \tag{9.122}$$

If the order is reversed, $uv^\dagger$ is an $m \times m$ matrix corresponding to the map $w \mapsto \langle v, w\rangle u$, and it is natural to write

$$|u\rangle\langle v| = u\, v^\dagger = \begin{pmatrix} u_1 \\ \vdots \\ u_m \end{pmatrix} (\overline{v}_1, \ldots, \overline{v}_m). \tag{9.123}$$

When $u = v$, this becomes the 1D projection onto the subspace spanned by u, i.e.,

$$P_u = \frac{1}{\|u\|^2} uu^\dagger = \frac{|u\rangle\langle u|}{\|u\|^2}. \tag{9.124}$$

In an abstract m-dimensional vector space, a "ket" vector $|u\rangle$ is analogous to a column vector. Its "dual" or "bra" vector $\langle u|$ is the analogue of a conjugated row vector. Moreover, this duality can be made completely rigorous by identifying $\langle u|$ with a vector in the usual Banach space dual via the Riesz representation theorem. The interpretation just given for $\langle v, u\rangle$ and $|u\rangle\langle v|$ then extend to general vector spaces in a natural way. (Note that it is natural to use the physicists' convention in which the inner product is linear in the second variable and antilinear in the first.)

In this notation, it is common to replace u by any convenient label, such as a (nondegenerate) eigenvalue, which identifies the vector u. Thus, one might write $|\lambda_k\rangle$ or even $|k\rangle$ for the eigenvector v_k associated with λ_k. In quantum computation, it is common to use $|0\rangle$ and $|1\rangle$ to label the two states of a qubit. As long as the convention used is clear, this should present no problem.

Appendix 9.B Trace and partial trace

The trace of a matrix (or operator) Q satisfies

$$\mathrm{Tr}Q = \sum_k q_{kk} = \sum_k \langle \phi_k, Q\phi_k \rangle, \qquad (9.125)$$

where $\{\phi_k\}$ is any orthonormal basis and q_{kk} denote the diagonal elements in a fixed matrix representation of A. One can define an inner product, known as the Hilbert-Schmidt inner product, on the bounded operators $\mathcal{B}(\mathcal{H})$ acting on any finite-dimensional space by

$$\langle A, B \rangle = \mathrm{Tr}\,(A^\dagger B). \qquad (9.126)$$

When Q is an operator on a tensor product space, $\mathcal{H}_A \otimes \mathcal{H}_B$, one often writes Q_{AB}. Formally, the partial trace Tr_B over $\mathcal{H}_B$ is defined by the requirement that $Q_A = \mathrm{Tr}_B Q_{AB}$ satisfies

$$\langle \chi, Q_A \psi \rangle = \sum_k \langle \chi \otimes \phi_k, Q_{AB} \psi \otimes \phi_k \rangle \qquad (9.127)$$

for any pair of vectors χ, ψ and any orthonormal basis $\{\phi_k\}$ for $\mathcal{H}_B$. There are several equivalent definitions that are somewhat easier to use. Any operator on $\mathcal{H}_A \otimes \mathcal{H}_B$ can be written in the form $Q_{AB} = \sum_j c_j S_j \otimes T_j$ with S_i and T_j operators on $\mathcal{H}_A$ and $\mathcal{H}_B$, respectively. Then

$$Q_A = \mathrm{Tr}_B Q_{AB} = \sum_j c_j (\mathrm{Tr} T_j) S_j, \qquad (9.128)$$

where Tr now denotes the usual trace on $\mathcal{H}_B$. In particular $\mathrm{Tr}_B S \otimes T = (\mathrm{Tr}T)S$. When the matrix M is written in block form

$$\begin{pmatrix} M_{11} & M_{12} & \dots & M_{1n} \\ M_{21} & M_{22} & \dots & M_{2n} \\ \vdots & \vdots & & \vdots \\ M_{n1} & M_{n2} & \dots & M_{nn} \end{pmatrix}, \qquad (9.129)$$

with the matrixes M_{jk} acting on $\mathcal{H}_B$, then $\mathrm{Tr}_B M$ is the matrix with elements $\mathrm{Tr} M_{jk}$, i.e, one takes the usual trace of each block, and

$$\mathrm{Tr}_A M = \sum_j M_{jj}, \qquad (9.130)$$

i.e., one sums over diagonal blocks.

Although less common, one can also formally define a "partial inner product," which was used in Sec. 9.3.4. If $\Psi = \sum_{jk} c_{jk} |\alpha_j\rangle \otimes |\beta_k\rangle$, then

$$\langle \phi, \Psi \rangle_A = \sum_k \Big(\sum_j c_{jk} \langle \phi, \alpha_j \rangle \Big) |\beta_k\rangle. \tag{9.131}$$

When the Hilbert spaces are chosen so that the inner product is given by an integral, this takes the familiar form

$$\langle \phi, \Psi \rangle_S = \int \overline{\phi(s)} \Psi(s, t)\, ds. \tag{9.132}$$

Appendix 9.C Singular value and Schmidt decompositions

On the tensor product of two Hilbert spaces with orthonormal bases $\{\phi_j\}$ and $\{\chi_k\}$, respectively, an arbitrary vector $|\Psi\rangle$ can be written as

$$|\Psi\rangle = \sum_{jk} b_{jk} |\phi_j\rangle \otimes |\chi_k\rangle \tag{9.133}$$

$$= \sum_j \alpha_j |\phi_j\rangle \otimes |\omega_j\rangle, \tag{9.134}$$

where $|\omega_j\rangle = \alpha_j^{-1} \sum_k b_{jk} |\chi_k\rangle$ and $\alpha_j = \sqrt{|b_{jk}|^2}$. In general, the vectors $\{\omega_j\}$ in Eq. (9.134) are not orthogonal. The so-called Schmidt decomposition is simply the statement that any vector on a tensor product space can be written in a form similar to Eq. (9.134) using orthonormal bases. It is an immediate consequence of the singular value decomposition (SVD) which is itself a corollary to the polar decomposition theorem.

Theorem 1 (Polar decomposition). *Any $m \times n$ matrix A can be written in the form $A = U|A|$, where the $n \times n$ matrix $|A| = \sqrt{A^\dagger A}$ is positive semidefinite and the $m \times n$ matrix U is a partial isometry.*

The term *partial isometry* means that $U^\dagger U$ (or, equivalently, $UU^\dagger$) is a projection. In general, U need not be unique but can be uniquely determined by the condition $\ker U = \ker A$. If A is a square $n \times n$ matrix, then U can instead be chosen (non-uniquely) to be unitary. Since $|A|$ is self-adjoint, it can be written as $|A| = VDV^\dagger$ where D is a diagonal matrix with nonnegative entries and V is unitary. Inserting this in Theorem 1 with U chosen to be unitary yields the SVD since $W = UV$ is also unitary.

Theorem 2 (Singular value decomposition). *Any $n \times n$ matrix A can be written in the form $A = WDV^\dagger$ with V and W unitary and D a positive semidefinite diagonal matrix.*

The nonzero elements of D are called the *singular values* of A. They are easily seen to be the eigenvalues of $|A|$ and, hence, their squares yield the nonzero eigenvalues of $A^\dagger A$. As an immediate corollary, one finds that $A^\dagger A$ and $AA^\dagger$ are unitarily equivalent and that V and W are, respectively, the unitary transformations that diagonalize $A^\dagger A$ and $AA^\dagger$. These results can be extended to nonsquare matrixes if the requirement that V and W be unitary is relaxed to partial isometry.

There are two ways to obtain the "Schmidt decomposition" from the SVD. One is to simply apply the SVD to the coefficient matrix b_{jk} in Eq. (9.133). The other is to observe that there is a one-to-one correspondence between vectors that have the form of Eq. (9.133) and operators of the form

$$K_\Psi = \sum_{jk} b_{jk} |\phi_j\rangle\langle\chi_k|. \tag{9.135}$$

Moreover, if $\rho_{AB} = |\Psi\rangle\langle\Psi|$, then

$$\rho_A \equiv T_B(\rho_{AB}) = K_\Psi K_\Psi^\dagger, \tag{9.136a}$$

$$\rho_B \equiv T_A(\rho_{AB}) = (K_\Psi^\dagger K_\Psi)^T, \tag{9.136b}$$

where ρ_A and ρ_B are the reduced density matrixes obtained by taking the indicated partial traces T_B and T_A. One then obtains the following result.

Theorem 3. *Any vector that has the form of* Eq. (9.133) *can be rewritten as*

$$\Psi = \sum_k \mu_k |\tilde{\psi}_k\rangle \otimes |\tilde{\chi}_k\rangle, \tag{9.137}$$

where μ_k are the singular values of the matrix B, the bases $\{\tilde{\psi}_k\}$ and $\{\tilde{\chi}_k\}$ are orthonormal and related by $\mu_k \tilde{\psi}_k = K_\Psi \tilde{\chi}_k$ with K_Ψ given by Eq. (9.135).

It follows immediately that the reduced density matrixes ρ_A and ρ_B have the same nonzero eigenvalues $\{{\mu_k}^2\}$ and $\{\tilde{\psi}_k\}$ and $\{\tilde{\chi}_k\}$ are the eigenvectors of ρ_A and ρ_B, respectively. Conversely, given a density matrix whose spectral decomposition is

$$\rho = \sum_{k=1}^{m} \lambda_k |\phi_k\rangle\langle\phi_k|, \tag{9.138}$$

one can define the pure state $|\Psi\rangle = \sum_{k=1}^m \sqrt{\lambda_k}|\phi_k\rangle \otimes |\psi_k\rangle$ on $\mathcal{H} \otimes \mathcal{H}$ with $\{\psi_k\}$ any m orthonormal vectors on $\mathcal{H}$. Then $|\Psi\rangle$ is called a purification of ρ since $\rho = \mathrm{Tr}_B|\Psi\rangle\langle\Psi|$.

Schmidt actually proved the SVD for integral kernels. For more about the history of the SVD and Schmidt decompositions, see Chap. 3 of Ref. 66 and Appendix A of Ref. 80.

Appendix 9.D A more complete description

9.D.1 Continuous variables

Most topics in QIT can be discussed using a model in which the underlying Hilbert space is finite dimensional and isomorphic to $\mathbf{C}^d$, particularly for $d = 2^n$. This enables one to avoid some delicate issues associated with operators, such as the position and momentum, with continuous spectrum. This can be refreshing, as some expositions of quantum theory leave the reader with the impression that the fact that particles do not have a definite position is the most fundamental feature of quantum theory. However, the very word "quantum" has quite a different meaning, originating with the observation that atoms emit and absorb light in a way that suggests they can have only certain allowed energies (the eigenvalues of the Hamiltonian).

Thus, observables with discrete spectra display fundamental quantum features. However, the commutator of the operators associated with a pair of observables limits the accuracy with which the two observables can be simultaneously measured. Indeed, the inequality

$$\Delta A \Delta B \geq \langle \phi, (AB - BA)\phi \rangle, \tag{9.139}$$

where $\Delta A = \sqrt{\langle \phi, A^2 \phi \rangle - |\langle \phi, A\phi \rangle|^2}$ gives a general uncertainty principle, regardless of whether the operators A and B have discrete or continuous spectra (or both).

9.D.2 The hidden spatial wave function

The standard description of qubits presented used in this chapter is incomplete. For example, the state of a electron is properly described by a vector in $L_2(\mathbf{R}^3) \otimes \mathbf{C}_2$, such as $|f(\boldsymbol{x}) \otimes \phi\rangle$, where $|\phi\rangle$ describes the spin and $\int_\Lambda |f(\boldsymbol{x})|^2 d^3\boldsymbol{x}$ is the probability of finding the electron in the region $\Lambda \subset \mathbf{R}^3$. The statement "qubit in state $|\phi\rangle$ in Sue's lab" should be interpreted as meaning that the full wave function of the qubit is $|f_S(\boldsymbol{x})\rangle \otimes |\phi\rangle$, where $f_S(\boldsymbol{x})$ has support in Sue's lab.

Thus, the full wave function for an entangled pair of particles in the Bell state $|\beta_1\rangle$ has the form

$$\frac{1}{\sqrt{2}}(|f, 0\rangle \otimes |g, 1\rangle + |f, 1\rangle \otimes |g, 0\rangle), \tag{9.140}$$

where we have written $|f, 0\rangle$ for $|f(\boldsymbol{x}, t)\rangle \otimes |0\rangle$ and the functions f and g may depend on time t as well as position $\boldsymbol{x}$. Sending the particle in Sue's lab to Tom means using a physical process so that $f(\boldsymbol{x}, t_0)$ has support in Sue's lab and at some later time $f(\boldsymbol{x}, t_1)$ has support in Tom's lab.

9.D.3 The Pauli principle

When there are n electrons, the full wave function must be antisymmetric with respect to exchange of particles. Thus, a simple product such as $|f,0\rangle \otimes |g,1\rangle$ must be replaced by an antisymmetrized product

$$\frac{1}{\sqrt{2}}\big(|f,0\rangle \otimes |g,1\rangle - |g,1\rangle \otimes |f,0\rangle\big). \tag{9.141}$$

In Eq. (9.141), the spin state $|0\rangle$ is always associated with the spatial state f and the spin state $|1\rangle$ is always associated with the spatial state g, regardless of whether it occurs as the first or second term in the product. This antisymmetry reflects the fact that electrons are identical particles that cannot be distinguished. The antisymmetrized wave function for an entangled state of the form of Eq. (9.140) can be written as

$$\begin{aligned}&\frac{1}{2}[(|f,0\rangle \otimes |g,1\rangle - |g,1\rangle \otimes |f,0\rangle) + (|f,1\rangle \otimes |g,0\rangle - |g,0\rangle \otimes |f,1\rangle)]\\ &\quad = \frac{1}{\sqrt{2}}(|fg\rangle - |gf\rangle)\frac{1}{\sqrt{2}}(|01\rangle + |10\rangle).\end{aligned} \tag{9.142}$$

The fact that the wave function can be factored into an antisymmetric spatial function times a symmetric spin function is rather atypical.

In general, the antisymmetry requirement applies only to the full wave function, not to the individual space and spin components. Indeed, for $n \geq 3$ there are no nontrivial antisymmetric functions on $\mathbf{C}_2^{\otimes n}$. In the general situation, Ψ has the form

$$\Psi(x_1, x_2, \ldots, x_N) = \sum_k F_k(\mathbf{r}_1, \mathbf{r}_2, \ldots, \mathbf{r}_N)\chi_k(s_1, s_2, \ldots, s_N), \tag{9.143}$$

where $x_k = (\mathbf{r}_k, s_k)$ with $\mathbf{r}$ a vector in $\mathbf{R}^3$ and s_k in $\mathbf{Z}_2$, χ_k are* elements of $\mathbf{C}^{2^n}$, and the "space functions" F_k are elements of $L^2(\mathbf{R}^{3N})$. When Ψ is antisymmetric, the sets $\{F_k\}$ and $\{\chi_k\}$ each transform as a representation of S_n. If these representations are irreducible, those for $\{F_k\}$ and $\{\chi_k\}$ have dual Young tableaux. For further discussion see Refs. 57, 87, 124 and 144.

Acknowledgment

This chapter has been partially supported by the National Security Agency (NSA) and Advanced Research and Development Activity (ARDA) under Army Research

*A spin state χ looks formally like a (possibly entangled) N-qubit state. However, unlike qubits that involve an implicit spatial component, we use here vectors in $\mathbf{C}^{2^n}$ itself.

Office (ARO) contract number DAAD19-02-1-0065, and by the National Science Foundation under contract number DMS-0314228.

Part of this chapter was written while I was a Walton visitor at the Communications Network Research Institute of Dublin Institute of Technology. I am grateful to John Lewis for the hospitable working environment there. It is also a pleasure to thank Professor Harriet Pollatsek for comments on earlier drafts and Dr. Christopher Fuchs for comments on Sec. 9.5.

While proofreading this chapter, I learned that John Lewis passed away on 21 January 2004. The concept of POVM was introduced in a 1970 paper he wrote with E. B. Davies, "An Operational Approach to Quantum Probability," *Commun. Math. Phys.* **17**, 239–260 (1970). His many contributions to quantum theory and mathematical physics have had a lasting impact on quantum information theory. This chapter is dedicated to his memory.

References

1. D. Aharonov and M. Ben-Or, "Fault-tolerant computation with constant error," in *Proc. 29th ACM Symposium on the Theory of Computing*, 176–188, ACM Press (1997).
2. K. M. R. Audenaert and S. L. Braunstein, "On strong superadditivity of the entanglement of formation," to appear in *Commun. Math. Phys.* **246**, 427–442 (2004).
3. G. G. Amosov, A. S. Holevo, and R. F. Werner, "On some additivity problems in quantum information theory," *Prob. Inf. Trans.* **36**, 305–313 (2000) (http://arxiv.org/math-ph/0003002).
4. D. Bacon, J. Kempe, D. A. Lidar, and K. B. Whaley, "Universal fault-tolerant computation on decoherence-free subspaces," *Phys. Rev. Lett.* **85**, 1758–1761 (2000).
5. D. Bacon, J. Kempe, D. P. DiVincenzo, D. A. Lidar, and K. B. Whaley, "Encoded universality in physical implementations of a quantum computer," in *Proceedings of the 1st International Conference on Experimental Implementations of Quantum Computation*, R. Clark, Ed., 257–264, Rinton Press, Princeton, NJ (2001) (quant-ph/0102140).
6. H. Barnum, C. Crépeau, D. Gottesman, A. Smith, and A. Tapp, "Authentication of quantum messages," in *Proc. 43rd IEEE Symposium on the Foundations of Computer Science*, 449–458 (2002).
7. H. Barnum, M. A. Nielsen, and B. Schumacher, "Information transmission through a noisy quantum channel," *Phys. Rev. A* **57**, 4153–4175 (1998).
8. J. S. Bell, *Speakable and Unspeakable in Quantum Mechanics*, Cambridge University Press, Cambridge (1989).
9. P. Benioff, "The computer as a physical system: a microscopic quantum mechanical Hamiltonian model of computers as represented by Turing machines," *J. Stat. Phys.* **22**, 563–591 (1980).

10. P. Benioff, "Quantum mechanical Hamiltonian models of Turing machines," *J. Stat. Phys.* **29**, 515–546 (1980).
11. P. Benioff, "Quantum mechanical models of Turing machines that dissipate no energy," *Phys. Rev. Lett.* **48**, 1581–1585 (1982).
12. C. H. Bennett, "Quantum cryptography using any two nonorthogonal states," *Phys. Rev. Lett.* **68**, 3121–3124 (1992).
13. C. H. Bennett and G. Brassard, "Quantum cryptography: public key distribution and coin tossing," in *Proc. IEEE Int. Conf. on Computers Systems and Signal Processing*, 175–179 (Bangalore India) (Dec. 1984).
14. C. H. Bennett, D. P. DiVincenzo, and J. A. Smolin, "Capacities of quantum erasure channels," *Phys. Rev. Lett.* **78**, 3217–3220 (1997).
15. C. H. Bennett, C. A. Fuchs, and J. A. Smolin, "Entanglement-enhanced classical communication on a noisy quantum channel," in *Quantum Communication, Computing and Measurement*, O. Hirota, A. S. Holevo, and C. M. Caves, Eds., 79–88, Plenum Press, New York (1997) (quant-ph/9611006).
16. C. H. Bennett and P. W. Shor, "Quantum information theory," *IEEE Trans. Inf. Theory* **44**, 2724–2742 (1998).
17. C. H. Bennett, D. P. DiVincenzo, J. A. Smolin, and W. K. Wootters, "Mixed-state entanglement and quantum error correction," *Phys. Rev. A* **54**, 3824–3851 (1996).
18. C. H. Bennett, P. W. Shor, J. A. Smolin, and A. V. Thapliyal, "Entanglement-assisted classical capacity of noisy quantum channels," *Phys. Rev. Lett.* **83**, 3081–3084 (1999).
19. C. H. Bennett, P. W. Shor, J. A. Smolin, and A. V. Thapliyal, "Entanglement-assisted capacity of a quantum channel and the reverse Shannon theorem," *IEEE Trans. Inf. Theory* **48**, 2637–2655 (2002).
20. E. Bernstein and U. Vazirani, "Quantum complexity theory," *SIAM J. Comput.* **26**, 1411–1473 (1997).
21. D. Bohm, *Quantum Theory*, Prentice Hall, Englewood Cliffs, New Jersey (1951); reprinted by Dover Press, New York (1989).
22. D. Bohm and Y. Aharonov, "Discussion of experimental proof for the paradox of Einstein, Rosen, and Podolsky," *Phys. Rev.* **108**, 1070–1076 (1957).
23. G. Brassard, C. Crépeau, D. Mayers, and L. Salvail, "A brief review on the impossibility of quantum bit commitment" (quant-ph/9712023).
24. H. J. Briegel, W. Dür, J. I. Cirac, and P. Zoller, "Quantum repeaters: the role of imperfect local operations in quantum communication," *Phys. Rev. Lett.* **81**, 5932–5935 (1998).
25. H. J. Briegel and R. Raussendorf, "Persistent entanglement in arrays of interacting particles," *Phys. Rev. Lett.* **86**, 910–913 (2001).
26. D. Bruss, "Characterizing entanglement," *J. Math. Phys.* **43**, 4237–4251 (2002).
27. A. R. Calderbank and P. W. Shor, "Good quantum error-correcting codes exist," *Phys. Rev. A* **54**, 1098–1105 (1996).

28. R. Calderbank, E. M. Rains, P. W. Shor, and N. J. A. Sloane, "Quantum error correction and orthogonal geometry," *Phys. Rev. Lett.* **78**, 405–408 (1997).
29. R. Calderbank, E. M. Rains, P. W. Shor, and N. J. A. Sloane, "Quantum error correction via codes over GF(4)," *IEEE Trans. Inf. Theory* **44**, 1369–1387 (1998).
30. R. Cleve, "An introduction to quantum complexity theory," in *Quantum Computation and Quantum Information Theory*, C. Macchiavello, G. M. Palma, and A. Zeilinger, Eds., World Scientific, Singapore, 103–127 (2000) (quant-ph/9906111).
31. M.-D. Choi, "Completely positive linear maps on complex matrixes," *Lin. Alg. Appl.* **10**, 285–290 (1975).
32. C. Crépeau, D. Gottesman, and A. Smith, "Secure multi-party quantum computing," in *Proc. 34th ACM Symp. on the Theory of Computing*, 643–652, ACM Press, New York (2002) (quant-ph/0206138).
33. C. Crépeau and J. Kilian, "Achieving oblivious transfer using weakened security assumptions," in *Proc. 29th FOCS*, 42–53, IEEE Press, Piscataway, NJ (1988).
34. D. Deutsch, "Quantum theory, the Church-Turing principle and the universal quantum computer," *Proc. R. Soc. Lond. A* **400**, 97–117 (1985).
35. D. Deutsch and R. Jozsa, "Rapid solution of problems by quantum computation," *Proc. R. Soc. Lond. A* **439**, 553–558 (1998).
36. D. P. DiVincenzo, "The physical implementation of quantum computation," *Forts. Phys.* **48**, 771–783, special issue, Experimental Proposals for Quantum Computation (2000) (quant-ph/0002077).
37. D. P. DiVincenzo, D. W. Leung, and B. M. Terhal, "Quantum data hiding," *IEEE Trans. Inf. Theory* **48**, 580–599 (2002).
38. A. Einstein, B. Podolsky, and N. Rosen, "Can quantum-mechanical description of physical reality be considered complete?" *Phys. Rev.* **47**, 777–780 (1935).
39. A. K. Ekert, "Quantum cryptography based on Bell's theorem," *Phys. Rev. Lett.* **67**, 661–663 (1991).
40. W. Faris, "Probability in quantum mechanics," appendix to *The Infamous Boundary*, Birkhauser, Boston (1995).
41. R. Feynman, "Simulating physics with computers," *Int. J. Theor. Phys.* **21**, 467–488 (1982).
42. R. Feynman, "Quantum mechanical computers," *Opt. News*, 11–20 (Feb. 1985); reprinted in *Found. Phys.* **16**, 507–531 (1986).
43. C. Fuchs, "Nonorthogonal quantum states maximize classical information capacity," *Phys. Rev. Lett.* **79**, 1162–1165 (1997).
44. N. Gisin, G. Ribordy, W. Tittel, and H. Zbinden, "Quantum cryptography," *Rev. Mod. Phys.* **74**, 145–195 (2002).
45. D. Gottesman, "Stabilizer codes and quantum error correction," PhD thesis, Caltech (1997) (quant-ph/9705052).

46. D. Gottesman, "A theory of fault-tolerant quantum computation," *Phys. Rev. A* **57**, 127–137 (1998).
47. D. Gottesman, "An introduction to quantum error correction," in *Proc. Symp. Appl. Math.* Vol. 58, 221–235, American Mathematical Society, Providence, RI (2000) (quant-ph/0004072).
48. D. Gottesman, "On the theory of quantum secret sharing," *Phys. Rev. A* **61**, 042311 (2000).
49. D. Gottesman, "Uncloneable encryption," *Quantum Inf. Comput.* **3**, 581–602 (2003) (quant-ph/0210062).
50. D. Gottesman and I. Chuang, "Quantum digital signatures" (quant-ph/0105032).
51. D. Gottesman, H.-K. Lo, N. Lütkenhaus, and J. Preskill, "Security of quantum key distribution with imperfect devices" (quant-ph/0212066).
52. L. Grover, "A fast quantum mechanical algorithm for database search," in *Proc. 28th Annual ACM Symp. on the Theory of Computing*, 212–219, ACM Press, New York (1996).
53. L. Grover, "From Schrödinger's equation to the quantum search algorithm," *Am. J. Phys.* **69**, 769–777 (2001).
54. S. Hallgren, "Polynomial-time quantum algorithms for Pell's equation and the principal ideal problem," in *Proc. 34th Annu. ACM Symp. on the Theory of Computing*, 653–658, ACM Press, New York (2002).
55. L. Hales and S. Hallgren, "An improved quantum Fourier transform algorithm and applications," in *Proc. 41st IEEE Symp. on Foundations of Computer Science*, 515–525, IEEE Press (2000).
56. P. R. Halmos, *Introduction to Hilbert Space and the Theory of Spectral Multiplicity*, Chelsea, New York (1957); reprinted by American Mathematical Society, Providence, RI (2000).
57. M. Hamermesh, *Group Theory*, Addison-Wesley, Reading, MA (1962); reprinted by Dover Press, New York (1990).
58. P. Hausladen, R. Jozsa, B. Schumacher, M. D. Westmoreland, and W. K. Wootters, "Classical information capacity of a quantum channel," *Phys. Rev. A* **54**, 1869–1876 (1996).
59. A. S. Holevo, "Information theoretical aspects of quantum measurement," *Prob. Inf. Transm. USSR* **9**, 31–42 (1973).
60. A. S. Holevo, "On the capacity of quantum communication channel," *Probl. Peredachi Inf.* **15**(4), 3–11 (1979); English translation: *Probl. Inf. Transm.* **15**(4), 247–253 (1979).
61. A. S. Holevo, "The capacity of a quantum channel with general signal states," *IEEE Trans. Inf. Theory* **44**, 269–273 (1998) (quant-ph/9611023).
62. A. S. Holevo, "Quantum coding theorems," *Russian Math. Surv.* **53**, 1295–1331 (1999); appeared as preprint "Coding theorem for quantum channels" (quant-ph/9809023).
63. A. S. Holevo, *Statistical Structure of Quantum Theory*, Springer, Berlin (2001).

64. A. S. Holevo, "On entanglement-assisted classical capacity," *J. Math. Phys.* **43**, 4326–4333 (2002).
65. R. A. Horn and C. R. Johnson, *Matrix Analysis*, Cambridge University Press, Cambridge (1985).
66. R. A. Horn and C. R. Johnson, *Topics in Matrix Analysis*, Cambridge University Press, Cambridge (1991).
67. M. Horodecki, P. Horodecki, and R. Horodecki, "Mixed-state entanglement and quantum communication," 151–195 in *Quantum Information: An Introduction to Basic Theoretical Concepts and Experiments*, Springer Tracts in Modern Physics, Vol. 173 (2001) (quant-ph/0109124).
68. M. Horodecki, "Limits for compression of quantum information carried by ensembles of mixed states," *Phys. Rev. A* **57**, 3364–3369 (1998).
69. R. J. Hughes, J. E. Nordholt, D. Derkacs, and C. G. Peterson, "Practical free-space quantum key distribution over 10 km in daylight and at night," *New J. Phys.* **4**, 43.1–43.14 (2004) (quant-ph/0206092).
70. H. Inamori, N. Lütkenhaus, and D. Mayers, "Unconditional security of practical quantum key distribution" (quant-ph/0107017).
71. ITP web site http://online.kitp.ucsb.edu/online/qinfo01/.
72. A. Jamiolkowski, "Linear transformations which preserve trace and positive semi-definiteness of operators," *Rep. Math. Phys.* **3**, 275–278 (1972).
73. R. Jozsa, "Quantum algorithms and the Fourier transform," *Proc. R. Soc. Lond. A* **454**, 323–337 (1998).
74. R. Jozsa, "Notes on Hallgren's efficient quantum algorithm for solving Pell's equation" (quant-ph/0302134).
75. R. Jozsa and N. Linden, "On the role of entanglement in quantum computational speed-up," *R. Soc. Lond. Proc. Ser. A Math. Phys. Eng. Sci.* **459**, 2011–2032 (2003) (quant-ph/0201143).
76. R. Jozsa and B. Schumacher, "A new proof of the quantum noiseless coding theorem," *J. Mod. Opt.* **41**, 2343–2349 (1994).
77. I. Kerenidis and R. de Wolf, "Exponential lower bound for 2-query locally decodable codes via a quantum argument," in *35th Annual ACM Symposium on Theory of Computing*, 106–115 (2003) (quant-ph/0208062).
78. Y.-H. Kim and Y. Shih, "Experimental realization of Popper's experiment: violation of the uncertainty principle?" *Found. Phys.* **29**, 1849–1861 (1999).
79. C. King, M. Nathanson, and M. B. Ruskai, "Qubit channels can require more than two inputs to achieve capacity," *Phys. Rev. Lett.* **88**, 057901 (2002).
80. C. King and M. B. Ruskai, "Minimal entropy of states emerging from noisy quantum channels," *IEEE Trans. Inf. Theory* **47**, 1–19 (2001).
81. C. King and M. B. Ruskai, "Capacity of quantum channels using product measurements," *J. Math. Phys.* **42**, 87–98 (2001).
82. A. Y. Kitaev, A. H. Shen, and M. N. Vyalyi, *Classical and Quantum Computation*, American Mathematical Society, Providence, RI (2002).
83. E. Knill and R. Laflamme, "A theory of quantum error-correcting codes," *Phys. Rev. A* **55**, 900–911 (1997).

84. E. Knill, R. Laflamme, and L. Viola, "Theory of quantum error correction for general noise," *Phys. Rev. Lett.* **84**, 2525–2528 (2000).
85. K. Kraus, "General state changes in quantum theory," *Ann. Phys.* **64**, 311–335 (1971).
86. K. Kraus, *States, Effects and Operations: Fundamental Notions of Quantum Theory*, Springer, Berlin (1983).
87. L. Landau and L. Lifshitz, *Quantum Mechanics*, 2nd ed. of English translation, Pergamon Press, Reading, MA (1965).
88. R. Landauer, "Irreversibility and heat generation in the computing process," *IBM J. Res. Develop.* **3**, 183–191 (1961); reprinted in *Maxwell's Demon*, H. S. Lef and A. F. Rex, Eds., Princeton (1990); 2nd ed. (2003).
89. R. Landauer, "Information is inevitably physical," in *Feynman and Computation*, J. G. Hey, Ed., 77–92, Perseus Press, Reading, MA (1999).
90. D. W. Leung, "Two-qubit projective measurements are universal for quantum computation" (quant-ph/0111122); see also "Quantum computation by measurements" (quant-ph/0310189).
91. D. Leung, "Choi's proof as a recipe for quantum process tomography," *J. Math. Phys.* **44**, 528–533 (2003).
92. D. Lidar and B. Whaley, "Decoherence-free subspaces and subsystems," in *Irreversible Quantum Dynamics*, F. Benatti and R. Floreanini, Eds., 83–120, Springer Lecture Notes in Physics, Vol. 622, Berlin (2003) (quant-ph/0301032).
93. E. Lieb, "The stability of matter: from atoms to stars," *Bull. Am. Math. Soc.* **22**, 1–49 (1990).
94. G. Lindblad, "Completely positive maps and entropy inequalities," *Commun. Math. Phys.* **40**, 147–151 (1975).
95. S. Lloyd, "Quantum search without entanglement," *Phys. Rev. A* **61**, 010301 (2000).
96. H.-K. Lo and H. F. Chau, "Why quantum bit commitment and ideal quantum coin tossing are impossible," *Phys. D* **120**, 177–187 (1998).
97. Y. Manin, *Computable and Uncomputable*, Sovetskoye Radio, Moscow (1980) (in Russian); see also "Classical computing, quantum computing, and Shor's factoring algorithm," *Astrisque* **266**, 375–404 (2000) (quant-ph/9903008).
98. MSRI web site http://www.msri.org/publications/video/index05.html.
99. K. Matsumoto, T. Shimono, and A. Winter, "Remarks on additivity of the Holevo channel capacity and of the entanglement of formation," to appear in *Commun. Math. Phys.* **246**, 443–452 (2004).
100. D. Mayers, "Unconditionally secure quantum bit commitment is impossible," *Phys. Rev. Lett.* **78**, 3414–3417 (1997).
101. M. Mehring, J. Mende, and W. Scherer, "Entanglement between an electron and a nuclear spin," *Phys. Rev. Lett.* **90**, 153001 (2003).
102. N. D. Mermin, *Boojums All the Way Through*, Cambridge University Press, Cambridge (1990).

103. N. D. Mermin, "Is the moon really there when nobody looks? Reality and the quantum theory," *Phys. Today* **38**(6), 38–47 (1990).
104. M. Nathanson, "Quantum guessing via Deutsch–Jozsa" (quant-ph/0301025).
105. A. W. Naylor and G. R. Sell, *Linear Operator Theory in Engineering and Science*, Springer, Berlin, New York (2000).
106. M. A. Nielsen, "Universal quantum computation using only projective measurement, quantum memory, and preparation of the 0 state" (quant-ph/0108020).
107. M. A. Nielsen and I. L. Chuang, *Quantum Computation and Quantum Information*, Cambridge University Press, Cambridge (2000).
108. M. Ohya and D. Petz, *Quantum Entropy and Its Use*, Springer, Berlin (1993).
109. M. Ohya, D. Petz, and Watanabe, "On capacities of quantum channels," *Probl. Math. Stats.* **17**, 170–196 (1997).
110. H. Pollatsek, "Quantum error correction: classic group theory meets a quantum challenge," *Am. Math. Monthly* **108**, 932–962 (Dec. 2001).
111. J. Preskill, http://theory.caltech.edu/~preskill/ph229/.
112. J. Preskill, "Reliable quantum computers," *Proc. R. Soc. Lond. A* **454**, 385–410 (1998).
113. J. Preskill, "Fault-tolerant quantum computation," in *Introduction to Quantum Computation*, H.-K. Lo, S. Popescu, and T. P. Spiller, Eds., 213–269, World Scientific, Singapore, (1996) (quant-ph/9712048).
114. J. Preskill, "Battling decoherence: the fault tolerant quantum computer," *Phys. Today* **52**(6), 24–30 (June 1999).
115. E. M. Rains, R. H. Hardin, P. W. Shor, and N. J. A. Sloane, "A nonadditive quantum code," *Phys. Rev. Lett.* **79**, 953–954 (1997).
116. V. P. Roychowdhury and F. Vatan, "On the existence of nonadditive quantum codes," *Quantum Computing and Quantum Communications*, Lecture Notes in Computer Science, Vol. 1509, Springer, Berlin, 325–336 (1999) (quant-ph/9710031).
117. R. Raussendorf and H. J. Briegel, "Persistent entanglement in arrays of interacting particles," *Phys. Rev. Lett.* **86**, 910–913 (2001).
118. R. Raussendorf and H. J. Briegel, "A one-way quantum computer," *Phys. Rev. Lett.* **86**, 5188–5191 (2001); see also "Quantum computing via measurements only" (quant-ph/0010033).
119. R. Raussendorf, D. E. Browne, and H. J. Briegel, "The one-way quantum computer—a non-network model of quantum computation," *J. Mod. Optic.* **49**, 1299–1306 (2002).
120. R. Raussendorf and H. J. Briegel, "Computational model underlying the one-way quantum computer," *Quantum Inf. Comput.* **2**, 443–486 (2002) see also "Computational model for the one-way quantum computer: concepts and summary" (quant-ph/0207183).
121. R. Raussendorf, D. E. Browne, and H. J. Briegel, "Measurement-based quantum computation with cluster states," *Phys. Rev. A* **68**, 022312 (2003) (quant-ph/0301052).

122. E. G. Rieffel and W. Polak, "An introduction to quantum computing for non-physicists," *ACM Comput. Surv.* **32**, 300–355 (2000) (quant-ph/9809016).
123. M. B. Ruskai, "Inequalities for quantum entropy: a review with conditions for equality," *J. Math. Phys.* **43**, 4358–4375 (2002).
124. M. B. Ruskai, "Pauli exchange errors in quantum computation," *Phys. Rev. Lett* **85**, 194–197 (2000); M. B. Ruskai, "Pauli-exchange errors and quantum error correction," in *Quantum Computation and Quantum Information Science*, S. Lomonaco, Ed., *Contemporary Math.* **305**, 251–263, AMS, Providence, RI (2002).
125. M. B. Ruskai, S. Szarek, and E. Werner, "An analysis of completely positive trace-preserving maps on $\mathcal{M}_2$," *Lin. Alg. Appl.* **347**, 159–187 (2002).
126. D. Schlingemann, "Cluster states, algorithms and graphs" (quant-ph/0305170); "Logical network implementation for graph codes and cluster states," *Quantum Inf. Comput.* **3**, 431–449 (2003).
127. B. Schumacher, "Quantum coding," *Phys. Rev. A* **51**, 2738–2747 (1995).
128. B. Schumacher, M. D. Westmoreland, and W. K. Wootters, "Limitation on the amount of accessible information in a quantum channel," *Phys. Rev. Lett.* **76**, 3452–3455 (1996).
129. B. Schumacher and M. D. Westmoreland, "Sending classical information via noisy quantum channels," *Phys. Rev. A* **56**, 131–138 (1997).
130. B. Schumacher and M. D. Westmoreland, "Optimal signal ensembles," *Phys. Rev. A* **63**, 022308 (2001).
131. P. W. Shor, "Algorithms for quantum computation: discrete logarithms and factoring," in *Proc. 35th IEEE Symp. on Foundations of Computer Science* (1994).
132. P. W. Shor, "Polynomial-time algorithms for prime factorization and discrete logarithms on a quantum computer," *SIAM J. Comput.* **26**, 1484–1509 (1997); reprinted and updated in *SIAM Rev.* **41**, 303–332 (1999).
133. P. W. Shor, "Scheme for reducing decoherence in quantum computer memory," *Phys. Rev. A* **52**, R2493–R2496 (1995).
134. P. W. Shor, "Capacities of quantum channels and how to find them," *Math. Program.* **97**, no. 1-2, ser. B, 311–335, ISMP, Copenhagen (2003) (quant-ph/0304102).
135. P. W. Shor, announced at *MSRI Workshop* (Nov. 2002), notes giving sketch of proof are available at www.msri.org/publications/ln/msri/2002/quantumcrypto/shor/1/index.html; for further developments, see I. Devetak and P. W. Shor, "The capacity of a quantum channel for simultaneous transmission of classical and quantum information" (quant-ph/0311131).
136. P. Shor, "Equivalence of additivity questions in quantum information theory," to appear in *Commun. Math. Phys.* **246**, 453–471 (2004).
137. P. W. Shor and J. Preskill, "Simple proof of security of the BB84 quantum key distribution protocol," *Phys. Rev. Lett.* **85**, 441–444 (2000).
138. Ch. Silberhorn, T. C. Ralph, N. Lütkenhaus, and G. Leuchs, "Continuous variable quantum cryptography—beating the 3 dB loss limit," *Phys. Rev. Lett.* **89**, 167901 (2002) (quant-ph/0204064).

139. D. Simon, "On the power of quantum computation," in *Proc. 35th IEEE Symp. on Foundations of Computer Science*, pp. 116–123 (1994); *SIAM J. Comput.* **26**, 1474–1483 (1997).
140. A. M. Steane, "Error correcting codes in quantum theory," *Phys. Rev. Lett.* **77**, 793–797 (1996).
141. A. M. Steane, "Multiple particle interference and quantum error correction," *Proc. R. Soc. Lond. A* **452**, 2551–25576 (1996).
142. A. M. Steane, "A quantum computer only needs one universe," *Stud. Hist. Philos. MP* **34**, 469–478 (2003) (quant-ph/0003084).
143. A. M. Steane, "Quantum computing and error correction," in *Decoherence and Its Implications in Quantum Computation and Information Transfer*, Gonis and Turchi, Eds., 284–298, IOS Press, Amsterdam (2001) (quant-ph/0304016).
144. S. Sternberg, *Group Theory and Physics*, Cambridge University Press, Cambridge (1995).
145. W. F. Stinespring, "Positive functions on C^*-algebras," *Proc. Am. Math. Soc.* **6**, 211–216 (1955).
146. *Virtual Journal of Quantum Information*, http://www.vjquantuminfo.org.
147. A. Wehrl, "General properties of entropy," *Rev. Mod. Phys.* **50**, 221–260 (1978).
148. R. F. Werner, "Quantum information theory—an invitation," in *Quantum Information: An Introduction to Basic Theoretical Concepts and Experiments*, 14–57, Springer Tracts in Modern Physics, Vol. 173, Springer (2001) (quant-ph/0101061).
149. R. F. Werner and M. M. Wolf, "Bell inequalities and entanglement," *Quantum Inf. Comput.* **1**(3), 1–25 (2001) (quant-ph/0107093).
150. D. Wick, *The Infamous Boundary*, Birkhauser, Boston (1995).
151. S. Wiesner, "Conjugate coding," *SIGACT News* **15**, 77 (1983).
152. A. Winter, A. C. A. Nascimento, and H. Imai, "Commitment capacity of discrete memoryless channels," preprint at (cs.CR/0304014).
153. J. A. Wheeler, "Polyelectrons," *Ann. New York Acad. Sci.* **48**, 219–238 (1946).
154. W. Wootters and W. Zurek, "A single quantum can not be cloned," *Nature* **299**, 802–803 (1982).
155. C. S. Wu and I. Shaknov, "The angular correlation of scattered annihilation radiation," *Phys. Rev.* **77**, 136 (1950).
156. H. P. Yuen and M. Ozawa, "Ultimate information carrying limit of quantum systems," *Phys. Rev. Lett.* **70**, 363–366 (1993).
157. T. F. Jordan, *Linear Operators for Quantum Mechanics*, Wiley, New York (1969); reissued by Krieger (1990).

List of acronyms

DFS	decoherence free subspace or subsystem
FFT	fast Fourier transform
POVM	positive operator valued measure
QC	quantum computer
QFT	quantum Fourier transform
QKD	quantum key distribution
QIP	quantum information processing
QIT	quantum information theory
SVD	singular value decomposition

Mary Beth Ruskai received her PhD degree in physical chemistry from the University of Wisconsin in 1969. She then became the Battelle Fellow at the Institut de Physique Théorique of the Université de Genéve. In 1971 and 1972, she was a postdoctoral fellow at MIT with Elliott Lieb, with whom she proved the strong subadditivity of quantum mechanical entropy. From 1973 to 1976, she was an assistant professor in the mathematics department at the University of Oregon. From 1977 to 2003 she was on the faculty of the University of Massachusetts–Lowell, from where she retired with professor emeritus rank. She has held visiting positions at many institutions, including the Rockefeller University, the University of Vienna, AT&T Bell Laboratories, the Courant Institute of New York University, Case Western Reserve University, the Technische Universität of Berlin. She is currently a research professor at Tufts University.

Prof. Ruskai has organized many conferences and workshops, including the 1990 CBMS conference on wavelets. She was coeditor of the September 2002 special issue of the *Journal of Mathematical Physics* on Quantum Information Theory, and now serves on the editoral board of *Communications in Mathematical Physics*. She has served on many committees of the American Mathematical Society, including its council, and on the Commission on Mathematical Physics (including a term as vice-chair) of the International Union of Pure and Applied Physics. In addition to quantum information theory, she has worked on the mathematical analysis of multiparticle quantum systems. In 1992, she was elected a fellow of the American Association for the Advancement of Science.

Index

2D electron gas (2DEG), 117

A

ab initio, 323–324
Abelian hidden subgroup problem, 427
absorption, 60–61, 192
acceptance probability, 223
accessible information, 436
adatom, 7–8
addition energy, 127–128
adsorption, 223, 326, 365–366
aggregate, surfactant, 225–226, 228–229
amphiphile, 227
ancilla, 444
angle of incidence, 47–48
anisotropy, density, 9
anisotropy, optical, 7
anomalous dispersion, 59
artificial atom, 109
association rate constant, 232
asymptotic capacity, 439
atomic elasticity (AE), 131
atomic force microscope (AFM), 112, 339, 344–347, 364–365
  magnetically activated, 346
  tapping mode, 344–345, 364–365
atomic orbital, 280
atomistic methods, 3, 208, 260, 274, 287
Auger transitions, 129
augmented continuum theories, 302–304
authentication, 433
automatic adaption, 301
average, statistical, 208, 221–222
average, temporal, 208, 214, 222
axial current, 153, 171
azimuthal current, 151

B

B92 protocol, 431
ballistic aggregation, 8
band gap, 51, 54, 60, 71–72, 75–76, 111, 150–151
basis, 67, 324
BB84 protocol, 431
beamsplitter, 29
Bell inequalities, 416
Bell measurement, 434
Bell states, 414–415
bianisotropy, 7, 10–11, 15, 32
biaxiality, 22
bideposition, 9
biexciton, 120
binding energy, 120
biochip, 30
biomedicine, 2
biopolymer, 230–231
bit flip, 442–443
Bloch equations, 151, 153, 183
Bloch functions, 65, 152, 178–179
Bloch sphere, 412
body force, 262
body-centered cubic (BCC) lattice, 73
Bohr magneton μ_B, 124
Born–Oppenheimer approximation, 275, 324
bosonic commutation, 171
boundary conditions, 233, 239, 335–337, 341
  absorbing, 240
  free, 210, 215
  helical, 231
  minimum image, 357
  no-slip, 321, 351–353

periodic, 209, 215, 235–236, 240, 333–334
stochastic, 333
zero shear, 352
boundary effects, 333
boundary value problem, 148, 159, 182, 270
Bragg phenomenon, 26
circular, 26–27, 30
Bragg soliton, 63
Brenner potential, 216
Brillouin zone, 54, 150–151, 153–154
brittle, 295
Bruggeman formalism, 15–16, 23
Buckingham potential, 216
buckyball, 217
bundle, 278

C

canonical ensemble, 290
carbon nanotube, 2, 30, 146–147, 259, 287, 293
armchair, 149
axial conductivity, 154–156
chiral, 149
conductivity law, 154
cross-sectional radius, 149
crystalline lattice, 148–149
dual index to characterize, 149
dynamical conductivity, 153, 171
edge effects, 159–162
electron transport in, 148–153
geometric chiral angle, 149
linear electrodynamics, 146
metallic conductivity, 151, 154
negative differential conductivity, 167–170
nonlinear effects, 146
quantum electrodynamics, 146
semi-classical conductivity, 155
zigzag, 149
Cauchy relation, 285–286
Cauchy's principle, 263
Cayley–Hamilton theorem, 56
cellular automata, 137
channel, 410
channel capacity, 435, 439
charge transfer, 216
chemical potential, 128, 154, 209, 223
chemical vapor deposition, 112
classical light, 192
cluster, 208, 220, 225–226, 228–229
CNOT gate, 404
coarse graining, 225, 233–234, 241
coherent information, 441
columnar morphology, 6–8
columnar thin film, 7, 10, 17, 22, 24
commutation relations, bosonic, 171
compatibility, 260, 268
compatibility equation, 268
completely positive maps, 410
compounded wavelet matrix (CWM), 235–236, 239
computational basis, 400
computational complexity, 404
concurrent multiscale method, 233
conductivity, negative absolute, 170
conductivity, negative differential, 146, 167–170
configurational bias, 224
configurational integral, 291
confinement energy, 118
confinement potential, 115
conservation law, 21, 187
constant pressure ensemble, 294
constitutive dyadic, 10, 12, 13
constitutive matrix, 15
constitutive relation, 11–13, 21, 147–148, 154, 260, 269–270
contact angle, 328, 348–349, 351, 353
contact line, 348, 353
continuous variables, 453
continuum, 7, 13, 16, 261
continuum mechanics, 131, 210
controlled phase gate, 429
correlation energy, 133
correlation functions, 327
Coulomb
blockade, 127, 167
effect, 123
energy, 127
interaction, 167–168, 179
screening, 156
coupling of properties, 311
CPT, 410
crack spalling, 273
creation and annihilation operators, for electron, 178
for electron-hole pairs, 179
for photons, 185–186
critical micelle concentration, 227
crystal, 2
CSS code, 443
current instability, 169

D

dark states, 126
data hiding, 433
data structures, 333
de Broglie thermal wavelength, 224
decoherence free subspace, 445
decoherence free subsystem (DFS), 445
defect modes, 84–88
deformation gradient tensor, 267
deformation map, 266
dense coding, 433
density functional theory, 217, 324–326
density matrix, 152, 407–408
density of modes, 54–55, 57, 110
dephasing, 130
depolarization, 15, 24, 183, 188, 190–192
  dyadic, 181
  field, 181, 185, 187, 189, 193
  Hamiltonian, 182, 185
  shift, 183–184, 188, 192–193
depolarizing channel, 445
desorption, 366
Deutsch–Jozsa problem, 420
diamagnetic shift, 125
dielectric, 7, 21
digital signatures, 433
dip-pen nanolithography, 346
dipole moment, electron-hole pair, 179, 181
dipole moment, atomic, 171, 177
Dirac delta function, 55
Dirac notation, 449
discrete model, 297
dislocations, 258
dispersion
  energy correction, 326
  equation, 54, 146
  interaction, 325
  of π-electrons, 148
  of π-electrons in carbon nanotube, 150–151
  of π-electrons in carbon nanotube, approximate law, 151
  of π-electrons in graphene, 150
  of π-electrons in quantum superlattice, 169
displacement field, 239
dissipative particle dynamics, 335, 338
  smoothed, 338
dissociation rate constant, 232
DNA, 320, 368–369
drag coefficient, 295
Drude-type conductivity, 175
ductile, 295
dyad, 15
dyadic, 7, 171
dynamic instability, 231

E

edge condition, 159
edge resonance, 162
edge scattering pattern, 161
effective
  boundary conditions, 156–157
  current, 156
  mass, 123
  mass model, 132
  parallelism, 398
eigenfunction symmetry, 71–73, 75–76
eigenstates, 112
eigenvalue equation, 56
elastic continuum, 234, 239, 243
elastic modulus, 269
elastic wave, 239
elastodynamics, 31
electric field operator, 170, 186, 189
electric field phasor, 16
electro-optic devices, 46
electrochromism, 31
electrodynamics
  classical, 3, 154, 192–194
  nonlinear, 163
  quantum, 147, 170, 184, 193–194
electroluminescence, 31
electromagnetic field operator, 170, 185
electromagnetic field quantization, 170
electron, 115
  affinity, 113, 115
  beam lithography, 112
  density, 285
  pump (EP), 128, 135
electron-hole effective mass concept, 192
electronic free pass in nanotubes, 154
element, 271
ellipsoid, 14
embedded atom method, 217, 285
embedding function, 286
emission, 192
empirical tight binding, 281
encoding information, 397
ensemble, 209, 211–213, 215–216, 221, 223–224, 227–228, 290, 436
entanglement, 414, 418
entanglement-assisted capacity, 440

enthalpy, 216
entropy exchange, 441
envelope wave function, 127
environment, 409
epitaxial growth, 2, 109, 111, 273
EPR experiment, 415, 417
equilibrium, 260, 264–265, 270, 290
equipartition theorem, 211
ergodicity, 213, 221
error threshold, 444
ethics, 1, 4
Euler–Cauchy law, 263
evanescence, 18, 121
evaporation, 7, 8, 32
Ewald summation, 284, 331
exchange errors, 446
exciton, 111, 119–120, 123
  resonance, 180
  transition frequency, 192
external current, 171

F

Fabry–Pérot etalon, 59
face-centered cubic (FCC) lattice, 73–74
far infrared, 121
far-zone scattered power density, 161
fast Fourier transform, 241, 399
fault-tolerant computation, 444
Fermi distribution, 111, 153–154
Fermi level, 150–151
Field-effect quantum dot (FEQD), 116
finite element method, 233–234
finite-difference time-domain (FDTD) method, 46, 82–88, 239
Floquet–Bloch theorem, 53
flow, 337
  charge, 358
  Couette, 337, 353–354, 356
  Hagen–Poiseuille, 354
  multiphase, 338
  Poiseuille, 337, 353, 356
  Taylor–Couette, 354
fluid mechanics, nanoscale, 322
Fock qubit, 189, 191
Fock state, 189, 193
force field, 327–329
free electron laser, 121
free space, 11, 18
free-space wave number, 156
free surfaces, 261, 306
frequency spectrum, 241–242
fullerene, 146, 148, 208, 217, 221, 281–282

G

gain band, 183
gap solitons, 63
gate, 115–116, 404
gradient of deformation, 267
graphene, 148, 150
  conductivity, 150, 154–155
  crystalline lattice, 148
graphite, 281–282, 324, 326, 370
grating, 29
Green function, classical dyadic, 171
  retarded, 184
  scalar, 172
Green–Lagrange strain tensor, 267
ground state, 275
group velocity, 55, 57
Grover oracle, 422
Grover's algorithm, 422
gyrotropy, 12

H

Hadamard transform, 413
Hallgren's algorithm, 413, 425
Hamiltonian, 210, 214, 323, 405
hardening modulus, 258
harmonic number, 164
Hartree–Fock method, 276
heat bath, 211–212, 243
Heaviside function, 64
helicoidal bianisotropic medium, thin-film, 10, 17, 22
Hellman-Feynman theorem, 289
Helmholtz free energy, 291
Hertz potential, scalar, 157, 159
heterostructure, 111
hidden subgroup problem, 427
high-order harmonic, 162–164, 166
Hilbert-Schmidt inner product, 450
Hohenberg–Kohn theorem, 276
Holevo bound, 436, 438
homeland security, 1
homogeneous
  homogeneous broadening, 111
  linewidth Γ, 130
homogeneous broadening, 193
homogenization, 13, 23
honeycomb cell, 149, 152
hybrid, 321, 336–337
hybrid methods, 3, 261
hydrogen bond, 325, 348, 356, 358, 360, 362

hydrophilic, 227, 322, 341–342, 346, 348, 351–352, 354, 358, 360
hydrophobic, 227, 322, 341, 344–345, 347–352, 354, 358–361, 365
  gating, 358
  pore, 360–361
hyperelastic, 269

I
ideal gas, 291
impedance mismatch, 239, 241
importance sampling, 223, 291
indistinguishability, 291
inhomogeneous broadening, 111, 126
insulator, 6
interaction energy, 324–325
interaction potential, 3, 321
interatomic potential, 216–217, 221, 233, 240, 243, 282
interband transition, 154, 162, 168, 175, 177, 179, 181
interferometry, 340
interlayer dielectric, 17, 31
intraband, 119
intraband motion, 154, 168, 179, 181
intraband transition, 175, 181
intrinsic length scale, 260, 304
ion bombardment, 8
ion channel, 355, 357
isolated systems, 405

J
Jacobi iteration technique, 16
Jellium, 326
joint density of states, 110

K
k · **p** method, 131
kinematics, 261, 265–268
kinetic energy, 210–212, 215
Knudsen number, 335–336
Kramers-Kronig relations, 48
Kraus representation, 411
Kronecker delta function, 67

L
Landau level, 125
laser action, 73
lateral cap model, 231–232
lattice basic vectors, 149
lattice mismatch, 117
lattice statics, 261, 287–289
lead, 128
Lennard–Jones, 330
  fluid, 337, 352–353
  interaction, 350
  molecules, 354
  potential, 216, 235, 240, 284, 330, 354
life time, 110
linear combination of atomic orbitals, 217, 280
liquid crystal, 7, 30
lithography, 2
local field, 148, 172, 179–180, 182, 184–185, 192–194
local-density approximation, 277
Lomer dislocation, 299
long-range interaction, 331, 341
long-time limit, 209, 234, 241
Lorentz-Berthelot mixing rule, 330
low-dimensional nanostructures, 146
luminescence, 30, 32, 122
Luttinger spinor, 124

M
macromolecules, 109
macroscopic approaches, 2
magnesium fluoride, 6
magnetic field operator, 170
magnetic field phasor, 16
magneto-optics, 12
magnetoelectricity, 11
many-body effects, 285
many-body potential, 216
Markov chain, 223
Markovian approximation, 171, 175
materials design, computer aided, 258
matrizant, 17
Maxwell's equations, 16, 48
  quantization of, 170
mean-field approximation, 185
mechanical behavior, 3, 32
memoryless channel, 430, 439
mesa, 119
mesoscale, 322, 337, 371
mesoscopic model, 335, 365
metal, 6
micelle, 226–229
microcanonical ensemble, 294
microcavity, 134, 172, 175
microphotoluminescence, 119
microscopic approach, 2
microscopic state, 3

microstructure, 235–236, 239
microtubule, 230
mixed methods, 261
mixed state, 407
modified embedded atom method, 217
modulation-doped heterojunction, 116
molecular beam epitaxy (MBE), 112
molecular dynamics, 3, 261, 287, 293–295
  ab initio, 217, 244, 276
  canonical, 211
  isobaric, 216
  microcanonical, 210
  non-equilibrium, 209
molecular gear, 357
molecular interaction potential, 3
molecular tagging, 342
monolayer, 112
monotonicity of relative entropy, 437
Monte Carlo method, 3, 221, 223, 225, 227, 287, 290–292
  kinetic, 209, 230–231, 243
morphology, 2, 9
  chevronic, 9
  chiral, 9
  columnar, 6–8
  helicoidal, 10
  nematic, 9, 12
  zigzag, 9
multibody contributions, 329
multipole expansion, 332
multiscale methods, 3, 233, 243, 258, 320–321, 334, 359, 371
multisection, 10, 12, 29
mutual information, 437

N

nanobubbles, 344–345
nanochannel, 339–340, 357, 369
nanocomputer, 4
nanocrystal, 112
nanoelectromechanical systems, 2, 256
nanoelectromagnetics, 2, 146
nanoelectronics, 4, 158
nanofluidics, 3, 320–321, 327, 341, 351, 362, 371
nanoindentation, 301
nanomechanics, 3, 256
nanopore, *see* pore
nanoscale sensor, 320
nanosieve, 31
nanostructures, 146, 258–259, 286, 303
nanotechnology, 46
nanowire, 278–279
natural linewidth, 119
no-cloning principle, 413
nodes, 271
noise, 409
noise current, 171
noncontact scanning force microscopy, 344
nonhomogeneity, 146–147, 157, 162, 170, 194
nonhomogeneous nanotube, 170
nonlinear
  composite, 162
  diffraction, 162
  optics, 3, 62, 148
  transport, 148
nonlinearity, 146, 153, 162, 168, 169
nonlocality, 260
nonorthogonal bases, 401
nonradiative decay, 173, 177
Nosé–Hoover thermostat, 294
nuclear magnetic resonance, 339–340

O

observable, 406
one-way quantum computer, 404
open system, 223, 409
optical
  activity, 10, 24, 27
  coatings, 8
  filter, 24, 26–28
  fluid sensor, 7, 17, 29
  interconnect, 30
  spectroscopy, 121
  switch, 7, 31
  transition band, 154
oracle, 420
orthogonal projection, 407
orthorhombicity, 9, 12
osculating plane, 11
overlap region, 336
overlapping integral, 150, 169

P

p-polarization, 48, 77, 81
pair potentials, 282, 284
partial inner product, 451
partial isometry, 451
partial trace, 450
particle annihilation, 224
particle creation, 224
particle-particle particle-mesh, 332
partition function, 224, 290

Pauli blocking, 121, 126
Pauli matrix, 412
Pauli's exclusion principle, 281
perfectly matched layer, 83
period finding, 425
periodicity, 12, 24, 30
permeability, 11
  relative, 47
permittivity, 11, 22
  relative, 47, 171, 175, 179
perturbative methods, 17
phase
  defect, 29
  diagram, 327
  error, 442
  space, 214–215, 222–225, 243, 293
  transformation, 216
phonon bottleneck, 129
photoabsorbtion, 112
photobleaching, 134
photocatalysis, 32
photoexcitation, 126
photoluminescence, 126
photon vacuum renormalization, 175
photonic band gap (PBG) structure, 2, 26, 29, 46
photostability, 114
physical vapor deposition, 2, 6–7, 32
piezoelectricity, 24, 118
plane wave, 18–19
plane wave methods, 63–77, 278
plasmon, 161
polar decomposition, 451
polarizability density, 15, 24
polarization, 402
  circular, 18, 26
  (-dependent) splitting, 183, 194
  linear, 18
  macroscopic, 182, 184, 186, 192
  operator, 178–181, 184
  power expansion, 165–166
  single-particle operator, 178
polydispersity, 228
polymer, 210, 215, 221, 225–226, 231
pore, 321, 335, 339, 353, 356, 358–361, 367, 369
porosity, 14, 24
positive operator-valued measure, 407
positivity-preserving maps, 410
potential, 327–329
  angle, 329
  box, 109
  Coulomb, 331
  dihedral angle, 330
  energy, 210, 222, 227, 327
  intramolecular, 329
  ionization, 326
  Morse, 329
  torsion, 330
Potts model, 235, 237–238
privacy amplification, 431
protein, 31, 230–231, 244
pseudo MD-FDTD coupling, 241
pseudopotential method, 131
pulse shaper, 30
Purcell effect, 175, 177
pure state, 400
purification, 452

Q

Q factor, 86, 88
QIP, 405
quadrature weights, 299
quanta, 126
quantization, 111
quantization electromagnetic field, 170, 182, 185
quantum
  bit commitment, 432
  circuit model, 404
  communication, 435
  computation, 420
  computer, 3, 397–398
  correlations, 408
  cryptography, 432
  dot (QD), 2, 3, 6, 109, 121, 146–147
  dot polarization, 183
  dot, dipole moment, 193
  efficiency, 116
  electrodynamics, 147–148, 170–171, 184, 186, 193–194
  entropy, 408
  error correction, 441
  Fourier transform, 399, 429
  gate, 404
  Hall effect, 112
  information, 130, 147, 170
  information processing, 3, 397
  information theory, 397
  key distribution, 397, 430
  light, 147, 184, 186, 191–193
  measurement, 406
  mechanics, 274, 323, 327
  optics of nonhomogeneous mediums, 185

oscillator, 3
teleportation, 434
theory, 406
well, 110
yield, 114
quantum-confined Stark effect, 124
quasi-continuum method, 233, 261, 295
quasi-momentum, 146, 150
quasi-particle, 133, 146, 167
qubit, 136, 189, 191, 397

R

radiation condition, 157, 159, 171
radiative
decay, 173, 177
life time, 127, 193
recombination, 115
Rahman-Stillinger potential, 216
rarefaction, 335–336
Rayleigh–Ritz method, 278
Rayleigh-Wood anomaly, 29
reactive empirical bond-order potential, 216
real-time MD-FDTD coupling, 241
reciprocal lattice, 66–67
reduced density matrix, 437
reference state, 297
reflectance, 17, 21, 26, 56
refractive index, 51, 57–59, 341–342
relative entropy, 437
relaxation, 153–154, 167, 183
relaxation time, 130, 154, 168–169, 183
relaxation-time approximation, 153, 167, 174
representative atom, 298
reptation, 225
reversibility, 403
rigidity, 258
rotation dyadic, 11–13, 22
rugate filter, 29

S

s-polarization, 48, 77, 81
scale parity, 243
scaled coordinates, 215
scanning near-field optical microscope, 121
scanning tunneling microscopy, 112, 257
scattered power density, 161
Schmidt decomposition, 451
Schrödinger equation, 274, 323, 405
sculptured nematic thin film, 10, 16, 22, 24, 29
sculptured thin film, 2, 10
chiral, 9–10, 12, 17, 24, 26, 31
second-harmonic generation, 32, 62–63
secret sharing, 433
secure distributed computing, 433
self-assembly, 113, 208, 210–221, 225–226, 230, 367
quantum dot, 113
semi-classical approximation, 167
semiconductor, 3, 6, 7, 30
semiempirical method, 323–324
semimetal, 150
serial multiscale method, 233
SETTLE, 330
SHAKE, 330
Shannon entropy, 409
shape function, 14, 24, 271
Shor's algorithm, 425
silicon oxide, 6
silicon-on-insulator (SOI) wafer, 116
Simon's algorithm, 428
simple cubic (SC) lattice, 73, 75–76, 80–81
simulation methods, deterministic, 208, 210
simulation methods, stochastic, 208, 221
single electron
pump, 128
transistor, 123, 128
single-photon source, 136
single-photon state, 192
singular value decomposition, 451
slip, 321
coefficient, 352
length, 352, 354
slow-wave coefficient, 157–158
slowly varying amplitude approximation, 61–62, 182
smoothing function, 331
solid body, 262
spatial confinement of charge carrier, 146
spatial dispersion, 153, 156
spatial wave function, 453
spectral
broadening, 130
hole, 28
line, 119
spin, 397, 403
spin-splitting, 125
spontaneous decay, 172–173, 175, 177
spontaneous emission, 171
spontaneous radiation, 171, 175, 177
sputtering, 8
square lattice, 72, 84
stabilizer code, 444

stabilizer groups, 420
stacking fault, 284
Stark effect, 124
Stark frequency, 164, 168
Stark harmonics, 169
statically determinate problems, 265
stationary state, 275
statistical mechanics, 290
steered molecular dynamics, 334
stiffness matrix, 272
Stillinger-Weber potential, 216
stop band, 47
strain, 234, 240, 262, 265
strain displacement relation, 270
strain-induced quantum dot, 118
Stranski-Krastanow mode, 117
strawberries, 256
streak camera, 119
stress, 210, 216, 234, 239–242, 262, 264–265
strong confinement regime, 179, 181, 193
structural handedness, 22, 23, 30
substrate, 7
superhelix, 10
superlattice, 111
superposition, 398
surface current density, 153, 154, 156, 163–168
surface force, 263, 339–341, 344, 347, 365
surface tension, 345, 348, 350, 357
surface wave, 157–159, 173, 175
  dispersion relation for, 157
  nanowaveguide, 158–159
surfactant, 210, 221, 224–229
susceptibility, nonlinear, 153
SWAP gate, 404, 446

T

tantalum oxide, 32
teleportation, 434
tensors, 261
Tersoff potential, 216–217, 220
Tersoff–Brenner potential, 287
thermodynamic limit, 208–209
thermodynamic potential, 290
thermodynamic property, 222–223
thermostat, 209, 215, 217
  Andersen, 212
  Hoover's constraint, 212, 218
  momentum rescaling, 212, 218
  Nosé-Hoover, 212, 217
thin film, 2
third harmonic, 164–166
third-order polarization, 165
three-body potential, 286
tight-binding approximation, 131, 151–152, 217–218
time-evolution, 405
titanium oxide, 26
trace, 450
trace-preserving maps, 410
traction, 263
transfer matrix, 19, 47, 53, 55–56, 77, 79–82
transition, 126, 223–224
transmission electron microscope, 112, 342
transmittance, 17, 21, 26, 56, 58
transverse electric (TE) modes, 84
transverse magnetic (TM) modes, 84, 86 87
transverse quantization, 152, 155
traveling wave, 156
triangular lattice, 67, 70–73, 84–88
tubulin, 230
tunneling, 111, 163, 167

U

unconditional security, 431
uncoupled modes, 71–73, 81
uniaxiality, 22
unit vector
  binormal, 11
  normal, 11
  tangential, 11
universal binding energy relation, 286
universal computation, 446
unsorted search, 422

V

vector wave equations, 48, 64
Verlet algorithm, 293–294
vertical quantum dot, 115
vertical-cavity surface-emitting quantum dot laser, 134
virial, 215
virus, 31
visco-elastic fluid, 131
void, 14
von Neumann entropy, 408
von Neumann measurement, 406–407

W

warping, 310
wave
  function, 112, 275, 323
  number, 18

packet, 240–241
propagation, electromagnetic, 16, 24
vector, 65–66, 71–72, 75–77, 86
wavelet transform, 234, 237
wetting, 347–349, 351–352, 354
Wiener–Hopf technique, 159
Wigner crystal, 132
Wronskian, 52–53

Z

Zeeman shift, 125